STEAM-H: Science, Technology, Engineering,
Agriculture, Mathematics & Health

STEAM-H: Science, Technology, Engineering, Agriculture, Mathematics & Health

Series Editor
Bourama Toni
Department of Mathematics
Howard University
Washington, DC, USA

This interdisciplinary series highlights the wealth of recent advances in the pure and applied sciences made by researchers collaborating between fields where mathematics is a core focus. As we continue to make fundamental advances in various scientific disciplines, the most powerful applications will increasingly be revealed by an interdisciplinary approach. This series serves as a catalyst for these researchers to develop novel applications of, and approaches to, the mathematical sciences. As such, we expect this series to become a national and international reference in STEAM-H education and research.

Interdisciplinary by design, the series focuses largely on scientists and mathematicians developing novel methodologies and research techniques that have benefits beyond a single community. This approach seeks to connect researchers from across the globe, united in the common language of the mathematical sciences. Thus, volumes in this series are suitable for both students and researchers in a variety of interdisciplinary fields, such as: mathematics as it applies to engineering; physical chemistry and material sciences; environmental, health, behavioral and life sciences; nanotechnology and robotics; computational and data sciences; signal/image processing and machine learning; finance, economics, operations research, and game theory.

The series originated from the weekly yearlong STEAM-H Lecture series at Virginia State University featuring world-class experts in a dynamic forum. Contributions reflected the most recent advances in scientific knowledge and were delivered in a standardized, self-contained and pedagogically-oriented manner to a multidisciplinary audience of faculty and students with the objective of fostering student interest and participation in the STEAM-H disciplines as well as fostering interdisciplinary collaborative research. The series strongly advocates multidisciplinary collaboration with the goal to generate new interdisciplinary holistic approaches, instruments and models, including new knowledge, and to transcend scientific boundaries.

More information about this series at http://www.springer.com/series/15560

Toka Diagana • Bourama Toni

Editors

Mathematical Structures and Applications

In Honor of Mahouton Norbert Hounkonnou

 Springer

Editors
Toka Diagana
Mathematical Sciences Department
The University of Alabama in Huntsville
Hunstville, AL, USA

Bourama Toni
Department of Mathematics
Howard University
Washington, DC, USA

ISSN 2520-193X   ISSN 2520-1948   (electronic)
STEAM-H: Science, Technology, Engineering, Agriculture, Mathematics & Health
ISBN 978-3-030-07316-9   ISBN 978-3-319-97175-9   (eBook)
https://doi.org/10.1007/978-3-319-97175-9

Mathematics Subject Classification: 42A16, 46B28, 46H35, 49S05, 53D30, 53D50, 53Z05, 57R57, 81T16, 82C22

This Springer imprint is published by the registered company Springer Nature Switzerland AG
The registered company address is: Gewerbestrasse 11, 6330 Cham, Switzerland

Dedication

Mahouton Norbert Hounkonnou is a full professor of Mathematics and Physics at the University of Abomey-Calavi, Cotonou, Benin. His research deals with non-commutative and nonlinear mathematics including differential equations, operator theory, coherent states, quantization techniques, orthogonal polynomials, special functions, graph theory, nonassociative algebras, nonlinear systems, noncommutative field theories, and geometric methods in physics.

Professor Hounkonnou has authored/coauthored and reviewed several books and refereed and served as an associate editor for renowned journals in mathematics, mathematical physics, and theoretical physics. He has published over 200 refereed research papers in outstanding ISI-ranked journals and international conference proceedings in the fields of mathematics, mathematical physics, and theoretical physics.

Norbert has been a visiting professor at several African, Asian, European, and North American universities. Together with his peers at the international level, he founded the International Chair in Mathematical Physics and Applications (ICMPA-UNESCO Chair) of the University of Abomey-Calavi offering multi-university master degrees and PhD programs in mathematics with connections, motivations, or applications to physics or in physics with important relationships to mathematics. The best African students from about 13 French- and English-speaking countries are selected to follow these graduate programs, which attracted prominent and leading mathematicians and mathematical physicists around the world who come to give lectures and supervise students' research, what has substantially increased international collaboration with African, Asian, American, European, and Indian scientists and mathematicians.

The ICMPA-UNESCO Chair presently hosts an International Conference Series (respectively, School) on Contemporary Problems in Mathematical Physics, which is held in Cotonou (Benin) every 2 years since 1999 (respectively, each year since 2005). These activities have led to a significant network of researchers connected with the ICMPA-UNESCO Chair. The ICMPA-UNESCO Chair gets its funding from various sources that are available in mathematics and mathematical physics and for the development of world-class mathematics and science in Africa. Professor Hounkonnou has directed/co-directed 32 PhD theses and 21 masters. His PhD and master students are from several countries including Belgium, Benin, Burkina Faso, Burundi, Cameroon, Democratic Republic of Congo, Niger, Nigeria, Senegal, Togo, and Zambia.

Professor Hounkonnou is the chair of the African Academy of Sciences Commission on Pan-African Science Olympiad (2014 to present), the chair of the African Academy of Sciences Membership Advisory Committee (MAC) on Mathematical Sciences (2013 to present), reviewer for the NANUM 2014 Award Committee Member of the International Congress of Mathematicians (ICM 2014), and TWAS research professor in Zambia and enjoys the membership of several important international scientific organizations.

Professor Hounkonnou is the current president of the Benin National Academy of Sciences, Arts and Letters. His membership extends to the International Association of Mathematical Physics, American Mathematical Society, African Academy of Sciences (AAS), The World Academy of Sciences (TWAS), UNESCO Scientific Board for International Basic Sciences Programme (IBSP), and to many others.

Among other things, Professor Hounkonnou is a Knight of the National Order of Benin (*Chevalier de l'Ordre National du Benin*). He has received a series of recognition for the excellence of his work such as the Prize of the Third World Academy of Sciences (TWAS) in 1996, the 2015 Tokyo University of Science President Award, and the 2016 World Academy of Sciences C.N.R. Rao Prize for Scientific Research "for his incisive work on noncommutative and nonlinear mathematics and his contributions to world-class mathematics education."

Preface

The multidisciplinary STEAM-H series (Science, Technology, Engineering, Agriculture, Mathematics, and Health) brings together leading researchers to present their work in the perspective to advance their specific fields and in a way to generate a genuine interdisciplinary interaction transcending disciplinary boundaries. All chapters therein were carefully edited and peer-reviewed; they are reasonably self-contained and pedagogically exposed for a multidisciplinary readership. Contributions are invited only and reflect the most recent advances delivered in a high standard, self-contained way. The goals of the series are as follows:

1. To enhance multidisciplinary understanding between the disciplines by showing how some new advances in a particular discipline can be of interest to the other discipline or how different disciplines contribute to a better understanding of a relevant issue at the interface of mathematics and the sciences
2. To promote the spirit of inquiry so characteristic of mathematics for the advances of the natural, physical, and behavioral sciences by featuring leading experts and outstanding presenters
3. To encourage diversity in the readers' background and expertise while structurally fostering genuine interdisciplinary interactions and networking

Current disciplinary boundaries do not encourage effective interactions between scientists; researchers from different fields usually occupy different buildings, publish in journals specific to their field, and attend different scientific meetings. Existing scientific meetings usually fall into either small gatherings specializing on specific questions, targeting specific and small group of scientists already aware of each other's work and potentially collaborating, or large meetings covering a wide field and targeting a diverse group of scientists but usually not allowing specific interactions to develop due to their large size and a crowded program. Traditional departmental seminars are becoming so technical as to be largely inaccessible to anyone who did not coauthor the research being presented. Here contributors focus on how to make their work intelligible and accessible to a diverse audience, which in the process enforces mastery of their own field of expertise.

This volume, as the previous ones, strongly advocates multidisciplinarity with the goal to generate new interdisciplinary approaches, instruments, and models including new knowledge, transcending scientific boundaries to adopt a more holistic approach. For instance, it should be acknowledged, following Nobel Laureate and president of the UK's Royal Society of Chemistry, Professor Sir Harry Kroto, "that the traditional chemistry, physics, biology departmentalised university infrastructures—which are now clearly out-of-date and a serious hindrance to progress—must be replaced by new ones which actively foster the synergy inherent in multidisciplinarity." The National Institutes of Health and the Howard Hughes Medical Institute have strongly recommended that undergraduate biology education should incorporate mathematics, physics, chemistry, computer science, and engineering until "interdisciplinary thinking and work become second nature." Young physicists and chemists are encouraged to think about the opportunities waiting for them at the interface with the life sciences. Mathematics is playing an ever more important role in the physical and life sciences, engineering, and technology, blurring the boundaries between scientific disciplines.

The series is to be a reference of choice for established interdisciplinary scientists and mathematicians and a source of inspiration for a broad spectrum of researchers and research students, graduate, and postdoctoral fellows; the shared emphasis of these carefully selected and refereed contributed chapters is on important methods, research directions, and applications of analysis including within and beyond mathematics. As such the volume promotes mathematical sciences, physical and life sciences, engineering, and technology education, as well as interdisciplinary, industrial, and academic genuine cooperation.

Toward such goals, the following chapters are featured in the current volume.

The present volume contains the contributions from the participants of the conference in honor of Professor Mahouton Norbert Hounkonnou on his 60th birthday held in Cotonou, Benin. It features the following chapters.

Chapter "Metric Operators, Generalized Hermiticity, and Partial Inner Product Spaces", by Jean-Pierre Antoine and Camillo Trapani, analyzes the structure of metric operators, bounded or unbounded, drawing from recent results on pseudo-Hermitian quantum mechanics.

In chapter "Beyond Frames: Semi-frames and Reproducing Pairs", Jean-Pierre Antoine and Camillo Trapani study semi-frames (upper and lower) and reproducing pairs which generate two Hilbert spaces conjugates of each other.

Chapter "On Hilbert-Schmidt Operator Formulation of Noncommutative Quantum Mechanics", by Isiaka Aremua, Ezinvi Baloitcha, Mahouton Norbert Hounkonnou, and Komi Sodoga, investigates a system of charged particle in a constant magnetic field as a way to emphasize the importance of Hilbert-Schmidt operators in the formulation of noncommutative quantum theory.

In chapter "Symplectic Affine Action and Momentum with Cocycle", Augustin Batubenge and Wallace Haziyu show that a symplectic structure can be defined on the orbit, a symplectic manifold, of a certain affine action.

Chapter "Some Difference Integral Inequalities", by Gaspard Bangerezako and Jean-Paul Nuwacu, uses the Lagrange method of linear difference equation of first order to establish different versions of some classical integral inequalities.

In chapter "Theoretical and Numerical Comparisons of the Parameter Estimator of the Fractional Brownian Motion", Jean-Marc Bardet presents theoretical and numerical comparisons of the most important methods of parameter estimators of the fractional Brownian motion.

Chapter "Minimal Lethal Disturbance for Finite Dimensional Linear Systems", by Abdes Samed Bernoussi, Mina Amharref, and Mustapha Ouardouz, considers the problem of robust viability giving some characterizations of the viability radius for finite dimensional disturbed linear systems leading to the determination of the so-called minimal lethal disturbance.

In chapter "Walker Osserman Metric of Signature (3, 3)", Abdoul Salam Diallo, Mouhamadou Hassirou, and Ousmane Toudou Issa investigate a torsion-free affine manifold and the related Riemann extension to produce an example of Walker Osserman metric of signature (3, 3).

Chapter "Conformal Symmetry Transformations and Nonlinear Maxwell Equations", by Gerald A. Goldin, Vladimir M. Shtelen, and Steven Duplij, explores ways to describe general, nonlinear Maxwell fields with conformal symmetry, making use of the conformal compactification of Minkowski spacetime.

Laure Gouba, in chapter "The Yukawa Model in One Space - One Time Dimensions", revisits the Yukawa model in one space-one time dimensions showing it as a constrained system at the classical level using the Dirac method and reformulating the model as a quantum level of scalar field by a bosonization procedure.

Chapter "Towards the Quantum Geometry of Saturated Quantum Uncertainty Relations: The Case of the (Q, P) Heisenberg Observables", by Jan Govaerts, outlines a program to identify geometric structures associated to the manifold in Hilbert space of the quantum states that saturate the Schrodinger-Robertson uncertainty relation to a specific set of quantum observables characterizing a given quantum system and its dynamics.

In chapter "The Role of the Jacobi Last Multiplier in Nonholonomic Systems and Locally Conformal Symplectic Structure", Partha Guha studies the geometric structure of nonholonomic system with almost symplectic structure in relation to Jacobi's last multiplier.

Chapter "Non-perturbative Renormalization Group of a $U(1)$ Tensor Model", by Vincent Lahoche and Dine Ousmane Samary, discusses the non-perturbative renormalization group of a U(1) tensor model.

Richard Kerner, in chapter "Ternary Z_2 and Z_3 Graded Algebras and Generalized Color Dynamics", studies cubic and ternary algebras as a direct generalization of Grassmann and Clifford algebras with Z_3 grading.

Using the fuel smuggling trade between Benin and Nigeria as a background, chapter "Pseudo-Solution of Weight Equations in Neural Networks: Application for Statistical Parameters Estimation", by Vincent J. M. Kiki, Villevo Adanhounme, and Mahouton Norbert Hounkonnou, presents a pseudo-solution to weight equations in a class of neural networks using an algebraic approach.

In chapter "A Note on Curvatures and Rank 2 Seiberg–Witten Invariants", Fortuné Massamba discusses lower bounds for certain curvature functionals on the space of Riemannian metrics of a smooth compact 4-manifold with nontrivial rank 2 Seiberg-Witten invariants.

In chapter "Shape Invariant Potential Formalism for Photon-Added Coherent State Construction", Komi Sodoga, Isiaka Aremua, and Mahouton Norbert Hounkonnou introduce the so-called shape invariant potential method, an algebro-operator approach to construct generalized coherent states for photon-added particle system, with illustrations on Pöschl-Teller potentials.

Mawoussi Todjro and Yaogan Mensah, in chapter "On the Fourier Analysis for L^2 Operator-Valued Functions", describe the construction of the Fourier transform of Hilbert-Schmidt operator-valued function on compact groups.

Chapter "Electrostatic Double Layers in a Magnetized Isothermal Plasma with two Maxwellian Electrons", by Odutayo Raji Rufai, discusses finite amplitude nonlinear ion-acoustic double layers in a magnetized plasma of warm isothermal ions fluid and two Boltzmann distributed electron species assuming the charge neutrality condition at equilibrium.

Finally Akira Yoshioka, in chapter "Star Products, Star Exponentials, and Star Functions", presents nonformal star products on polynomials with positive deformation parameter, star exponentials in the star product algebra, leading to the so-called star functions in the algebra with some noncommutative identities.

The book as a whole certainly enhances the overall objective of the series, that is, to foster the readership interest and enthusiasm in the STEAM-H disciplines (Science, Technology, Engineering, Agriculture, Mathematics, and Health), stimulate graduate and undergraduate research, and generate collaboration among researchers on a genuine interdisciplinary basis.

The STEAM-H series is now hosted at Howard University, Washington, DC, USA, an area that is socially, economically, intellectually very dynamic and home to some of the most important research centers in the USA. This series, by now well established and published by Springer, a world-renowned publisher, is expected to become a national and international reference in interdisciplinary education and research.

Washington, DC, USA   Bourama Toni

Acknowledgments

We would like to express our sincere appreciation to all the contributors and to all the anonymous referees for their professionalism. They all made this volume a reality for the greater benefice of the community of Science, Engineering, and Mathematics and in honor of Professor Mahouton Norbert Hounkonnou for his own outstanding, inspiring, and everlasting contributions to mathematics and the sciences.

It has indeed been a great pleasure and a privilege to edit such a volume!

Contents

Contributors

Villévo Adanhounme University of Abomey-Calavi, International Chair in Mathematical Physics and Applications (ICMPA), Cotonou, Benin

Mina Amharref GAT, Tangier, Morocco

Jean-Pierre Antoine Institut de Recherche en Mathématique et Physique, Université catholique de Louvain, Louvain-la-Neuve, Belgium

Isiaka Aremua Université de Lomé, Faculté des Sciences, Département de Physique, Laboratoire de Physique des Matériaux et de Mécanique Appliquée, Lomé, Togo

University of Abomey-Calavi, International Chair in Mathematical Physics and Applications (ICMPA), Cotonou, Benin

Ezinvi Baloïtcha University of Abomey-Calavi, International Chair in Mathematical Physics and Applications (ICMPA), Cotonou, Benin

G. Bangerezako University of Burundi, Faculty of Sciences, Department of Mathematics, Bujumbura, Burundi

Jean-Marc Bardet University Paris 1 Pantheon-Sorbonne, Paris, France

Augustin Batubenge Départment de Mathématiques et Statistique, Université de Montréal, Montréal, QC, Canada

University of Zambia, Lusaka, Zambia

Abdes Samed Bernoussi GAT, Tangier, Morocco

Abdoul Salam Diallo Université Alioune Diop de Bambey, UFR SATIC, Département de Mathématiques, Bambey, Senegal

Steven Duplij University of Münster, Münster, Germany

Gerald A. Goldin Department of Mathematics, Rutgers University, New Brunswick, NJ, USA

Department of Physics, Rutgers University, New Brunswick, NJ, USA

Laure Gouba The Abdus Salam International Centre for Theoretical Physics (ICTP), Trieste, Italy

Jan Govaerts Centre for Cosmology, Particle Physics and Phenomenology (CP3), Institut de Recherche en Mathématique et Physique (IRMP), Université catholique de Louvain (U.C.L.), Louvain-la-Neuve, Belgium

National Institute for Theoretical Physics (NITheP), Stellenbosch, Republic of South Africa

International Chair in Mathematical Physics and Applications (ICMPA–UNESCO Chair), University of Abomey-Calavi, Cotonou, Republic of Benin

Partha Guha IFSC, Universidade de São Paulo, São Carlos, SP, Brazil

S.N. Bose National Centre for Basic Sciences, Salt Lake City, Kolkata, India

Mouhamadou Hassirou Département de Mathématiques et Informatique, Faculté des Sciences et Techniques, Université Abdou Moumouni, Niamey, Niger

Wallace Haziyu Department of Mathematics and Statistics, University of Zambia, Lusaka, Zambia

Mahouton Norbert Hounkonnou University of Abomey-Calavi, International Chair in Mathematical Physics and Applications (ICMPA), Cotonou, Benin

Ousmane Toudou Issa Département de l'Environnement, Université de Tillaberi, Tillaberi, Niger

Richard Kerner Laboratoire de Physique Théorique de la Matière Condensée, Sorbonne-Universités, CNRS UMR, Paris, France

Vincent J. M. Kiki Ecole Nationale d'Economie Appliquée et de Management, Université d'Abomey-Calavi, Cotonou, Benin

Vincent Lahoche LaBRI, Univ. Bordeaux 351 cours de la Libération, Talence, France

Fortuné Massamba School of Mathematics, Statistics and Computer Science, University of KwaZulu-Natal, Scottsville, South Africa

Yaogan Mensah Department of Mathematics, University of Lomé, Lomé, Togo

J. P. Nuwacu University of Burundi, Faculty of Sciences, Department of Mathematics, Bujumbura, Burundi

Mustapha Ouardouz MMC Team, Faculty of Sciences and Techniques, Tangier, Morocco

Dine Ousmane Samary Max Planck Institute for Gravitational Physics, Albert Einstein Institute, Potsdam, Germany

Faculté des Sciences et Techniques/ICMPA-UNESCO Chair, Université d'Abomey-Calavi, Abomey-Calavi, Benin

Odutayo Raji Rufai Department of Physics and Astronomy, University of the Western Cape, Bellville, Cape Town, South Africa

Mawoussi Todjro Department of Mathematics, University of Lomé, Lomé, Togo

Camillo Trapani Dipartimento di Matematica e Informatica, Università di Palermo, Palermo, Italy

Vladimir M. Shtelen Department of Mathematics, Rutgers University, New Brunswick, NJ, USA

Komi Sodoga Université de Lomé, Faculté des Sciences, Département de Physique, Laboratoire de Physique des Matériaux et de Mécanique Appliquée, Lomé, Togo

University of Abomey-Calavi, International Chair in Mathematical Physics and Applications (ICMPA), Cotonou, Benin

Akira Yoshioka Tokyo University of Science, Tokyo, Japan

Metric Operators, Generalized Hermiticity, and Partial Inner Product Spaces

Jean-Pierre Antoine and Camillo Trapani

Abstract A quasi-Hermitian operator is an operator in a Hilbert space that is similar to its adjoint in some sense, via a metric operator, i.e., a strictly positive self-adjoint operator. Motivated by the recent developments of pseudo-Hermitian quantum mechanics, we analyze the structure of metric operators, bounded or unbounded. We introduce several generalizations of the notion of similarity between operators and explore to what extent they preserve spectral properties.

Next we consider canonical lattices of Hilbert space s generated by unbounded metric operators. Since such lattices constitute the simplest case of a partial inner product space (PIP-space), we can exploit the technique of PIP-space operators. Thus we apply some of the previous results to operators on a particular PIP-space, namely, the scale of Hilbert space s generated by a single metric operator. Finally, we reformulate the notion of pseudo-hermitian operators in the preceding formalism.

Keywords Metric operators · Quasi-Hermitian operators · Similar operators · Lattices and scales of Hilbert spaces · Partial inner product spaces (PIP spaces)

1 Introduction

Non-self-adjoint operators with real spectrum appear in different contexts: the so-called $\mathcal{PT}$-symmetric quantum mechanics [10], pseudo-Hermitian quantum mechanics [19, 20], three-Hilbert-space formulation of quantum mechanics [27],

---

Based on a talk given at the COPROMAPH8 conference [3].

J.-P. Antoine (✉)
Institut de Recherche en Mathématique et Physique, Université catholique de Louvain, Louvain-la-Neuve, Belgium
e-mail: jean-pierre.antoine@uclouvain.be

C. Trapani
Dipartimento di Matematica e Informatica, Università di Palermo, Palermo, Italy
e-mail: camillo.trapani@unipa.it

nonlinear pseudo-bosons [9], nonlinear supersymmetry, and so on. In addition, they appear under various names: pseudo-Hermitian, quasi-Hermitian, cryptohermitian operators.

The $\mathcal{PT}$-symmetric Hamiltonians, that is, Hamiltonians invariant under the joint action of space reflection ($\mathcal{P}$) and complex conjugation ($\mathcal{T}$), are usually pseudo-Hermitian operators. This term was introduced a long time ago by Dieudonné [14] (under the name "quasi-Hermitian") for characterizing those bounded operators A which satisfy a relation of the form

$$GA = A^*G, \tag{1.1}$$

where G is a *metric operator*, i.e., a strictly positive self-adjoint operator. This operator G then defines a new metric (hence the name) and a new Hilbert space (sometimes called physical) in which A is symmetric and possesses a self-adjoint extension. For a systematic analysis of pseudo-Hermitian QM, we may refer to the review of Mostafazadeh [19] and the special issues [11, 12], which contain a variety of concrete applications in quantum physics.

According to (1.1), the generic structure of these operators is $A^* = GAG^{-1}$. Thus A^* is *similar* to A, in some sense, via a metric operator G, i.e., a strictly positive self-adjoint operator $G > 0$, thus invertible, with (possibly unbounded) inverse G^{-1}. Now, in most of the literature, the metric operators are assumed to be bounded. In some recent works, however, unbounded metric operators are introduced [7–9, 20].

On the other hand, if G^{-1} is bounded, (1.1) implies that A is similar to a self-adjoint operator, thus it is a spectral operator of scalar type and real spectrum, in the sense of Dunford [15]. This is the case treated by Scholtz et al. [25] and Geyer et al. [17], who introduced the concept in the physics literature.

The aim of this chapter is to study in a rigorous way the problem of *operator similarity* under a metric operator, bounded or unbounded. In particular, we will formulate the analysis in the framework of partial inner product spaces (PIP-spaces), since the latter appear naturally in this context. Most of the information contained here comes from our papers [4–6].

To conclude, we fix our notations. The framework is a separable Hilbert space $\mathcal{H}$, with inner product $\langle\cdot|\cdot\rangle$, linear in the first entry. Then, for any operator A in $\mathcal{H}$, we denote its domain by $D(A)$, its range by $R(A)$ and, if A is positive, its form domain by $Q(A) := D(A^{1/2})$.

2 Metric Operators

By a *metric operator*, in a Hilbert space $\mathcal{H}$, we mean a strictly positive self-adjoint operator G, that is, $G > 0$ or $\langle G\xi|\xi\rangle \geq 0$ for every $\xi \in D(G)$ and $\langle G\xi|\xi\rangle = 0$ if and only if $\xi = 0$.

Of course, G is densely defined and invertible, but need not be bounded; its inverse G^{-1} is also a metric operator, bounded or not (in this case, in fact, 0 belongs to the continuous spectrum of G).

Let G, G_1, G_2 be metric operators. Then

(1) If G_1 and G_2 are both bounded, then $G_1 + G_2$ is a bounded metric operator;
(2) λG is a bounded metric operator for every $\lambda > 0$;
(3) if G_1 and G_2 commute, their product $G_1 G_2$ is also a bounded metric operator;
(4) $G^{1/2}$ and, more generally, $G^\alpha (\alpha \in \mathbb{R})$ are bounded metric operators.

Given a bounded metric operator G, define $\langle \xi | \eta \rangle_G := \langle G\xi | \eta \rangle, \xi, \eta \in \mathcal{H}$. This is a positive definite inner product on $\mathcal{H}$ with corresponding norm $\|\xi\|_G = \|G^{1/2}\xi\|$. We denote by $\mathcal{H}(G)$ the completion of $\mathcal{H}$ in this norm. Thus we get $\mathcal{H} \subseteq \mathcal{H}(G)$. If $G^{-1/2}$ is bounded, $\mathcal{H}$ and $\mathcal{H}(G)$ are the same as vector spaces and they carry different, but equivalent, norms.

Clearly, the conjugate dual space $\mathcal{H}(G)^\times$ of $\mathcal{H}(G)$ is a subspace of $\mathcal{H}$ and $\mathcal{H}(G)^\times \equiv \mathcal{H}(G^{-1}) = D(G^{-1/2})$ with inner product $\langle \xi | \eta \rangle_{G^{-1}} = \langle G^{-1}\xi | \eta \rangle$. The upshot is a triplet of Hilbert spaces

$$\mathcal{H}(G^{-1}) \hookrightarrow \mathcal{H} \hookrightarrow \mathcal{H}(G), \tag{2.1}$$

where $\hookrightarrow$ denotes a continuous embedding with dense range. If G^{-1} is bounded, $\mathcal{H}(G^{-1}) = \mathcal{H}(G) = \mathcal{H}$ with norms equivalent to (but different from) the norm of $\mathcal{H}$. In the triplet (2.1), $G^{-1/2}$ is a unitary operator from $\mathcal{H}(G^{-1})$ onto $\mathcal{H}$ and from $\mathcal{H}$ onto $\mathcal{H}(G)$. In the same way, $G^{1/2}$ is a unitary operator from $\mathcal{H}$ onto $\mathcal{H}(G^{-1})$ and from $\mathcal{H}(G)$ onto $\mathcal{H}$.

Now, the triplet (2.1) is the central part of the infinite scale of Hilbert spaces built on the powers of $G^{-1/2}$, $V_I := \{\mathcal{H}_n, n \in \mathbb{Z}\}$, where $\mathcal{H}_n = D(G^{-n/2}), n \in \mathbb{N}$, with a norm equivalent to the graph norm, and $\mathcal{H}_{-n} = \mathcal{H}_n^\times$:

$$\ldots \subset \mathcal{H}_2 \subset \mathcal{H}_1 \subset \mathcal{H} \subset \mathcal{H}_{-1} \subset \mathcal{H}_{-2} \subset \ldots \tag{2.2}$$

The obvious question is how to identify the end spaces of the scale:

$$\mathcal{H}_\infty(G^{-1/2}) := \bigcap_{n \in \mathbb{Z}} \mathcal{H}_n, \qquad \mathcal{H}_{-\infty}(G^{-1/2}) := \bigcup_{n \in \mathbb{Z}} \mathcal{H}_n. \tag{2.3}$$

By *quadratic interpolation* [13], one may build the continuous scale $\mathcal{H}_t, 0 \le t \le 1$, between $\mathcal{H}_1$ and $\mathcal{H}$, where $\mathcal{H}_t = D(G^{-t/2})$, with norm $\|\xi\|_t = \|G^{-t/2}\xi\|$. Next, defining $\mathcal{H}_{-t} = \mathcal{H}_t^\times$ and iterating, one obtains the full continuous scale $V_{\tilde{I}} := \{\mathcal{H}_t, t \in \mathbb{R}\}$, a simple example of a PIP-space [2]. Then, of course, one can replace $\mathbb{Z}$ by $\mathbb{R}$ in the definition (2.3) of the end spaces of the scale.

3   Similar and Quasi-Similar Operators

Before proceeding to our main topic, we quote two easy properties. Let A be a linear operator in the Hilbert space $\mathcal{H}$, with domain $D(A)$. Then, (i) if $D(A)$ is dense in $\mathcal{H}$, it is also dense in $\mathcal{H}(G)$; (ii) if A is closed in $\mathcal{H}(G)$, it is also closed in $\mathcal{H}$.

Now we introduce the central definitions.

Definition 3.1

(1) Let $\mathcal{H}$ and $\mathcal{K}$ be Hilbert spaces, A and B densely defined linear operators in $\mathcal{H}$, resp. $\mathcal{K}$. A bounded operator $T : \mathcal{H} \to \mathcal{K}$ is called a *bounded intertwining operator* for A and B if

   (io$_1$) $T : D(A) \to D(B)$;
   (io$_2$) $BT\xi = TA\xi, \ \forall \xi \in D(A)$.

   If T is a bounded intertwining operator for A and B, then $T^* : \mathcal{K} \to \mathcal{H}$ is a bounded intertwining operator for B^* and A^*.
(2) A and B are *similar*, which is denoted $A \sim B$, if there exists a bounded intertwining operator T for A and B with bounded inverse $T^{-1} : \mathcal{K} \to \mathcal{H}$, which is intertwining for B and A. In addition, A and B are *metrically similar* if T is a metric operator.
(3) A and B are *unitarily equivalent* $(A \overset{u}{\sim} B)$ if $A \sim B$ and $T : \mathcal{H} \to \mathcal{K}$ is unitary.

Obviously, $\sim$ and $\overset{u}{\sim}$ are equivalence relations.

The following properties are immediate. Let $A \sim B$. Then:

 (i) $TD(A) = D(B)$.
 (ii) A is closed iff B is closed.
(iii) $A \sim B$ iff $B^* \sim A^*$.
(iv) A^{-1} exists iff B^{-1} exists; in that case, $B^{-1} \sim A^{-1}$.

In the sequel, we will examine to what extent the spectral properties of operators behave under the similarity relation. In order to do that, it is worth recalling the basic definitions, especially since we are dealing with closed, non-self-adjoint operators.

Given a closed operator A in $\mathcal{H}$, consider $A - \lambda I : D(A) \to \mathcal{H}$ and the resolvent $R_A(\lambda) := (A - \lambda I)^{-1}$. Then one defines:

- The resolvent set $\rho(A) := \{\lambda \in \mathbb{C} : A - \lambda I$ is one-to-one and $(A - \lambda I)^{-1}$ is bounded$\}$.
- The spectrum $\sigma(A) := \mathbb{C} \setminus \rho(A)$.
- The point spectrum $\sigma_p(A) := \{\lambda \in \mathbb{C} : A - \lambda I$ is not one-to-one$\}$, that is, the set of eigenvalues of A.
- The continuous spectrum $\sigma_c(A) := \{\lambda \in \mathbb{C} : A - \lambda I$ is one-to-one and has dense range, different from $\mathcal{H}\}$, hence $(A - \lambda I)^{-1}$ is densely defined, but unbounded.
- The residual spectrum $\sigma_r(A) := \{\lambda \in \mathbb{C} : A - \lambda I$ is one-to-one, but its range is not dense$\}$, hence $(A - \lambda I)^{-1}$ is not densely defined.

With these definitions, the three sets $\sigma_p(A), \sigma_c(A), \sigma_r(A)$ are disjoint and

$$\sigma(A) = \sigma_p(A) \cup \sigma_c(A) \cup \sigma_r(A). \tag{3.1}$$

We note also that $\sigma_r(A) = \overline{\sigma_p(A^*)} = \{\bar{\lambda} : \lambda \in \sigma_p(A^*)\}$. Indeed, for any $\lambda \in \sigma_r(A)$, there exists $\eta \neq 0$ such that

$$0 = \langle (A - \lambda I)\xi | \eta \rangle = \langle \xi | (A^* - \bar{\lambda} I)\eta \rangle, \ \forall \, \xi \in D(A),$$

which implies $\bar{\lambda} \in \sigma_p(A^*)$. Also $\sigma_r(A) = \emptyset$ if A is self-adjoint.

Note that here we follow Dunford–Schwartz [16], but other authors give a different definition of the continuous spectrum, implying that it is no longer disjoint from the point spectrum, for instance, Reed–Simon [22] or Schmüdgen [24]. This alternative definition allows for eigenvalues embedded in the continuous spectrum, a situation common in many physical situations.

The answer to the question raised above is given in the following proposition.

Proposition 3.2 *Let A, B be closed operators such that $A \sim B$ with the bounded intertwining operator T. Then,*

(i) $\rho(A) = \rho(B)$.
(ii) $\sigma_p(A) = \sigma_p(B)$. *Moreover if $\xi \in D(A)$ is an eigenvector of A corresponding to the eigenvalue λ, then $T\xi$ is an eigenvector of B corresponding to the same eigenvalue. Conversely, if $\eta \in D(B)$ is an eigenvector of B corresponding to the eigenvalue λ, then $T^{-1}\eta$ is an eigenvector of A corresponding to the same eigenvalue. Moreover, the multiplicity of λ as eigenvalue of A is the same as its multiplicity as eigenvalue of B.*
(iii) $\sigma_c(A) = \sigma_c(B)$ *and* $\sigma_r(A) = \sigma_r(B)$.
(iv) If A is self-adjoint, then B has real spectrum and $\sigma_r(B) = \emptyset$.

The property (iv) means that B is then a spectral operator of scalar type with real spectrum, a notion introduced by Dunford [15].

In conclusion, similarity preserves the various parts of the spectra, but it does *not* preserve self-adjointness. This means we are on the good track, since we are seeking a form of similarity that transforms a non-self-adjoint operator into a self-adjoint one.

However, the notion of similarity just defined is too strong in many situations. A natural step is to drop the boundedness of T^{-1}.

Definition 3.3 We say that A is *quasi-similar* to B, and write $A \dashv B$, if there exists a bounded intertwining operator T for A and B which is invertible, with inverse T^{-1} densely defined (but not necessarily bounded).

Even if T^{-1} is bounded, A and B need not be similar, unless T^{-1} is also an intertwining operator. Indeed, T^{-1} does not necessarily map $D(B)$ into $D(A)$, unless of course if $T D(A) = D(B)$. Note that one can always suppose that T is a metric operator

Actually there is a great confusion in the literature about the terminology of (quasi-)similarity. We refer to our paper [6] for a detailed discussion.

We proceed now to show the stability of the different parts of the spectrum under the quasi-similarity relation $\dashv$, following mostly [4] and [6].

Proposition 3.4 *Let A and B be closed operators and assume that $A \dashv B$, with the bounded intertwining operator T. Then the following statements hold.*

(i) *$\sigma_p(A) \subseteq \sigma_p(B)$ and for every $\lambda \in \sigma_p(A)$ one has $m_A(\lambda) \le m_B(\lambda)$, where $m_A(\lambda)$, resp. $m_B(\lambda)$, denotes the multiplicity of λ as eigenvalue of the operator A, resp. B.*

(ii) *$\sigma_r(B) \subseteq \sigma_r(A)$.*

(iii) *If $T D(A) = D(B)$, then $\sigma_p(B) = \sigma_p(A)$.*

(iv) *If T^{-1} is bounded and $T D(A)$ is a core for B, then $\sigma_p(B) \subseteq \sigma(A)$.*

(v) *If T^{-1} is everywhere defined and bounded, then $\rho(A) \setminus \sigma_p(B) \subseteq \rho(B)$ and $\rho(B) \setminus \sigma_r(A) \subseteq \rho(A)$.*

(vi) *Assume that T^{-1} is everywhere defined and bounded and $T D(A)$ is a core for B. Then*

$$\sigma_p(A) \subseteq \sigma_p(B) \subseteq \sigma(B) \subseteq \sigma(A).$$

The situation described in Proposition 3.4 (vi) is quite important for possible applications. Even if the spectra of A and B may be different, it gives a certain number of information on $\sigma(B)$ once $\sigma(A)$ is known. For instance, if A has a pure point spectrum, then B is isospectral to A. More generally, if A is self-adjoint, then any operator B quasi-similar to A by means of an intertwining operator T with bounded inverse T^{-1} has real spectrum.

We will illustrate the previous proposition by two examples, both taken from [4]. In the first one, $A \dashv B$, A, B and T are all bounded, and the two spectra, which are pure point, coincide.

Example 3.5 In $\mathcal{H} = L^2(\mathbb{R}, \mathrm{d}x)$, take the operator Q of multiplication by x, on the dense domain

$$D(Q) = \left\{ f \in L^2(\mathbb{R}) : \int_{\mathbb{R}} x^2 |f(x)|^2 \, \mathrm{d}x < \infty \right\},$$

and define the two operators

- $P_\varphi := |\varphi\rangle\langle\varphi|$, for $\varphi \in L^2(\mathbb{R})$, with $\|\varphi\| = 1$,
- $A_\varphi f = \langle (I + Q^2) f | \varphi \rangle (I + Q^2)^{-1} \varphi, \quad \varphi \in D(A_\varphi)$.

Then

(i) $P_\varphi \dashv A_\varphi$ with the bounded intertwining operator $T = (I + Q^2)^{-1}$.
(ii) P_φ is everywhere defined and bounded, but the operator A_φ is closable iff $\varphi \in D(Q^2)$.
(iii) If $\varphi \in D(Q^2)$, A_φ is bounded and everywhere defined, and $\sigma(A_\varphi) = \sigma(P_\varphi) = \{0, 1\}$.

In the second example, A and B are both unbounded. In that case, the two spectra coincide as a whole, but not their individual parts. In particular, A has a nonempty residual spectrum, whereas B does not.

Example 3.6 In $\mathcal{H} = L^2(\mathbb{R}, dx)$, define the two operators

- $(Af)(x) = f'(x) - \dfrac{2x}{1 + x^2} f(x)$, $\quad f \in D(A) = W^{1,2}(\mathbb{R})$
- $(Bf)(x) = f'(x)$, $\quad f \in D(B) = W^{1,2}(\mathbb{R})$

Then

(i) $A \dashv B$ with the bounded intertwining operator $T = (I + Q^2)^{-1}$.
(ii) $\sigma(A) = \sigma(B)$.
(iii) $\sigma_p(A) = \emptyset$, $\sigma_r(A) = \{0\}$, but $\sigma(B) = \sigma_c(B) = i\mathbb{R}$.

It is easy to generalize the preceding analysis to the case of an unbounded intertwining operator, but we have to adapt the definition.

Definition 3.7 Let A, B two densely defined linear operators on the Hilbert spaces $\mathcal{H}, \mathcal{K}$, respectively. A closed (densely defined) operator $T : D(T) \subseteq \mathcal{H} \to \mathcal{K}$ is called an *intertwining operator* for A and B if

(io$_0$) $D(TA) = D(A) \subset D(T)$;
(io$_1$) $T : D(A) \to D(B)$;
(io$_2$) $BT\xi = TA\xi$, $\forall \xi \in D(A)$.

The first part of condition (io$_0$) means that $\xi \in D(A)$ implies $A\xi \in D(T)$.

Then we say again that A is *quasi-similar* to B, $A \dashv B$, if there exists a (possibly unbounded) intertwining operator T for A and B which is invertible, with inverse T^{-1} densely defined. Note that $A \dashv B$ does not imply $B^* \dashv A^*$, since (io$_0$) may fail for B^*. Furthermore, we say that A and B are *mutually quasi-similar* if we have both $A \dashv B$ and $B \dashv A$, which we denote by $A \dashv\vdash B$. Clearly, $\dashv\vdash$ is an equivalence relation and $A \dashv\vdash B$ implies $A^* \dashv\vdash B^*$.

We add that quasi-similarity with an unbounded intertwining operator may occur only under *singular*, even pathological, circumstances. For instance, one may note that, if $A \dashv B$ with the intertwining operator T and the resolvent set $\rho(A)$ is not empty, then T is necessarily bounded.

At this point, one may examine to what extent some parts of Proposition 3.4 survive when the intertwining operator T is no longer bounded, and also what happens if $A \dashv\vdash B$. We refer to [6] for a thorough analysis.

4   The Lattice Generated by a Single Metric Operator

Now we turn to the general case, where G and G^{-1} are both possibly unbounded.

Define $\mathcal{H}(R_G)$ as $D(G^{1/2})$ equipped with the graph norm $\|\xi\|_{R_G}^2 := \|\xi\|^2 + \|G^{1/2}\xi\|^2$. Then define $\mathcal{H}(G)$ as the completion of $\mathcal{H}(R_G)$ in the norm $\|\xi\|_G^2 := \|G^{1/2}\xi\|^2$. It follows that $\mathcal{H}(R_G) = \mathcal{H} \cap \mathcal{H}(G)$, with the projective norm [2, Sec. I.2.1].

Now, since $D(G^{1/2}) = Q(G)$, the form domain of G, we may write

$$\|\xi\|_{R_G}^2 = \langle (1+G)\xi|\xi\rangle = \langle R_G\xi|\xi\rangle, \quad \|\xi\|_G^2 = \langle G\xi|\xi\rangle, \quad \text{with } R_G = 1+G,$$

which justifies the notation $\mathcal{H}(R_G)$.

Next, the conjugate dual $\mathcal{H}(R_G)^\times = \mathcal{H}(R_G^{-1})$, so that

$$\mathcal{H}(R_G) \subset \mathcal{H} \subset \mathcal{H}(R_G^{-1}) = \mathcal{H} + \mathcal{H}(G^{-1}),$$

with the inductive norm [2, Sec. I.2.1]. Putting everything together, we get the lattice shown in Fig. 1.

To give a concrete example, take $G = x^2$ in $L^2(\mathbb{R}, dx)$, so that $R_G = 1 + x^2$. Then all the spaces in the diagram are weighted L^2 spaces, as shown in Fig. 2.

Actually one can go further, following a construction made in [1]. If G is unbounded, $R_G = 1 + G > 1$ and R_G^{-1} bounded, so that we have the triplet $\mathcal{H}(R_G) \subset \mathcal{H} \subset \mathcal{H}(R_G^{-1})$.

Iterating as before, we get the infinite Hilbert scale built on powers of $R_G^{1/2}$, $\mathcal{H}_n = D(R_G^{n/2}), n \in \mathbb{N}$, and $\mathcal{H}_{-n} = \mathcal{H}_n^\times$:

$$\ldots \subset \mathcal{H}_2 \subset \mathcal{H}_1 \subset \mathcal{H} \subset \mathcal{H}_{-1} \subset \mathcal{H}_{-2} \subset \ldots \tag{4.1}$$

Fig. 1 The lattice of Hilbert space s generated by a metric operator

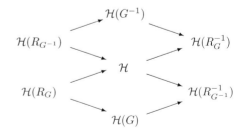

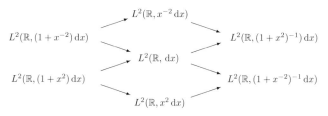

Fig. 2 The lattice of Hilbert space s generated by $G = x^2$

Taking $\mathcal{H} = L^2(\mathbb{R}, \, dx)$, we find familiar examples, namely,

- $G_x = (1 + x^2)^{1/2}$, so that $\mathcal{H}_\infty(G_x^{1/2})$ consists of fast decreasing L^2 functions.
- $G_p = (1 - d^2/dx^2)^{1/2} = \mathcal{F}G_x\mathcal{F}-1$, so that the scale consists of the Sobolev spaces $W^{n,2}(\mathbb{R})$.

An interesting variant of the last example is the LHS of analytic functions described in [2, Sec. 4.6.3], in which the order parameter is the opening angle of a sector, instead of the rate of growth at infinity. This LHS simplifies considerably the formulation of scattering theory, in the form presented by van Winter, as explained in [2, Sec. 7.2]. Let us give some details.

Define $G(a, b)$ $(-\pi < a < b < \pi)$ as the space of all functions $f(z)$, $z = re^{i\varphi}$, which are analytic in the open sector $S_{a,b} := \{z = re^{i\varphi}, \, a < \varphi < b\}$, and such that the integral $\int_0^\infty |f(re^{i\varphi})|^2 \, dr < \infty$ is uniformly bounded in $\varphi \in (a, b)$. It turns out that the family $\{G(a, b), -\frac{\pi}{2} \leqslant a < b \leqslant \frac{\pi}{2}\}$ may be identified, via a Mellin transform, with a part of an LHS of weighted L^2 spaces. First, for $-\frac{\pi}{2} \leqslant a \leqslant \frac{\pi}{2}$, define the Hilbert space

$$L^2(a) := \{f : \int_{-\infty}^{+\infty} e^{ax} |f(x)|^2 \, dx < \infty\} = L^2(r_a), \quad \text{with } r_a(t) = e^{-ax}.$$
(4.2)

Then consider the lattice generated by the family $L^2(a)$, $L^2(0) = L^2$ and $L^2(-a)$, following the construction described previously. The infimum is $L^2(a) \wedge L^2(b) = L^2(a) \cap L^2(b) = L^2(a \wedge b)$ and the supremum is $L^2(a) \vee L^2(b) = L^2(a) + L^2(b) = L^2(a \vee b)$, with $r_{a \wedge b}(x) = \min(r_a(x), r_b(x))$ and $r_{a \vee b}(x) = \max(r_a(x), r_b(x))$. As usual, these norms are equivalent to the projective, resp. inductive, norms. For instance, the following two norms are equivalent

$$\|f\|_{L^2(r_{a \wedge -a})}^2 = \int_{-\infty}^{+\infty} e^{a|x|}|f(x)|^2 \, dx \; \asymp \; \int_{-\infty}^{+\infty} (e^{ax} + e^{-ax})|f(x)|^2 \, dx.$$

Next, the discrete lattice of nine spaces may be converted into a continuous one by interpolation. This yields $\{L^2(a), -\frac{\pi}{2} \leqslant a \leqslant \frac{\pi}{2}\}$. Thus we obtain an LHS, with extreme spaces $V^\# = L^2(-\frac{\pi}{2}) \cap L^2(\frac{\pi}{2})$, $V = L^2(-\frac{\pi}{2}) + L^2(\frac{\pi}{2})$, which are

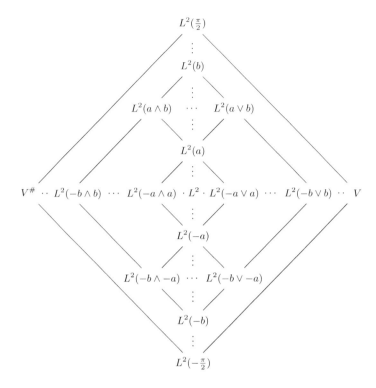

Fig. 3 The van Winter LHS (from [2])

themselves Hilbert space s. In addition, all spaces are obtained at the first generation, i.e., they are all of the form $L^2(c \wedge d)$ or $L^2(c \vee d)$.

In the case $0 < a < b$, one gets the picture shown in Fig. 3. Duality corresponds to symmetry with respect to the center (i.e., L^2): $a \wedge b \iff -b \vee -a$.

5 Quasi-Hermitian Operators

According to Dieudonné [14], a bounded operator A is called quasi-Hermitian if there exists a metric operator G such that $GA = A^*G$. However, this definition is too restrictive for applications, hence we generalize it in order to cover unbounded operators.

Definition 5.1 A closed operator A is called *quasi-Hermitian* if there exists a metric operator G such that $D(A) \subset D(G)$ and

$$\langle A\xi | G\eta \rangle = \langle G\xi | A\eta \rangle, \quad \xi, \eta \in D(A) \tag{5.1}$$

Let us consider first a bounded quasi-Hermitian operator A. If in addition, the metric operator G is bounded with bounded inverse, then (5.1) implies immediately that GA is self-adjoint in $\mathcal{H}$. Actually, there is more.

Proposition 5.2 *Let A be bounded. Then the following statements are equivalent.*

(i) *A is quasi-Hermitian.*
(ii) *There exists a bounded metric operator G, with bounded inverse, such that $GA (= A^*G)$ is self-adjoint.*
(iii) *A is metrically similar to a self-adjoint operator K.*

We turn now to unbounded quasi-Hermitian operators. The following results are easy.

Proposition 5.3 *Let A be an unbounded quasi-Hermitian operator and G a bounded metric operator. Then (i) A is quasi-Hermitian iff GA is symmetric in $\mathcal{H}$; (ii) If A is self-adjoint in $\mathcal{H}(G)$, then GA is symmetric in $\mathcal{H}$. If G^{-1} is also bounded, A is self-adjoint in $\mathcal{H}(G)$ iff GA is self-adjoint in $\mathcal{H}$.*

Now we turn the problem around. Namely, given the closed densely defined operator A, we seek whether there is a metric operator G that makes A quasi-Hermitian and self-adjoint in $\mathcal{H}(G)$. We first obtain a metric operator with bounded inverse.

Proposition 5.4 *Let A be closed and densely defined. Then the following statements are equivalent:*

(i) *There exists a bounded metric operator G, with bounded inverse, such that A is self-adjoint in $\mathcal{H}(G)$.*
(ii) *There exists a bounded metric operator G, with bounded inverse, such that $GA = A^*G$, i.e., A is similar to its adjoint A^*, with intertwining operator G.*
(iii) *There exists a bounded metric operator G, with bounded inverse, such that $G^{1/2}AG^{-1/2}$ is self-adjoint.*
(iv) *A is a spectral operator of scalar type with real spectrum, i.e., $A = \int_{\mathbb{R}} \lambda \, dX(\lambda)$, where $\{X(\lambda)\}$ is a spectral family (not necessarily self-adjoint).*

Instead of requiring that A be similar to A^*, we may ask that they be only quasi-similar. The price to pay is that now G^{-1} is no longer bounded and, therefore, the equivalences stated above are no longer true. Instead Proposition 5.4 is replaced by the following weaker result [5].

Proposition 5.5 *Let A be closed and densely defined. Consider the statements*

(i) *There exists a bounded metric operator G such that $GD(A) = D(A^*)$, $A^*G\xi = GA\xi$, for every $\xi \in D(A)$, in particular, A is quasi-similar to its adjoint A^*, with intertwining operator G.*
(ii) *There exists a bounded metric operator G, such that $G^{1/2}AG^{-1/2}$ is self-adjoint.*
(iii) *There exists a bounded metric operator G such that A is self-adjoint in $\mathcal{H}(G)$; then we say that A is quasi-self-adjoint.*

(iv) *There exists a bounded metric operator G such that $GD(A) = D(G^{-1}A^*)$,*
$A^*G\xi = GA\xi$, for every $\xi \in D(A)$, in particular, A is quasi-similar to its
adjoint A^*, with intertwining operator G.

Then, the following implications hold:

$$(i) \Rightarrow (ii) \Rightarrow (iii) \Rightarrow (iv).$$

If the range $R(A^*)$ of A^* is contained in $D(G^{-1})$, then the four conditions (i)-(iv)
are equivalent.

When G is unbounded, we say that A is *strictly quasi-Hermitian* if it is quasi-
Hermitian, in the sense of Definition 5.1, and $AD(A) \subset D(G)$ or, equivalently,
$D(GA) = D(A)$. Therefore, A is strictly quasi-Hermitian iff $A \dashv A^*$.

More results may be obtained if one uses the PIP-*space formalism*, as we shall
see below.

6   The LHS Generated by Metric Operators

Denote by $\mathcal{M}(\mathcal{H})$ the set of all metric operators and by $\mathcal{M}_b(\mathcal{H})$ the set of bounded
ones. There is a natural order on $\mathcal{M}(\mathcal{H})$

$$G_1 \preceq G_2 \Longleftrightarrow \exists \gamma > 0 \text{ such that } G_2 \leq \gamma G_1$$

$$\Longleftrightarrow \mathcal{H}(G_1) \subset \mathcal{H}(G_2), \text{ where the embedding is continuous and has a}$$

$$\text{dense range.}$$

As a consequence, we have

$$G_2^{-1} \preceq G_1^{-1} \Longleftrightarrow G_1 \preceq G_2 \text{ if } G_1, G_2 \in \mathcal{M}(\mathcal{H})$$

$$G^{-1} \preceq I \preceq G, \ \forall \, G \in \mathcal{M}_b(\mathcal{H})$$

Thus, given $X, Y \in \mathcal{M}(\mathcal{H})$, one has $X \preceq Y \Leftrightarrow \mathcal{H}(X) \hookrightarrow \mathcal{H}(Y)$. We will show
that the spaces $\{\mathcal{H}(X) : X \in \mathcal{M}(\mathcal{H})\}$ constitute a *lattice of Hilbert spaces (LHS)*.

Let $\mathcal{O} \subset \mathcal{M}(\mathcal{H})$ be a family of metric operators, containing I and at least one
unbounded element, and assume that

$$\mathcal{D} := \bigcap_{G \in \mathcal{O}} D(G^{1/2})$$

is a dense subspace of $\mathcal{H}$. Since every operator $G \in \mathcal{O}$ is self-adjoint and invertible,
one can define on $\mathcal{D}$, the graph topology $t_{\mathcal{O}}$ by means of the norms

$$\xi \in \mathcal{D} \mapsto \|G^{1/2}\xi\|, \quad G \in \mathcal{O}.$$

Let $\mathcal{D}^\times$ denote the conjugate dual of $\mathcal{D}[t_\mathcal{O}]$, with strong dual topology $t_\mathcal{O}^\times$. Then the triplet

$$\mathcal{D}[t_\mathcal{O}] \hookrightarrow \mathcal{H} \hookrightarrow \mathcal{D}^\times[t_\mathcal{O}^\times]$$

is the Rigged Hilbert Space associated with $\mathcal{O}$. We will show that $\mathcal{O}$ generates a canonical lattice of Hilbert spaces (LHS) interpolating between $\mathcal{D}$ and $\mathcal{D}^\times$.

On the family $\{\mathcal{H}(X) : X \in \mathcal{O}^{-1}\}$ define the lattice operations as

$$\mathcal{H}(X \wedge Y) := \mathcal{H}(X) \cap \mathcal{H}(Y),$$

$$\mathcal{H}(X \vee Y) := \mathcal{H}(X) + \mathcal{H}(Y),$$

equipped, respectively, with the projective and the inductive norms, namely,

$$\|\xi\|_{X \wedge Y}^2 = \|\xi\|_X^2 + \|\xi\|_Y^2,$$

$$\|\xi\|_{X \vee Y}^2 = \inf_{\xi = \eta + \zeta} \left(\|\eta\|_X^2 + \|\zeta\|_Y^2 \right), \quad \eta \in \mathcal{H}(X), \zeta \in \mathcal{H}(Y).$$

The corresponding operators read as

$$X \wedge Y := X \dotplus Y,$$

$$X \vee Y := (X^{-1} \dotplus Y^{-1})^{-1}.$$

Here $\dotplus$ stands for the *form sum* and $X, Y \in \mathcal{O}^{-1}$: given two positive operators, $T := T_1 \dotplus T_2$ is the positive operator associated with the quadratic form $t = t_1 + t_2$, where t_1, t_2 are the quadratic forms of T_1, T_2, respectively [18, §VI.2.5]. Note that both $X \vee Y$ and $X \wedge Y$ are inverses of a metric operator, but they need not belong to $\mathcal{O}^{-1}$. In particular, for $\mathcal{O} = \mathcal{M}(\mathcal{H})$, the corresponding family $\mathcal{M}(\mathcal{H})^{-1}$ is a lattice by itself, but the domain $\mathcal{D}$ usually fails to be dense.

Define $\mathcal{R} = \{G^{\pm 1/2}, G \in \mathcal{O}\}$ and the domain $\mathcal{D}_\mathcal{R} := \bigcap_{X \in \mathcal{R}} D(X)$. Let Σ be the minimal set of self-adjoint operators containing $\mathcal{O} \cup \mathcal{O}^{-1}$, stable under inversion and form sums, such that $\mathcal{D}_\mathcal{R}$ is dense in $H(Z)$, for every $Z \in \Sigma$.

Then $\mathcal{O}$ generates a lattice of Hilbert space s $\mathcal{J}_\Sigma := \{\mathcal{H}(X) : X \in \Sigma\}$ and a PIP-space V_Σ with central Hilbert space $\mathcal{H} = \mathcal{H}(I)$. The total space is $V = \sum_{G \in \Sigma} \mathcal{H}(G)$ (algebraic inductive limit, in general) and the "smallest" space is $V^\# = \mathcal{D}_\mathcal{R}$. The compatibility and the partial inner product read, respectively, as

$$\xi \# \eta \iff \exists G \in \Sigma \text{ such that } \xi \in \mathcal{H}(G), \eta \in \mathcal{H}(G^{-1}),$$

$$\langle \xi | \eta \rangle_\Sigma = \langle G^{1/2} \xi | G^{-1/2} \eta \rangle_\mathcal{H}.$$

For simplicity, we write $\langle \xi | \eta \rangle_\Sigma = \langle \xi | \eta \rangle$.

For instance, for $\mathcal{O} = \{I, G\}$, the lattice Σ consists of the nine operators shown in Fig. 4.

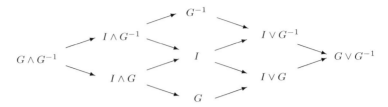

Fig. 4 The lattice Σ generated by the metric operator G

7 (Quasi-)Similarity for PIP-Space Operators

7.1 General PIP-Space Operators

Given the PIP-space V_Σ, an *operator* A on V_Σ is a map from a subset $\mathcal{D}(A) \subset V$ into V, such that

(i) $\mathcal{D}(A) = \bigcup_{X \in \mathsf{d}(A)} \mathcal{H}(X)$, where $\mathsf{d}(A)$ is a nonempty subset of Σ;
(ii) For every $X \in \mathsf{d}(A)$, there exists $Y \in \Sigma$ such that the restriction of A to $\mathcal{H}(X)$ is a continuous linear map into $\mathcal{H}(Y)$ (we denote this restriction by A_{YX});
(iii) A has no proper extension satisfying (i) and (ii).

We denote by $\mathrm{Op}(V_\Sigma)$ the set of all operators on V_Σ. The continuous linear operator $A_{YX} : \mathcal{H}(X) \to \mathcal{H}(Y)$ is called a *representative* of A.

The properties of the operator A are encoded in the set $\mathsf{j}(A)$ of couples $(X, Y) \in \Sigma \times \Sigma$ such that $A : \mathcal{H}(X) \to \mathcal{H}(Y)$, continuously. Thus the operator A may be identified with the collection of its representatives, $A \simeq \{A_{YX} : (X, Y) \in \mathsf{j}(A)\}$. This is a *coherent* family, that is, if $\mathcal{H}(W) \subset \mathcal{H}(X)$ and $\mathcal{H}(Y) \subset \mathcal{H}(Z)$, then one has $A_{ZW} = E_{ZY} A_{YX} E_{XW}$ ($E_{..} \simeq$ identity). More generally, $(X, Y) \in \mathsf{j}(A)$ if $Y^{1/2} A X^{-1/2}$ is bounded in $\mathcal{H}$.

Every operator has an *adjoint* $A^\times$ defined as follows: $(X, Y) \in \mathsf{j}(A)$ implies $(Y^{-1}, X^{-1}) \in \mathsf{j}(A^\times)$ and

$$\langle A^\times \eta | \xi \rangle = \langle \eta | A \xi \rangle, \quad \text{for } \xi \in \mathcal{H}(X), \eta \in \mathcal{H}(Y^{-1}).$$

In particular, $(X, X) \in \mathsf{j}(A)$ implies $(X^{-1}, X^{-1}) \in \mathsf{j}(A^\times)$.

An operator A is *symmetric* if $A = A^\times$. Therefore, $(X, X) \in \mathsf{j}(A)$ implies $(X^{-1}, X^{-1}) \in \mathsf{j}(A)$. Then, by interpolation, $(I, I) \in \mathsf{j}(A)$, that is, A has a bounded representative $A_{II} : \mathcal{H} \to \mathcal{H}$.

We will now examine the quasi-similarity properties of PIP-space operators.

(1) Let first $(G, G) \in \mathsf{j}(A)$, for some $G \in \mathcal{M}(\mathcal{H})$. Then the operator $\mathsf{B} = G^{1/2} A_{GG} G^{-1/2}$ is bounded on $\mathcal{H}$ and $A_{GG} \dashv \mathsf{B}$.
(2) Next, let $(G, G) \in \mathsf{j}(A)$, with G bounded and G^{-1} unbounded, so that

$$\mathcal{H}(G^{-1}) \subset \mathcal{H} \subset \mathcal{H}(G).$$

Consider the restriction A of A_{GG} to $\mathcal{H}$ and assume that $D(A) = \{\xi \in \mathcal{H} : A\xi \in \mathcal{H}\}$ is dense in $\mathcal{H}$. Then $G^{1/2} : D(A) \to D(B)$ and $B\,G^{1/2}\eta = G^{1/2}A\,\eta$, $\forall\,\eta \in D(A)$, i.e. $A \dashv B$, where the two operators act in $\mathcal{H}$.

Now we have $B\,G^{1/2}\eta = G^{1/2}A\,\eta$, $\forall\,\eta \in \mathcal{H}(G)$ and $G^{1/2} : \mathcal{H}(G) \to \mathcal{H}$ is a unitary operator. Therefore, A and B are unitarily equivalent (but acting in different Hilbert spaces).

(3) Let finally $(G, G) \in \mathsf{j}(A)$ with G unbounded and G^{-1} bounded, so that

$$\mathcal{H}(G) \subset \mathcal{H} \subset \mathcal{H}(G^{-1}).$$

Then $A : \mathcal{H}(G) \to \mathcal{H}(G)$ is a densely defined operator in $\mathcal{H}$. Since $B = G^{1/2}A_{GG}G^{-1/2}$ is bounded and everywhere defined on $\mathcal{H}$, one has $G^{-1/2}B\xi = A_{GG}G^{-1/2}\xi$, $\forall\,\xi \in \mathcal{H}$, i.e., $B \dashv A_{GG}$. However, by (1), $A_{GG} \dashv B$, hence we have $A_{GG} \dashv\vdash B$. In addition $A_{GG} \overset{u}{\sim} B$, since $G^{\pm 1/2}$ are unitary between $\mathcal{H}$ and $\mathcal{H}(G)$.

7.2 The Case of Symmetric PIP-Space Operators

If $A \in \mathrm{Op}(V_\Sigma)$ is symmetric, $A = A^\times$, there is a possibility of self-adjoint *restrictions* to $\mathcal{H}$, that is, candidates for quantum observables.

However, if $A = A^\times$, then $(G, G) \in \mathsf{j}(A)$ iff $(G^{-1}, G^{-1}) \in \mathsf{j}(A)$, which implies $(I, I) \in \mathsf{j}(A)$. Thus, every symmetric operator $A \in \mathrm{Op}(V_\Sigma)$ such that $(G, G) \in \mathsf{j}(A)$, with $G \in \mathcal{M}(\mathcal{H})$, has a *bounded* restriction A_{II} to $\mathcal{H}$.

Therefore, we conclude that the assumption $(G, G) \in \mathsf{j}(A)$ is too strong for applications ! Thus we will assume instead that $(G^{-1}, G) \in \mathsf{j}(A)$, where G is bounded with unbounded inverse, so that

$$\mathcal{H}(G^{-1}) \subset \mathcal{H} \subset \mathcal{H}(G).$$

In that case, one can apply the *KLMN theorem*,[1] namely,

Given a symmetric operator $A = A^\times$, assume there is a metric operator $G \in \mathcal{M}_b(\mathcal{H})$ with an unbounded inverse, for which there exists a $\lambda \in \mathbb{R}$ such that $A - \lambda I$ has a boundedly invertible representative $(A - \lambda I)_{GG^{-1}} : \mathcal{H}(G^{-1}) \to \mathcal{H}(G)$. Then $A_{GG^{-1}}$ has a unique restriction to a self-adjoint operator A in the Hilbert space $\mathcal{H}$, with dense domain $D(A) = \{\xi \in \mathcal{H} : A\xi \in \mathcal{H}\}$. In addition, $\lambda \in \rho(A)$.

If there is no bounded G as before, i.e. $(G^{-1}, G) \in \mathsf{j}(A)$, one can still use the KLMN theorem, but in the Hilbert scale V_G built on the powers of $G^{-1/2}$ or $(R_G)^{-1/2}$.

---

[1]KLMN stands for Kato, Lax, Lions, Milgram, Nelson.

Fig. 5 The semi-similarity
scheme

Let $V_{\mathcal{G}} = \{\mathcal{H}_n, n \in \mathbb{Z}\}$ be the Hilbert scale built on the powers of the operator $G^{\pm 1/2}$ or $(R_G)^{-1/2}$, depending on the (un)boundedness of $G^{\pm 1} \in \mathcal{M}(\mathcal{H})$ and let $A = A^{\times}$ be a symmetric operator in $V_{\mathcal{G}}$.

(i) Assume there is a $\lambda \in \mathbb{R}$ such that $A - \lambda I$ has a boundedly invertible representative $(A - \lambda I)_{nm} : \mathcal{H}_m \to \mathcal{H}_n$, with $\mathcal{H}_m \subset \mathcal{H}_n$. Then A_{nm} has a unique restriction to a self-adjoint operator A in the Hilbert space $\mathcal{H}$, with dense domain $D(\mathsf{A}) = \{\xi \in \mathcal{H} : A\xi \in \mathcal{H}\}$. In addition, $\lambda \in \rho(\mathsf{A})$.

(ii) If the natural embedding $\mathcal{H}_m \to \mathcal{H}_n$ is compact, the operator A has a purely point spectrum of finite multiplicity, thus $\sigma(\mathsf{A}) = \sigma_p(\mathsf{A})$, $m_{\mathsf{A}}(\lambda_j) < \infty$ for every $\lambda_j \in \sigma_p(\mathsf{A})$ and $\sigma_c(\mathsf{A}) = \emptyset$.

Note, however, that there is so far no known (quasi-)similarity relation between $A_{GG^{-1}}$ or A and another operator! On the contrary, under the previous assumption $A : \mathcal{H}(G^{-1}) \to \mathcal{H}(G)$, $\mathsf{B} = G^{1/2}A_{GG^{-1}}G^{1/2}$ is bounded on $\mathcal{H}$, but $A_{GG^{-1}} \not\dashv \mathsf{B}$. Indeed, (io_1) imposes $T = G^{-1/2}$, hence unbounded, but then conditions (io_0) and (io_2) cannot be satisfied.

8 Semi-Similarity of PIP-Space Operators

So far we have considered only the case of one metric operator G in relation to A. Assume now we take two different metric operators $G_1, G_2 \in \mathcal{M}(\mathcal{H})$. What can be said concerning A if it maps $\mathcal{H}(G_1)$ into $\mathcal{H}(G_2)$?

One possibility is to introduce, following [4], a notion slightly more general than quasi-similarity, called *semi-similarity*.

Definition 8.1 Let $\mathcal{H}, \mathcal{K}_1$, and $\mathcal{K}_2$ be three Hilbert spaces, A a closed, densely defined operator from $\mathcal{K}_1$ to $\mathcal{K}_2$, B a closed, densely defined operator on $\mathcal{H}$. Then A is said to be *semi-similar* to B, which we denote by $A \dashv\vdash B$, if there exist two bounded operators $T : \mathcal{K}_1 \to \mathcal{H}$ and $S : \mathcal{K}_2 \to \mathcal{H}$ such that (see Fig. 5):

(i) $T : D(A) \to D(B)$;
(ii) $BT\xi = SA\xi$, $\forall \xi \in D(A)$.

The pair (T, S) is called an *intertwining couple*.

Of course, if $\mathcal{K}_1 = \mathcal{K}_2$ and $S = T$, we recover the notion of quasi-similarity and $A \dashv B$ (with a bounded intertwining operator).

Assume there exist two bounded metric operators G_1, G_2 such that $A : \mathcal{H}(G_1) \to \mathcal{H}(G_2)$ continuously. Then $\mathsf{B}_0 := G_2^{1/2}A_{G_2G_1}G_1^{-1/2}$ has a bounded extension B to $\mathcal{H}$ (its closure) and $A_{G_2 G_1} \dashv\vdash B$, with respect to the intertwining couple $T = G_1^{1/2}, S = G_2^{1/2}$.

Take now $A = A^\times$ symmetric. Then $A : \mathcal{H}(G_1) \to \mathcal{H}(G_2)$ implies $A : \mathcal{H}(G_2^{-1}) \to \mathcal{H}(G_1^{-1})$. Assume that $G_1 \preceq G_2$, that is, $\mathcal{H}(G_1) \subset \mathcal{H}(G_2)$. Then we have

$$\mathcal{H}(G_2^{-1}) \subset \mathcal{H}(G_1^{-1}) \subset \mathcal{H} \subset \mathcal{H}(G_1) \subset \mathcal{H}(G_2).$$

It follows that the KLMN theorem applies. Assume indeed there exists $\lambda \in \mathbb{R}$ such that $A - \lambda I$ has an invertible representative $(A - \lambda I)_{G_2 G_2^{-1}} : \mathcal{H}(G_2^{-1}) \to \mathcal{H}(G_2)$. Then $A_{G_2 G_2^{-1}}$ has a unique restriction to a *self-adjoint* operator A in $\mathcal{H}$, hence $A_{G_2 G_2^{-1}} \dashv\vdash B$ and $A \dashv\vdash B$. A question remains open, namely, A is self-adjoint, but is the spectrum of B real?

In conclusion, there are three cases: if $A : \mathcal{H}(G_1) \to \mathcal{H}(G_2)$, then

(i) G_1 is unbounded and G_2 is bounded: then

$$\mathcal{H}(G_1) \subset \mathcal{H} \subset \mathcal{H}(G_2),$$

and A maps the small space into the large one, thus the KLMN theorem applies.
(ii) G_1 and G_2 are both unbounded, with $\mathcal{H}(G_1) \subset \mathcal{H}(G_2)$; then the KLMN theorem applies.
(iii) G_1 is bounded and G_2 is unbounded: then

$$\mathcal{H}(G_2) \subset \mathcal{H} \subset \mathcal{H}(G_1) \quad \text{and} \quad \mathcal{H}(G_1^{-1}) \subset \mathcal{H} \subset \mathcal{H}(G_2^{-1}),$$

so that, in both cases, A maps the large space into the small one; hence, the KLMN theorem does *not* apply.

9 Pseudo-Hermitian Hamiltonians

Non-self-adjoint Hamiltonians appear in *Pseudo-Hermitian quantum mechanics* [19, 20]. In general, they are $\mathcal{PT}$-symmetric operators, that is, invariant under the joint action of space reflection ($\mathcal{P}$) and complex conjugation ($\mathcal{T}$). Typical examples are $H = p^2 + ix^3$ and $H = p^2 - x^4$, which are both $\mathcal{PT}$-symmetric, but non-self-adjoint, and have both a purely point spectrum, real and positive.

Now, the usual assumption is that H is pseudo-Hermitian in the sense of Dieudonné [14], that is, there exists an (unbounded) metric operator G satisfying the relation $H^* G = G H$.

Assume instead that H is *pseudo-Hermitian*, that is, $D(H) \subset D(G)$ and

$$\langle H\xi | G\eta \rangle = \langle G\xi | H\eta \rangle, \quad \forall \xi, \eta \in \mathcal{D}(H).$$

Then, if G is bounded, one gets $H \dashv H^*$ and $G^{1/2}HG^{-1/2}$ is self-adjoint. If G is unbounded and H is strictly quasi-Hermitian, then $H \dashv H^*$. If, in addition, G^{-1} is bounded, then $G^{-1}H^*G\eta = H\eta$, $\forall \eta \in \mathcal{D}(H)$, which is a restrictive form of similarity.

Finally, assume that H is a quasi-Hermitian operator which possesses a (large) set of vectors, $\phi \in \mathcal{D}_G^\omega(H)$, *analytic* in the norm $\|\cdot\|_G$ and contained in $D(G)$ [21], that is,

$$\sum_{n=0}^\infty \frac{\|H^n \phi\|_G}{n!} t^n < \infty, \text{ for some } t \in \mathbb{R}.$$

Thus, $\mathcal{D}_G^\omega(H) \subset D(H) \subset D(G) \subset D(G^{1/2}) \subset \mathcal{H}$.

Under this assumption, we can proceed to the construction of the physical system, following [4, Sec.6]. Define $\mathcal{H}_G$ as the completion of $\mathcal{D}_G^\omega(H)$ in the norm $\|\cdot\|_G$. This is a closed subspace of $\mathcal{H}(G)$ and one has

$$\langle \phi | H\psi \rangle_G = \langle H\phi | \psi \rangle_G, \quad \forall \phi, \psi \in \mathcal{D}_G^\omega(H).$$

Thus H is a densely defined symmetric operator in $\mathcal{H}_G$, with a dense set of analytic vectors. Therefore, H essentially self-adjoint, according to Nelson's theorem [21].

Then the closure $\overline{H}$ of H is self-adjoint in $\mathcal{H}_G$. The pair $(\mathcal{H}_G, \overline{H})$ may be interpreted as the physical quantum system.

Next, $W_\mathcal{D} = G^{1/2} \upharpoonright \mathcal{D}_G^\omega(H)$ is isometric from $\mathcal{D}_G^\omega(H)$ into $\mathcal{H}$, hence it extends to an isometry $W = \overline{W_\mathcal{D}} : \mathcal{H}_G \to \mathcal{H}$. The range of W is a closed subspace of $\mathcal{H}$, denoted by $\mathcal{H}_\text{phys}$, and the operator W is unitary from $\mathcal{H}_G$ to $\mathcal{H}_\text{phys}$. Therefore, the operator $h = W \overline{H} W^{-1}$ is self-adjoint in $\mathcal{H}_\text{phys}$. This operator h is interpreted as the genuine Hamiltonian of the system, acting in the physical Hilbert space $\mathcal{H}_\text{phys}$.

The situation becomes simpler if $\mathcal{D}_G^\omega(H)$ is dense in $\mathcal{H}$. Then, indeed, $W(\mathcal{D}_G^\omega(H))$ is also dense, $\mathcal{H}_G = \mathcal{H}(G)$, $\mathcal{H}_\text{phys} = \mathcal{H}$ and $W = G^{1/2}$ is unitary from $\mathcal{H}(G)$ onto $\mathcal{H}$.

Now, every eigenvector of an operator is automatically analytic, hence this construction generalizes that of [20]. This applies, for instance, to the example given there, namely, the $\mathcal{PT}$-symmetric operator $H = \frac{1}{2}(p-i\alpha)^2 + \frac{1}{2}\omega^2 x^2$ in $\mathcal{H} = L^2(\mathbb{R})$, for any $\alpha \in \mathbb{R}$, which has an orthonormal basis of eigenvectors.

A beautiful example of the situation just analyzed has been given recently by Samsonov [23], namely, the second derivative on the positive half-line, with special boundary conditions at the origin (this example stems from Schwartz [26]).

References

1. J.-P. Antoine, P. Balazs, Frames, semi-frames, and Hilbert scales. Numer. Funct. Anal. Optim. **33**, 1–34 (2012)
2. J.-P. Antoine, C. Trapani, *Partial Inner Product Spaces: Theory and Applications*. Lecture Notes in Mathematics, vol. 1986 (Springer, Berlin, 2009)
3. J.-P. Antoine, C. Trapani, Metric operators, generalized hermiticity and partial inner product spaces. Presented at *Contemporary Problems in Mathematical Physics (Eighth Int. Workshop Cotonou, Bénin, Nov. 2013)*. ICMPA-UNESCO Chair, Univ. of Abomey-Calavi, Bénin
4. J.-P. Antoine, C. Trapani, Partial inner product spaces, metric operators and generalized hermiticity. J. Phys. A: Math. Theor. **46**, 025204 (2013); Corrigendum, J. Phys. A: Math. Theor. **46**, 272703 (2013)
5. J.-P. Antoine, C. Trapani, Some remarks on quasi-Hermitian operators. J. Math. Phys. **55**, 013503 (2014)
6. J.-P. Antoine, C. Trapani, Metric operators, generalized hermiticity and lattices of Hilbert spaces (Chap. 7), in *Non-Selfadjoint Operators in Quantum Physics: Mathematical Aspects*, ed. by F. Bagarello, J.-P. Gazeau, F.H. Szafraniec, M. Znojil (Wiley, Hoboken, 2015), pp. 345–402
7. F. Bagarello, From self-adjoint to non-self-adjoint harmonic oscillators: physical consequences and mathematical pitfalls. Phys. Rev. A **88**, 0321120 (2013)
8. F. Bagarello, A. Fring, Non-self-adjoint model of a two-dimensional noncommutative space with an unbounded metric. Phys. Rev. A **88**, 0421119 (2013)
9. F. Bagarello, M. Znojil, Nonlinear pseudo-bosons versus hidden hermiticity. II. The case of unbounded operators. J. Phys. A: Math. Theor. **45**, 115311 (2012)
10. C.M. Bender, Making sense of non-Hermitian Hamiltonians. Rep. Prog. Phys. **70**, 947–1018 (2007)
11. C.M. Bender, A. Fring, U. Günther, H. Jones, Quantum physics with non-Hermitian operators. J. Phys. A: Math. Theor. **45**, 440301 (2012)
12. C.M. Bender, M. DeKieviet, S.P. Klevansky, $\mathcal{PT}$ quantum mechanics. Phil. Trans. R. Soc. Lond. **371**, 20120523 (2013)
13. J. Bergh, J. Löfström, *Interpolation Spaces* (Springer, Berlin, 1976)
14. J. Dieudonné, Quasi-Hermitian operators, in *Proc. Int. Symposium on Linear Spaces, Jerusalem 1960* (Pergamon Press, Oxford, 1961), pp. 115–122
15. N. Dunford, A survey of the theory of spectral operators. Bull. Amer. Math. Soc. **64**, 217–274 (1958)
16. N. Dunford, J.T. Schwartz, *Linear Operators. Part I: General Theory; Part II: Spectral Theory; Part III: Spectral Operators* (Interscience, New York, 1957/1963/1971)
17. H.B. Geyer, W.D. Heiss, F.G. Scholtz, Non-Hermitian Hamiltonians, metric, other observables and physical implications. arXiv:0710.5593v1 (2007)
18. T. Kato, *Perturbation Theory for Linear Operators* (Springer, Berlin, 1976)
19. A. Mostafazadeh, Pseudo-Hermitian representation of quantum mechanics. Int. J. Geom. Methods Mod. Phys. **7**, 1191–1306 (2010)
20. A. Mostafazadeh, Pseudo–Hermitian quantum mechanics with unbounded metric operators. Phil. Trans. R. Soc. Lond. **371**, 20120050 (2013)
21. E. Nelson, Analytic vectors, Ann. Math. **70**, 572–615 (1959)
22. M. Reed, B. Simon, *Methods of Modern Mathematical Physics. I. Functional Analysis* (Academic, New York/London, 1972/1980)
23. B.F. Samsonov, Hermitian Hamiltonian equivalent to a given non-Hermitian one: manifestation of spectral singularity. Phil. Trans. R. Soc. Lond. **371**, 20120044 (2013)

24. K. Schmüdgen, *Unbounded Self-Adjoint Operators on Hilbert Space* (Springer, Dordrecht/Heidelberg, 2012)
25. F.G. Scholtz, H.B. Geyer, F.J.W. Hahne, Quasi-Hermitian operators in quantum mechanics and the variational principle. Ann. Phys. NY **213**, 74–101 (1992)
26. J. Schwartz, Some non-selfadjoint operators. Commun. Pure Appl. Math. **13**, 609–639 (1960)
27. M. Znojil, Three-Hilbert space formulation of quantum mechanics. Symm. Integr. Geom. Methods Appl. (SIGMA) **5**, 001 (2009)

Beyond Frames: Semi-frames and Reproducing Pairs

Jean-Pierre Antoine and Camillo Trapani

Abstract Frames are nowadays a standard tool in many areas of mathematics, physics, and engineering. However, there are situations where it is difficult, even impossible, to design an appropriate frame. Thus there is room for generalizations, obtained by relaxing the constraints. A first case is that of semi-frames, in which one frame bound only is satisfied. Accordingly, one has to distinguish between upper and lower semi-frames. We will summarize this construction. Even more, one may get rid of both bounds, but then one needs two basic functions and one is led to the notion of reproducing pair. It turns out that every reproducing pair generates two Hilbert spaces, conjugate dual of each other. We will discuss in detail their construction and provide a number of examples, both discrete and continuous. Next, we notice that, by their very definition, the natural environment of a reproducing pair is a partial inner product space (PIP-space) with an L^2 central Hilbert space. A first possibility is to work in a rigged Hilbert space. Then, after describing the general construction, we will discuss two characteristic examples, namely, we take for the partial inner product space a Hilbert scale or a lattice of L^p spaces.

Keywords Frames · Upper and lower semi-frames · Reproducing Pairs · Rigged Hilbert space · Partial inner product spaces

J.-P. Antoine (✉)
Institut de Recherche en Mathématique et Physique, Université catholique de Louvain,
Louvain-la-Neuve, Belgium
e-mail: jean-pierre.antoine@uclouvain.be

C. Trapani
Dipartimento di Matematica e Informatica, Università di Palermo, Palermo, Italy
e-mail: camillo.trapani@unipa.it

© Springer Nature Switzerland AG 2018
T. Diagana, B. Toni (eds.), *Mathematical Structures and Applications*,
STEAM-H: Science, Technology, Engineering, Agriculture,
Mathematics & Health, https://doi.org/10.1007/978-3-319-97175-9_2

1 Introduction

Representing functions in terms of simple ones, preferably with a small number of them, is a recurrent problem in analysis. It is particularly acute in signal and image processing, where transmission imposes severe constraints. Such signals are usually taken as square integrable functions on some manifold, hence they constitute a Hilbert space.

More generally, given a separable Hilbert space $\mathcal{H}$, one seeks to expand an arbitrary element $f \in \mathcal{H}$ in a sequence of simple, basic elements (atoms) $\Psi = (\psi_k)$, $k \in \Gamma$, with Γ a countable index set:

$$f = \sum_{k \in \Gamma} c_k \psi_k, \tag{1.1}$$

where the sum converges in an adequate fashion (e.g., in norm or unconditionally) and the coefficients c_k are (preferably) unique and easy to compute. There are several possibilities for obtaining that result. Namely, we can require that Ψ be:

(i) an orthonormal basis: the coefficients are unique, namely, $c_k = \langle \psi_k | f \rangle$, the convergence is unconditional;

(ii) a Riesz basis, i.e., $\psi_k = V e_k$, where (e_k) is an orthonormal basis and V is bounded bijective operator; the coefficients are unique, namely, $c_k = \langle \phi_k | f \rangle$, where (ϕ_k) is a unique Riesz basis dual to $(V e_k)$; the convergence is unconditional.

These two notions solve the problem, but they are very rigid and often not very manageable, leading mostly to infinite expansions. Thus frames were introduced for ensuring a better flexibility, originally in 1952 by Duffin and Schaeffer [19] in the context of nonharmonic analysis. The notion was revived by Daubechies, Grossmann, and Meyer [18] in the early stages of wavelet theory and then became a very popular topic, in particular in Gabor and wavelet analysis [14, 16, 17, 24]. The reason is that a good frame in a Hilbert space is almost as good as an orthonormal basis for expanding arbitrary elements (albeit non-uniquely) and is often easier to construct. In order to put the present work in perspective, we recall that a sequence $\Psi = (\psi_k)$ is a frame for a Hilbert space $\mathcal{H}$ if there exist constants $0 < \mathsf{m} \leqslant \mathsf{M} < \infty$ (the frame bounds) such that

$$\mathsf{m} \|f\|^2 \leq \sum_{k \in \Gamma} |\langle \psi_k | f \rangle|^2 \leq \mathsf{M} \|f\|^2, \forall f \in \mathcal{H}. \tag{1.2}$$

Actually frames are most often considered in the discrete case, for instance in signal processing [16]. However, continuous frames have also been studied and offer interesting mathematical problems. They have been introduced originally by Ali, Gazeau, and one of us [1, 2] and also, independently, by Kaiser [25]. Since then, several papers dealt with various aspects of the concept, see, for instance,

[13, 21, 22] or [27]. The next step towards numerical applications will be, of course, discretization, but this is not our purpose in this chapter.

However, there may occur situations where it is impossible to satisfy both frame bounds at the same time. Therefore, several generalizations of frames have been introduced. Semi-frames [4, 5], for example, are obtained when functions only satisfy one of the two frame bounds. It turns out that a large portion of frame theory can be extended to this larger framework, in particular the notion of duality.

More recently, a new generalization of frames was introduced by Balazs and Speckbacher [30], namely, reproducing pairs. Here, given a measure space (X, μ), one considers a couple of weakly measurable functions (ψ, ϕ), instead of a single mapping, and one studies the correlation between the two (a precise definition is given below). This definition also includes the original definition of a continuous frame [1, 2] to which it reduces when $\psi = \phi$. The increase of freedom in choosing the mappings ψ and ϕ, however, leads to the problem of characterizing the range of the analysis operators, which in general need no more be contained in $L^2(X, \mathrm{d}\mu)$, as in the frame case. Therefore, it is natural to extend the theory to the case where the weakly measurable functions take their values in a partial inner product space (PIP-space), for instance, a rigged Hilbert space or a Hilbert scale.

The paper is organized as follows. In Sect. 2, we review the notions of frames and semi-frames and recall their salient properties. In Sect. 3 we introduce reproducing pairs, in particular their duality properties. Then, in Sect. 4, we discuss briefly the existence and uniqueness of reproducing partners. In Sect. 6, we motivate the link between reproducing pairs and PIP-spaces, first a Rigged Hilbert space (RHS) in Sect. 7, then a general PIP-space, more precisely a lattice of Banach spaces (LBS) or a lattice of Hilbert spaces (LHS), in Sect. 8. Finally, in Sects. 9 and 10, respectively, we examine two particular cases, namely, a Hilbert scale and a lattice of L^p spaces.

2   Preliminaries: Frames and Semi-frames

2.1   Frames

Before proceeding, we list our definitions and conventions. The framework is a (separable) Hilbert space $\mathcal{H}$, with the inner product $\langle \cdot | \cdot \rangle$ linear in the first factor. Given an operator A on $\mathcal{H}$, we denote its domain by $D(A)$, its range by Ran (A) and its kernel by Ker (A). $GL(\mathcal{H})$ denotes the set of all invertible bounded operators on $\mathcal{H}$ with bounded inverse. Throughout the paper, we will consider weakly measurable functions $\psi : X \to \mathcal{H}$, where (X, μ) is a locally compact space with a Radon measure μ, that is, $\langle f | \psi_x \rangle$ is μ-measurable for every $f \in \mathcal{H}$.

The weakly measurable function ψ is a *continuous frame* if there exist constants $\mathsf{m} > 0$ and $\mathsf{M} < \infty$ (the frame bounds) such that

$$\mathsf{m} \, \|f\|^2 \leqslant \int_X |\langle f|\psi_x\rangle|^2 \, \mathrm{d}\mu(x) \leq \mathsf{M} \, \|f\|^2 , \forall \, f \in \mathcal{H}. \tag{2.1}$$

Given the continuous frame ψ, the *analysis* operator $C_\psi : \mathcal{H} \rightarrow L^2(X, d\mu)$ is defined[1] as

$$(C_\psi f)(x) = \langle f | \psi_x \rangle, \quad f \in \mathcal{H}, \tag{2.2}$$

and the corresponding *synthesis* operator $C_\psi^* : L^2(X, d\mu) \rightarrow \mathcal{H}$ as (the integral being understood in the weak sense, as usual)

$$C_\psi^* \xi = \int_X \xi(x) \, \psi_x \, d\mu(x), \quad \text{for } \xi \in L^2(X, d\mu). \tag{2.3}$$

We set $S_\psi := C_\psi^* C_\psi$, i.e.,

$$\langle f | S_\psi f \rangle = \int_X |\langle f | \psi_x \rangle|^2 \, d\mu(x). \tag{2.4}$$

Thus the so-called *frame or resolution operator* S_ψ is self-adjoint, invertible, bounded with bounded inverse S_ψ^{-1}, that is, $S_\psi \in GL(\mathcal{H})$.

In particular, if X is a discrete set with μ being a counting measure, we recover the standard definition (1.2) of a (discrete) frame [14, 16, 19].

An important concept in frame theory is that of duality. Given a frame $\Psi = \{\psi_x\}$, one says that a frame $\Phi = \{\phi_x\}$ is *dual* to the frame $\{\psi_x\}$ if one has

$$\langle f | g \rangle = \int_X \langle f | \phi_x \rangle \langle \psi_x | g \rangle \, d\mu(x), \quad \forall \, f, g \in \mathcal{H}. \tag{2.5}$$

Then Ψ is dual to Φ as well. The dual of a given frame Ψ is not unique in general, but one of them is distinguished, namely, the canonical dual $\widetilde{\psi}_x := S^{-1}\psi_x$.

2.2 Semi-frames

In practice, there are situations where the notion of frame is too restrictive, in the sense that one cannot satisfy *both* frame bounds simultaneously. Thus there is room for two natural generalizations, namely, we say that a family Ψ is an *upper (resp. lower) semi-frame*, if

 (i) Ψ is total in $\mathcal{H}$;
(ii) Ψ satisfies the upper (resp. lower) frame inequality in (2.1).

Note that the lower frame inequality automatically implies that the family is total, i.e., (ii) $\Rightarrow$ (i) for a lower semi-frame.

---

[1] As usual, we identify a function f with its residue class in $L^2(X, d\mu)$.

Let first Ψ be a (continuous) *upper semi-frame*, that is, there exists a constant $0 < M < \infty$ such that

$$0 < \int_X |\langle f|\psi_x\rangle|^2 \, d\mu(x) \leqslant M \|f\|^2, \ \forall f \in \mathcal{H}, \ f \neq 0. \tag{2.6}$$

In this case, Ψ is a total set in $\mathcal{H}$, the operators C_ψ and S_ψ are bounded, S_ψ is injective and self-adjoint. Therefore $\mathsf{Ran}\,(S_\psi)$ is dense in $\mathcal{H}$ and S_ψ^{-1} is also self-adjoint. Thus, if Ψ is an upper semi-frame and not a frame, S_ψ is bounded and S_ψ^{-1} is unbounded, as follows immediately from (2.6).

Note that, if a family Ψ verifies the upper frame bound only, the map $x \mapsto \psi_x$ is often called a Bessel mapping. More precisely, Bessel maps are those for which $\int_X |\langle f|\psi_x\rangle|^2 \, d\mu(x) < \infty$. By a standard argument based on the closed graph theorem, one gets the inequality on the right of (2.6).

We notice that an upper semi-frame Ψ is a frame if and only if there exists another upper semi-frame Φ which is dual to Ψ, in the sense of (2.5) [22].

Next, we say that a family $\Phi = \{\phi_x\}$ is a *lower semi-frame* if it satisfies the lower frame condition, that is, there exists a constant $m > 0$ such that

$$m \|f\|^2 \leq \int_X |\langle \phi_x|f\rangle|^2 \, d\mu(x), \ \forall f \in \mathcal{H}. \tag{2.7}$$

Clearly, (2.7) implies that the family Φ is total in $\mathcal{H}$.

Following the terminology of Young [34] in the discrete case, we may call *moment space* of a measurable function ψ the range of its analysis operator, $C_\psi(\mathcal{H})$. Then one may say that a measurable function ψ is Bessel or an upper semi-frame if its moment space is contained in $L^2(X, d\mu)$. On the contrary, a measurable function ϕ is a lower semi-frame if its moment space contains $L^2(X, d\mu)$.

In the lower case, the definition of S_ϕ must be changed, since C_ϕ need not be densely defined, so that C_ϕ^* may not be well-defined. Instead, following [4, Sec.2] one defines the analysis operator (2.2) on the domain

$$D(C_\phi) = \{f \in \mathcal{H} : \int_X |\langle f|\phi_x\rangle|^2 \, d\mu(x) < \infty\},$$

which need not be dense. As for the synthesis operator, we put

$$D_\phi F = \int_X F(x) \phi_x \, d\mu(x), \quad F \in L^2(X, d\mu), \tag{2.8}$$

on the domain of all elements F for which the integral in (2.8) converges weakly in $\mathcal{H}$. Defining $S_\phi := D_\phi C_\phi$, it is shown in [4, Sec.2] that S_ϕ is unbounded and S_ϕ^{-1} is bounded.

With these definitions, we obtain a nice duality property between upper and lower semi-frames. In the discrete case, the role of upper, resp. lower semi-frame, is played by Bessel, resp. Riesz-Fischer sequences [34]. A Riesz-Fischer sequence is a sequence for which, for every sequence $\{a_n\} \in \ell^2$, there is a solution of the equation $\langle f | \phi_n \rangle = a_n$. One knows that every total Riesz-Fischer sequence satisfies the lower frame condition, which is equivalent to the existence of a Bessel sequence dual to it [15]. The same result holds here.

Proposition 2.1

(i) *Let* $\Psi = \{\psi_x\}$ *be an upper semi-frame, with upper frame bound* M *and let* $\Phi = \{\phi_x\}$ *be a total family dual to* Ψ. *Then* Φ *is a lower semi-frame, with lower frame bound* M^{-1}.

(ii) *Conversely, if* $\Phi = \{\phi_x\}$ *is a lower semi-frame, there exists an upper semi-frame* $\Psi = \{\psi_x\}$ *dual to* Φ, *that is, one has, in the weak sense,*

$$f = \int_X \langle f | \phi_x \rangle \, \psi_x \, d\mu(x), \quad \forall \, f \in D(C_\Phi).$$

A proof may be found in [4, Lemma 2.5 and Proposition 2.6]. However, the latter is slightly incomplete. Here is a corrected proof (M. Speckbacher, private communication).

Proof of (ii) The 'if' part is Lemma 2.5 of [4]. Let Φ be a lower semi-frame. Then $\mathsf{Ran}\,(C_\phi)$ is a closed subspace of $L^2(X, d\mu)$, in virtue of the lower frame bound condition, and it is a reproducing kernel Hilbert space (RKHS), since one has, for $F = C_\phi f \in \mathsf{Ran}\,(C_\phi)$,

$$|F(x)| = |C_\phi f(x)| \leqslant \|f\| \, \|\phi_x\| \leqslant \sqrt{1/m} \, \|\phi_x\| \, \|C_\phi f\|_2 = C_x \, \|F\|_2 \,,$$

where m is the lower frame bound of Φ, i.e., the evaluation functional is bounded. Let P be the orthogonal projection on $\mathsf{Ran}\,(C_\phi)$ and let $\{e_n\}_{n \in \mathbb{N}}$ be an arbitrary orthonormal basis of $L^2(X, d\mu)$.

Define a linear operator $V : L^2(X, d\mu) \to \mathcal{H}$ by $V = C_\phi^{-1}$ on $\mathsf{Ran}\,(C_\phi)$, by $V = 0$ on $\mathsf{Ran}\,(C_\phi)^\perp$ and extending by linearity. Then V is bounded, since $C_\phi^{-1} : \mathsf{Ran}\,(C_\phi) \to \mathcal{H}$ is bounded. Then, for all $f \in D(C_\phi)$, $g \in \mathcal{H}$, we have

$$\langle f | g \rangle = \langle V C_\phi f | g \rangle = \langle C_\phi f | V^* g \rangle_2 = \langle C_\phi f | V^* (\sum_{n \in \mathbb{N}} \langle g | e_n \rangle e_n) \rangle_2$$

$$= \langle C_\phi f | \sum_{n \in \mathbb{N}} \langle g | e_n \rangle V^* e_n \rangle_2 = \langle C_\phi f | \sum_{n \in \mathbb{N}} \langle g | e_n \rangle P V^* e_n \rangle_2 = \langle C_\phi f | C_\psi g \rangle_2,$$

where we have put $\psi_x := \sum_{n \in \mathbb{N}} e_n \overline{(P V^* e_n)}(x)$. It remains to show that ψ_x is well defined for every $x \in X$. This is the case if and only if

$$\sum_{n \in \mathbb{N}} |(P V^* e_n)(x)|^2 < \infty, \quad \forall \, x \in X.$$

But this follows from the fact that $\{PV^*e_n\}_{n\in\mathbb{N}}$ is a Bessel sequence in the RKHS Ran (C_ϕ). One has indeed, for any $F \in$ Ran (C_ϕ),

$$\sum_{n\in\mathbb{N}} |\langle F|PV^*e_n\rangle_2|^2 = \sum_{n\in\mathbb{N}} |\langle VPF|e_n\rangle_2|^2 = \|VF\|^2 \leqslant C \|F\|_2^2,$$

since V is bounded and $PF = F$. □

In the same paper [4], concrete examples are presented, namely an upper semi-frame of affine coherent states and a lower semi-frame of wavelets on the 2-sphere. We will come back to these examples in Sect. 5.2.

In conclusion, if two semi-frames are in duality, either they are both frames, or else at least one of them is a lower semi-frame.

3 Reproducing Pairs

Quite recently, a new generalization of frames was introduced by Balazs and Speckbacher [30], namely, reproducing pairs. Here, given a measure space (X, μ), one considers a couple of weakly measurable functions (ψ, ϕ), instead of a single mapping. The advantage is that no further conditions are imposed on these functions, which results in an increased flexibility.

More precisely, the couple of weakly measurable functions (ψ, ϕ) is called a *reproducing pair* if [12]

(a) The sesquilinear form

$$\Omega_{\psi,\phi}(f, g) = \int_X \langle f|\psi_x\rangle\langle\phi_x|g\rangle \,d\mu(x) \tag{3.1}$$

is well-defined and bounded on $\mathcal{H} \times \mathcal{H}$, that is, $|\Omega_{\psi,\phi}(f, g)| \leqslant c \|f\| \|g\|$, for some $c > 0$.
(b) The corresponding bounded (resolution) operator $S_{\psi,\phi}$ belongs to $GL(\mathcal{H})$.

Under these hypotheses, one has

$$S_{\psi,\phi}f = \int_X \langle f|\psi_x\rangle\phi_x \,d\mu(x), \ \forall f \in \mathcal{H}, \tag{3.2}$$

the integral on the r.h.s. being defined in weak sense. If $\psi = \phi$, we recover the notion of continuous frame, as introduced in [1, 2], so that we have indeed a genuine generalization of the latter.

Notice that $S_{\psi,\phi}$ is in general neither positive nor self-adjoint, since $S_{\psi,\phi}^* = S_{\phi,\psi}$. However, if (ψ, ϕ) is a reproducing pair, then $(\psi, S_{\psi,\phi}^{-1}\phi)$ is also a reproducing pair, for which the corresponding resolution operator is the identity, that is, ψ and ϕ are in duality. Therefore, there is no restriction of generality to assume that $S_{\phi,\psi} = I$ [30]. The worst that can happen is to replace some norms by equivalent ones.

3.1 The Hilbert Spaces Generated by a Reproducing Pair

It has been shown in [12] that each weakly measurable function ϕ generates an intrinsic pre-Hilbert space $V_\phi(X, \mu)$ and, moreover, a reproducing pair (ψ, ϕ) generates two Hilbert spaces, $V_\psi(X, \mu)$ and $V_\phi(X, \mu)$, conjugate dual of each other with respect to the $L^2(X, \mu)$ inner product. Let us sketch that construction, following closely [12]. Further generalizations will follow.

Given a weakly measurable function ϕ, let us denote by $\mathcal{V}_\phi(X, \mu)$ the space of all measurable functions $\xi : X \to \mathbb{C}$ such that the integral $\int_X \xi(x)\langle\phi_x|g\rangle \, d\mu(x)$ exists for every $g \in \mathcal{H}$ (in the sense that $\xi\langle\phi.|g\rangle \in L^1(X, d\mu)$) and defines a bounded conjugate linear functional on $\mathcal{H}$, i.e., $\exists \, c > 0$ such that

$$\left|\int_X \xi(x)\langle\phi_x|g\rangle \, d\mu(x)\right| \leqslant c \, \|g\|, \ \forall g \in \mathcal{H}. \tag{3.3}$$

Clearly, if (ψ, ϕ) is a reproducing pair, all functions $\xi(x) = \langle f|\psi_x\rangle$ belong to $\mathcal{V}_\phi(X, \mu)$.

By the Riesz lemma, we can define a linear map $T_\phi : \mathcal{V}_\phi(X, \mu) \to \mathcal{H}$ by the following weak relation

$$\langle T_\phi\xi|g\rangle = \int_X \xi(x)\langle\phi_x|g\rangle \, d\mu(x), \ \forall \xi \in \mathcal{V}_\phi(X, \mu), g \in \mathcal{H}. \tag{3.4}$$

Next, we define the vector space

$$V_\phi(X, \mu) = \mathcal{V}_\phi(X, \mu)/\mathsf{Ker} \, T_\phi$$

and equip it with the norm

$$\left\|[\xi]_\phi\right\|_\phi := \sup_{\|g\|\leqslant 1}\left|\int_X \xi(x)\langle\phi_x|g\rangle \, d\mu(x)\right| = \sup_{\|g\|\leqslant 1}\left|\langle T_\phi\xi|g\rangle\right|, \tag{3.5}$$

where we have put $[\xi]_\phi = \xi + \mathsf{Ker} \, T_\phi$ for $\xi \in \mathcal{V}_\phi(X, \mu)$. Clearly, $V_\phi(X, \mu)$ is a normed space. However, the norm $\|\cdot\|_\phi$ is in fact Hilbertian, that is, it derives from an inner product, as can be seen as follows. First, it turns out that the map $\widehat{T}_\phi : V_\phi(X, \mu) \to \mathcal{H}, \, \widehat{T}_\phi[\xi]_\phi := T_\phi\xi$ is a well-defined isometry of $V_\phi(X, \mu)$ into $\mathcal{H}$. Next, one may define on $V_\phi(X, \mu)$ an inner product by setting

$$\langle[\xi]_\phi|[\eta]_\phi\rangle_{(\phi)} := \langle\widehat{T}_\phi[\xi]_\phi|\widehat{T}_\phi[\eta]_\phi\rangle, \ [\xi]_\phi, [\eta]_\phi \in V_\phi(X, \mu),$$

and one shows that the norm defined by $\langle\cdot|\cdot\rangle_{(\phi)}$ coincides with the norm $\|\cdot\|_\phi$ defined in (3.5). One has indeed

$$\left\|[\xi]_\phi\right\|_{(\phi)} = \left\|\widehat{T}_\phi[\xi]_\phi\right\| = \left\|T_\phi\xi\right\| = \sup_{\|g\|\leqslant 1} \left|\langle T_\phi\xi|g\rangle\right| = \left\|[\xi]_\phi\right\|_\phi.$$

Thus we may state:

Proposition 3.1 *Let ϕ be a weakly measurable function. Then $V_\phi(X, \mu)$ is a pre-Hilbert space with respect to the norm $\|\cdot\|_\phi$ and the map $\widehat{T}_\phi : V_\phi(X, \mu) \to \mathcal{H}$, $\widehat{T}_\phi[\xi]_\phi := T_\phi\xi$ is a well-defined isometry of $V_\phi(X, \mu)$ into $\mathcal{H}$.*

Let us denote by $V_\phi(X, \mu)^*$ the Hilbert dual space of $V_\phi(X, \mu)$, that is, the set of continuous linear functionals on $V_\phi(X, \mu)$. The norm $\|\cdot\|_{\phi^*}$ of $V_\phi(X, \mu)^*$ is defined, as usual, by

$$\|F\|_{\phi^*} = \sup_{\|[\xi]_\phi\|_\phi \leqslant 1} |F([\xi]_\phi)|, \quad F \in V_\phi(X, \mu)^*.$$

Now we define a conjugate linear map $\mathsf{C}_\phi : \mathcal{H} \to V_\phi(X, \mu)^*$ by

$$(\mathsf{C}_\phi f)([\xi]_\phi) := \int_X \xi(x)\langle\phi_x|f\rangle \, d\mu(x), \quad f \in \mathcal{H}, \tag{3.6}$$

which will take the role of the analysis operator C_ϕ of Sect. 2. Notice that C_ϕ is a *linear* map, whereas C_ϕ is conjugate linear. The discrepancy is explained in Remark 3.11 below.

Of course, (3.6) means that $(\mathsf{C}_\phi f)([\xi]_\phi) = \langle T_\phi\xi|f\rangle = \langle\widehat{T}_\phi[\xi]_\phi|f\rangle$, for every $f \in \mathcal{H}$. Thus $\mathsf{C}_\phi = \widehat{T}_\phi^*$, the adjoint map of $\widehat{T}_\phi$. By (3.3) it follows that C_ϕ is continuous. This implies that

$$\mathcal{H} = [\mathsf{Ran}\, \widehat{T}_\phi]^\sim \oplus \mathsf{Ker}\, \mathsf{C}_\phi, \tag{3.7}$$

where the first summand denotes the closure of $\mathsf{Ran}\, \widehat{T}_\phi$. Hence $\mathsf{C}_\phi^* = \widehat{T}_\phi^{**} = \widehat{T}_\phi$, if $V_\phi(X, \mu)$ is complete.

By modifying in an obvious way the definition given in Sect. 2, we say that ϕ is μ-*total* if $\mathsf{Ker}\, \mathsf{C}_\phi = \{0\}$, that is $\overline{\mathsf{Ran}\, \widehat{T}_\phi} = \mathcal{H}$.

Remark 3.2 Whenever no confusion may arise, we will omit the explicit indication of residues classes and write simply, for instance, $\xi \in V_\phi(X, \mu)$ instead of $[\xi]_\phi \in V_\phi(X, \mu)$. Similarly, for the operator C_ϕ introduced in (3.6), we will often identify $\mathsf{C}_\phi f$, $f \in \mathcal{H}$, with $\langle\phi_x|f\rangle$, as a shortcut to $(\mathsf{C}_\phi f)(\xi) = \int_X \xi(x)\langle\phi_x|f\rangle \, d\mu(x)$.

It is easy to see that $V_\phi(X, \mu)[\langle\cdot|\cdot\rangle_{(\phi)}]$ is complete, i.e., it is a Hilbert space if and only if $\widehat{T}_\phi$ has closed range [12].

As a consequence of (3.7) we get

Corollary 3.3 *The following statements hold.*

(i) A weakly measurable function ϕ is μ-total if and only if $\operatorname{Ran} \widehat{T}_\phi$ is dense in $\mathcal{H}$.
(ii) If $V_\phi(X, \mu)$ is a Hilbert space, $\operatorname{Ran} \widehat{T}_\phi$ is equal to $\mathcal{H}$ if and only if ϕ is μ-total.

Now, things get simpler in the case of a reproducing pair. Namely,

Lemma 3.4 *If (ψ, ϕ) is a reproducing pair, then $\operatorname{Ran} \widehat{T}_\phi = \mathcal{H}$.*

Proof Since $S_{\psi,\phi} \in GL(\mathcal{H})$, for every $h \in \mathcal{H}$, there exists a unique $f \in \mathcal{H}$ such that $S_{\psi,\phi} f = h$. But, by (3.2), we get

$$\langle h|g \rangle = \int_X \langle f|\psi_x\rangle\langle\phi_x|g\rangle \, d\mu(x), \ \forall f, g \in \mathcal{H},$$

that is, $h = \widehat{T}_\phi \overline{[\mathsf{C}_\psi f]_\phi}$, where the overbar denotes complex conjugation, as usual.   □

Notice that, if (ψ, ϕ) is a reproducing pair, both functions are necessarily μ-total.

3.2   Duality Properties of the Spaces $V_\phi(X, \mu)$

When the space $V_\phi(X, \mu)$ is a Hilbert space, it is conjugate isomorphic to its dual, via the Riesz operator. In addition, if (ψ, ϕ) is a reproducing pair, the dual of $V_\phi(X, \mu)$ can be identified with $V_\psi(X, \mu)$ as we shall prove below. We emphasize that the duality is taken with respect to the sesquilinear form

$$\langle\xi|\eta\rangle_\mu := \int_X \xi(x)\overline{\eta(x)} \, d\mu(x), \tag{3.8}$$

which coincides with the inner product of $L^2(X, \mu)$ whenever the latter makes sense.

Theorem 3.5 *Let ϕ be a weakly measurable function. If F is a continuous linear functional on $V_\phi(X, \mu)$, then there exists a unique $g \in [\mathcal{M}_\phi]\widetilde{\,}$, the closure of the range of $\widehat{T}_\phi$, such that*

$$F([\xi]_\phi) = \int_X \xi(x)\langle\phi_x|g\rangle \, d\mu(x), \ \forall \xi \in \mathcal{V}_\phi(X, \mu) \tag{3.9}$$

and $\|F\|_{\phi^*} = \|g\|$, where $\|\cdot\|_{\phi^*}$ denotes the (dual) norm on $V_\phi(X, \mu)^*$. Moreover, every $g \in \mathcal{H}$ defines a continuous linear functional F on $V_\phi(X, \mu)$ with $\|F\|_{\phi^*} \leqslant \|g\|$, by (3.9). In particular, if $g \in \operatorname{Ran} \widehat{T}_\phi$, then $\|F\|_{\phi^*} = \|g\|$.

Proof Let $F \in V_\phi(X, \mu)^*$. Then, there exists $c > 0$ such that

$$|F([\xi]_\phi)| \leqslant c \left\| [\xi]_\phi \right\|_\phi = c \left\| T_\phi \xi \right\|, \ \forall \xi \in V_\phi(X, \mu).$$

Let $\mathcal{M}_\phi := \{T_\phi \xi : \xi \in V_\phi(X, \mu)\} = \mathsf{Ran}\, \widehat{T}_\phi$. Then $\mathcal{M}_\phi$ is a vector subspace of $\mathcal{H}$, with closure $[\mathcal{M}_\phi]\widetilde{\,}$.

Let $\widetilde{F}$ be the linear functional defined on $\mathcal{M}_\phi$ by

$$\widetilde{F}(T_\phi \xi) := F([\xi]_\phi), \ \xi \in V_\phi(X, \mu).$$

We notice that $\widetilde{F}$ is well-defined. Indeed, if $T_\phi \xi = T_\phi \xi'$, then $\xi - \xi' \in \mathsf{Ker}\, T_\phi$. Hence, $[\xi]_\phi = [\xi']_\phi$ and $F([\xi]_\phi) = F([\xi']_\phi)$.

Hence, $\widetilde{F}$ is a continuous linear functional on $\mathcal{M}_\phi$. Thus there exists a unique $g \in [\mathcal{M}_\phi]\widetilde{\,}$ such that

$$\widetilde{F}(T_\phi \xi) = \langle \widehat{T}_\phi [\xi]_\phi | g \rangle = \int_X \xi(x) \langle \phi_x | g \rangle \, d\mu(x)$$

and $\|g\| = \|\widetilde{F}\|$.

In conclusion,

$$F([\xi]_\phi) = \int_X \xi(x) \langle \phi_x | g \rangle \, d\mu(x), \ \forall \xi \in V_\phi(X, \mu).$$

and $\|F\|_{\phi^*} = \|g\|$.

Moreover, every $g \in \mathcal{H}$ obviously defines a continuous linear functional F by (3.9) as $|F([\xi]_\phi)| \leqslant \|g\| \left\| [\xi]_\phi \right\|_\phi$. This inequality implies that $\|F\|_{\phi^*} \leqslant \|g\|$. In particular, if $g \in \mathsf{Ran}\, \widehat{T}_\phi$, then there exists $[\xi]_\phi \in V_\phi(X, \mu)$, $\|[\xi]_\phi\|_\phi = 1$, such that $\widehat{T}_\phi [\xi]_\phi = g\|g\|^{-1}$. Hence $F([\xi]_\phi) = \langle \widehat{T}_\phi [\xi]_\phi | g \rangle = \|g\|$. $\qquad\square$

Corollary 3.6 *Let ϕ be a μ-total weakly measurable function, then $\mathsf{C}_\phi : \mathcal{H} \to V_\phi(X, \mu)^*$ is a conjugate linear isometric isomorphism.*

Proof C_ϕ is surjective by Theorem 3.5. As ϕ is μ-total, it follows by Corollary 3.3 that $\mathsf{Ran}\, \widehat{T}_\phi$ is dense in $\mathcal{H}$. Consequently, for $f \in \mathcal{H}$ it follows that

$$\left\| \mathsf{C}_\phi f \right\|_{\phi^*} = \sup_{\|[\xi]_\phi\|_\phi = 1} \left| \int_X \xi(x) \langle \phi_x | f \rangle \, d\mu(x) \right|$$

$$= \sup_{\|[\xi]_\phi\|_\phi = 1} |\langle \widehat{T}_\phi \xi | f \rangle| = \sup_{\|g\|=1, \ g \in \mathsf{Ran}\, \widehat{T}_\phi} |\langle g | f \rangle| = \|f\|. \qquad\square$$

Theorem 3.7 *If (ψ, ϕ) is a reproducing pair, then every continuous linear functional F on $V_\phi(X, \mu)$, i.e., $F \in V_\phi(X, \mu)^*$, can be represented as*

$$F([\xi]_\phi) = \int_X \xi(x)\overline{\eta(x)}\, d\mu(x), \ \forall [\xi]_\phi \in V_\phi(X, \mu), \tag{3.10}$$

with $\eta \in \mathcal{V}_\psi(X, \mu)$. The residue class $[\eta]_\psi \in V_\psi(X, \mu)$ is uniquely determined.

Proof By Theorem 3.5, we have the representation

$$F(\xi) = \int_X \xi(x)\langle \phi_x | g \rangle\, d\mu(x).$$

It is easily seen that $\eta(x) = \langle g | \phi_x \rangle \in \mathcal{V}_\psi(X, \mu)$. Uniqueness is easy. □

The lesson of the previous statements is that the map

$$j : F \in V_\phi(X, \mu)^* \mapsto [\eta]_\psi \in V_\psi(X, \mu) \tag{3.11}$$

is well-defined and conjugate linear. On the other hand, $j(F) = j(F')$ implies easily $F = F'$. Therefore $V_\phi(X, \mu)^*$ can be identified with a closed subspace of $\overline{V_\psi(X, \mu)} := \{\overline{[\xi]_\psi} : \xi \in \mathcal{V}_\psi(X, \mu)\}$.

Let (ψ, ϕ) be a reproducing pair. We want to prove that the spaces $V_\phi(X, \mu)^*$ and $\overline{V_\psi(X, \mu)}$ can be identified. This is the first essential result of [12], to which we refer for a proof, see [12, 12, Lemmas 3.11 and 3.12]. Corresponding to $\widehat{T}_\phi$, we introduce the linear operator $\widehat{C}_{\psi,\phi} : \mathcal{H} \to V_\phi(X, \mu)$ by $\widehat{C}_{\psi,\phi}f := [C_\psi f]_\phi$. We note that $\widehat{C}_{\psi,\phi}f = \widehat{C}_{\psi,\phi}f'$ implies $f = f'$, as can be seen easily.

Thus we state:

Theorem 3.8 *If (ψ, ϕ) is a reproducing pair, the map j defined in (3.11) is surjective. Hence $V_\phi(X, \mu)^* \simeq \overline{V_\psi(X, \mu)}$, where $\simeq$ denotes a bounded isomorphism and the norm $\|\cdot\|_\psi$ is the dual norm of $\|\cdot\|_\phi$. Moreover, $\mathsf{Ran}\,\widehat{C}_{\psi,\phi}[\|\cdot\|_\phi] = V_\phi(X, \mu)[\|\cdot\|_\phi]$ and $\mathsf{Ran}\,\widehat{C}_{\phi,\psi}[\|\cdot\|_\psi] = V_\psi(X, \mu)[\|\cdot\|_\psi]$.*

Proof Let (ψ, ϕ) be a reproducing pair. First one shows that $\mathsf{Ran}\,\widehat{C}_{\psi,\phi}$ is closed in $V_\phi(X, \mu)[\|\cdot\|_\phi]$. Moreover, every $[\eta]_\psi \in V_\psi(X, \mu)$ defines a continuous linear functional on the closed subspace $\mathsf{Ran}\,\widehat{C}_{\psi,\phi}[\|\cdot\|_\phi]$, which implies that the map j is surjective. Next, it turns out that $\mathsf{Ran}\,\widehat{C}_{\psi,\phi}$ is dense in $V_\phi(X, \mu)$. Hence, $\mathsf{Ran}\,\widehat{C}_{\psi,\phi}[\|\cdot\|_\phi]$ and $V_\phi(X, \mu)[\|\cdot\|_\phi]$ coincide, and similarly for the other pair. □

The first statement of the theorem implies that there exist $0 < \mathsf{m} \leqslant \mathsf{M} < \infty$ such that

$$\mathsf{m}\|f\| \leqslant \|\widehat{C}_{\psi,\phi}f\|_\phi \leqslant \mathsf{M}\|f\|, \ \forall f \in \mathcal{H}, \tag{3.12}$$

a relation that may have an independent interest. The inequalities (3.12) are, of course, very similar to the ones defining a frame, viz. (1.2) or (2.1). Yet they are more general, since they are satisfied for any reproducing pair, be it a frame or not.

By Theorems 3.7 and 3.8, it follows that, if (ψ, ϕ) is a reproducing pair, then for every $\eta \in V_\psi(X, \mu)$, there exists $g \in \mathcal{H}$ such that $\eta = \langle \phi. | g \rangle$.

In conclusion, we may state

Theorem 3.9 *If (ψ, ϕ) is a reproducing pair, the spaces $V_\phi(X, \mu)$ and $V_\psi(X, \mu)$ are both Hilbert spaces, conjugate dual of each other with respect to the sesquilinear form* (3.8).

Corollary 3.10 *If (ψ, ϕ) is a reproducing pair and $\phi = \psi$, then ψ is a continuous frame and $V_\psi(X, \mu)$ is a closed subspace of $L^2(X, \mu)$.*

Proof Since the duality takes place with respect to the L^2 inner product, $V_\psi(X, \mu)$ is a subspace of $L^2(X, \mu)$. The equality $\mathsf{Ran}\,\widehat{C}_{\psi,\psi} = V_\psi(X, \mu)$ and the fact that $\widehat{C}_{\psi,\psi}$ is bounded from below with respect to the L^2-norm, by (3.12), imply that it is closed. ☐

Remark 3.11 The operator C_ϕ defined by (2.2) is linear, but the operator C_ϕ given in (3.6) is conjugate linear. However the latter maps $\mathcal{H}$ into $V_\phi(X, \mu)^*$, which is identified with $\overline{V_\psi(X, \mu)}$, thus C_ϕ maps $\mathcal{H}$ linearly into $V_\psi(X, \mu)$.

Actually Theorem 3.9 has an inverse. Indeed:

Theorem 3.12 *Let ϕ and ψ be weakly measurable and μ-total. Then, the couple (ψ, ϕ) is a reproducing pair if and only if $V_\phi(X, \mu)$ and $V_\psi(X, \mu)$ are Hilbert spaces, conjugate dual of each other with respect to the sesquilinear form* (3.8).

Proof The "if" part is Theorem 3.9. Let now $V_\phi(X, \mu)$ and $V_\psi(X, \mu)$ be Hilbert spaces in conjugate duality. Consider the sesquilinear form

$$\Omega_{\psi,\phi}(f, g) = \int_X \langle f | \psi_x \rangle \langle \phi_x | g \rangle \, d\mu(x), \quad f, g \in \mathcal{H}.$$

By the definition of the norms $\|\cdot\|_\phi$, $\|\cdot\|_\psi$ and the duality condition, we have, for every $f, g \in \mathcal{H}$, the two inequalities

$$|\Omega_{\psi,\phi}(f, g)| \leqslant \left\| [\langle f | \psi. \rangle]_\phi \right\|_\phi \|g\|,$$

$$|\Omega_{\psi,\phi}(f, g)| \leqslant \left\| [\langle g | \phi. \rangle]_\psi \right\|_\psi \|f\|.$$

This means, the form $\Omega_{\psi,\phi}$ is separately continuous, hence jointly continuous. Therefore there exists a bounded operator $S_{\psi,\phi}$ such that $\Omega_{\psi,\phi}(f, g) = \langle S_{\psi,\phi} f | g \rangle$. First the operator $S_{\psi,\phi}$ is injective. Indeed, since $\mathsf{C}_\phi^* = \widehat{T}_\phi$, we have

$$\langle S_{\psi,\phi} f | g \rangle = \langle C_\psi f | C_\phi g \rangle = \langle \widehat{C}_{\psi,\phi} f | C_\phi g \rangle = \langle \widehat{T}_\phi \widehat{C}_{\psi,\phi} f | g \rangle, \quad \forall f, g \in \mathcal{H}.$$

Now $\widehat{T}_\phi$ is isometric and $\widehat{C}_{\psi,\phi}$ is injective, hence $\widehat{T}_\phi \widehat{C}_{\psi,\phi} f = 0$ implies $f = 0$. Next, $S_{\psi,\phi}$ is also surjective, by Corollary 3.3. Hence $S_{\psi,\phi}$ belongs to $GL(\mathcal{H})$. ☐

In addition to Lemma 3.8, there is another characterization of the space $V_\psi(X, \mu)$, in terms of an eigenvalue equation, based on the fact that $\langle S_{\psi,\phi}^{-1}\phi_y|\psi_x\rangle$ is a reproducing kernel [30, Prop.3].

Proposition 3.13 *Let (ψ, ϕ) be a reproducing pair. Let $\xi \in V_\psi(X, \mu)$ and consider the eigenvalue equation*

$$\int_X \xi(y)\langle S_{\psi,\phi}^{-1}\phi_y|\psi_x\rangle \, d\mu(y) = \lambda\,\xi(x). \tag{3.13}$$

Then $\xi \in \mathsf{Ran}\,C_\phi$ if and only if $\lambda = 1$, and $\xi \in \mathsf{Ker}\,T_\psi$ if and only if $\lambda = 0$. Moreover, there are no other eigenvalues.

4 Existence and Nonuniqueness of Reproducing Partners

Given a weakly measurable function ψ, it is not obvious that there exists another function ϕ such that (ψ, ϕ) is a reproducing pair. Here is a criterion towards the existence of a specific dual partner.

Theorem 4.1 *Let ϕ be a weakly measurable function and $e = \{e_n\}_{n\in\mathbb{N}}$ an orthonormal basis of $\mathcal{H}$. There exists another measurable function ψ, such that (ψ, ϕ) is a reproducing pair if and only if $\mathsf{Ran}\,\widehat{T_\phi} = \mathcal{H}$ and there exists a family $\{\xi_n\}_{n\in\mathbb{N}} \subset V_\phi(X, \mu)$ such that*

$$[\xi_n]_\phi = [\widehat{T}_\phi^{-1}e_n]_\phi, \ \forall n \in \mathbb{N}, \quad and \quad \sum_{n\in\mathbb{N}}|\xi_n(x)|^2 < \infty, \ for\ a.e.\ x \in X. \tag{4.1}$$

The proof of this theorem is quite technical and may be found in [12, Sec.4]. Note that, if ϕ is a frame, then the reproducing partner ψ given by the proof of Theorem 4.1 is also a frame.

Actually, given the weakly measurable function ϕ, the fact that (ψ, ϕ) is a reproducing pair does not determine the function ψ uniquely. Indeed we have:

Theorem 4.2 *Let (ψ, ϕ) be a reproducing pair. Then (θ, ϕ) is a reproducing pair if and only if $\theta = A\psi + \theta_0$, where $A \in GL(\mathcal{H})$ and $[\langle f|\theta_0(\cdot)\rangle]_\phi = [0]_\phi$, $\forall f \in \mathcal{H}$, i.e., $\widehat{C}_{\theta_0,\phi}f = 0, \forall f \in \mathcal{H}$.*

Proof If $\theta = A\psi + \theta_0$ as above, then $S_{\theta,\phi}f = \widehat{T}_\phi(\widehat{C}_{A\psi,\phi} + \widehat{C}_{\theta_0,\phi})f = \widehat{T}_\phi(\widehat{C}_{A\psi,\phi}f) = \widehat{T}_\phi\widehat{C}_{\psi,\phi}A^*f = S_{\psi,\phi}A^*f$, hence $S_{\theta,\phi} = S_{\psi,\phi}A^* \in GL(\mathcal{H})$.

Conversely, assume that (θ, ϕ) is a reproducing pair. By Theorem 3.8, we have $V_\phi(X, \mu) = \mathsf{Ran}\,C_\psi/\mathsf{Ker}\,T_\phi = \mathsf{Ran}\,C_\theta/\mathsf{Ker}\,T_\phi$, i.e., for every $f \in \mathcal{H}$ there exists $g \in \mathcal{H}$ such that $[C_\theta f]_\phi = [C_\psi g]_\phi$. Then, using successively the definition of $S_{\phi,\theta}$, the relation above and the reproducing kernel (3.13), we obtain

$$\langle f | S_{\phi,\theta}(S_{\psi,\phi}^{-1})^* \psi(\cdot) \rangle = \int_X \langle f | \theta_x \rangle \langle \phi_x | (S_{\psi,\phi}^{-1})^* \psi. \rangle \, d\mu(x)$$

$$= \int_X \langle g | \psi_x \rangle \langle \phi_x | (S_{\psi,\phi}^{-1})^* \psi. \rangle \, d\mu(x) = \langle g | \psi. \rangle = \langle f | \theta. \rangle \,, \ \forall f \in \mathcal{H}.$$

This means that, for all $f \in \mathcal{H}$, we have $[C_\theta f]_\phi = [C_{A\psi} f]_\phi$ or, equivalently, $\widehat{C}_{\theta,\phi} = \widehat{C}_{A\psi,\phi}$, where $A := S_{\phi,\theta}(S_{\psi,\phi}^{-1})^* \in GL(\mathcal{H})$. Moreover, $C_\theta f(x) = C_{A\psi} f(x) + F(f, x)$ for a.e. $x \in X$ and every $f \in \mathcal{H}$, where $F(f, \cdot) \in \mathrm{Ker}\, T_\phi$, i.e., $F(f, x) = \langle f | (\theta - A\psi)_x \rangle =: \langle f | \theta_{0x} \rangle$. □

In addition, the existence of a reproducing partner to a given function ϕ is preserved if one replaces $V_\phi(X, \mu)$ by an isomorphic space $V_\phi'(X, \mu)$. Indeed:

Corollary 4.3 *Let ϕ, ϕ' be weakly measurable functions and $V_\phi(X, \mu) \simeq V_{\phi'}(X, \mu)$, where $\simeq$ denotes a bounded isomorphism. There exists ψ such that (ψ, ϕ) is a reproducing pair if and only if there exists ψ' such that (ψ', ϕ') is a reproducing pair.*

Proof Suppose that (ψ, ϕ) is a reproducing pair. Let j be the bounded isomorphism $j : V_\phi(X, \mu) \to V_{\phi'}(X, \mu)$. Then $j^* : V_{\phi'}(X, \mu)^* \to V_\phi(X, \mu)^*$ is also a bounded isomorphism. Since $C_{\phi'}$ is bounded by (3.3), in order to verify that there exists ψ' such that $S_{\psi',\phi'} \in GL(\mathcal{H})$, we only need to check that $C_{\phi'}^{-1}$ is bounded. For every $f \in \mathcal{H}$, there exists $g \in \mathcal{H}$ such that $C_\phi f = j^* C_{\phi'} g$. Hence, $g \mapsto C_\phi^{-1} j^* C_{\phi'} g$ is surjective and bounded. It is moreover injective since μ-totality of ϕ' implies injectivity of $C_{\phi'}$. Thus, the bounded inverse theorem implies that $(C_\phi^{-1} j^* C_{\phi'})^{-1}$ and consequently $C_{\phi'}^{-1}$ are bounded. Since the statement is symmetric in (ψ, ϕ) and (ψ', ϕ'), the converse implication holds as well. □

5 Examples of Reproducing Pairs

In this section, we present a few concrete examples of the construction of Sect. 3. More details may be found in [12]. We begin with discrete examples, that is, $X = \mathbb{N}$ with the counting measure.

5.1 Discrete Examples

5.1.1 Orthonormal Basis

Let $e = \{e_n\}_{n \in \mathbb{N}}$ be an orthonormal basis, then e is a frame and $V_e(\mathbb{N}) = V_e(\mathbb{N}) = \ell^2(\mathbb{N})$. Indeed, for $\xi \in V_e(\mathbb{N})$, we have

$$\left| \sum_{n \in \mathbb{N}} \xi_n \langle e_n | g \rangle \right| = \left| \sum_{n \in \mathbb{N}} \xi_n \overline{g_n} \right| \leqslant c \, \|g\| = c \, \|\{g_n\}_{n \in \mathbb{N}}\|_{\ell^2} \,, \ \forall g \in \mathcal{H},$$

where $g_n := \langle g | e_n \rangle$. Hence, $\xi \in \ell^2(\mathbb{N})^* = \ell^2(\mathbb{N})$. Moreover, since $\mathsf{Ker}\, T_e = \{0\}$, it follows that $\mathcal{V}_e(\mathbb{N}) = V_e(\mathbb{N})$ and $\|\cdot\|_{\ell^2} = \|\cdot\|_e$.

5.1.2 Riesz Basis

Now consider a Riesz basis $r = \{r_n\}_{n \in \mathbb{N}}$. Then $r_n = Ae_n$ for some $A \in GL(\mathcal{H})$ [16]. Therefore $\mathcal{V}_r(\mathbb{N}) = V_r(\mathbb{N}) = \ell^2(\mathbb{N})$ as sets, but with equivalent (not necessary equal) norms, as can be seen easily. Hence r is a frame.

5.1.3 Discrete Upper and Lower-Semi Frames

Let $\theta = \{\theta_n\}_{n \in \mathbb{N}}$ be a discrete frame, $m = \{m_n\}_{n \in \mathbb{N}} \subset \mathbb{C} \backslash \{0\}$ and define $\phi := \{m_n \theta_n\}_{n \in \mathbb{N}}$. If $\{|m_n|\}_{n \in \mathbb{N}} \in c_0$, then ϕ is an upper semi-frame and if $\{|m_n|^{-1}\}_{n \in \mathbb{N}} \in c_0$, then ϕ is a lower semi-frame. Observe that in both cases ϕ is not a frame.

It can easily be seen that $V_\phi(\mathbb{N}) = M_{1/m}(V_\theta(\mathbb{N})) = M_{1/m}(\mathsf{Ran}\, C_\theta)$ as sets, where M_m is the multiplication operator defined by $(M_m \xi)_n = m_n \xi_n$. Moreover, $\|\cdot\|_\phi \asymp \|\cdot\|_{\ell^2_m}$, where $\|\xi\|_{\ell^2_m} := \sum_{n \in \mathbb{N}} |\xi_n m_n|^2$. Also, $V_\phi(\mathbb{N})^* = M_m(V_\theta(\mathbb{N})^*) = M_m(V_\theta(\mathbb{N})) = M_m(\mathsf{Ran}\, C_\theta)$ as sets.

Using Theorem 4.1, one may show that there exists ψ such that (ψ, ϕ) is a reproducing pair [12]. A natural choice of a reproducing partner is $\psi := \{(1/\overline{m_n})\theta_n\}_{n \in \mathbb{N}}$ as $S_{\psi,\phi} = S_\theta \in GL(\mathcal{H})$.

5.2 Continuous Examples

5.2.1 Continuous Frames

If ϕ is a continuous frame, Corollary 3.10 implies that $V_\phi(X, \mu)$ is a closed subspace of $L^2(X, \mu)$. Now, since $L^2(X, \mu) = \mathsf{Ran}\, C_\phi \oplus \mathsf{Ker}\, D_\phi$, it follows that $V_\phi(X, \mu)[\|\cdot\|_\phi] \simeq \mathsf{Ran}\, C_\phi[\|\cdot\|_{L^2}]$.

5.2.2 1D Continuous Wavelets

Let $\psi, \phi \in L^2(\mathbb{R}, \mathrm{d}x)$ and consider the continuous wavelet systems $\phi_{x,a} = T_x D_a \phi$, where, as usual, T_x denotes the translation operator and D_a the dilation operator. If

$$\int_\mathbb{R} |\widehat{\psi}(\omega)\widehat{\phi}(\omega)| \frac{\mathrm{d}\omega}{|\omega|} < \infty \tag{5.1}$$

then (ψ, ϕ) is a reproducing pair for $L^2(\mathbb{R}, dx)$ with $S_{\psi,\phi} = c_{\psi,\phi}I$ [24, Theorem 10.1], where

$$c_{\psi,\phi} := \int_{\mathbb{R}} \overline{\widehat{\psi}(\omega)}\widehat{\phi}(\omega)\frac{d\omega}{|\omega|}.$$

Actually the relation $S_{\psi,\phi} = c_{\psi,\phi}I$ simply expresses the well-known orthogonality relations of wavelet transforms. A similar result holds true for D-dimensional continuous wavelets [3] and, more generally, for all coherent states associated with square integrable group representations [3, Chaps. 8 and 12].

For $\psi = \phi$, the cross-admissibility condition (5.1) reduces to the classical admissibility condition

$$c_\phi := \int_{\mathbb{R}} |\widehat{\phi}(\omega)|^2 \frac{d\omega}{|\omega|} < \infty. \tag{5.2}$$

Considering the obvious inequalities

$$|c_{\psi,\phi}| \leqslant \int_{\mathbb{R}} |\widehat{\psi}(\omega)\widehat{\phi}(\omega)|\frac{d\omega}{|\omega|} \leqslant c_\phi^{1/2}c_\psi^{1/2},$$

we see that condition (5.1) is automatically satisfied whenever ϕ and ψ are both admissible. However, it is possible to choose a mother wavelet ϕ that does not satisfy the admissibility condition (5.2) and still obtain a reproducing pair (ψ, ϕ). Consider, for example, the Gaussian window $\phi(x) = e^{-\pi x^2}$, then $c_\phi = \infty$, which implies that ϕ is not a continuous wavelet frame. However, if one defines $\psi \in L^2(\mathbb{R}, dx)$ in the Fourier domain via $\widehat{\psi}(\omega) = |\omega|\widehat{\phi}(\omega)$, it follows that $0 < c_{\psi,\phi} = \|\phi\|_2^2 < \infty$. Hence (ψ, ϕ) is a reproducing pair. This example clearly shows the increasing flexibility obtained when replacing continuous frames by reproducing pairs.

5.2.3 A Continuous Upper Semi-frame: Affine Coherent States

In [4, Sec. 2.6] the following example of an upper semi-frame is investigated. Define $\mathcal{H}^{(n)} := L^2(\mathbb{R}^+, r^{n-1} dr)$, $n \in \mathbb{N}$, and the measure space $(X, \mu) = (\mathbb{R}, dx)$. Let $\psi \in \mathcal{H}^{(n)}$ and define the affine coherent state

$$\psi_x(r) = e^{-ixr}\psi(r), \quad r \in \mathbb{R}^+.$$

Then ψ is admissible if $\sup_{r\in\mathbb{R}^+} \mathfrak{s}(r) = 1$, where $\mathfrak{s}(r) := 2\pi r^{n-1}|\psi(r)|^2$, and $|\psi(r)| \neq 0$, for a.e. $r \in \mathbb{R}^+$. The frame operator is given by the multiplication operator on $\mathcal{H}^{(n)}$

$$(S_\psi f)(r) = \mathfrak{s}(r)f(r),$$

and, more generally,

$$(S_\psi^m f)(r) = [\mathfrak{s}(r)]^m f(r), \ \forall\, m \in \mathbb{Z}.$$

Hence S_ψ is bounded and S_ψ^{-1} is unbounded.

The function ψ enjoys the interesting property that we can characterize the space $V_\psi(\mathbb{R}, dx)$ and its norm. First, we show that $T_\phi \xi = \widehat{\xi}\psi$, which in turn implies that $\widehat{\xi}$ has to be given by an almost everywhere defined function which satisfies $\widehat{\xi}\psi \in \mathcal{H}^{(n)}$. Hence $\xi \in V_\psi(\mathbb{R}, dx)$ provided $\widehat{\xi}\psi \in \mathcal{H}^{(n)}$ and then $\|\xi\|_\psi = \|\widehat{\xi}\psi\|$.

When looking for a reproducing partner for ψ, we first ask whether there exists an affine coherent state $\phi_x(r) = e^{-ixr}\phi(r)$, $r \in \mathbb{R}^+$, $\phi \in \mathcal{H}^{(n)}$, such that (ψ, ϕ) forms a reproducing pair. The answer is negative. Indeed, since ψ is Bessel and not a frame, its dual ϕ is by necessity a lower semi-frame, whereas an affine coherent state must be Bessel, but can never satisfy the lower frame bound. Hence, there is no pair of affine coherent states forming a reproducing pair. This fact can also be proven by an explicit calculation.

More generally, we may look for a reproducing partner which is not an affine coherent state. First C_ψ is an isometry by Corollary 3.6, but $\mathsf{Ran}\,\widehat{T}_\psi \neq \mathcal{H}$. Indeed, if we had $\mathsf{Ran}\,\widehat{T}_\psi = \mathcal{H}$, an arbitrary element $h \in \mathcal{H}^{(n)} = L^2(\mathbb{R}^+, r^{n-1}\,dr)$ could be written as $h = \widehat{T}_\psi \xi = \widehat{\xi}\psi$ for some $\xi \in V_\psi(\mathbb{R}, dx)$. This applies, in particular, to ψ itself, which also belongs to $\mathcal{H}^{(n)}$. This in turn implies that there exists ξ, such that $\widehat{\xi}(r) = 1$ for a.e. $r \geqslant 0$. But there is no function that satisfies this condition (however, the δ-distribution does the job).

This discussion has two major consequences. First, it shows that $V_\psi(\mathbb{R}, dx)$ is *not* a Hilbert space, since it is not complete. Second, ψ has *no* reproducing partner at all.

5.2.4 Continuous Wavelets on the Sphere

Next we consider the continuous wavelet transform on the 2-sphere $\mathbb{S}^2$ [3, 10]. For a mother wavelet $\phi \in \mathcal{H} = L^2(\mathbb{S}^2, d\mu)$, define the family of spherical wavelets

$$\phi_{\varrho,a} := R_\varrho D_a \phi, \ \text{where } (\varrho, a) \in X := SO(3) \times \mathbb{R}^+.$$

Here, D_a denotes the stereographic dilation operator and R_ϱ the unitary rotation on $\mathbb{S}^2$.

It has been shown in [10, Theorem 3.3] that the operator S_ϕ is diagonal in Fourier space (harmonic analysis on the 2-sphere reduces to expansions in spherical harmonics Y_l^m, $l \in \mathbb{N}$, $m = -l, \ldots, l$), thus it is a Fourier multiplier $\widehat{S_\phi f}(l, m) = s_\phi(l)\widehat{f}(l, m)$ with the symbol s_ϕ given by

$$s_\phi(l) := \frac{8\pi^2}{2l+1} \sum_{|m| \leqslant l} \int_0^\infty |\widehat{D_a\phi}(l, m)|^2 \frac{da}{a^3}, \ l \in \{0\} \cup \mathbb{N},$$

where $\widehat{D_a\phi}(l, m) := \langle Y_l^m | D_a\phi \rangle$ is the Fourier coefficient of $D_a\phi$.

The result of the analysis is twofold. First, the wavelet $\phi \in L^2(\mathbb{S}^2, d\mu)$ is admissible if and only if there exists a constant $c > 0$ such that $s_\phi(l) \leqslant c$, $\forall l \in \mathbb{N}$, equivalently, if the frame operator S_ϕ is bounded. In addition, for any admissible axisymmetric wavelet ϕ, there exists a constant $d > 0$ such that $d \leqslant s_\phi(l) \leqslant c$, $\forall l \in \mathbb{N}$. Equivalently, S_ϕ and S_ϕ^{-1} are both bounded, i.e., the family of spherical wavelets $\{\phi_{a,\varrho}, (\varrho, a) \in X = SO(3) \times \mathbb{R}_+^*\}$ is a continuous frame. One notices, however, that the upper frame bound, which is implied by the constant c, does depend on ϕ, whereas the lower frame bound, which derives from d, does not, it follows from the asymptotic behavior of the function Y_l^m for large l.

However, it turns out [33] that the reconstruction formula converges if $d \leqslant s_\phi(l) < \infty$ for all $l \in \{0\} \cup \mathbb{N}$, and this implies that ϕ (which is *not* admissible) is in fact a lower semi-frame and S_ϕ is unbounded, but densely defined.

We may now apply Theorem 4.1 to investigate the existence of a reproducing partner for ϕ. First, we show that $\mathrm{Ran}\,\widehat{T_\phi} = \mathcal{H}$. The operator M_ϕ defined by $\widehat{M_\phi f}(l, m) = s_\phi(l)^{-1}\widehat{f}(l, m)$ is bounded and constitutes a right inverse to S_ϕ. Hence, for every $f \in \mathcal{H}$, it holds

$$f = S_\phi M_\phi f = \widehat{T_\phi}[C_\phi M_\phi f]_\phi \in \mathrm{Ran}\,\widehat{T_\phi}.$$

Choosing $\xi_{l,m}(\varrho, a) := C_\phi(S_\phi^{-1}Y_l^m)(\varrho, a) = \langle S_\phi^{-1}Y_l^m | \phi_{\varrho,a}\rangle$ as a representative of $[\widehat{T_\phi}^{-1}Y_l^m]_\phi$ yields for every $(\varrho, a) \in \mathbb{R} \times \mathbb{R}^+$:

$$\sum_{l=0}^\infty \sum_{|m|\leqslant l} |\xi_{l,m}(\varrho, a)|^2 = \sum_{l=0}^\infty \sum_{|m|\leqslant l} |\langle S_\phi^{-1}Y_l^n | \phi_{\varrho,a}\rangle|^2 = \sum_{l=0}^\infty \sum_{|m|\leqslant l} |s_\phi(l)^{-1}\widehat{\phi}_{\varrho,a}(l, m)|^2$$

$$\leqslant \frac{1}{d}\sum_{l=0}^\infty \sum_{|m|\leqslant l} |\widehat{\phi}_{\varrho,a}(l, m)|^2 = \frac{1}{d}\left\|\phi_{\varrho,a}\right\|^2 < \infty.$$

Thus there exists (at least one) function $\psi \in L^2(\mathbb{S}^2, d\mu)$ such that (ψ, ϕ) is a reproducing pair.

Moreover, as for the wavelets on $\mathbb{R}^d$, it is possible to choose another continuous wavelet system $\psi_{\varrho,a}$ as reproducing partner if the symbol $s_{\psi,\phi}$, defined by

$$s_{\psi,\phi}(l) := \frac{8\pi^2}{2l + 1}\sum_{|m|\leqslant l}\int_0^\infty \widehat{D_a\psi}(l, m)\,\overline{\widehat{D_a\phi}(l, m)}\,\frac{da}{a^3}.$$

satisfies $m \leqslant |s_{\psi,\phi}(l)| \leqslant M$ for all $l \in \{0\} \cup \mathbb{N}$.

5.2.5 Genuine Reproducing Pairs, Applications

Explicit examples of reproducing pairs, some of them containing neither frames, nor semi-frames, have been given in the original paper [30]. An interesting class of such objects arises in the context of the Gabor transform, or rather the so-called continuous nonstationary Gabor transform. The latter relies on representations of the Weyl-Heisenberg group, pioneered by Torrésani [32]. These techniques, and in particular their discretized versions, seem to have a rich future in signal analysis. The same can be said of the α-modulation transform and the attending α-modulation frames [20, 31]. Obviously the whole analysis described in the present chapter, including semi-frames and reproducing pairs, could and should be extended to these more general frameworks. This offers interesting perspectives, both from the mathematical point of view and towards applications in signal processing. But this is another story...

6   Interlude: Reproducing Pairs and PIP-Spaces

Let (ψ, ϕ) be a reproducing pair. By definition,

$$\langle S_{\psi,\phi} f | g \rangle = \int_X \langle f | \psi_x \rangle \langle \phi_x | g \rangle \, d\mu(x) = \int_X C_\psi f(x) \, \overline{C_\phi g(x)} \, d\mu(x) \tag{6.1}$$

is well-defined for all $f, g \in \mathcal{H}$ (here we revert to the *linear* maps C_ψ, C_ϕ defined in (2.2)). The r.h.s. coincides with the sesquilinear form (3.8), that is, the L^2 inner product, but generalized, since in general $C_\psi f, C_\phi g$ need not belong to $L^2(X, d\mu)$.

This fact clearly indicates that the analysis should be performed in the framework of a partial inner product space (PIP-space) of measurable functions on X [6]. The question is, how to embed $\mathsf{Ran}\,(C_\psi)$ and $\mathsf{Ran}\,(C_\phi)$ into the corresponding assaying subspaces. Next we have to determine how the Hilbert spaces V_ψ and V_ϕ are related to the latter. Following [9], we will examine successively the cases of a rigged Hilbert space (RHS) and a genuine PIP-space. Then we particularize the results to a Hilbert scale and to a PIP-space of L^p spaces. The motivation for the last case is the following. If, following [30], we make the innocuous assumption that the map $x \mapsto \psi_x$ is bounded, i.e., $\sup_{x \in X} \|\psi_x\|_{\mathcal{H}} \leqslant c$ for some $c > 0$ (often $\|\psi_x\|_{\mathcal{H}} = $ const., e.g. for wavelets or coherent states), then $(C_\psi f)(x) = \langle f | \psi_x \rangle \in L^\infty(X, d\mu)$ so that a PIP-space based on the lattice generated by the family $\{L^p(X, d\mu), 1 \leqslant p \leqslant \infty, \}$ may be a good solution.

7 Reproducing Pairs and RHS

We begin with the simplest example of a PIP-space, namely, a rigged Hilbert space (RHS). Let indeed $\mathcal{D}[t] \subset \mathcal{H} \subset \mathcal{D}^\times[t^\times]$ be an RHS, where $\mathcal{D}^\times[t^\times]$ denotes the space of continuous conjugate linear functionals on $\mathcal{D}$, equipped with the strong dual topology $t^\times$. We assume that $\mathcal{D}[t]$ is reflexive, so that t and $t^\times$ coincide with the respective Mackey topologies [29]. Given a measure space (X, μ), we denote by $\langle \cdot, \cdot \rangle$ the sesquilinear form expressing the duality between $\mathcal{D}^\times$ and $\mathcal{D}$. As usual, we suppose that this sesquilinear form extends the inner product of $\mathcal{D}$ (and $\mathcal{H}$). This allows to build the triplet above.

Let $x \in X \mapsto \psi_x$, $x \in X \mapsto \phi_x$ be weakly measurable functions from X into $\mathcal{D}^\times$. Instead of (3.1), we consider the sesquilinear form

$$\Omega^{\mathcal{D}}_{\psi,\phi}(f, g) = \int_X \langle f, \psi_x \rangle \langle \phi_x, g \rangle \, d\mu(x), \quad f, g \in \mathcal{D}. \tag{7.1}$$

For short, we put $\Omega^{\mathcal{D}} := \Omega^{\mathcal{D}}_{\psi,\phi}$ and we assume that $\Omega^{\mathcal{D}}$ is jointly continuous on $\mathcal{D} \times \mathcal{D}$, that is, $\Omega^{\mathcal{D}} \in \mathsf{B}(\mathcal{D}, \mathcal{D})$ in the notation of [11, Sec.10.2]. Then the relation

$$\langle S_{\psi,\phi} f, g \rangle := \int_X \langle f, \psi_x \rangle \langle \phi_x, g \rangle \, d\mu(x), \quad \forall f, g \in \mathcal{D}, \tag{7.2}$$

tells us that the operator $S_{\psi,\phi}$ belongs to $\mathcal{L}(\mathcal{D}, \mathcal{D}^\times)$, the space of all continuous linear maps from $\mathcal{D}$ into $\mathcal{D}^\times$.

7.1 A Hilbertian Approach

We first assume that the sesquilinear form $\Omega^{\mathcal{D}}$ is well-defined and bounded on $\mathcal{D} \times \mathcal{D}$ in the topology of $\mathcal{H}$. Then $\Omega^{\mathcal{D}}$ extends to a bounded sesquilinear form on $\mathcal{H} \times \mathcal{H}$, denoted by the same symbol.

The definition of the space $V_\phi(X, \mu)$ must be modified as follows. Instead of (3.3), we suppose that the integral below exists and defines a conjugate linear functional on $\mathcal{D}$, bounded in the topology of $\mathcal{H}$, i.e.,

$$\left| \int_X \xi(x) \langle \phi_x, g \rangle \, d\mu(x) \right| \leqslant c \, \|g\|, \quad \forall g \in \mathcal{D}. \tag{7.3}$$

Then the functional extends to a bounded conjugate linear functional on $\mathcal{H}$, since $\mathcal{D}$ is dense in $\mathcal{H}$. Hence, for every $\xi \in V_\phi(X, \mu)$, there exists a unique vector $h_{\phi,\xi} \in \mathcal{H}$ such that

$$\int_X \xi(x) \langle \phi_x, g \rangle \, d\mu(x) = \langle h_{\phi,\xi} | g \rangle, \quad \forall g \in \mathcal{D}.$$

Next, we define a linear map $T_\phi : \mathcal{V}_\phi(X, \mu) \to \mathcal{H}$ by

$$T_\phi \xi = h_{\phi,\xi} \in \mathcal{H}, \ \forall \xi \in \mathcal{V}_\phi(X, \mu), \tag{7.4}$$

in the following weak sense

$$\langle T_\phi \xi | g \rangle = \langle h_{\phi,\xi} | g \rangle = \int_X \xi(x) \langle \phi_x, g \rangle \, d\mu(x), \ g \in \mathcal{D}, \xi \in \mathcal{V}_\phi(X, \mu).$$

The rest proceeds as before. We consider the space $V_\phi(X, \mu) = \mathcal{V}_\phi(X, \mu)/\mathrm{Ker}\, T_\phi$, with the norm $\left\| [\xi]_\phi \right\|_\phi = \left\| T_\phi \xi \right\|$, where, for $\xi \in \mathcal{V}_\phi(X, \mu)$, we have put $[\xi]_\phi = \xi + \mathrm{Ker}\, T_\phi$. Then $V_\phi(X, \mu)$ is a pre-Hilbert space for that norm.

Assume, in addition, that the corresponding bounded operator $S_{\psi,\phi}$ is an element of $GL(\mathcal{H})$. Then (ψ, ϕ) is a reproducing pair and Theorem 3.9 remains true, that is,

Theorem 7.1 *If (ψ, ϕ) is a reproducing pair, the spaces $V_\phi(X, \mu)$ and $V_\psi(X, \mu)$ are both Hilbert spaces, conjugate dual of each other with respect to the sesquilinear form (3.8) or (3.10), namely,*

$$\langle [\xi]_\phi | [\eta]_\psi \rangle = \langle \xi | \eta \rangle_\mu = \int_X \xi(x) \overline{\eta(x)} \, d\mu(x), \ \forall \xi \in V_\phi(X, \mu), \ \eta \in V_\psi(X, \mu). \tag{7.5}$$

Example 7.2 To give a trivial example, consider the Schwartz rigged Hilbert space $\mathcal{S}(\mathbb{R}) \subset L^2(\mathbb{R}, dx) \subset \mathcal{S}^\times(\mathbb{R})$, $(X, \mu) = (\mathbb{R}, dx)$, $\psi_x(t) = \phi_x(t) = \frac{1}{\sqrt{2\pi}} e^{ixt}$. Then $C_\phi f = \widehat{f}$, the Fourier transform, so that $\langle f | \phi(\cdot) \rangle \in L^2(\mathbb{R}, dx)$. In this case

$$\Omega^{\mathcal{D}}_{\psi,\phi}(f, g) = \int_\mathbb{R} \langle f, \psi_x \rangle \langle \phi_x, g \rangle \, dx = \langle \widehat{f} | \widehat{g} \rangle = \langle f | g \rangle, \ \forall f, g \in \mathcal{S}(\mathbb{R}),$$

and $V_\psi(\mathbb{R}, dx) = V_\phi(\mathbb{R}, dx) = L^2(\mathbb{R}, dx)$.

7.2   The General Case

In the general case, we only assume that the form $\Omega^{\mathcal{D}}$ is jointly continuous on $\mathcal{D} \times \mathcal{D}$, with no other regularity requirement. In that case, the vector space $\mathcal{V}_\phi(X, \mu)$ must be defined differently. Let the topology of $\mathcal{D}$ be given by a directed family $\mathfrak{P}$ of seminorms. Given a weakly measurable function ϕ, we denote again by $\mathcal{V}_\phi(X, \mu)$ the space of all measurable functions $\xi : X \to \mathbb{C}$ such that the integral $\int_X \xi(x) \langle \phi_x, g \rangle \, d\mu(x)$ exists for every $g \in \mathcal{D}$ and defines a continuous conjugate linear functional on $\mathcal{D}$, that is, there exists a constant $c > 0$ and a seminorm $\mathsf{p} \in \mathfrak{P}$ such that

$$\left| \int_X \xi(x) \langle \phi_x, g \rangle \, d\mu(x) \right| \leqslant c \, \mathsf{p}(g), \ \forall g \in \mathcal{D}.$$

This in turn determines a linear map $T_\phi : \mathcal{V}_\phi(X, \mu) \to \mathcal{D}^\times$ by the following relation

$$\langle T_\phi \xi, g \rangle = \int_X \xi(x) \langle \phi_x, g \rangle \, d\mu(x), \quad \forall \xi \in \mathcal{V}_\phi(X, \mu), g \in \mathcal{D}. \tag{7.6}$$

Next, we define as before the vector space

$$V_\phi(X, \mu) = \mathcal{V}_\phi(X, \mu)/\mathsf{Ker}\, T_\phi,$$

and we put again $[\xi]_\phi = \xi + \mathsf{Ker}\, T_\phi$ for $\xi \in \mathcal{V}_\phi(X, \mu)$.

Now we define the topology of $V_\phi(X, \mu)$ by means of the strong dual topology $t^\times$ of $\mathcal{D}^\times$, which we recall is defined by the seminorms

$$\|F\|_Q = \sup_{g \in Q} |\langle F | g \rangle|, \quad F \in \mathcal{D}^\times,$$

where Q runs over the family of bounded subsets of $\mathcal{D}[t]$. As said above, the reflexivity of $\mathcal{D}$ entails that $t^\times$ is equal to the Mackey topology $\tau(\mathcal{D}^\times, \mathcal{D})$. Define the following seminorm on $V_\phi(X, \mu)$:

$$\widehat{\mathsf{p}}_Q([\xi]_\phi) := \sup_{g \in Q} \left| \langle T_\phi \xi, g \rangle \right|, \tag{7.7}$$

where Q is a bounded subset of $\mathcal{D}[t]$. Then we may state

Lemma 7.3 *The map* $\widehat{T}_\phi : V_\phi(X, \mu) \to \mathcal{D}^\times$, $\widehat{T}_\phi[\xi]_\phi := T_\phi \xi$ *is a well-defined linear map of* $V_\phi(X, \mu)$ *into* $\mathcal{D}^\times$ *and, for every bounded subset* Q *of* $\mathcal{D}[t]$, *one has*

$$\widehat{\mathsf{p}}_Q([\xi]_\phi) = \|T_\phi \xi\|_Q, \quad \forall \xi \in \mathcal{V}_\phi(X, \mu)$$

The latter equality obviously implies the continuity of T_ϕ.

Next we investigate the dual $V_\phi(X, \mu)^*$ of the space $V_\phi(X, \mu)$, that is, the set of continuous linear functionals on $V_\phi(X, \mu)$. We equip $V_\phi(X, \mu)^*$ with the strong dual topology, which is defined by the family of seminorms

$$\mathsf{q}_\mathcal{R}(F) := \sup_{[\xi]_\phi \in \mathcal{R}} |F([\xi]_\phi)|,$$

where $\mathcal{R}$ runs over the bounded subsets of $V_\phi(X, \mu)$.

Theorem 7.4 *Assume that* $\mathcal{D}[t]$ *is a reflexive space and let* ϕ *be a weakly measurable function. If* F *is a continuous linear functional on* $V_\phi(X, \mu)$, *then there exists a unique* $g \in \mathcal{D}$ *such that*

$$F([\xi]_\phi) = \int_X \xi(x) \langle \phi_x, g \rangle \, d\mu(x), \quad \forall \xi \in \mathcal{V}_\phi(X, \mu) \tag{7.8}$$

Moreover, every $g \in \mathcal{H}$ *defines a continuous linear functional* F *on* $V_\phi(X, \mu)$ *with* $\|F\|_{\phi^*} \leqslant \|g\|$, *by* (7.8).

The proof of this theorem follows closely that of Theorem 3.5, replacing Hilbertian norms by appropriate seminorms and using the reflexivity of $\mathcal{D}$. Details may be found in [9].

In the present context, the analysis operator C_ϕ is defined in the usual way, given in (2.2). Then, particularizing the discussion of Theorem 3.5 to the functional $\langle \cdot, C_\phi g \rangle$, one can interpret the analysis operator C_ϕ as a continuous operator from $\mathcal{D}$ to $V_\phi(X, \mu)^*$. As in the case of frames or semi-frames, one may characterize the synthesis operator in terms of the analysis operator.

Proposition 7.5 *For a weakly measurable function ϕ, $\widehat{T}_\phi \subseteq C_\phi^*$. If, in addition, $V_\phi(X, \mu)$ is reflexive, then $\widehat{T}_\phi^* = C_\phi$. Moreover, ϕ is μ-total (i.e., $\mathsf{Ker}\, C_\phi = \{0\}$) if and only if $\mathsf{Ran}\, \widehat{T}_\phi$ is dense in $\mathcal{D}^\times$.*

Proof As $C_\phi : \mathcal{D} \to V_\phi(X, \mu)^*$ is a continuous operator, it has a continuous adjoint $C_\phi^* : V_\phi(X, \mu)^{**} \to \mathcal{H}$ [29, Sec.IV.7.4]. Let $C_\phi^\sharp := C_\phi^* \lceil V_\phi(X, \mu)$. Then $C_\phi^\sharp = \widehat{T}_\phi$ since, for every $f \in \mathcal{D}$, $[\xi]_\phi \in V_\phi(X, \mu)$,

$$\langle C_\phi f, [\xi]_\phi \rangle = \int_X \langle f, \phi_x \rangle \overline{\xi(x)}\, d\mu(x) = \langle f, \widehat{T}_\phi [\xi]_\phi \rangle. \tag{7.9}$$

If $V_\phi(X, \mu)$ is reflexive, we have, of course, $C_\phi^\sharp = C_\phi^* = \widehat{T}_\phi$.

If ϕ is not μ-total, then there exists $f \in \mathcal{D}$, $f \neq 0$ such that $(C_\phi f)(x) = 0$ for a.e. $x \in X$. Hence, $f \in (\mathsf{Ran}\, \widehat{T}_\phi)^\perp := \{f \in \mathcal{D} : \langle F, f \rangle = 0, \forall F \in \mathsf{Ran}\, \widehat{T}_\phi\}$ by (7.9). Conversely, if ϕ is μ-total, as $(\mathsf{Ran}\, \widehat{T}_\phi)^\perp = \mathsf{Ker}\, C_\phi = \{0\}$, by the reflexivity of $\mathcal{D}$ and $\mathcal{D}^\times$, it follows that $\mathsf{Ran}\, \widehat{T}_\phi$ is dense in $\mathcal{D}^\times$. $\qquad\square$

In a way similar to what we have done above, we can define the space $V_\psi(X, \mu)$, its topology, the residue classes $[\eta]_\psi$, the operator T_ψ, etc., replacing ϕ by ψ. Then, $V_\psi(X, \mu)$ is a locally convex space.

Theorem 7.6 *Assume that the form (7.1) is jointly continuous on $\mathcal{D} \times \mathcal{D}$. Then, every continuous linear functional F on $V_\phi(X, \mu)$, i.e., $F \in V_\phi(X, \mu)^*$, can be represented as*

$$F([\xi]_\phi) = \int_X \xi(x) \overline{\eta(x)}\, d\mu(x), \ \forall [\xi]_\phi \in V_\phi(X, \mu), \tag{7.10}$$

with $\eta \in \mathcal{V}_\psi(X, \mu)$. The residue class $[\eta]_\psi \in V_\psi(X, \mu)$ is uniquely determined.

Proof By Theorem 7.4, we have the representation

$$F(\xi) = \int_X \xi(x) \langle \phi_x, g \rangle\, d\mu(x).$$

It is easily seen that $\eta(x) := \langle g, \phi_x \rangle \in \mathcal{V}_\psi(X, \mu)$.

It remains to prove uniqueness. Suppose that

$$F(\xi) = \int_X \xi(x)\overline{\eta'(x)} \, d\mu(x).$$

Then

$$\int_X \xi(x)(\overline{\eta'(x)} - \overline{\eta(x)}) \, d\mu(x) = 0.$$

Now the function $\xi(x)$ is arbitrary. Hence, taking in particular for $\xi(x)$ the functions $\langle \psi_x, f \rangle$, $f \in \mathcal{D}$, we get $[\eta]_\psi = [\eta']_\psi$.   □

The lesson of the previous statements is that the map

$$j : F \in V_\phi(X, \mu)^* \mapsto [\eta]_\psi \in V_\psi(X, \mu) \tag{7.11}$$

is well-defined and conjugate linear. On the other hand, $j(F) = j(F')$ implies easily $F = F'$. Therefore $V_\phi(X, \mu)^*$ can be identified with a closed subspace of $\overline{V_\psi}(X, \mu) := \{[\overline{\xi}]_\psi : \xi \in V_\psi(X, \mu)\}$.

Working in the framework of Hilbert spaces, as in Sect. 7.1, we proved in Theorem 3.9 that the spaces $V_\phi(X, \mu)^*$ and $\overline{V_\psi}(X, \mu)$ can be identified. The conclusion was that if (ψ, ϕ) is a reproducing pair, the spaces $V_\phi(X, \mu)$ and $V_\psi(X, \mu)$ are both Hilbert spaces, conjugate dual of each other with respect to the sesquilinear form (3.10). And if ϕ and ψ are also μ-total, then the converse statement holds true.

In the present situation, however, a result of this kind cannot be proved with techniques similar to those of Sect. 3.2, which are specific of Hilbert spaces. In particular, the condition (b), $S_{\psi, \phi} \in GL(\mathcal{H})$, which was essential in the proof of [12, Lemma 3.11], is now missing, and it is not clear by what regularity condition it should replaced.

8 Reproducing Pairs and Genuine PIP-Spaces

In this section, we will consider the case where our measurable functions take their values in a genuine PIP-space. However, for simplicity, we will restrict ourselves to a lattice of Banach spaces (LBS) or a lattice of Hilbert spaces (LHS). For the convenience of the reader, we have summarized in the Appendix the basic notions concerning LBSs and LHSs. Further information may be found in our monograph [6].

Let (X, μ) be a locally compact, σ-compact measure space. Let $V_J = \{V_p, p \in J\}$ be an LBS or an LHS of measurable functions on X. Thus the central Hilbert space is $\mathcal{H} := V_o = L^2(X, \mu)$ and the spaces V_p, $V_{\overline{p}}$ are reflexive Banach spaces or Hilbert spaces, conjugate dual of each other with respect to the L^2 inner product, as follows from (A.2). The partial inner product, which extends that of $L^2(X, \mu)$, is denoted again by $\langle \cdot | \cdot \rangle$. As usual we put $V = \sum_{p \in J} V_p$ and $V^\# = \bigcap_{p \in J} V_p$. Thus $\psi : X \to V$ really means that $\psi : X \to V_p$ for some $p \in J$, since V is the algebraic inductive limit of $\{V_p, p \in J\}$ [29] (see the Appendix).

Example 8.1 A typical example is the lattice generated by the Lebesgue spaces $L^p(\mathbb{R}, dx)$, $1 \leqslant p \leqslant \infty$, with $\frac{1}{p} + \frac{1}{\overline{p}} = 1$ [6]. We shall discuss it in detail in Sect. 10.

Two approaches are possible, depending whether the functions ψ_x themselves belong to V or rather the scalar functions $C_\psi f$. However, the first possibility is the exact generalization of the one used in the RHS case in Sect. 7. Hence it does not exploit the PIP-space structure, only the RHS $V^\# \subset \mathcal{H} \subset V$! Thus we turn to the second strategy.

Let ψ, ϕ be weakly measurable functions from X into $\mathcal{H}$. In view of (6.1), (A.2) and the definition of V, we assume that the following condition holds:

(p) $\exists\, p \in J$ such that $C_\psi f = \langle f | \psi. \rangle \in V_p$ and $C_\phi g = \langle g | \phi. \rangle \in V_{\overline{p}}, \forall\, f, g \in \mathcal{H}$.

Notice that, in Condition (p), the index p cannot depend on f, g. We need some uniformity, in the form $C_\psi(\mathcal{H}) \subset V_p$ and $C_\phi(\mathcal{H}) \subset V_{\overline{p}}$. This is fully in line with the philosophy of PIP-spaces: the building blocks are the (assaying) subspaces V_p, not individual vectors.

Since $V_{\overline{p}}$ is the conjugate dual of V_p, the relation

$$\Omega_{\psi,\phi}(f, g) := \int_X \langle f | \psi_x \rangle \langle \phi_x | g \rangle \, d\mu(x), \quad f, g \in \mathcal{H},$$

defines a sesquilinear form on $\mathcal{H} \times \mathcal{H}$ and one has

$$|\Omega_{\psi,\phi}(f, g)| \leqslant \|C_\psi f\|_p \, \|C_\phi g\|_{\overline{p}}, \quad \forall\, f, g \in \mathcal{H}. \tag{8.1}$$

If $\Omega_{\psi,\phi}$ is bounded as a form on $\mathcal{H} \times \mathcal{H}$ (this is not automatic, see Proposition 8.2), there exists a bounded operator $S_{\psi,\phi}$ in $\mathcal{H}$ such that

$$\int_X \langle f | \psi_x \rangle \langle \phi_x | g \rangle \, d\mu(x) = \langle S_{\psi,\phi} f | g \rangle, \quad \forall\, f, g \in \mathcal{H}. \tag{8.2}$$

Then (ψ, ϕ) is a *reproducing pair* if $S_{\psi,\phi} \in GL(\mathcal{H})$.

Let us suppose that the spaces V_p have the following property:

(k) If $\xi_n \to \xi$ in V_p, then, for every compact subset $K \subset X$, there exists a subsequence $\{\xi_n^K\}$ of $\{\xi_n\}$ which converges to ξ almost everywhere in K.

We note that condition (k) is satisfied by the L^p-spaces [28].

As seen before, $C_\psi : \mathcal{H} \to V$, in general. This means, given $f \in \mathcal{H}$, there exists $p \in J$ such that $C_\psi f = \langle f | \psi. \rangle \in V_p$. We define

$$D_r(C_\psi) = \{ f \in \mathcal{H} : C_\psi f \in V_r \}, \ r \in J.$$

In particular, $D_r(C_\psi) = \mathcal{H}$ is equivalent to $C_\psi(\mathcal{H}) \subset V_r$.

Proposition 8.2 *Assume that (k) holds. Then*

(i) $C_\psi : D_r(C_\psi) \to V_r$ is a closed linear map.
(ii) If, for some $r \in J$, $C_\psi(\mathcal{H}) \subset V_r$, then $C_\psi : \mathcal{H} \to V_r$ is continuous.

Proof

(i) Let $f_n \to f$ in $\mathcal{H}$ and $\{C_\psi f_n\}$ be Cauchy in V_r. Since V_r is complete, there exists $\xi \in V_r$ such that $\| C_\psi f_n - \xi \|_r \to 0$. By (k), for every compact subset $K \subset X$, there exists a subsequence $\{ f_n^K \}$ of $\{ f_n \}$ such that $(C_\psi f_n^K)(x) \to \xi(x)$ a.e. in K. On the other hand, since $f_n \to f$ in $\mathcal{H}$, we get

$$\langle f_n | \psi_x \rangle \to \langle f | \psi_x \rangle, \quad \forall x \in X,$$

and the same holds true, of course, for $\{ f_n^K \}$. From this we conclude that $\xi(x) = \langle f | \psi_x \rangle$ almost everywhere. Thus, $f \in D_r(C_\psi)$ and $\xi = C_\psi f$.

(ii) As for the continuity of $C_\psi : \mathcal{H} \to V_r$ it follows from (i) and the closed graph theorem. □

Combining Proposition 8.2(ii) with (8.1), we get

Corollary 8.3 *Assume that (k) holds. If $C_\psi(\mathcal{H}) \subset V_p$ and $C_\phi(\mathcal{H}) \subset V_{\bar{p}}$, the form Ω is bounded on $\mathcal{H} \times \mathcal{H}$, that is, $|\Omega_{\psi,\phi}(f, g)| \leqslant c \| f \| \, \| g \|$.*

If $C_\psi(\mathcal{H}) \subset V_r$, we will assume that $C_\psi : \mathcal{H} \to V_r$ is continuous. According to Proposition 8.2, this is automatic if condition (k) holds.

If $C_\psi : \mathcal{H} \to V_r$ continuously, then $C_\psi^* : V_{\bar{r}} \to \mathcal{H}$ exists and it is continuous. By definition, if $\xi \in V_{\bar{r}}$,

$$\langle C_\psi f | \xi \rangle = \int_X \langle f | \psi_x \rangle \overline{\xi(x)} \, d\mu(x) = \langle f | \int_X \psi_x \, \xi(x) \, d\mu(x) \rangle, \ \forall f \in \mathcal{H}. \qquad (8.3)$$

Thus,

$$C_\psi^* \xi = \int_X \psi_x \, \xi(x) \, d\mu(x).$$

Assume now that for some $p \in J$, $C_\psi : \mathcal{H} \to V_p$ and $C_\phi : \mathcal{H} \to V_{\bar{p}}$ continuously. Then, $C_\phi^* : V_p \to \mathcal{H}$ so that $C_\phi^* C_\psi$ is a well-defined bounded operator in $\mathcal{H}$. As before, we have

$$C_\phi^* \eta = \int_X \eta(x) \phi_x \, d\mu(x), \ \forall \eta \in V_p.$$

Hence,

$$C_\phi^* C_\psi f = \int_X \langle f | \psi_x \rangle \phi_x \, d\mu(x) = S_{\psi,\phi} f, \quad \forall f \in \mathcal{H},$$

the last equality following also from (8.2) and Corollary 8.3. Of course, this does not yet imply that $S_{\psi,\phi} \in GL(\mathcal{H})$, thus we don't know whether (ψ, ϕ) is a reproducing pair.

According to (3.3), the pre-Hilbert space $V_\phi(X, \mu)$ consists of the measurable functions ξ such that

$$\left| \int_X \xi(x) \overline{(C_\phi g)(x)} \, d\mu(x) \right| \leqslant c \, \|g\| , \quad \forall g \in \mathcal{H}. \tag{8.4}$$

Since $C_\phi : \mathcal{H} \to V_{\overline{p}}$, the integral is well-defined for all $\xi \in V_p$. This means, the inner product on the l.h.s. is in fact the partial inner product of V, which coincides with the L^2 inner product whenever the latter makes sense. Thus we may rewrite the relation (8.4) as

$$|\langle \xi | C_\phi g \rangle| \leqslant c \, \|g\| , \forall g \in \mathcal{H}, \ \xi \in V_p ,$$

where $\langle \cdot | \cdot \rangle$ denotes the partial inner product. Next, by (A.2), one has, for $\xi \in V_p$, $g \in \mathcal{H}$,

$$|\langle \xi | C_\phi g \rangle| \leqslant \|\xi\|_p \, \big\| C_\phi g \big\|_{\overline{p}} \leqslant c \, \|\xi\|_p \, \|g\| ,$$

where the last inequality follows from Proposition 8.2 or the assumption of continuity of C_ϕ. Hence $\xi \in V_\phi(X, \mu)$, so that $V_p \subset V_\phi(X, \mu)$.

As for the adjoint operator, we have $C_\phi^* : V_p \to \mathcal{H}$. Then we may write, for $\xi \in V_p, g \in \mathcal{H}$, $\langle \xi | C_\phi g \rangle = \langle T_\phi \xi | g \rangle$, thus C_ϕ^* is the restriction from $V_\phi(X, \mu)$ to V_p of the operator $T_\phi : V_\phi \to \mathcal{H}$ introduced in Sect. 2, which reads now as

$$\langle T_\phi \xi | g \rangle = \int_X \xi(x) \langle \phi_x | g \rangle \, d\mu(x), \ \forall \xi \in V_p, g \in \mathcal{H}. \tag{8.5}$$

Thus $C_\phi^* \subset T_\phi$.

From now on, the construction proceeds as in Sect. 7. The space $V_\phi(X, \mu) = V_\phi(X, \mu)/\text{Ker } T_\phi$, with the norm $\big\| [\xi]_\phi \big\|_\phi = \big\| T_\phi \xi \big\|$, is a pre-Hilbert space. Then Theorem 3.9 and the other results from Sect. 3.2 remain true. In particular, we have:

Theorem 8.4 *If (ψ, ϕ) be a reproducing pair, the spaces $V_\phi(X, \mu)$ and $V_\psi(X, \mu)$ are both Hilbert spaces, conjugate dual of each other with respect to the sesquilinear form (3.8), namely,*

$$\langle\!\langle \xi | \eta \rangle\!\rangle_\mu := \int_X \xi(x) \overline{\eta(x)} \, d\mu(x).$$

Note the form (3.8) coincides with the inner product of $L^2(X, \mu)$ whenever the latter makes sense.

Let (ψ, ϕ) is a reproducing pair. Assume again that $C_\phi : \mathcal{H} \to V_{\overline{p}}$ continuously, which me may write $\widehat{C}_{\phi,\psi} : \mathcal{H} \to V_{\overline{p}}/\mathrm{Ker}\, T_\psi$, where $\widehat{C}_{\phi,\psi} : \mathcal{H} \to V_\psi(X, \mu)$ is the operator defined by $\widehat{C}_{\phi,\psi} f := [C_\phi f]_\psi$, already introduced in Sect. 3.2. In addition, by Theorem 3.8, one has $\mathsf{Ran}\,\widehat{C}_{\psi,\phi}[\|\cdot\|_\phi] = V_\phi(X, \mu)[\|\cdot\|_\phi]$ and $\mathsf{Ran}\,\widehat{C}_{\phi,\psi}[\|\cdot\|_\psi] = V_\psi(X, \mu)[\|\cdot\|_\psi]$.

Putting everything together, we get

Corollary 8.5 *Let (ψ, ϕ) be a reproducing pair. Then, if $C_\psi : \mathcal{H} \to V_p$ and $C_\phi : \mathcal{H} \to V_{\overline{p}}$ continuously, one has*

$$\widehat{C}_{\phi,\psi} : \mathcal{H} \to V_{\overline{p}}/\mathsf{Ker}\, T_\psi = V_\psi(X, \mu) \simeq \overline{V_\phi}(X, \mu)^*, \qquad (8.6)$$

$$\widehat{C}_{\psi,\phi} : \mathcal{H} \to V_p/\mathsf{Ker}\, T_\phi = V_\phi(X, \mu) \simeq \overline{V_\psi}(X, \mu)^*. \qquad (8.7)$$

In these relations, the equality sign means an isomorphism of vector spaces, whereas $\simeq$ denotes an isomorphism of Hilbert spaces.

Proof On the one hand, we have $\mathsf{Ran}\,\widehat{C}_{\phi,\psi} = V_\psi(X, \mu)$. On the other hand, under the assumption $C_\phi(\mathcal{H}) \subset V_{\overline{p}}$, one has $V_{\overline{p}} \subset \mathcal{V}_\psi(X, \mu)$, hence $V_{\overline{p}}/\mathsf{Ker}\, T_\psi = \{\xi + \mathsf{Ker}\, T_\psi, \ \xi \in V_{\overline{p}}\} \subset V_\psi(X, \mu)$. Thus we get $V_\psi(X, \mu) = V_{\overline{p}}/\mathsf{Ker}\, T_\psi$ as vector spaces. Similarly $V_\phi(X, \mu) = V_p/\mathsf{Ker}\, T_\phi$. $\qquad\square$

9 The Case of a Hilbert Triplet or a Hilbert Scale

9.1 The General Construction

We have derived in the previous section the relations $V_p \subset \mathcal{V}_\phi(X, \mu)$, $V_{\overline{p}} \subset \mathcal{V}_\psi(X, \mu)$, and their equivalent ones (8.6)–(8.7). Then, since $V_\psi(X, \mu)$ and $V_\phi(X, \mu)$ are both Hilbert spaces, it seems natural to take for $V_p, V_{\overline{p}}$ Hilbert spaces as well, that is, take for V an LHS. The simplest case is then a Hilbert chain, for instance, the scale (A.4) $\{\mathcal{H}_k, k \in \mathbb{Z}\}$ built on the powers of a self-adjoint operator $A > I$. This situation is quite interesting, since in that case one may get results about spectral properties of symmetric operators (in the sense of PIP-space operators) [6, 8].

Thus, let (ψ, ϕ) be a reproducing pair. For simplicity, we assume that $S_{\psi,\phi} = I$, that is, ψ, ϕ are dual to each other.

If ψ and ϕ are both frames, there is nothing to say, since then $C_\psi(\mathcal{H}), C_\phi(\mathcal{H}) \subset L^2(X, \mu) = \mathcal{H}_o$, so that there is no need for a Hilbert scale. Thus we assume that ψ is an upper semi-frame and ϕ is a lower semi-frame, dual to each other. It follows that $C_\psi(\mathcal{H}) \subset L^2(X, \mu)$. Hence Condition (p) becomes: There is an index $k \geqslant 1$

such that $C_\psi : \mathcal{H} \to \mathcal{H}_k$ and $C_\phi : \mathcal{H} \to \mathcal{H}_{\bar{k}}$ continuously, thus $V_p \equiv \mathcal{H}_k$ and $V_{\bar{p}} \equiv \mathcal{H}_{\bar{k}}$. This means we are working in the Hilbert triplet

$$V_p \equiv \mathcal{H}_k \subset \mathcal{H}_o = L^2(X, \mu) \subset \mathcal{H}_{\bar{k}} \equiv V_{\bar{p}}. \tag{9.1}$$

Next, according to Corollary 8.5, we have $V_\psi(X, \mu) = \mathcal{H}_{\bar{k}}/\mathrm{Ker}\, T_\psi$ and $V_\phi(X, \mu) = \mathcal{H}_k/\mathrm{Ker}\, T_\phi$, as vector spaces.

In addition, since ϕ is a lower semi-frame, we know that C_ϕ has closed range in $L^2(X, \mu)$ and is injective [4, Lemma 2.1]. However its domain

$$D(C_\phi) := \{f \in \mathcal{H} : \int_X |\langle f|\phi_x\rangle|^2 \, \mathrm{d}\mu(x) < \infty\}$$

need not be dense, it could be $\{0\}$. Thus C_ϕ maps its domain $D(C_\phi)$ onto a closed subspace of $L^2(X, \mu)$, possibly trivial, and the whole of $\mathcal{H}$ into the larger space $\mathcal{H}_{\bar{k}}$.

9.2 Examples

As for concrete examples of such Hilbert scales, we might mention two. First the Sobolev spaces $H^k(\mathbb{R})$, $k \in \mathbb{Z}$, in $\mathcal{H}_0 = L^2(\mathbb{R}, dx)$, which is the scale generated by the powers of the positive self-adjoint operator $A^{1/2}$, where $A := 1 - \frac{d^2}{dx^2}$. The other one corresponds to the quantum harmonic oscillator, with Hamiltonian $A_{\mathrm{osc}} := x^2 - \frac{d^2}{dx^2}$. The spectrum of A_{osc} is $\{2n + 1, n = 0, 1, 2, \ldots\}$ and it gets diagonalized on the basis of Hermite functions. It follows that A_{osc}^{-1}, which maps every $\mathcal{H}_k$ onto $\mathcal{H}_{k-1}$, is a Hilbert-Schmidt operator. Therefore, the end space of the scale $\mathcal{D}^\infty(A_{\mathrm{osc}}) := \bigcap_k \mathcal{H}_k$, which is simply Schwartz' space $\mathcal{S}$ of C^∞ functions of fast decrease, is a nuclear space.

Actually one may give an explicit example, using a Sobolev-type scale. Let $\mathcal{H}_K$ be a reproducing kernel Hilbert space (RKHS) of (nice) functions on a measure space (X, μ), with kernel function k_x, $x \in X$, that is, $f(x) = \langle f|k_x\rangle_K$, $\forall f \in \mathcal{H}_K$. The corresponding reproducing kernel is $K(x, y) = k_y(x) = \langle k_y|k_x\rangle_K$. Choose the weight function $m(x) > 1$, the analog of the weight $(1 + |x|^2)$ considered in the Sobolev case. Define the Hilbert scale $\mathcal{H}_l$, $l \in \mathbb{Z}$, determined by the multiplication operator $Af(x) = m(x)f(x)$, $\forall x \in X$. Hence, for each $l \geqslant 1$,

$$\mathcal{H}_l \subset \mathcal{H}_0 \equiv \mathcal{H}_K \subset \mathcal{H}_{\bar{l}}.$$

Then, for some $n \geqslant 1$, define the measurable functions $\phi_x = k_x m^n(x)$, $\psi_x = k_x m^{-n}(x)$, so that $C_\psi : \mathcal{H}_K \to \mathcal{H}_n$, $C_\phi : \mathcal{H}_K \to \mathcal{H}_{\bar{n}}$ continuously, where $\mathcal{H}_n \subset \mathcal{H}_K \subset \mathcal{H}_{\bar{n}}$, and ψ, ϕ are dual of each other. One has indeed $\langle \phi_x|g\rangle_K = \langle k_x m^n(x)|g\rangle_K = \langle k_x|g\, m^n(x)\rangle_K = \overline{g(x)}\, m^n(x) \in \mathcal{H}_{\bar{n}}$ and

$\langle \psi_x | g \rangle_K = \overline{g(x)} \, m^{-n}(x) \in \mathcal{H}_n$, which implies duality. Thus (ψ, ϕ) is a reproducing pair with $S_{\psi,\phi} = I$, ψ is an upper semi-frame and ϕ a lower semi-frame.

In this case, one can compute the operators T_ψ, T_ϕ explicitly. The definition (8.5) reads as

$$\langle T_\phi \xi | g \rangle_K = \int_X \xi(x) \langle \phi_x | g \rangle_K \, d\mu(x), \ \forall \xi \in \mathcal{H}_n, g \in \mathcal{H}_K,$$

$$= \int_X \xi(x) \, \overline{g(x)} \, m^n(x) \, d\mu,$$

that is, $(T_\phi \xi)(x) = \xi(x) \, m^n(x)$ or $T_\phi \xi = \xi \, m^n$. However, since the weight $m(x) > 1$ is invertible, $\overline{g} \, m^n$ runs over the whole of $\mathcal{H}_{\overline{n}}$ whenever g runs over H_K. Hence $\xi \in \text{Ker } T_\phi \subset \mathcal{H}_n$ means that $\langle T_\phi \xi | g \rangle_K = 0$, $\forall g \in H_K$, which implies $\xi = 0$, since the duality between $\mathcal{H}_n$ and $\mathcal{H}_{\overline{n}}$ is separating. The same reasoning yields $\text{Ker } T_\psi = \{0\}$. Therefore $V_\phi(X, \mu) \simeq \text{Ran } C_\psi = \mathcal{H}_n$ and $V_\psi(X, \mu) \simeq \text{Ran } C_\phi = \mathcal{H}_{\overline{n}}$.

A more general situation may be derived from the discrete example of Section 6.1.3 of [12]. Take a sequence of weights $m := \{|m_n|\}_{n \in \mathbb{N}} \in c_0, m_n \neq 0$, and consider the space ℓ_m^2 with norm $\|\xi\|_{\ell_m^2} := \sum_{n \in \mathbb{N}} |m_n \xi_n|^2$. Then we have the following triplet replacing (9.1)

$$\ell_{1/m}^2 \subset \ell^2 \subset \ell_m^2. \tag{9.2}$$

Next, for each $n \in \mathbb{N}$, define $\psi_n = m_n \theta_n$, where θ is a frame or an orthonormal basis in ℓ^2. Then ψ is an upper semi-frame. Moreover, $\phi := \{(1/\overline{m_n}) \theta_n\}_{n \in \mathbb{N}}$ is a lower semi-frame, dual to ψ, thus (ψ, ϕ) is a reproducing pair. Hence, by [12, Theorem 3.13] (see also Sect. 5.1.3), $V_\psi \simeq \text{Ran } C_\phi = M_{1/m}(V_\theta(\mathbb{N})) = \ell_m^2$ and $V_\phi \simeq \text{Ran } C_\psi = M_m(V_\theta(\mathbb{N})) = \ell_{1/m}^2$ (here we take for granted that $\text{Ker } T_\psi = \text{Ker } T_\phi = \{0\}$).

For making contact with the situation of (9.1), consider in ℓ^2 the diagonal operator $A := \text{diag}[n], n \in \mathbb{N}$ (the number operator), that is $(A\xi)_n = n \xi_n, n \in \mathbb{N}$, which is obviously self-adjoint and larger than 1. Then $\mathcal{H}_k = D(A^k)$ with norm $\|\xi\|_k = \|A^k \xi\| \equiv \ell_{r^{(k)}}^2$, where $(r^{(k)})_n = n^k$ (note that $1/r^{(k)} \in c_0$). Hence we have

$$\mathcal{H}_k = \ell_{r^{(k)}}^2 \subset \mathcal{H}_o = \ell^2 \subset \mathcal{H}_{\overline{k}} = \ell_{1/r^{(k)}}^2, \tag{9.3}$$

where $(1/r^{(k)})_n = n^{-k}$. In addition, as in the continuous case discussed above, the end space of the scale, $\mathcal{D}^\infty(A) := \bigcap_k \mathcal{H}_k$, is simply Schwartz's space s of fast decreasing sequences, with dual $\mathcal{D}_{\overline{\infty}}(A) := \bigcup_k \mathcal{H}_k = s'$, the space of slowly increasing sequences. Here too, this construction shows that the space s is nuclear, since every embedding $A^{-1} : \mathcal{H}_{k+1} \to \mathcal{H}_k$ is a Hilbert-Schmidt operator.

However, the construction described above yields a much more general family of examples, since the weight sequences m are not ordered.

10   The Case of L^p Spaces

Following the suggestion made at the end of Sect. 2, we present now several possibilities of taking Ran C_ψ in the context of the Lebesgue spaces $L^p(\mathbb{R}, \, dx)$.

As it is well-known, these spaces don't form a chain, since two of them are never comparable. We have only

$$L^p \cap L^q \subset L^s, \text{ for all } s \text{ such that } p < s < q.$$

Take the lattice $\mathcal{J}$ generated by $\mathcal{I} = \{L^p(\mathbb{R}, \, dx), 1 \leqslant p \leqslant \infty\}$, with lattice operations [6, Sec.4.1.2]:

- $L^p \wedge L^q = L^p \cap L^q$ is a Banach space for the projective norm $\|f\|_{p \wedge q} = \|f\|_p + \|f\|_q$
- $L^p \vee L^q = L^p + L^q$ is a Banach space for the inductive norm
$$\|f\|_{p \vee q} = \inf_{f = g + h} \left\{ \|g\|_p + \|h\|_q; g \in L^p, h \in L^q \right\}$$
- For $1 < p, q < \infty$, both spaces $L^p \wedge L^q$ and $L^p \vee L^q$ are reflexive and $(L^p \wedge L^q)^\times = L^{\bar{p}} \vee L^{\bar{q}}$.

Moreover, no additional spaces are obtained by iterating the lattice operations to any finite order. Thus we obtain an involutive lattice and an LBS, denoted by V_J with elements denoted generically by $L^{(s)}$, $s = (p, q)$.

Following [6, Sec.4.1.2], we represent the space $L^{(p,q)}$ by the point $(1/p, 1/q)$ of the unit square $J = [0, 1] \times [0, 1]$. In this representation, the spaces L^p are on the main diagonal, intersections $L^p \cap L^q$ above it and sums $L^p + L^q$ below, the duality is $[L^{(s)}]^\times = L^{(\bar{s})}$, where $s = (p, q)$ and $\bar{s} = (\bar{p}, \bar{q})$, that is, symmetry with respect to L^2. Hence, $L^{(p,q)} \subset L^{(p',q')}$ if $(1/p, 1/q)$ is on the left and/or above $(1/p', 1/q')$ The extreme spaces are

$$V_J^\# = L^\infty \cap L^1 \quad \text{and} \quad V_J = L^1 + L^\infty.$$

Note that the space $L^1 + L^\infty$ has been considered by Gould [23]. For a full picture, see [6, Fig.4.1].

There are three possibilities for using the L^p lattice for controlling reproducing pairs

(1) Exploit the *full lattice* $\mathcal{J}$, that is, find (p, q) such that $\forall f, g \in \mathcal{H}$, $C_\psi f \# C_\phi g$ in the PIP-space V_J, that is, $C_\psi f \in L^{(p,q)}$ and $C_\phi g \in L^{(\bar{p},\bar{q})}$.
(2) Select in V_J a self-dual *Banach chain* V_I, centered around L^2, symbolically.

$$\ldots L^{(s)} \subset \ldots \subset L^2 \subset \ldots \subset L^{(\bar{s})} \ldots, \tag{10.1}$$

such that $C_\psi f \in L^{(s)}$ and $C_\phi g \in L^{(\bar{s})}$ (or vice-versa). Here are three examples of such Banach chains. Of course, in each case, one may also select a symmetric subset of the chain.

- The *anti-diagonal* chain : $q = \bar{p}$

$$L^\infty \cap L^1 \subset \ldots \subset L^{\bar{q}} \cap L^q \subset \ldots \subset L^2 \subset \ldots \subset L^q + L^{\bar{q}}$$
$$= (L^{\bar{q}} \cap L^q)^\times \subset \ldots \subset L^1 + L^\infty.$$

- The *horizontal* chain $q = 2$:

$$L^\infty \cap L^2 \subset \ldots \subset L^2 \subset \ldots \subset L^1 + L^2.$$

- The *vertical* chain $p = 2$:

$$L^2 \cap L^1 \subset \ldots \subset L^2 \subset \ldots \subset L^2 + L^\infty.$$

All three chains are presented in Fig. 1. In each case, the full chain belongs to the second and fourth quadrants (top left and bottom right). A typical point is then $s = (p, q)$ with, $2 \leqslant p \leqslant \infty$, $1 \leqslant q \leqslant 2$, so that one has the situation depicted

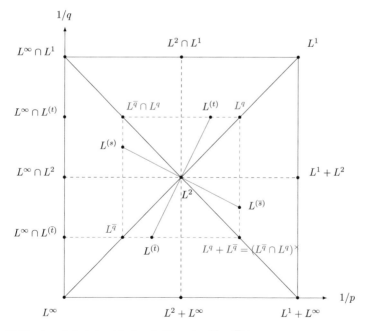

Fig. 1 (1) The three full chains (black); (2) The pair $L^{(s)}$, $L^{(\bar{s})}$ for s in the second quadrant (blue); (3) The pair $L^{(t)}$, $L^{(\bar{t})}$ for t in the first quadrant (blue) (from [9])

in (10.1), that is, the spaces $L^{(s)}$, $L^{(\bar{s})}$ to which $C_\psi f$, resp. $C_\phi g$, belong, are necessarily comparable to each other and to L^2. In particular, one of them is necessarily contained in L^2 (see Fig. 1).

(3) Choose a dual pair in the first and third quadrant (top right, bottom left). A typical point is then $t = (p', q')$, with $1 < p', q' < 2$, so that the spaces $L^{(t)}$, $L^{(\bar{t})}$ are never comparable to each other, nor to L^2.

Let us now add the boundedness condition already mentioned in Sect. 2, $\sup_{x \in X} \|\psi_x\|_{\mathcal{H}} \leqslant c$ and $\sup_{x \in X} \|\phi_x\|_{\mathcal{H}} \leqslant c'$ for some $c, c' > 0$. Then $C_\psi f(x) = \langle f | \psi_x \rangle \in L^\infty(X, \mathrm{d}\mu)$ and $C_\phi f(x) = \langle f | \phi_x \rangle \in L^\infty(X, \mathrm{d}\mu)$. Therefore, the third case reduces to the second one, since we have now (in the situation of Fig. 1).

$$L^\infty \cap L^{(t)} \subset L^\infty \cap L^2 \subset L^\infty \cap L^{(\bar{t})}. \tag{10.2}$$

Note that none of these spaces is reflexive.

Following the pattern of Hilbert scales, we choose a (Gel'fand) triplet of Banach spaces. One could have, for instance, a triplet of reflexive Banach spaces such as

$$L^{(s)} \subset L^2 \subset L^{(\bar{s})}, \tag{10.3}$$

corresponding to a point s inside of the second quadrant, as shown in Fig. 1. In this case, according to (8.6) and (8.7), $V_\psi = L^{(\bar{s})}/\text{Ker}\, T_\psi$ and $V_\phi = L^{(s)}/\text{Ker}\, T_\phi$.

On the contrary, if we choose a point t in the second quadrant, case (3) above, it seems that no triplet arises. However, if (ψ, ϕ) is a nontrivial reproducing pair, with $S_{\psi,\phi} = I$, that is, ψ, ϕ are dual to each other, one of them, say ψ, is an upper semi-frame and then necessarily ϕ is a lower semi-frame (see the remark at the end of Sect. 2.2). Therefore $C_\psi(\mathcal{H}) \subset L^2(X, \mu)$, that is, case (3) cannot be realized.

In conclusion, the only acceptable solution is a triplet of the type (10.3), with s strictly inside of the second quadrant, that is, $s = (p, q)$ with, $2 \leqslant p < \infty, 1 < q \leqslant 2$.

A word of explanation is in order here, concerning the relations $V_\psi = L^{(\bar{s})}/\text{Ker}\, T_\psi$ and $V_\phi = L^{(s)}/\text{Ker}\, T_\phi$. Both $L^{(s)}$ and $L^{(\bar{s})}$ are reflexive Banach spaces, with their usual norm, and so are the quotients by T_ψ, resp. T_ϕ. On the other hand, $V_\psi(X, \mu)[\| \cdot \|_\psi]$ and $V_\phi(X, \mu)[\| \cdot \|_\phi]$ are Hilbert spaces. However, there is no contradiction, since the equality sign $=$ denotes an isomorphism of vector spaces only, without reference to any topology. Moreover, the two norms, Banach and Hilbert, *cannot* be comparable, lest they are equivalent [26, Coroll. 1.6.8], which is impossible in the case of L^p, $p \neq 2$. The same is true for any LBS where the spaces V_p are not Hilbert spaces.

Although we don't have an explicit example of a reproducing pair, we indicate a possible construction towards one. Let $\theta^{(1)} : \mathbb{R} \to L^2$ be a measurable function such that $\langle h | \theta_x^{(1)} \rangle \in L^q$, $1 < q < 2$, $\forall h \in L^2$, and let $\theta^{(2)} : \mathbb{R} \to L^2$ be a measurable function such that $\langle h | \theta_x^{(2)} \rangle \in L^{\overline{q}}$, $\forall h \in L^2$. Define $\psi_x := \min(\theta_x^{(1)}, \theta_x^{(2)}) \equiv \theta_x^{(1)} \wedge \theta_x^{(2)}$ and $\phi_x := \max(\theta_x^{(1)}, \theta_x^{(2)}) \equiv \theta_x^{(1)} \vee \theta_x^{(2)}$. Then we have $\psi_x \leqslant \phi_x$ for almost all $x \in \mathbb{R}$, and

$$(C_\psi h)(x) = \langle h | \psi_x \rangle \in L^q \cap L^{\overline{q}}, \ \forall h \in L^2$$

$$(C_\phi h)(x) = \langle h | \phi_x \rangle \in L^q + L^{\overline{q}}, \ \forall h \in L^2$$

and we have indeed $L^q \cap L^{\overline{q}} \subset L^2 \subset L^q + L^{\overline{q}}$. It remains to guarantee that ψ and ϕ are dual to each other, that is,

$$\int_X \langle f | \psi_x \rangle \langle \phi_x | g \rangle \, d\mu(x) = \int_X C_\psi f(x) \, \overline{C_\phi g(x)} \, d\mu(x) = \langle f | g \rangle, \ \forall f, g \in L^2.$$

11 Concluding Remarks

Starting with the well-known notion of frame, both discrete and continuous, we have introduced a first natural generalization, namely, semi-frames, both upper and lower ones. The main result is that the two types are dual of each other. Indeed, if two semi-frames are in duality, either they are both frames, or else at least one of them is a lower semi-frame. Take, for example, $\phi = \{n e_n\} \cup \{\frac{1}{k} e_k\}_k$ and $\psi = \{\frac{1}{2n} e_n\}_n \cup \{\frac{k}{2} e_k\}_k$ with e_n an orthonormal basis, then ψ, ϕ are in duality and both are lower semi-frames but not Bessel (M. Speckbacher, private communication).

Then the next step is to drop the restriction imposed by the frame bounds on the two measurable functions in duality, and this leads to the notion of reproducing pair. We have seen that the latter is quite rich. It generates a whole mathematical structure, which ultimately leads to a pair of Hilbert spaces, conjugate dual to each other with respect to the $L^2(X, \mu)$ inner product. We have given several concrete examples in Sect. 5. These, and additional ones, should allow one to better specify the best assumptions to be made on the measurable functions or, more precisely, on the nature of the range of the analysis operators C_ψ, C_ϕ.

This is clearly seen in the definition (6.1), which immediately suggests to perform the analysis in the context of PIP-spaces [6], as already remarked in [30]. In particular, a natural choice is a scale, or simply a triplet, of Hilbert spaces, the two extreme spaces being conjugate duals of each other with respect to the $L^2(X, \mu)$ inner product. Another possibility consists of exploiting the lattice of all $L^p(\mathbb{R}, dx)$ spaces, or a subset thereof, in particular a (Gel'fand) triplet of Banach spaces. Some examples have been described above, but obviously more work along these lines is in order.

Another interesting direction consists in considering a whole family $\mathcal{G}$ of μ-total, weakly measurable functions $\phi : X \to \mathcal{H}$, instead of only one. To each $\phi \in \mathcal{G}$ we can associate the pre-Hilbert space $V_\phi(X, \mu)[\|\cdot\|_\phi]$ and take its completion

$\widetilde{V}_\phi(X,\mu)[\|\cdot\|_\phi]$. If ϕ has a partner $\psi \in \mathcal{G}$ such that (ψ, ϕ) is a reproducing pair, both spaces $V_\phi(X,\mu) = \widetilde{V}_\phi(X,\mu)[\|\cdot\|_\phi]$ and $V_\psi(X,\mu) = \widetilde{V}_\psi(X,\mu)[\|\cdot\|_\phi]$ are Hilbert spaces, conjugate dual to each other. In the general case, however, the question of completeness of $V_\phi(X,\mu)[\|\cdot\|_\phi]$ is open. Can one find conditions under which it holds? Also once might study the relationship between different pre-Hilbert spaces $V_\phi(X,\mu)$. When is one contained in another one?

Acknowledgements This work was partly supported by the Istituto Nazionale di Alta Matematica (GNAMPA project "Proprietà spettrali di quasi *-algebre di operatori"). JPA acknowledges gratefully the hospitality of the Dipartimento di Matematica e Informatica, Università di Palermo, whereas CT acknowledges that of the Institut de Recherche en Mathématique et Physique, Université catholique de Louvain. We also thank Michael Speckbacher for supplying the new proof of Proposition 2.1.

Appendix: Lattices of Banach or Hilbert Spaces

For the convenience of the reader, we summarize in this Appendix the basic facts concerning PIP-spaces and operators on them. However, we will restrict the discussion to the simpler case of a lattice of Banach (LBS) or Hilbert spaces (LHS). Further information may be found in our monograph [6] or our review paper [7].

Let thus $\mathcal{J} = \{V_p, \ p \in I\}$ be a family of Hilbert spaces or reflexive Banach spaces, partially ordered by inclusion. Then $\mathcal{I}$ generates an involutive lattice $\mathcal{J}$, indexed by J, through the operations $(p, q, r \in I)$:

- involution: $V_r \leftrightarrow V_{\bar{r}} = V_r^\times$, the conjugate dual of V_r
- infimum: $V_{p \wedge q} := V_p \wedge V_q = V_p \cap V_q$
- supremum: $V_{p \vee q} := V_p \vee V_q = V_p + V_q$.

It turns out that both $V_{p \wedge q}$ and $V_{p \vee q}$ are Hilbert spaces, resp. reflexive Banach spaces, under appropriate norms (the so-called projective, resp. inductive norms). Assume that the following conditions are satisfied:

(1) $\mathcal{I}$ contains a unique self-dual, Hilbert subspace $V_o = V_{\bar{o}}$.
(2) for every $V_r \in \mathcal{I}$, the norm $\|\cdot\|_{\bar{r}}$ on $V_{\bar{r}} = V_r^\times$ is the conjugate of the norm $\|\cdot\|_r$ on V_r.

In addition to the family $\mathcal{J} = \{V_r, \ r \in J\}$, it is convenient to consider the two spaces $V^\#$ and V defined as

$$V = \sum_{q \in I} V_q, \quad V^\# = \bigcap_{q \in I} V_q. \tag{A.1}$$

These two spaces themselves usually do *not* belong to $\mathcal{I}$. According to the general theory of PIP-spaces [6], V is the algebraic inductive limit of the V_p's, and $V^\#$ is the projective limit of the V_p's.

We say that two vectors $f, g \in V$ are *compatible* if there exists $r \in J$ such that $f \in V_r, g \in V_{\bar{r}}$. Then a *partial inner product* on V is a Hermitian form $\langle \cdot | \cdot \rangle$ defined exactly on compatible pairs of vectors. In particular, the partial inner product $\langle \cdot | \cdot \rangle$ coincides with the inner product of V_o on the latter. A *partial inner product space* (PIP-space) is a vector space V equipped with a partial inner product. Clearly LBSs and LHSs are particular cases of PIP-spaces.

We will assume that our PIP-space $(V, \langle \cdot | \cdot \rangle)$ is *nondegenerate*, that is, $\langle f | g \rangle = 0$ for all $f \in V^{\#}$ implies $g = 0$. As a consequence, $(V^{\#}, V)$ and every couple $(V_r, V_{\bar{r}})$, $r \in J$, are a dual pair in the sense of topological vector spaces [29]. In particular, the original norm topology on V_r coincides with its Mackey topology $\tau(V_r, V_{\bar{r}})$, so that indeed its conjugate dual is $(V_r)^{\times} = V_{\bar{r}}$, $\forall r \in J$. Then, $r < s$ implies $V_r \subset V_s$, and the embedding operator $E_{sr} : V_r \to V_s$ is continuous and has dense range. In particular, $V^{\#}$ is dense in every V_r. In the sequel, we also assume the partial inner product to be positive definite, $\langle f | f \rangle > 0$ whenever $f \neq 0$.

Then we have the familiar (Schwarz) inequality

$$\xi \in V_p, \ \eta \in V_{\bar{p}} \quad \text{implies} \quad \xi \bar{\eta} \in L^1(X, \mu) \quad \text{and}$$

$$\left| \int_X \xi(x) \overline{\eta(x)} \, d\mu(x) \right| \leqslant \|\xi\|_p \, \|\eta\|_{\bar{p}}. \tag{A.2}$$

A standard, albeit trivial, example is that of a rigged Hilbert space (RHS) $\Phi \subset \mathcal{H} \subset \Phi^{\#}$ (it is trivial because the lattice $\mathcal{I}$ contains only three elements).

Familiar concrete examples of PIP-spaces are sequence spaces, with $V = \omega$ the space of *all* complex sequences $x = (x_n)$, and spaces of locally integrable functions with $V = L^1_{\text{loc}}(\mathbb{R}, dx)$, the space of Lebesgue measurable functions, integrable over compact subsets.

Among LBSs, the simplest example is that of a chain of reflexive Banach spaces. The prototype is the chain $\mathcal{I} = \{L^p := L^p([0, 1]; dx), \ 1 < p < \infty\}$ of Lebesgue spaces over the interval $[0, 1]$.

$$L^{\infty} \subset \ldots \subset L^{\bar{q}} \subset L^{\bar{r}} \subset \ldots \subset L^2 \subset \ldots \subset L^r \subset L^q \subset \ldots \subset L^1, \tag{A.3}$$

where $1 < q < r < 2$ (of course, L^{∞} and L^1 are not reflexive). Here L^q and $L^{\bar{q}}$ are dual to each other ($1/q + 1/\bar{q} = 1$), and similarly $L^r, L^{\bar{r}}$ ($1/r + 1/\bar{r} = 1$).

As for an LHS, the simplest example is the Hilbert scale generated by a self-adjoint operator $A > I$ in a Hilbert space $\mathcal{H}_o$. Let $\mathcal{H}_n$ be $D(A^n)$, the domain of A^n, equipped with the graph norm $\|f\|_n = \|A^n f\|$, $f \in D(A^n)$, for $n \in \mathbb{N}$ or $n \in \mathbb{R}^+$, and $\mathcal{H}_{\bar{n}} := \mathcal{H}_{-n} = \mathcal{H}_n^{\times}$ (conjugate dual):

$$\mathcal{D}^{\infty}(A) := \bigcap_n \mathcal{H}_n \subset \ldots \subset \mathcal{H}_2 \subset \mathcal{H}_1 \subset \mathcal{H}_0 \subset \mathcal{H}_{\bar{1}} \subset \mathcal{H}_{\bar{2}} \ldots \subset \mathcal{D}_{\overline{\infty}}(A) := \bigcup_n \mathcal{H}_n. \tag{A.4}$$

Note that here the index n may be integer or real, the link between the two cases being established by the spectral theorem for self-adjoint operators. Here again the inner product of $\mathcal{H}_0$ extends to each pair $\mathcal{H}_n$, $\mathcal{H}_{-n}$, but on $\mathcal{D}_{\overline{\infty}}(A)$ it yields only a *partial* inner product. A standard example is the scale of Sobolev spaces $H^s(\mathbb{R})$, $s \in \mathbb{Z}$, in $\mathcal{H}_0 = L^2(\mathbb{R}, dx)$.

References

1. S.T. Ali, J.-P. Antoine, J.-P. Gazeau, Square integrability of group representations on homogeneous spaces I. Reproducing triples and frames. Ann. Inst. H. Poincaré **55**, 829–856 (1991)
2. S.T. Ali, J.-P. Antoine, J.-P. Gazeau, Continuous frames in Hilbert space. Ann. Phys. **222**, 1–37 (1993)
3. S.T. Ali, J.-P. Antoine, J.-P. Gazeau, *Coherent States, Wavelets and Their Generalizations*, 2nd edn. (Springer, New York, 2014)
4. J.-P. Antoine, P. Balazs, Frames and semi-frames. J. Phys. A: Math. Theor. **44**, 205201 (2011); Corrigendum, ibid. **44**, 479501 (2011)
5. J.-P. Antoine, P. Balazs, Frames, semi-frames, and Hilbert scales. Numer. Funct. Anal. Optim. **33**, 736–769 (2012)
6. J.-P. Antoine, C. Trapani, *Partial Inner Product Spaces: Theory and Applications*. Lecture Notes in Mathematics, vol. 1986 (Springer, Berlin, 2009)
7. J.-P. Antoine, C. Trapani, The partial inner product space method: a quick overview. Adv. Math. Phys. **2010**, 457635 (2010); Erratum, ibid. **2011**, 272703 (2010)
8. J.-P. Antoine, C. Trapani, Operators on partial inner product spaces: towards a spectral analysis. Mediterr. J. Math. **13**, 323–351 (2016)
9. J.-P. Antoine, C. Trapani, Reproducing pairs of measurable functions and partial inner product spaces. Adv. Oper. Theory **2**, 126–146 (2017)
10. J.-P. Antoine, P. Vandergheynst, Wavelets on the 2-sphere: a group theoretical approach. Appl. Comput. Harmon. Anal. **7**, 262–291 (1999)
11. J.-P. Antoine, A. Inoue, C. Trapani, *Partial *-Algebras and Their Operator Realizations*. Mathematics and Its Applications, vol. 553 (Kluwer, Dordrecht, 2002)
12. J.-P. Antoine, M. Speckbacher, C. Trapani, Reproducing pairs of measurable functions. Acta Appl. Math. **150**, 81–101 (2017)
13. A. Askari-Hemmat, M.A. Dehghan, M. Radjabalipour, Generalized frames and their redundancy. Proc. Am. Math. Soc. **129**, 1143–1147 (2001)
14. P.G. Casazza, The art of frame theory. Taiwan. J. Math. **4**, 129–202 (2000)
15. P. Casazza, O. Christensen, S. Li, A. Lindner, Riesz-Fischer sequences and lower frame bounds. Z. Anal. Anwend. **21**, 305–314 (2002)
16. O. Christensen, *An Introduction to Frames and Riesz Bases* (Birkhäuser, Boston, 2003)
17. I. Daubechies, *Ten Lectures On Wavelets*. CBMS-NSF Regional Conference Series in Applied Mathematics (SIAM, Philadelphia, 1992)
18. I. Daubechies, A. Grossmann, Y. Meyer, Painless nonorthogonal expansions. J. Math. Phys. **27**, 1271–1283 (1986)
19. R.J. Duffin, A.C. Schaeffer, A class of nonharmonic Fourier series. Trans. Am. Math. Soc. **72**, 341–366 (1952)
20. M. Fornasier, Banach frames for α-modulation spaces. Appl. Comput. Harmon. Anal. **22**, 157–175 (2007)
21. M. Fornasier, H. Rauhut, Continuous frames, function spaces, and the discretization problem. J. Fourier Anal. Appl. **11**, 245–287 (2005)
22. J.-P. Gabardo, D. Han, Frames associated with measurable spaces. Adv. Comput. Math. **18**, 127–147 (2003)

23. G.G. Gould, On a class of integration spaces. J. Lond. Math. Soc. **34**, 161–172 (1959)
24. K. Gröchenig, *Foundations of Time-Frequency Analysis* (Birkhäuser, Boston, 2001)
25. G. Kaiser, *A Friendly Guide to Wavelets* (Birkhäuser, Boston, 1994)
26. R.E. Megginson, *An Introduction to Banach Space Theory* (Springer, New York-Heidelberg-Berlin, 1998)
27. A. Rahimi, A. Najati, Y.N. Dehghan, Continuous frames in Hilbert spaces. Methods Funct. Anal. Topol. **12**, 170–182 (2006)
28. W. Rudin, *Real and Complex Analysis*, Int. edn. (McGraw Hill, New York, 1987); p.73, from Ex.18
29. H.H. Schaefer, *Topological Vector Spaces* (Springer, New York-Heidelberg-Berlin, 1971)
30. M. Speckbacher, P. Balazs, Reproducing pairs and the continuous nonstationary Gabor transform on LCA groups. J. Phys. A: Math. Theor. **48**, 395201 (2015)
31. M. Speckbacher, D. Bayer, S. Dahlke, P. Balazs, The α-modulation transform: admissibility, coorbit theory and frames of compactly supported functions. Monatsh. Math. **184**, 133–169 (2017)
32. B. Torrésani, Wavelets associated with representations of the Weyl-Heisenberg group. J. Math. Phys. **32**, 1273–1279 (1991)
33. Y. Wiaux, L. Jacques, P. Vandergheynst, Correspondence principle between spherical and Euclidean wavelets. Astrophys. J. **632**, 15–28 (2005)
34. R.M. Young, *An Introduction to Nonharmonic Fourier Series*, Rev. 1st edn. (Academic Press, San Diego, 2001)

On Hilbert-Schmidt Operator Formulation of Noncommutative Quantum Mechanics

Isiaka Aremua, Ezinvi Baloïtcha, Mahouton Norbert Hounkonnou, and Komi Sodoga

Abstract This work gives value to the importance of Hilbert-Schmidt operators in the formulation of noncommutative quantum theory. A system of charged particle in a constant magnetic field is investigated in this framework.

Keywords Hilbert spaces · Operator theory · Hilbert-Schmidt operators · von Neumann algebra · Modular theory · Density matrix · Coherent states · Noncommutative quantum mechanics

1 Introduction

The theory of Hilbert-Schmidt operators plays a key role in the formulation of the noncommutative quantum mechanics. In the past three decades, the von Neumann algebras [19, 27] underwent a vigorous growth after the discovery of a natural infinite family of pairwise nonisomorphic factors, and the advent of Tomita-Takesaki theory [25] and Connes noncommutative geometry [11]. The latter was initiated with the classification theorems for von Neumann algebras and the extensions of C^*-algebras [10]. The modular theory of von Neumann algebras was created by Tomita [26] in 1967 and perfected by M. Takesaki around 1970.

I. Aremua (✉) · K. Sodoga
Université de Lomé, Faculté des Sciences, Département de Physique, Laboratoire de Physique des Matériaux et de Mécanique Appliquée, Lomé, Togo

University of Abomey-Calavi, International Chair in Mathematical Physics and Applications (ICMPA), Cotonou, Benin
e-mail: iaremua@univ-lome.tg; ksodoga@univ-lome.tg

E. Baloïtcha · M. N. Hounkonnou
University of Abomey-Calavi, International Chair in Mathematical Physics and Applications (ICMPA), Cotonou, Benin
e-mail: ezinvi.baloitcha@cipma.uac.bj; norbert.hounkonnou@cipma.uac.bj

61

From physical point of view, a charged particle interacting with a constant magnetic field is one of the important problems in quantum mechanics described by the Hamiltonian $\mathcal{H} = \frac{1}{2M}(\mathbf{p} + \frac{e}{c}\mathbf{A})^2$, inspired by condensed matter physics, quantum optics, etc. The Landau problem [18] is related to the motion of a charged particle on the flat plane xy in the presence of a constant magnetic field along the z-axis. In metals, the electrons occupy many Landau levels [13] $E_n = \hbar\omega_c(n + \frac{1}{2})$, each level being infinitely degenerate, with $\omega_c = eB/Mc$, the cyclotron frequency, which are those of the one-dimensional harmonic oscillator, and correspond to the kinetic energy levels of electrons.

This physical model represents an interesting application [1] of the Tomita-Takesaki modular theory [25, 26]. Taking into account the sense of the magnetic field, one obtains a pair of commuting Hamiltonians. Both these Hamiltonians can be written in terms of two pairs of mutually commuting oscillator-type creation and annihilation operators, which then generate two mutually commuting von Neumann algebras, commutants of each other. The associated von Neumann algebra of observables displays a modular structure in the sense of the Tomita-Takesaki theory, with the algebra and its commutant referring to the two orientations of the magnetic field.

Hilbert spaces, at the mathematical side, realize the skeleton of quantum theories. Coherent states (CS), defined as a specific overcomplete family of vectors in the Hilbert space describing quantum phenomena [3, 12, 17, 20, 23], constitute an important tool of investigation. In the studies and understanding of noncommutative geometry, CS were proved to be useful objects [14]. Based on the approach developed in [22], Gazeau-Klauder CS were constructed in noncommutative quantum mechanics [6]. Besides, in studying, in the noncommutative plane [15], the behavior of an electron in an external uniform electromagnetic background coupled to a harmonic potential, matrix vector coherent states (MVCS) as well as quaternionic vector coherent states (QVCS) were constructed and discussed.

Our present contribution paper is organized as follows:

- First, we formulate the Hilbert-Schmidt operators and the Tomita-Takesaki modular theories in the framework of noncommutative quantum mechanics.
- Detailed proofs are given for main frequently used statements in the study of modular theory and Hilbert-Schmidt operators. As application, a construction of CS from the thermal state is achieved as in a previous work [1]. Relevant properties are discussed. Then, a light is put on the Wigner map as an interplay between the noncommutative quantum mechanics formalism [22] and the modular theory based on Hilbert-Schmidt operators.
- Finally, the motion of a charged particle on the flat plane xy in the presence of a constant magnetic field along the z-axis with a harmonic potential is studied.

2 von Neumann Algebras: Modular Theory, Hilbert-Schmidt Operators, and Coherent States

This section recapitulates fundamental notions and main ingredients of the modular theory used in the sequel. More details on these mathematical structures and their applications may be found in a series of works [1, 2, 7, 8, 10, 19, 21, 25–27] (and references therein), widely exploited to write this review section.

2.1 Basics on von Neumann Algebras

In this paragraph, $\mathfrak{H}$ denotes a Hilbert space over $\mathbb{C}$. $\mathfrak{H}$ is assumed to be separable, of dimension N, which could be finite or infinite. Denote by $\mathcal{L}(\mathfrak{H})$ the C^*-algebra of all bounded operators on $\mathfrak{H}$. The following definitions are in order:

Definition 2.1 Let $\mathfrak{G}$ be an algebra. A mapping $A \in \mathfrak{G} \mapsto A^* \in \mathfrak{G}$ is called an *involution*, or *adjoint operation*, of the algebra $\mathfrak{G}$, if it has the following properties:

1. $A^{**} = A$
2. $(AB)^* = B^*A^*$, with $A, B \in \mathfrak{G}$, $A^*, B^* \in \mathfrak{G}$
3. $(\alpha A + \beta B)^* = \bar{\alpha} A^* + \bar{\beta} B^*$, $\alpha, \beta \in \mathbb{C}$.
   ($\bar{\alpha}$ is the complex conjugate of α.)

Definition 2.2 *-algebra
An algebra with an involution is called *-*algebra* and a subset $\mathfrak{B}$ of $\mathfrak{G}$ is called self-adjoint if $A \in \mathfrak{B}$ implies that $A^* \in \mathfrak{B}$.

The algebra $\mathfrak{G}$ is a normed algebra if to each $A \in \mathfrak{G}$ there is associated a real number $||A||$, the norm of A, satisfying the requirements

1. $||A|| \geqslant 0$ and $||A|| = 0$ if, and only if, $||A|| = 0$,
2. $||\alpha A|| = |\alpha| ||A||$,
3. $||A + B|| \leqslant ||A|| + ||B||$,
4. $||AB|| \leqslant ||A|| ||B||$.

The third of these conditions is called the triangle inequality and the fourth the product inequality. The norm defines a metric topology on $\mathfrak{G}$ which is referred to as the uniform topology. The neighborhoods of an element $A \in \mathfrak{G}$ in this topology are given by

$$\mathcal{U}(A; \varepsilon) = \{B; B \in \mathfrak{G}, ||B - A|| < \varepsilon\}, \tag{2.1}$$

where $\varepsilon > 0$. If $\mathfrak{G}$ is complete with respect to the uniform topology, then it is called a Banach algebra. A normed algebra with involution which is complete and has the property $||A|| = ||A^*||$ is called a *Banach* *-algebra*. Then, follows the definition:

Definition 2.3 A C^*-algebra is a Banach *-algebra $\mathfrak{G}$ with the property

$$||A^*A|| = ||A||^2 \tag{2.2}$$

for all $A \in \mathfrak{G}$.

Before going further, let us deal, in the following, with some notions about representations and states.

Definition 2.4 *-Morphism between two *-algebras

Let $\mathfrak{G}$ and $\mathfrak{B}$ be two *-algebras. The *-morphism between $\mathfrak{G}$ and $\mathfrak{B}$ is given by the mapping $\pi : A \in \mathfrak{G} \mapsto \pi(A) \in \mathfrak{B}$, satisfying:

1. $\pi(\alpha A + \beta B) = \alpha\pi(A) + \beta\pi(B)$
2. $\pi(AB) = \pi(A)\pi(B)$
3. $\pi(A^*) = \pi(A)^*$

for all $A, B \in \mathfrak{G}, \alpha \in \mathbb{C}$.

Remark 2.5 Each *-automorphism π between two *-algebras $\mathfrak{G}$ and $\mathfrak{B}$ is positive because if $A \geqslant 0$, then $A = B^*B$ for some $B \in \mathfrak{G}$. Hence,

$$\pi(A) = \pi(B^*B) = \pi(B)^*\pi(B) \geqslant 0. \tag{2.3}$$

Definition 2.6 Representation of a C^*-algebra

A *representation of a C^*-algebra* $\mathfrak{G}$ is defined to be a pair $(\mathfrak{H}, \pi)$, where $\mathfrak{H}$ is a complex Hilbert space and π is a *-morphism of $\mathfrak{G}$ into $\mathcal{L}(\mathfrak{H})$. The representation $(\mathfrak{H}, \pi)$ is said to be *faithful* if, and only if, π is a *-isomorphism between $\mathfrak{G}$ and $\pi(\mathfrak{G})$, i.e., if, and only if, $\ker \pi = \{0\}$.

Each representation $(\mathfrak{H}, \pi)$ of a C^*-algebra $\mathfrak{G}$ defines a faithful representation of the quotient algebra $\mathfrak{G}_\pi = \mathfrak{G}/\ker \pi$.

Then, follows the proposition on the criteria for faithfulness:

Proposition 2.7 ([8], p. 44)
Let $(\mathfrak{H}, \pi)$ be a representation of the C^*-algebra $\mathfrak{G}$. The representation is faithful if, and only if, it satisfies each of the following equivalent conditions:

1. *$\ker \pi = \{0\}$;*
2. *$||\pi(A)|| = ||A||$ for all $A \in \mathfrak{G}$;*
3. *$\pi(A) > 0$ for all $A > 0$.*

The proof of this proposition is achieved by the following proposition:

Proposition 2.8 ([8], pp. 42–43)
Let $\mathfrak{G}$ be a Banach *-algebra with identity, $\mathfrak{B}$ a C^*-algebra, and π a *-morphism of $\mathfrak{G}$ into $\mathfrak{B}$. Then π is continuous and

$$||\pi(A)|| \leqslant ||A|| \tag{2.4}$$

for all $A \in \mathfrak{G}$. Moreover, if $\mathfrak{G}$ is a C^*-algebra, then the range $\mathfrak{B}_\pi = \{\pi(A); A \in \mathfrak{G}\}$ of π is a C^*-subalgebra of $\mathfrak{B}$.

Proof (See [8], p. 43)

First assume $A = A^*$. Then since $\mathfrak{B}$ is a C^*-algebra and $\pi(A) \in \mathfrak{B}$, one has

$$||\pi(A)|| = \sup\{|\lambda|; \lambda \in \sigma(\pi(A))\} \tag{2.5}$$

by Theorem 2.2.5(a) (see [8, p. 29]). Next, define $P = \pi(\mathbb{1}_{\mathfrak{G}})$ where $\mathbb{1}_{\mathfrak{G}}$ denotes the identity of $\mathfrak{G}$. It follows from the definition of π that P is a projection in $\mathfrak{B}$.

Hence replacing $\mathfrak{B}$ by the C^*-algebra $P\mathfrak{B}P$, the projection P becomes the identity $\mathbb{1}_{\mathfrak{B}}$ of the new algebra $\mathfrak{B}$. Moreover, $\pi(\mathfrak{G}) \subseteq \mathfrak{B}$. Now it follows from the definitions of a morphism and of the spectrum that $\sigma_{\mathfrak{B}}(\pi(A)) \subseteq \sigma_{\mathfrak{G}}(A)$. Therefore,

$$||\pi(A)|| \leqslant \sup\{|\lambda|; \lambda \in \sigma_{\mathfrak{G}}(A)\} \leqslant ||A|| \tag{2.6}$$

by the following Proposition:

Proposition 2.9 ([8], p. 26)

Let A be an element of a Banach algebra with identity and define the spectral radius $\rho(A)$ of A by

$$\rho(A) = \sup\{|\lambda|; \lambda \in \sigma_{\mathfrak{G}}(A)\}. \tag{2.7}$$

It follows that

$$\rho(A) = \lim_{n \to \infty} ||A^n||^{1/n} = \inf_n ||A^n||^{1/n} \leqslant ||A||. \tag{2.8}$$

In particular, the limit exists. Thus the spectrum of A is a nonempty compact set.

Proof (See [8], p. 26) Let $|\lambda|^n > ||A^n||$ for some $n > 0$. As each $m \in \mathbb{Z}$ can be decomposed as $m = pn + q$ with $p, q \in \mathbb{Z}$ and $0 \leqslant q < n$ one again establishes that the series

$$\lambda^{-1} \sum_{m \geqslant 0} \left(\frac{A}{\lambda}\right)^m \tag{2.9}$$

is Cauchy in the uniform topology and defines $(\lambda\mathbb{1} - A)^{-1}$. Therefore,

$$\rho(A) \leqslant ||A^n||^{1/n} \tag{2.10}$$

for all $n > 0$, and consequently

$$\rho(A) \leqslant \inf_n ||A^n||^{1/n} \leqslant \lim_{n \to \infty} \inf ||A^n||^{1/n}. \tag{2.11}$$

Thus to complete the proof it suffices to establish that $\rho(A) \geqslant r_A$, where

$$r_A = \lim_{n \to \infty} \sup ||A^n||^{1/n}. \tag{2.12}$$

There are two cases.

Firstly, assume $0 \in r_{\mathfrak{G}}(A)$, i.e., A is invertible. Then $1 = ||A^n A^{-n}|| \leqslant ||A^n|| \, ||A^{-n}||$ and hence $1 \leqslant r_A r_{A^{-1}}$. This implies $r_A > 0$. Consequently, if $r_A = 0$, one must have $0 \in r_{\mathfrak{G}}(A)$ and $\rho(A) \geqslant r_A$.

Secondly, we may assume $r_A > 0$. We will need the following observation. If A_n is any sequence of elements such that $R_n = (\mathbb{1} - A_n)^{-1}$ exists, then $\mathbb{1} - R_n = -A_n(\mathbb{1} - A_n)^{-1}$ and $A_n = -(\mathbb{1} - R_n)(\mathbb{1} - (\mathbb{1} - R_n))^{-1}$. Therefore, $||\mathbb{1} - R_n|| \to 0$ is equivalent to $||A_n|| \to 0$ by power series expansion.

Define $S_A = \{\lambda; \lambda \in \mathbb{C}, |\lambda| \geqslant r_A\}$. We assume that $S_A \subseteq r_{\mathfrak{G}}(A)$ and obtain a contradiction. Let ω be a primitive nth root of unity. By assumption

$$R_n(A; \lambda) = n^{-1} \sum_{k=1}^{n} \left(\mathbb{1} - \frac{\omega^k A}{\lambda} \right)^{-1} \tag{2.13}$$

is well defined for all $\lambda \in S_A$. But an elementary calculation shows that

$$R_n(A; \lambda) = \left(\mathbb{1} - \frac{A^n}{\lambda^n} \right)^{-1} . \tag{2.14}$$

Next one has the continuity estimate

$$\left\| \left(\mathbb{1} - \frac{\omega^k A}{r_A} \right)^{-1} - \left(\mathbb{1} - \frac{\omega^k A}{\lambda} \right)^{-1} \right\|$$

$$= \left\| \left(\mathbb{1} - \frac{\omega^k A}{r_A} \right)^{-1} \omega^k A \left(\frac{1}{\lambda} - \frac{1}{r_A} \right) \left(\mathbb{1} - \frac{\omega^k A}{\lambda} \right)^{-1} \right\|$$

$$\leqslant |\lambda - r_A| \, ||A|| \sup_{\gamma \in S_A} ||(\gamma \mathbb{1} - A)^{-1}||^2, \tag{2.15}$$

which is uniform in k. The supremum is finite since $\lambda \mapsto ||(\lambda \mathbb{1} - A)^{-1}||$ is continuous on $r_{\mathfrak{G}}(A)$ and for $|\lambda| > ||A||$ one has

$$||(\lambda \mathbb{1} - A)^{-1}|| \leqslant |\lambda|^{-1} \sum_{n \geqslant 0} ||A||^n / |\lambda|^n = (|\lambda| - ||A||)^{-1}. \tag{2.16}$$

It follows then that for each $\varepsilon > 0$ there is a $\lambda > r_A$ such that

$$\left\| \left(\mathbb{1} - \frac{A^n}{r_A^n} \right)^{-1} - \left(\mathbb{1} - \frac{A^n}{\lambda^n} \right)^{-1} \right\| < \varepsilon \tag{2.17}$$

uniformly in n. But $||A^n|| / \lambda^n \to 0$ and by the above observation $||(\mathbb{1} - A^n/\lambda^n)^{-1} - \mathbb{1}|| \to 0$. This implies that $||(\mathbb{1} - A^n/r_A^n)^{-1} - \mathbb{1}|| \to 0$ and $||A^n|| / r_A^n \to 0$ by another application of the same observation. This last statement contradicts, however, the definition of r_A and hence the proof is complete. $\qquad \square$

Finally, if A is not self-adjoint one can combine this inequality with the C^*-norm property and the product inequality to deduce that

$$||\pi(A)||^2 = ||\pi(A^*A)|| \leqslant ||A^*A|| \leqslant ||A||^2. \qquad (2.18)$$

Thus $||\pi(A)|| \leqslant ||A||$ for all $A \in \mathfrak{G}$ and π is continuous.

The range $\mathfrak{B}_\pi$ is a $*$-subalgebra of $\mathfrak{B}$ by definition and to deduce that it is a C^*-subalgebra we must prove that it is closed, under the assumption that $\mathfrak{G}$ is a C^*-algebra.

Now introduce the kernel $\ker \pi$ of π by

$$\ker \pi = \{A \in \mathfrak{G}; \pi(A) = 0\} \qquad (2.19)$$

then $\ker \pi$ is closed two-sided $*$-ideal. Given $A \in \mathfrak{G}$ and $B \in \ker \pi$ then $\pi(AB) = \pi(A)\pi(B) = 0, \pi(BA) = \pi(B)\pi(A) = 0$, and $\pi(B^*) = \pi(B) = 0$. The closedness follows from the estimate $||\pi(A)|| \leqslant ||A||$. Thus we can form the quotient algebra $\mathfrak{G}_\pi = \mathfrak{G}/\ker \pi$ and $\mathfrak{G}_\pi$ is a C^*-algebra. The elements of $\mathfrak{G}_\pi$ are the classes $\hat{A} = \{A+I; I \in \ker \pi\}$ and the morphism π induces a morphism $\hat{\pi}$ from $\mathfrak{G}_\pi$ onto $\mathfrak{B}_\pi$ by the definition $\hat{\pi}(\hat{A}) = \pi(A)$. The kernel of $\hat{\pi}$ is zero by construction and hence $\hat{\pi}$ is an isomorphism between $\mathfrak{G}_\pi$ and $\mathfrak{B}_\pi$. Therefore, one can define a morphism $\hat{\pi}^{-1}$ from the $*$-algebra $\mathfrak{B}_\pi$ onto the C^*-algebra $\mathfrak{G}_\pi$ by $\hat{\pi}^{-1}(\hat{\pi}(\hat{A})) = \hat{A}$ and then applying the first statement of the proposition to $\hat{\pi}^{-1}$ and $\hat{\pi}$ successively one obtains

$$||\hat{A}|| = ||\hat{\pi}^{-1}(\hat{\pi}(\hat{A}))|| \leqslant ||\hat{\pi}(\hat{A})|| \leqslant ||\hat{A}||. \qquad (2.20)$$

Thus $||\hat{A}|| = ||\hat{\pi}(\hat{A})|| = ||\pi(A)||$. Consequently, if $\pi(A_n)$ converges uniformly in $\mathfrak{B}$ to an element A_π then $\hat{A}_n$ converges in $\mathfrak{G}_\pi$ to an element $\hat{A}$ and $A_\pi = \hat{\pi}(\hat{A}) = \pi(A)$ where A is any element of the equivalence class $\hat{A}$. Thus $A_\pi \in \mathfrak{B}_\pi$ and $\mathfrak{B}_\pi$ is closed. □

Proof of Proposition 2.7 *(See [8], p. 44)* The equivalence of condition (1) and faithfulness is by definition. Prove that $(1) \Rightarrow (2) \Rightarrow (3) \Rightarrow (1)$.
$(1) \Rightarrow (2)$ Since $\ker \pi = \{0\}$, we can define a morphism π^{-1} from the range of π into $\mathfrak{G}$ by $\pi^{-1}(\pi(A)) = A$ and then applying Proposition 2.8 to π^{-1} and π successively one has

$$||A|| = ||\pi^{-1}(\pi(A))|| \leqslant ||\pi(A)|| \leqslant ||A||. \qquad (2.21)$$

$(2) \Rightarrow (3)$ If $A > 0$, then $||A|| > 0$ and hence $||\pi(A)|| > 0$, or $\pi(A) \neq 0$. But $\pi(A) \geqslant 0$ by Proposition 2.8 and therefore $\pi(A) > 0$. $(3) \Rightarrow (1)$ If condition (1) is false, then there is a $B \in \ker \pi$ with $B \neq 0$ and $\pi(B^*B) = 0$. But $||B^*B|| \geqslant 0$ and as $||B^*B|| = ||B||^2$ one has $B^*B > 0$. Thus condition (3) is false. □

Definition 2.10 Cyclic representation of a C^*-algebra

A *cyclic representation of a C^*-algebra* $\mathfrak{G}$ is defined to be a triplet $(\mathfrak{H}, \pi, \Omega)$, where $(\mathfrak{H}, \pi)$ is a representation of $\mathfrak{G}$ and Ω is a vector in $\mathfrak{H}$ which is cyclic for π, in $\mathfrak{H}$. Ω is called *cyclic vector* or *cyclic vector for π*. If $\mathfrak{K}$ is a closed subspace of $\mathfrak{H}$ then $\mathfrak{K}$ is called a *cyclic subspace* for $\mathfrak{H}$ whenever the set

$$\left\{ \sum_i \pi(A_i)\psi_i; \ A_i \in \mathfrak{G}, \ \psi_i \in \mathfrak{K} \right\} \tag{2.22}$$

is dense in $\mathfrak{H}$.

Definition 2.11 State over a C^*-algebra

A linear functional ω over the C^*-algebra $\mathfrak{G}$ is defined to be positive if

$$\omega(A^*A) \geqslant 0 \tag{2.23}$$

for all $A \in \mathfrak{G}$. A positive linear functional ω over a C^*-algebra $\mathfrak{G}$ with $||\omega|| = 1$ is called a *state*.

Remark 2.12

1. Every positive element of a C^*-algebra is of the form A^*A and hence positivity of ω is equivalent to ω being positive on positive elements.
2. Considering a representation $(\mathfrak{H}, \pi)$ of the C^*-algebra $\mathfrak{G}$, taking $\Omega \in \mathfrak{H}$ being a nonzero vector and define ω_Ω by

$$\omega_\Omega(A) = (\Omega, \pi(A)\Omega) \tag{2.24}$$

for all $A \in \mathfrak{G}$. It follows that ω_Ω is a linear function over $\mathfrak{G}$, it is also positive since

$$\omega_\Omega(A^*A) = ||\pi(A)\Omega||^2 \geqslant 0. \tag{2.25}$$

$||\omega_\Omega|| = 1$ whenever $||\Omega|| = 1$ and then, π is nondegenerate. In this case ω_Ω is a state, and is usually called *vector state* for the representation $(\mathfrak{H}, \pi)$.

Definition 2.13 The cyclic representation $(\mathfrak{H}_\omega, \pi_\omega, \Omega_\omega)$, constructed from the state ω over the C^*-algebra $\mathfrak{G}$, is defined as the *canonical cyclic representation of $\mathfrak{G}$ associated with ω*.

Next it will be demonstrated that the notions of purity of a state ω and irreducibility of the representation associated with ω are intimately related.

Theorem 2.14 ([8], p. 57)
Let ω be a state over the C^*-algebra $\mathfrak{G}$ and $(\mathfrak{H}_\omega, \pi_\omega, \Omega_\omega)$ the associated cyclic representation. The following conditions are equivalent:

1. *$(\mathfrak{H}_\omega, \pi_\omega)$ is irreducible;*
2. *ω is pure;*
3. *ω is an extremal point of the set $E_\mathfrak{G}$ of states over $\mathfrak{G}$. Furthermore, there is one-to-one correspondence*

$$\omega_T(A) = (T\Omega_\omega, \pi_\omega(A)\Omega_\omega) \tag{2.26}$$

between positive functionals ω_T, over $\mathfrak{G}$, majorized by ω and positive operators T in the commutant π'_ω, of π_ω, with $||T|| \leqslant 1$.

Proof (See [8], pp. 57–58)
(1) $\Rightarrow$ (2) Assume that (2) is false. Thus there exists a positive functional ρ such that $\rho(A^*A) \leqslant \omega(A^*A)$ for all $A \in \mathfrak{G}$. But applying the Cauchy-Schwarz inequality one then has

$$\begin{aligned}
|\rho(B^*A)|^2 &\leqslant \rho(B^*B)\rho(A^*A) \\
&\leqslant \omega(B^*B)\omega(A^*A) \\
&= ||\pi_\omega(B)\Omega_\omega||^2 ||\pi_\omega(A)\Omega_\omega||^2.
\end{aligned}$$

Thus $\pi_\omega(B)\Omega_\omega \times \pi_\omega(A)\Omega_\omega \longmapsto \rho(B^*A)$ is a densely defined, bounded, sesquilinear functional, over $\mathfrak{H}_\omega \times \mathfrak{H}_\omega$, and there exists a unique bounded operator T, on $\mathfrak{H}_\omega$, such that

$$(\pi_\omega(B)\Omega_\omega, T\pi_\omega(A)\Omega_\omega) = \rho(B^*A).$$

As ρ is not a multiple of ω operator T is not a multiple of the identity. Moreover,

$$\begin{aligned}
0 &\leqslant \rho(A^*A) \\
&= (\pi_\omega(A)\Omega_\omega, T\pi_\omega(A)\Omega_\omega) \\
&\leqslant \omega(A^*A) = (\pi_\omega(A)\Omega_\omega, \pi_\omega(A)\Omega_\omega)
\end{aligned}$$

and hence $0 \leqslant T \leqslant \mathbb{1}$. But

$$\begin{aligned}
(\pi_\omega(B)\Omega_\omega, T\pi_\omega(C)\pi_\omega(A)\Omega_\omega) &= \rho(B^*CA) \\
&= \rho((C^*B)^*A) = (\pi_\omega(B)\Omega_\omega, \pi_\omega(C)T\pi_\omega(A)\Omega_\omega)
\end{aligned}$$

and therefore $T \in \pi'_\omega$. Thus condition (1) is false. (2) $\Rightarrow$ (1) Assume that (1) is false. If $T \in \pi'_\omega$, then $T^* \in \pi'_\omega$ and $T + T^*$, $(T - T^*)/i$ are also elements of the commutant. Thus there exists a self-adjoint element S of π'_ω which is not a multiple of the identity. Therefore there exists a spectral projector P of S such that $0 < P < \mathbb{1}$ and $P \in \pi'_{\omega}$. Consider the functional

$$\rho(A) = (P\Omega_{\omega}, \pi_{\omega}(A)\Omega_{\omega}).$$

This is certainly positive since

$$\rho(A^*A) = (P\pi_{\omega}(A)\Omega_{\omega}, P\pi_{\omega}(A)\Omega_{\omega}) \geqslant 0.$$

Moreover,

$$\omega(A^*A) - \rho(A^*A) = (\pi_{\omega}(A)\Omega_{\omega}, (\mathbb{1} - P)\pi_{\omega}(A)\Omega_{\omega})$$
$$\geqslant 0.$$

Thus ω majorizes ρ. It is verified that ρ is not a multiple of ω and hence (2) is false. This proves the equivalence of the first two conditions stated in the theorem and simultaneously establishes the correspondence described by the last statement.

The equivalence of conditions (2) and (3) is performed as follows. Suppose that ω is an extremal point of $E_{\mathfrak{G}}$ and $\omega \neq 0$. Then, we must have $||\omega|| = 1$. Thus ω is a state and we must deduce that it is pure. Suppose the contrary; then there is a state $\omega_1 \neq \omega$ and a λ with $0 < \lambda < 1$ such that $\omega \geqslant \lambda \omega_1$. Define ω_2 by $\omega_2 = (\omega - \lambda \omega_1)/(1 - \lambda)$; then $||\omega_2|| = (||\omega|| - \lambda ||\omega_1||)/(1 - \lambda) = 1$ and ω_2 is also a state. But $\omega = \lambda \omega_1 + (1 - \lambda)\omega_2$ and ω is not extremal, which is a contradiction. $\square$

In the following, some notions on von Neumann algebra are provided. To specify the Hilbert space upon which a von Neumann algebra $\mathfrak{A}$ acts, one often uses the notation $\{\mathfrak{A}, \mathfrak{H}\}$ to denote the von Neumann algebra $\mathfrak{A}$.

Definition 2.15 von Neumann algebra

Let $\mathfrak{H}$ be a Hilbert space. For each subset $\mathfrak{A}$ of $\mathcal{L}(\mathfrak{H})$, let $\mathfrak{A}'$ denote the set of all bounded operators on $\mathfrak{H}$ commuting with every operator in $\mathfrak{A}$. Clearly, $\mathfrak{A}'$ is a Banach algebra of operators containing the identity operator $I_{\mathfrak{H}}$ on $\mathfrak{H}$.

(i) A *von Neumann algebra* is a *-subalgebra $\mathfrak{A}$ of $\mathcal{L}(\mathfrak{H})$ such that $\mathfrak{A} = \mathfrak{A}''$.
(ii) $\mathfrak{A}'$ denotes the *commutant* of $\mathfrak{A}$, the set of all elements in $\mathcal{L}(\mathfrak{H})$ which commute with every element of $\mathfrak{A}$.
(iii) A von Neumann algebra always contains the identity operator $I_{\mathfrak{H}}$ on $\mathfrak{H}$. It is called a *factor* if $\mathfrak{A} \cap \mathfrak{A}' = \mathbb{C}I_{\mathfrak{H}}$.
(iv) If a subset S of $\mathcal{L}(\mathfrak{H})$ is invariant under the *-operation, then S'', the double commutant of S, is the smallest von Neumann algebra containing S, and it is called the von Neumann algebra *generated* by S.

We also have the following definition:

Definition 2.16 A von Neumann algebra $\mathfrak{A} \subset \mathcal{L}(\mathfrak{H})$ is a C^*-algebra acting on the Hilbert space $\mathfrak{H}$ that is closed under the *weak-operator topology*: $A_n \overset{n \to +\infty}{\longrightarrow} A$ iff $\langle \xi | A_n \eta \rangle \overset{n \to +\infty}{\longrightarrow} \langle \xi | A\eta \rangle, \forall \xi, \eta \in \mathfrak{H}$, or equivalently under the σ-*weak topology*:

$A_n \overset{n \to +\infty}{\longrightarrow} A$ iff for all sequences (ξ_k), (ζ_k) in $\mathfrak{H}$ such that $\sum_{k=1}^{+\infty} ||\xi_k||^2 < +\infty$ and $\sum_{k=1}^{+\infty} ||\zeta_k||^2 < +\infty$ we have $\sum_{k=1}^{+\infty} \langle \xi_k | A_n \zeta_k \rangle \overset{n \to +\infty}{\longrightarrow} \sum_{k=1}^{+\infty} \langle \xi_k | A \zeta_k \rangle$.

Case of $\mathcal{L}(\mathfrak{H})$

(i) $\mathcal{L}(\mathfrak{H})$ is a von Neumann algebra and even a factor since $\mathcal{L}(\mathfrak{H})' = \mathbb{C}\mathbb{1}$ ($\mathbb{1} = I_{\mathfrak{H}}$).
(ii) The Hilbert space adjoint operation defines an involution on $\mathcal{L}(\mathfrak{H})$ and with respect to these operations and this norm, $\mathcal{L}(\mathfrak{H})$ is a C^*-algebra. In particular, the C^*-norm property follows from $||A||^2 \leqslant ||A^*|| ||A|| = ||A||^2$.
(iii) Any uniformly closed subalgebra $\mathfrak{M}$ of $\mathcal{L}(\mathfrak{H})$ which is self-adjoint is also a C^*-algebra.

Next, it comes the following definition:

Definition 2.17 Closure-Orthogonal projection
If $\mathfrak{M}$ is a subset of $\mathcal{L}(\mathfrak{H})$ and $\mathfrak{K}$ is a subset of $\mathfrak{H}$, $[\mathfrak{M}\mathfrak{K}]$ denotes the closure of the linear span of elements of the form $A\xi$, where $A \in \mathfrak{M}$ and $\xi \in \mathfrak{K}$. $[\mathfrak{M}\mathfrak{K}]$ also denotes the orthogonal projection onto $[\mathfrak{M}\mathfrak{K}]$.

(iv) A *-subalgebra $\mathfrak{M} \subseteq \mathcal{L}(\mathfrak{H})$ is said to be nondegenerate if $[\mathfrak{M}\mathfrak{H}] = \mathfrak{H}$.
(v) If $\mathfrak{M} \subseteq \mathcal{L}(\mathfrak{H})$ contains the identity operator, then, it is automatically nondegenerate.

A nondegenerate *-algebra contains the identity operator; If a subalgebra of $\mathcal{L}(\mathfrak{H})$ is invariant under the *-operation, then it is called a *-subalgebra of $\mathcal{L}(\mathfrak{H})$ or a *-algebra of operators on $\mathfrak{H}$.

We have the following proposition (see [25, pp. 72–73]):

Proposition 2.18 ([25], p. 72)
The subset $\mathfrak{M}$ of $\mathcal{L}(\mathfrak{H})$ is a von Neumann algebra on $\mathfrak{H}$.

Proof (See [25], p. 72)
Let $\{\mathfrak{M}_i, \mathfrak{H}_i\}_{i \in I}$ be a family of von Neumann algebras. Let $\mathfrak{H}$ denote the direct sum $\sum_{i \in I}^{\oplus} \mathfrak{H}_i$ of Hilbert spaces $\{\mathfrak{H}_i\}_{i \in I}$. Each vector $\xi = \{\xi_i\}_{i \in I}$ in $\mathfrak{H}$ is denoted by $\sum_{i \in I}^{\oplus} \xi_i$. For each bounded sequence $\{x_i\}_{i \in I}$ in $\prod_{i \in I} \mathfrak{M}_i$, one defines an operator x on $\mathfrak{H}$ by

$$x \sum_{i \in I}^{\oplus} \xi_i = \sum_{i \in I}^{\oplus} x_i \xi_i \tag{2.27}$$

Then, x is a bounded operator on $\mathfrak{H}$ denoted by $\sum_{i \in I}^{\oplus} x_i$. Let $\mathfrak{M}$ be the set of all such x.

Particularly, taking $\mathfrak{M}$ as a subset of $\mathcal{L}(\mathfrak{H})$, the proof is completed. $\square$

Definition 2.19 Cyclic and separating vector

The modular theory of von Neumann algebras is such that to every von Neumann algebra $\mathfrak{M} \subset \mathcal{L}(\mathfrak{H})$, and to every vector $\xi \in \mathfrak{H}$ that is *cyclic*

$$\overline{(\mathfrak{M}\xi)} = \mathfrak{H} \tag{2.28}$$

i.e., the set $\{\Psi \text{ by } \xi\}$ ($\mathfrak{M}$ denoting a set of bounded operators on $\mathfrak{H}$) is dense in $\mathfrak{H}$; and *separating* i.e. for $A \in \mathfrak{M}$,

$$A\xi = 0 \Rightarrow A = 0. \tag{2.29}$$

Moreover, a vector $\psi \in \mathfrak{H}$ is said separating for a von Neumann algebra $\mathfrak{A}$ if $A\psi = B\psi$, $A, B \in \mathfrak{A}$, if and only if $A = B$.

We have the following definitions:

Definition 2.20 Separating subset.

Let $\mathfrak{A}$ be a von Neumann algebra on a Hilbert space $\mathfrak{H}$. A subset $\mathfrak{K} \subseteq \mathfrak{H}$ is separating for $\mathfrak{A}$ if for any $A \in \mathfrak{A}$, $A\xi = 0$ for all $\xi \in \mathfrak{K}$ implies $A = 0$.

Definition 2.21 Cyclic and separating subset of a von Neumann algebra.

Let $\{\mathfrak{A}, \mathfrak{H}\}$ be a von Neumann algebra. A subset $\mathfrak{M}$ of $\mathfrak{H}$ is called *separating* (resp.*cyclic*) for $\mathfrak{A}$ if $a\xi = 0$, $a \in \mathfrak{A}$, for every $\xi \in \mathfrak{M}$ implies $a = 0$ (resp. the smallest invariant subspace $[\mathfrak{A}\mathfrak{M}]$ under $\mathfrak{A}$ containing $\mathfrak{M}$ in the whole space $\mathfrak{H}$).

Recall that a subset $\mathfrak{K} \subseteq \mathfrak{H}$ is cyclic for $\mathfrak{M}$ if $[\mathfrak{M}\mathfrak{K}] = \mathfrak{H}$. There is a dual relation between the properties of cyclic for the algebra and separating for the commutant.

We have the following propositions:

Proposition 2.22 ([8], p. 85)
Let $\mathfrak{A}$ be a von Neumann algebra on $\mathfrak{H}$ and $\mathfrak{K} \subseteq \mathfrak{H}$ a subset. The following conditions are equivalent:

(1) $\mathfrak{K}$ is cyclic for $\mathfrak{A}$;
(2) $\mathfrak{K}$ is separating for $\mathfrak{A}'$.

Proof (See [8], p. 85)
(1) $\Rightarrow$ (2) Assume that $\mathfrak{K}$ is cyclic for $\mathfrak{A}$ and choose $A' \in \mathfrak{A}'$ such that $A'\mathfrak{K} = \{0\}$. Then, for any $B \in \mathfrak{A}$ and $\xi \in \mathfrak{K}$, $A'B\xi = BA'\xi = 0$, hence $A'[\mathfrak{A}\mathfrak{K}] = 0$ and $A' = 0$.

(2) $\Rightarrow$ (1) Suppose that $\mathfrak{K}$ is separating for $\mathfrak{A}'$ and set $P' = [\mathfrak{A}\mathfrak{K}]$. P' is then a projection in $\mathfrak{A}'$ and $(\mathbb{1} - P')\mathfrak{K} = \{0\}$. Hence $\mathbb{1} - P' = 0$ and $[\mathfrak{A}\mathfrak{K}] = \mathfrak{H}$. $\square$

Definition 2.23 The weak and σ-weak topologies

If $\xi, \eta \in \mathfrak{H}$, then $A \mapsto |(\xi, A\eta)|$ is a seminorm on $\mathcal{L}(\mathfrak{H})$. The locally convex topology on $\mathcal{L}(\mathfrak{H})$ defined by these seminorms is called the *weak topology*. The seminorms defined by the vector states $A \mapsto |(\xi, A\xi)|$ suffice to define this topology because $\mathfrak{H}$ is complex and one has the polarization identity

$$4(\xi, A\eta) = \sum_{n=0}^{3} i^{-n}(\xi + i^n \eta, A(\xi + i^n \eta)). \tag{2.30}$$

Let $\{\xi_n\}, \{\eta_n\}$ be two sequences from $\mathfrak{H}$ such that

$$\sum_n ||\xi_n||^2 < \infty, \qquad \sum_n ||\eta_n||^2 < \infty. \tag{2.31}$$

Then for $A \in \mathcal{L}(\mathfrak{H})$

$$\left| \sum_n (\xi_n, A\eta_n) \right| \leqslant \sum_n ||\xi_n|| \, ||A|| \, ||\eta_n||$$

$$\leqslant ||A|| \left(\sum_n ||\xi_n||^2 \right)^{1/2} \left(\sum_n ||\eta_n||^2 \right)^{1/2}$$

$$\leqslant \infty. \tag{2.32}$$

Hence $A \mapsto \left| \sum_n (\xi_n, A\eta_n) \right|$ is a seminorm on $\mathcal{L}(\mathfrak{H})$. The locally convex topology on $\mathcal{L}(\mathfrak{H})$ induced by these seminorms is called the σ-weak topology.

Notations: In the sequel,

- $\mathfrak{A}_+$ denotes the *positive part* of the von Neumann algebra $\mathfrak{A}$ or the set of positive elements of the von Neumann algebra $\mathfrak{A}$;
- $\mathfrak{A}_*$ denotes the *predual of a von Neumann algebra*. It is the space of all σ-weakly continuous linear functionals on $\mathfrak{A}$;
- $\mathcal{L}(\mathfrak{H})_1$ denotes the unit ball of $\mathcal{L}(\mathfrak{H})$. $\mathcal{L}(\mathfrak{H})_1$ is norm dense in the unit ball of the norm closure of $\mathcal{L}(\mathfrak{H})$, and it is taken as a C^*-algebra (see [8, p. 74]).

Definition 2.24 Let $\varphi : \mathfrak{A} \to \mathbb{C}$ be a bounded linear functional on $\mathfrak{A}$, which is denoted by $\langle \varphi; A \rangle$, $A \in \mathfrak{A}$.

φ is called a state on this algebra if it also satisfies the two conditions:

(a) $\langle \varphi; A^*A \rangle \geqslant 0$, $\quad \forall A \in \mathfrak{A}$
(b) $\langle \varphi; I_{\mathfrak{H}} \rangle = 1$.

The state φ is called a *vector state* if there exists a vector $\phi \in \mathfrak{H}$ such that

$$\langle \varphi; A \rangle = \langle \phi | A\phi \rangle, \quad \forall A \in \mathfrak{A}. \tag{2.33}$$

Such a state is also *normal*.

Definition 2.25 A state ω on a von Neumann algebra $\mathfrak{A}$ is *faithful* if $\omega(A) > 0$ for all nonzero $A \in \mathfrak{A}_+$.

Remark 2.26 (See [8], Example 2.5.5 p. 85)
Let $\mathfrak{A} = \mathcal{L}(\mathfrak{H})$ with $\mathfrak{H}$ separable. Every normal state ω over $\mathfrak{A}$ is of the form

$$\omega(A) = \text{Tr}(\rho A), \tag{2.34}$$

where ρ is a density matrix. If ω is faithful, then $\omega(E) > 0$ for each rank one projector, i.e., $||\rho^{1/2}\psi|| > 0$ for each $\psi \in \mathfrak{H} \setminus \{0\}$. Thus ρ is invertible (in the densely defined self-adjoint operators on $\mathfrak{H}$). Conversely, if ω is not faithful, then $\omega(A^*A) = 0$ for some nonzero A and hence $||\rho^{1/2}A^*\psi|| = 0$ for all $\psi \in \mathfrak{H}$, i.e., ρ is not invertible. This establishes that ω is faithful if, and only if, ρ is invertible.

Lemma 2.27 ([8], p. 76) *Let $\{A_\alpha\}$ be an increasing set in $\mathcal{L}(\mathfrak{H})_+$ with an upper bound in $\mathcal{L}(\mathfrak{H})_+$. Then $\{A_\alpha\}$ has a least upper bound (l.u.b.) A, and the net converges σ-strongly to A.*

Proof (See [8], p. 76) Let $\mathfrak{K}_\alpha$ be the weak closure of the set of A_β with $\beta > \alpha$. Since $\mathcal{L}(\mathfrak{H})_1$ is weakly compact, there exists an element A in $\bigcap_\alpha \mathfrak{K}_\alpha$. For all A_α the set of $B \in \mathcal{L}(\mathfrak{H})_+$ such that $B \geqslant A_\alpha$ is σ-weakly closed and contains $\mathfrak{K}_\alpha$, hence $A \geqslant A_\alpha$. Thus, A majorizes $\{A_\alpha\}$ and lies in the weak closure of $\{A_\alpha\}$. If B is another operator majorizing $\{A_\alpha\}$, then it majorizes its weak closure; thus, $B \geqslant A$ and A is the least upper bound of $\{A_\alpha\}$. Finally, if $\xi \in \mathfrak{H}$, then

$$\begin{aligned}
||(A - A_\alpha)\xi||^2 &\leqslant ||A - A_\alpha|| \, ||(A - A_\alpha)^{1/2}\xi||^2 \\
&\leqslant ||A||(\xi, (A - A_\alpha)\xi) \\
&\xrightarrow[\alpha]{} 0.
\end{aligned} \tag{2.35}$$

Since the strong and σ-strong topology coincide on $\mathcal{L}(\mathfrak{H})_1$, this ends the proof. □

Proposition 2.28 ([8], p. 68) *Let Tr be the usual trace on $\mathcal{L}(\mathfrak{H})$, and let $\mathcal{T}(\mathfrak{H})$ be the Banach space of trace-class operators on $\mathfrak{H}$ equipped with the trace norm $T \mapsto \text{Tr}(|T|) = ||T||_{Tr}$. Then it follows that $\mathcal{L}(\mathfrak{H})$ is the dual $\mathcal{T}(\mathfrak{H})^*$ of $\mathcal{T}(\mathfrak{H})$ by the duality*

$$A \times T \in \mathcal{L}(\mathfrak{H}) \times \mathcal{T}(\mathfrak{H}) \mapsto \text{Tr}(AT). \tag{2.36}$$

The weak topology on $\mathcal{L}(\mathfrak{H})$ arising from this duality is just the σ-weak topology.*

Proof (See [8], pp. 68–69) Due to the inequality $|\text{Tr}(AT)| \leqslant ||A|| \, ||T||_{Tr}$, $\mathcal{L}(\mathfrak{H})$ is the subspace of $\mathcal{T}(\mathfrak{H})^*$ by the duality described in the proposition. Conversely, assume $\omega \in \mathcal{T}(\mathfrak{H})^*$ and consider a rank one operator $E_{\varphi,\psi}$ defined for $\varphi, \psi \in \mathfrak{H}$ by

$$E_{\varphi,\psi}\chi = \varphi(\psi, \chi). \tag{2.37}$$

One has $E_{\varphi,\psi}^{*} = E_{\psi,\varphi}$ and $E_{\varphi,\psi}E_{\psi,\varphi} = ||\psi||^2 E_{\varphi,\varphi}$. Hence

$$||E_{\varphi,\psi}||_{\mathrm{Tr}} = ||\psi||\,\mathrm{Tr}(E_{\varphi,\varphi})^{1/2} = ||\psi||\,||\varphi||. \tag{2.38}$$

It follows that

$$|\omega(E_{\varphi,\psi})| \leqslant ||\omega||\,||\varphi||\,||\psi||. \tag{2.39}$$

Hence there exists, by the Riesz representation theorem, an $A \in \mathcal{L}(\mathfrak{H})$ with $||A|| \leqslant ||\omega||$ such that

$$\omega(E_{\varphi,\psi}) = (\psi, A\varphi). \tag{2.40}$$

Consider $\omega_0 \in \mathcal{T}(\mathfrak{H})^*$ defined by

$$\omega_0(T) = \mathrm{Tr}(AT) \tag{2.41}$$

then

$$\begin{aligned}
\omega_0(E_{\varphi,\psi}) &= \mathrm{Tr}(AE_{\varphi,\psi}) \\
&= (\psi, A\varphi) \\
&= \omega(E_{\varphi,\psi}).
\end{aligned} \tag{2.42}$$

Now for any $T \in \mathcal{T}(\mathfrak{H})$ there exist bounded sequences $\{\psi_n\}$ and $\{\varphi_n\}$ and a sequence $\{\alpha_n\}$ of complex numbers such that

$$\sum_n |\alpha_n| < \infty \tag{2.43}$$

and

$$T = \sum_n \alpha_n E_{\varphi_n,\psi_n}. \tag{2.44}$$

The latter series converges with respect to the trace norm and hence

$$\begin{aligned}
\omega(T) &= \sum_n \alpha_n \omega(E_{\varphi_n,\psi_n}) \\
&= \sum_n \alpha_n \omega_0(E_{\varphi_n,\psi_n}) = \omega_0(T) = \mathrm{Tr}(AT).
\end{aligned} \tag{2.45}$$

Thus $\mathcal{L}(\mathfrak{H})$ is just the dual of $\mathcal{T}(\mathfrak{H})$. The weak* topology on $\mathcal{L}(\mathfrak{H})$ arising from this duality is given by the seminorms

$$A \in \mathcal{L}(\mathfrak{H}) \mapsto |\mathrm{Tr}(AT)|. \tag{2.46}$$

Now, for T as in (2.44), one has

$$\mathrm{Tr}(AT) = \sum_n \alpha_n \mathrm{Tr}(E_{\varphi_n, \psi_n} A)$$
$$= \sum_n \alpha_n (\psi_n, A\varphi_n). \tag{2.47}$$

Thus the seminorms are equivalent to the seminorms defining the σ-weak topology.
□

It follows the theorem below:

Theorem 2.29 *([8, p. 76])*
Let ω be a state on a von Neumann algebra $\mathfrak{A}$ acting on a Hilbert space $\mathfrak{H}$. The following conditions are equivalent:

(1) ω is normal;
(2) ω is σ-weakly continuous;
(3) there exists a density matrix ρ, i.e., a positive trace-class operator ρ on $\mathfrak{H}$ with $Tr(\rho) = 1$, such that

$$\omega(A) = Tr(\rho A). \tag{2.48}$$

Proof (See [8, pp. 76 to 78])
$(3) \Rightarrow (2)$ follows from Proposition 2.28 and $(2) \Rightarrow (1)$ from Lemma 2.27. Next show $(2) \Rightarrow (3)$. If ω is σ-weakly continuous there exist sequences $\{\xi_n\}$, $\{\eta_n\}$ of vectors such that $\sum_n \|\xi_n\|^2 < \infty$, $\sum_n \|\eta_n\|^2 < \infty$, and $\omega(A) = \sum_n (\xi_n, A\eta_n)$. Define $\tilde{\mathfrak{H}} = \bigoplus_{n=1}^{\infty} \mathfrak{H}_n$ and introduce a representation π of $\mathfrak{A}$ on $\tilde{\mathfrak{H}}$ by $\pi(A)(\bigoplus_n \psi_n) = \bigoplus_n (A\psi_n)$. Let $\xi = \bigoplus_n \xi_n$, $\eta = \bigoplus_n \eta_n$ and then $\omega(A) = (\xi, \pi(A)\eta)$. Since $\omega(A)$ is real for $A \in \mathfrak{A}_+$ (with $\mathfrak{A}_+$ denoting the *positive part* of the von Neumann algebra $\mathfrak{A}$ or the set of positive elements of the von Neumann algebra $\mathfrak{A}$), we have

$$\begin{aligned}
4\omega(A) &= 2(\xi, \pi(A)\eta) + 2(\xi, \pi(A^*)\eta) \\
&= 2(\xi, \pi(A)\eta) + 2(\eta, \pi(A)\xi) \\
&= (\xi + \eta, \pi(A)(\xi + \eta)) - (\xi - \eta, \pi(A)(\xi - \eta)) \\
&\leqslant (\xi + \eta, \pi(A)(\xi + \eta)). \tag{2.49}
\end{aligned}$$

Hence, by Theorem 2.14 there exists a positive $T \in \pi(\mathfrak{A})'$ with $0 \leqslant T \leqslant \mathbb{1}/2$ such that

$$\begin{aligned}
(\xi, \pi(A)\eta) &= (T(\xi + \eta), \pi(A)T(\xi + \eta)) \\
&= (\psi, \pi(A)\psi). \tag{2.50}
\end{aligned}$$

Now $\psi \in \tilde{\mathfrak{H}}$ has the form $\psi = \bigoplus_n \psi_n$, and therefore

$$\omega(A) = \sum_n (\psi_n, A\psi_n). \tag{2.51}$$

The right side of this relation can be used to extend ω to a σ-weakly continuous positive linear functional $\tilde{\omega}$ on $\mathcal{L}(\mathfrak{H})$. Since $\tilde{\omega}(\mathbb{1}) = 1$, it is a state. Thus, by Proposition 2.28 there exists a trace-class operator ρ with $Tr(\rho) = 1$ such that

$$\tilde{\omega}(A) = Tr(\rho A). \tag{2.52}$$

Let P be the rank one projector with range ξ; then

$$(\xi, \rho\xi) = Tr(P\rho P) = Tr(\rho P) = \tilde{\omega}(P) \geqslant 0. \tag{2.53}$$

Thus ρ is positive. Turn now to the proof of (1) $\Rightarrow$ 2. Assume that ω is a normal state on $\mathfrak{A}$. Let $\{B_\alpha\}$ be an increasing net of elements in $\mathfrak{A}_+$ such that $||B_\alpha|| \leqslant 1$ for all α and such that $A \mapsto \omega(AB_\alpha)$ is σ-strongly continuous for all α. One can use Lemma 2.27 to define B by

$$B = \text{l.u.b.}_\alpha\, B_\alpha = \sigma\text{-strong} \lim_\alpha B_\alpha. \tag{2.54}$$

Then $0 \leqslant B \leqslant \mathbb{1}$ and $B \in \mathfrak{A}$. But for all $A \in \mathfrak{A}$ we have

$$\begin{aligned}
|\omega(AB - AB_\alpha)|^2 &= |\omega(A(B - B_\alpha)^{1/2}(B - B_\alpha)^{1/2})|^2 \\
&\leqslant \omega(A(B - B_\alpha)A^*)\omega(B - B_\alpha) \\
&\leqslant ||A||^2 \omega(B - B_\alpha).
\end{aligned} \tag{2.55}$$

Hence

$$||\omega(\cdot B) - \omega(\cdot B_\alpha)|| \leqslant (\omega(B - B_\alpha))^{1/2}. \tag{2.56}$$

But ω is normal. Therefore $\omega(B - B_\alpha) \to 0$ and $\omega(\cdot B_\alpha)$ tends to $\omega(\cdot B)$ in norm. As $\mathfrak{A}_*$ is a Banach space, $\omega(\cdot B) \in \mathfrak{A}_*$. Now, applying Zorn's lemma, we can find a maximal element $P \in \mathfrak{A}_+ \cap \mathfrak{A}_1$ such that $A \mapsto \omega(AP)$ is σ-strongly continuous. If $P = \mathbb{1}$, the theorem is proved. Assume *ad absurdum* that $P \neq \mathbb{1}$. Put $P' = \mathbb{1} - P$ and choose $\xi \in \mathfrak{H}$ such that $\omega(P') < (\xi, P'\xi)$. If $\{B\}$ is an increasing net in $\mathfrak{A}_+$ such that $B_\alpha \leqslant P'$, $\omega(B_\alpha) \geqslant (\xi, B_\alpha\xi)$, and $B = \text{l.u.b.}_\alpha\, B_\alpha = \sigma\text{-strong} \lim_\alpha B_\alpha$, then $B \in \mathfrak{A}_+$, $B \leqslant P'$, and $\omega(B) = \sup \omega(B_\alpha) \geqslant \sup (\xi, B_\alpha\xi) = (\xi, B\xi)$. Hence, by Zorn's lemma, there exists a maximal $B \in \mathfrak{A}_+$ such that $B \leqslant P'$ and $\omega(B) \geqslant (\xi, B\xi)$. Take $Q = P' - B$. Then, $Q \in \mathfrak{A}_+$, $Q \neq 0$, since $\omega(P') < (\xi, P'\xi)$, and if $A \in \mathfrak{A}_+$, $A \leqslant Q$, $A \neq 0$, then $\omega(A) < (\xi, A\xi)$ by the maximality of B.

For any $A \in \mathfrak{A}$ one has

$$QA^*AQ \leqslant ||A||^2 Q^2 \leqslant ||A||^2 ||Q||Q. \tag{2.57}$$

Hence $(QA^*AQ)/||A||^2 ||Q|| \leqslant Q$ and $\omega(QA^*AQ) < (\xi, QA^*AQ\xi)$. Combining this with the Cauchy-Schwartz inequality one finds

$$\begin{aligned} |\omega(AQ)|^2 &\leqslant \omega(\mathbb{1})\omega(QA^*AQ) \\ &< (\xi, QA^*AQ\xi) = ||AQ\xi||^2. \end{aligned} \tag{2.58}$$

Thus both $A \mapsto \omega(AQ)$ and $A \mapsto \omega(A(P+Q))$ are σ-strongly continuous. Since $P + Q \leqslant \mathbb{1}$, this contradicts the maximality of P. $\qquad\square$

Proposition 2.30 ([8], p. 86)
Let $\mathfrak{A}$ be a von Neumann algebra on a Hilbert space $\mathfrak{H}$. Then the following four conditions are equivalent:

(1) $\mathfrak{A}$ is σ-finite;
(2) there exists a countable subset of $\mathfrak{H}$ which is separating for $\mathfrak{A}$;
(3) there exists a faithful normal state on $\mathfrak{A}$;
(4) $\mathfrak{A}$ is isomorphic with a von Neumann algebra $\pi(\mathfrak{A})$ which admits a separating and cyclic vector.

Proof (See [8, p. 86])
$(1) \Rightarrow (2)$ Let $\{\xi_\alpha\}$ be a maximal family of vectors in $\mathfrak{H}$ such that $[\mathfrak{A}'\xi_\alpha]$ and $[\mathfrak{A}'\xi_{\alpha'}]$ are orthogonal whenever $\alpha \neq \alpha'$. Since $[\mathfrak{A}'\xi_\alpha]$ is a projection in $\mathfrak{A}$ (in fact the smallest projection in $\mathfrak{A}$ containing ξ_α), $\{\xi_\alpha\}$ is countable. But by the maximality,

$$\sum_\alpha [\mathfrak{A}'\xi_\alpha] = \mathbb{1}. \tag{2.59}$$

Thus $\{\xi_\alpha\}$ is cyclic for $\mathfrak{A}'$. Hence $\{\xi_\alpha\}$ is separating for $\mathfrak{A}$ by Proposition 2.22. $(2) \Rightarrow (3)$ Choose a sequence ξ_n such that the set $\{\xi_n\}$ is separating for $\mathfrak{A}$ and such that $\sum_n ||\xi_n||^2 = 1$. Define ω by

$$\omega(A) = \sum_n (\xi_n, A\xi_n). \tag{2.60}$$

ω is σ-weakly continuous, hence normal, by using Theorem 2.29. If $\omega(A^*A) = 0$, then $0 = (\xi_n, A^*A\xi_n) = ||A\xi_n||^2$ for all n, hence $A = 0$. $(3) \Rightarrow (4)$ Let ω be a faithful normal state on $\mathfrak{A}$ and $(\mathfrak{H}, \pi, \Omega)$ the corresponding cyclic representation. Since $\pi(\mathfrak{A})$ is a von Neumann algebra, if $\pi(A)\Omega = 0$ for an $A \in \mathfrak{A}$, then $\omega(A^*A) = ||\pi(A)\Omega||^2 = 0$, hence $A^*A = 0$ and $A = 0$. This proves that π is faithful and Ω separating for $\pi(\mathfrak{A})$. $(4) \Rightarrow (1)$ Let Ω be the separating (and cyclic) vector for $\pi(\mathfrak{A})$, and let $\{E_\alpha\}$ be a family of mutually orthogonal projections in $\mathfrak{A}$. Set $E = \sum_\alpha E_\alpha$.

Then

$$
\begin{aligned}
||\pi(E)\Omega||^2 &= (\pi(E)\Omega, \pi(E)\Omega) \\
&= \sum_{\alpha,\alpha'} (\pi(E_\alpha)\Omega, \pi(E_{\alpha'})\Omega) \\
&= \sum_{\alpha} ||\pi(E_\alpha)\Omega||^2
\end{aligned}
\tag{2.61}
$$

by Lemma 2.27. Since $\sum_{\alpha} ||\pi(E_\alpha)\Omega||^2 < +\infty$, only a countable number of the $\pi(E_\alpha)\Omega$ is nonzero, and thus the same is true for the E_α. □

2.2 Hilbert Space of Hilbert-Schmidt Operators

Here we recall some definitions provided in [1, 2, 21] (and references therein).

Definition 2.31 The trace of a linear operator.

A linear operator A defined on the separable Hilbert space $\mathfrak{H}$ is said to be of *trace class* if the series $\sum_k \langle e_k | A e_k \rangle$ converges and has the same value in any orthonormal basis $\{e_k\}$ of $\mathfrak{H}$. The sum

$$
\text{Tr} A = \sum_k \langle e_k | A e_k \rangle
\tag{2.62}
$$

is called the *trace* of A.

Definition 2.32 Trace norm

Consider the class of Hilbert-Schmidt operators. For every such operator A, the *trace norm* is given by

$$
\text{Tr}[\sqrt{A^* A}] = \text{Tr}[\sum_k |e_k\rangle \lambda_k \langle e_k|] = \sum_k \lambda_k < +\infty.
\tag{2.63}
$$

Remark 2.33 If A is any operator of trace class, then A^* is also of trace class:

$$
\text{Tr} A^* = \sum_k \langle e_k | A^* e_k \rangle = \sum_k \langle e_k | A e_k \rangle^* = (\text{Tr} A)^*.
\tag{2.64}
$$

Definition 2.34 Hilbert-Schmidt operator.

Given a bounded operator, having the decomposition $A = \sum_k |\phi_k\rangle \lambda_k \langle \phi_k|$, where $\{\phi_k\}$ is an orthonormal basis of $\mathfrak{H}$, and $\lambda_1, \lambda_2, \ldots$ positive numbers, A is called a *Hilbert-Schmidt operator* if

$$
\text{Tr}[A A^*] = \sum_k \langle \phi_k | A^* A \phi_k \rangle = \sum_k \lambda_k^2 < +\infty.
\tag{2.65}
$$

Remark 2.35 If the series (2.63) is infinite, its convergence implies that $\lambda_k \to 0$ when $k \to +\infty$. Consequently, $\lambda_k^2 \geqslant \lambda_k$ for sufficiently large value of k. Hence, $\sum_k \lambda_k^2$ converges when $\sum_k \lambda_k$ converges. This shows that any completely continuous operator A satisfying (2.63) is a Hilbert-Schmidt operator.

Definition 2.36 Hilbert-Schmidt norm.

For any Hilbert-Schmidt operator A, the quantity

$$||A||_2 = \sqrt{\mathrm{Tr}[A^*A]} \tag{2.66}$$

exists, and is called *Hilbert-Schmidt norm* of A.

Definition 2.37 Let $\mathcal{B}_2(\mathfrak{H})$, $\mathcal{B}_2(\mathfrak{H}) \subset \mathcal{L}(\mathfrak{H})$ the set of all bounded operators on $\mathfrak{H}$, be the Hilbert space of Hilbert-Schmidt operators on $\mathfrak{H} = L^2(\mathbb{R})$, with the scalar product

$$\langle X|Y\rangle_2 = Tr[X^*Y] = \sum_k \langle \Phi_k|X^*Y\Phi_k\rangle, \tag{2.67}$$

where $\{\Phi_k\}_{k=0}^\infty$ is an orthonormal basis of $\mathfrak{H}$.

$\mathcal{B}_2(\mathfrak{H}) \simeq \mathfrak{H} \otimes \bar{\mathfrak{H}}$ (where $\bar{\mathfrak{H}}$ denotes the dual of $\mathfrak{H}$) and basis vectors of $\mathcal{B}_2(\mathfrak{H})$ are given by

$$\Phi_{nl} := |\Phi_n\rangle\langle\Phi_l|, \quad n, l = 0, 1, 2, \ldots, \infty. \tag{2.68}$$

Remark 2.38 In the notation $\mathcal{B}_2(\mathfrak{H}) \simeq \mathfrak{H} \otimes \bar{\mathfrak{H}}$, $\mathfrak{H} \otimes \bar{\mathfrak{H}}$ is taken here as the completion of the algebraic tensor product of $\mathfrak{H}$ by $\bar{\mathfrak{H}}$ which is a pre-Hilbert space containing finite sums of the type $\sum_{j,k=0}^n \lambda_{jk}|\phi_j\rangle \otimes |\overline{\phi_k}\rangle$, where a basis of $\mathfrak{H} \otimes \bar{\mathfrak{H}}$ is $\{|\phi_j\rangle \otimes |\overline{\phi_k}\rangle\}_{j,k=0}^\infty$. Then, $\mathcal{B}_2(\mathfrak{H})$, being the Hilbert space of Hilbert-Schmidt operators on $\mathfrak{H}$ is isomorphic to $\mathfrak{H} \otimes \bar{\mathfrak{H}}$, since the separable Hilbert spaces are taken two by two isomorphic each other. Setting $|\phi_j\rangle \otimes |\overline{\phi_k}\rangle = |\phi_j\rangle\langle\phi_k|$, $\mathcal{B}_2(\mathfrak{H})$ admits for orthonormal basis $\{\phi_{jk}\}_{j,k=0}^\infty$ such that $\phi_{jk} := |\phi_j\rangle\langle\phi_k|$.

Definition 2.39 Let A and B be two operators on $\mathfrak{H}$. The operator $A \vee B$ is such that

$$A \vee B(X) = AXB^*, \quad X \in B_2(\mathfrak{H}). \tag{2.69}$$

For bounded linear operators A and B, $A \vee B$ defines a linear operator on $\mathcal{B}_2(\mathfrak{H})$. Indeed, $\forall A, B \in \mathcal{L}(\mathfrak{H})$, (the space of bounded linear operators on $\mathfrak{H}$), since $\mathcal{B}_2(\mathfrak{H}) \subset \mathcal{L}(\mathfrak{H})$, we get $\forall X \in \mathcal{B}_2(\mathfrak{H})$, $X \in \mathcal{L}(\mathfrak{H})$. Then, $AXB^* \in \mathcal{L}(\mathfrak{H})$, i.e. $(A \vee B) \in \mathcal{L}(\mathfrak{H})$. Thus, $A \vee B$ defines a bounded linear operator on $\mathcal{B}_2(\mathfrak{H})$.

From the scalar product in $B_2(\mathfrak{H})$,

$$\langle X|Y\rangle_2 = Tr[X^*Y], \quad X, Y \in B_2(\mathfrak{H}), \tag{2.70}$$

it comes

$$Tr[X^*(AYB^*)] = Tr[(A^*XB)^*Y] \Rightarrow (A \vee B)^* = A^* \vee B^*. \tag{2.71}$$

Since for any $X \in B_2(\mathfrak{H})$,

$$(A_1 \vee B_1)(A_2 \vee B_2)(X) = A_1[(A_2 \vee B_2)(X)]B_1^* = A_1 A_2 X B_2^* B_1^*, \tag{2.72}$$

we have

$$(A_1 \vee B_1)(A_2 \vee B_2) = (A_1 A_2) \vee (B_1 B_2). \tag{2.73}$$

2.3   Modular Theory and Hilbert-Schmidt Operators

2.3.1   Modular Theory

This paragraph is devoted to Tomita-Takesaki modular theory [24–26] of von Neumann algebras [19, 27]. Recall that the origins of Tomita-Takesaki modular theory lie in two unpublished papers of M. Tomita in 1967 [26] and a slim volume by M. Takesaki. As one of the most important contributions in the operator algebras, this theory finds many applications in mathematical physics.

We provide some key ingredients from [1, 2, 7, 8, 25] as needed for this section. First, let us deal with some notions from [8]:

1. Let $\mathfrak{A}$ be a von Neumann algebra on a Hilbert space $\mathfrak{H}$ and $\mathfrak{A}'$ its commutant. Let $\Phi \in \mathfrak{H}$ be a unit vector which is cyclic and separating for $\mathfrak{A}$. This is the case, if $\mathfrak{A}$ is a σ-finite von Neumann algebra, and by applying Proposition 2.22.
2. The mapping $A \in \mathfrak{A} \mapsto A\Omega \in \mathfrak{H}$ then establishes a one-to-one linear correspondence between $\mathfrak{A}$ and a dense subspace $\mathfrak{A}\Omega$ of $\mathfrak{H}$. Let S_0 and F_0 be two antilinear operators on $\mathfrak{A}\Omega$ and $\mathfrak{A}'\Omega$, respectively. By Proposition 2.22, Ω is cyclic and separating for $\mathfrak{A}$ and $\mathfrak{A}'$. Therefore the two antilinear operators S_0 and F_0, given by

$$\begin{aligned} S_0 A\Omega &= A^*\Omega, \quad \text{for} \quad A \in \mathfrak{A} \\ F_0 A'\Omega &= A'^*\Omega \quad \text{for} \quad A' \in \mathfrak{A}' \end{aligned} \tag{2.74}$$

are both well defined on the dense domains on $D(S_0) = \mathfrak{A}\Omega$ and $D(F_0) = \mathfrak{A}'\Omega$. Then follows the definition:

Definition 2.40 Define S and F as the closures of S_0 and F_0, respectively, i.e.,

$$S = \bar{S}_0, \quad F = \bar{F}_0 \tag{2.75}$$

where the bar denotes the closure. Let Δ be the unique, positive, self-adjoint operator and J the unique antiunitary operator occurring in the polar decomposition

$$S = J\Delta^{1/2} \tag{2.76}$$

of S. Δ is called the *modular operator* associated with the pair $\{\mathfrak{A}, \Omega\}$ and J the *modular conjugation*.

The following proposition provides connections between S, F, Δ, and J:

Proposition 2.41 *([8, p. 89]) The following relations are valid:*

$$\begin{aligned}
\Delta &= FS, & \Delta^{-1} &= SF \\
S &= J\Delta^{1/2}, & F &= J\Delta^{-1/2} \\
J &= J^*, & J^2 &= I_{\mathfrak{H}} \\
\Delta^{-1/2} &= J\Delta^{1/2}J.
\end{aligned} \tag{2.77}$$

Proof (See [8, pp. 89–90]) $\Delta = S^*S = FS$, and $S = J\Delta^{1/2}$ by Definition 2.40. Using the fact that for any $\psi \in D(\bar{S}_0)$ there exists a closed operator Q on $\mathfrak{H}$, with $S_0^* = \bar{F}_0$, $F_0^* = \bar{S}_0$, such that

$$Q\Omega = \psi, \quad Q^*\Omega = \bar{S}_0\psi \tag{2.78}$$

where $\mathfrak{A}'D(Q) \subseteq D(Q)$, $QQ' \supseteq Q'Q$ for all $Q' \in \mathfrak{A}'$, with $S_0 = S_0^{-1}$, it follows by closure that $S = S^{-1}$, and hence

$$J\Delta^{1/2} = S = S^{-1} = \Delta^{1/2}J^*, \tag{2.79}$$

so that $J^2\Delta^{1/2} = J\Delta^{-1/2}J^*$. Since $J\Delta^{-1/2}J^*$ is a positive operator, and by the uniqueness of the polar decomposition one deduces that

$$J^2 = I_{\mathfrak{H}} \tag{2.80}$$

and then

$$J^* = J, \quad \Delta^{-1/2} = J\Delta^{1/2}J. \tag{2.81}$$

But this implies that

$$F = S^* = (\Delta^{-1/2}J)^* = J\Delta^{-1/2} \tag{2.82}$$

and

$$SF = \Delta^{-1/2}JJ\Delta^{-1/2} = \Delta^{-1}. \tag{2.83}$$

$\square$

3. The principal result of the Tomita-Takesaki theory [8, 25] is that the following relations

$$J \mathfrak{A} J = \mathfrak{A}', \qquad \Delta^{it} \mathfrak{A} \Delta^{-it} = \mathfrak{A} \qquad (2.84)$$

hold for all $t \in \mathbb{R}$.

4. **Definition 2.42** Modular automorphism group.

Let $\mathfrak{A}$ be a von Neumann algebra, ω a faithful, normal state on $\mathfrak{A}$, $(\mathfrak{H}_\omega, \pi_\omega, \Omega_\omega)$ the corresponding cyclic representation, and Δ the modular operator associated with the pair $(\omega(\mathfrak{A}), \Omega_\omega)$. The Tomita-Takesaki theorem establishes the existence of a σ-weakly continuous one-parameter group $t \mapsto \sigma_t^\omega$ of *-automorphisms of $\mathfrak{A}$ through the definition

$$\sigma_t^\omega(A) = \pi_\omega^{-1}(\Delta^{it} \pi_\omega(A) \Delta^{-it}). \qquad (2.85)$$

The group $t \mapsto \sigma_t^\omega$ is called *the modular automorphism group associated with the pair* $(\mathfrak{A}, \omega)$.

5. **Definition 2.43** *C*-dynamical system.*

A C^*-dynamical system $(\mathfrak{G}, \alpha)$ is a C^*-algebra $\mathfrak{G}$ equipped with a group homomorphism $\alpha : G \to \mathrm{Aut}(\mathfrak{G})$ that is strongly continuous, i.e., $g \mapsto \|\alpha_g(x)\|$ is a continuous map for all $x \in \mathfrak{G}$. A *von Neumann dynamical system* $(\mathfrak{A}, \alpha)$ is a von Neumann algebra acting on the Hilbert space $\mathfrak{H}$ equipped with a group homomorphism $\alpha : G \to \mathrm{Aut}(\mathfrak{A})$ that is weakly continuous, i.e., $g \mapsto \langle \xi | \alpha_g(x) \eta \rangle$ is continuous for all $x \in \mathfrak{A}$ and all $\xi, \eta \in \mathfrak{H}$.

Definition 2.44 A state ω on a one-parameter C^*-dynamical system $(\mathfrak{G}, \alpha)$ is a (α, β)-*KMS state*, for $\beta \in \mathbb{R}$, if for all pairs of elements x, y in a norm dense α-invariant *-subalgebra of α-analytic elements of $\mathfrak{G}$, then $\omega(x \alpha_{i\beta}(y)) = \omega(yx)$.

Remark 2.45 In the case of a von Neumann dynamical system $(\mathfrak{A}, \alpha)$, a (α, β)-KMS state must be normal (i.e., for every increasing bounded net of positive elements $x_\lambda \to x$, we have $\omega(x_\lambda) \to \omega(x)$). Besides, given $\alpha : \mathbb{R} \to \mathrm{Aut}(A)$, an element $x \in \mathfrak{G}$ is α-*analytic* if there exists a holomorphic extension of the map $t \mapsto \alpha_t(x)$ to an open horizontal strip $\{z \in \mathbb{C} | |\mathrm{Im}\, z| < r\}$, with $r > 0$, in the complex plane. The set of α-analytic elements is always α-invariant (i.e., for all x is analytic, $\alpha(x)$ is analytic)*-subalgebra of $\mathfrak{G}$ that is norm dense in the C^* case and weakly dense in the von Neumann case.

6. The modular automorphism group associated with ω is only the one parameter automorphism group that satisfies the Kubo-Martin-Schwinger (KMS)-condition with respect to the state ω, at inverse temperature β, i.e.,

$$\omega(\sigma_t^\omega(x)) = \omega(x), \qquad \forall x \in \mathfrak{A} \qquad (2.86)$$

and for all $x, y \in \mathfrak{A}$, there exists a function $F_{x,y} : \mathbb{R} \times [0, \beta] \to \mathbb{C}$ such that:

$$
\begin{array}{ll}
F_{x,y} & \text{is holomorphic on} \quad \mathbb{R} \times]0, \beta[, \\
F_{x,y} & \text{is bounded continuous on} \quad \mathbb{R} \times [0, \beta], \\
F_{x,y}(t) = \omega(\sigma_t^{\omega}(y)x), & t \in \mathbb{R}, \\
F_{x,y}(i\beta + t) = \omega(x\sigma_t^{\omega}(y)), & t \in \mathbb{R}.
\end{array}
\tag{2.87}
$$

2.3.2 Vector State and von Neumann Algebras

Let α_i, $i = 1, 2, \ldots, N$ be a sequence of nonzero, positive numbers, satisfying :
$\sum_{i=1}^{N} \alpha_i = 1$. Let

$$
\Phi = \sum_{i=1}^{N} \alpha_i^{\frac{1}{2}} \mathbb{P}_i = \sum_{i=1}^{N} \alpha_i^{\frac{1}{2}} X_{ii} \in \mathcal{B}_2(\mathfrak{H}) \quad \text{with} \quad X_{ii} = |\zeta_i\rangle\langle\zeta_i|
\tag{2.88}
$$

$\{\zeta_i\}_{i=0}^{\infty}$ being an orthonormal basis of $\mathfrak{H}$, and the vectors $\{X_{ij} = |\zeta_i\rangle\langle\zeta_j|, i, j = 1, 2, \ldots, N\}$ forming an orthonormal basis of $\mathcal{B}_2(\mathfrak{H})$,

$$
\langle X_{ij} | X_{kl}\rangle_2 = \delta_{ik}\delta_{lj}.
\tag{2.89}
$$

In particular, the vectors

$$
\mathbb{P}_i = X_{ii} = |\zeta_i\rangle\langle\zeta_i|
\tag{2.90}
$$

are one-dimensional projection operators on $\mathfrak{H}$. Then, we have the following properties:

(i) **Proposition 2.46** Φ *defines a vector state φ on the von Neumann algebra $\mathfrak{A}_l$ corresponding to the operators given with A in the left of the identity operator $I_{\mathfrak{H}}$ on $\mathfrak{H}$, i.e., $\mathfrak{A}_l = \{A_l = A \vee I | A \in \mathcal{L}(\mathfrak{H})\}$.*

Proof Indeed, for any $A \vee I \in \mathfrak{A}_l$, since $\mathcal{B}_2(\mathfrak{H}) \subset \mathcal{L}(\mathfrak{H})$ and $\mathfrak{A} \subset \mathcal{L}(\mathfrak{H})$, from the Remark 2.26 and the equality (2.34) together, the state φ on $\mathfrak{A}_l$ may be defined by

$$
\langle\varphi; A \vee I\rangle = \langle\Phi|(A \vee I)(\Phi)\rangle_2 = \mathrm{Tr}[\Phi^* A\Phi] = \mathrm{Tr}[\rho_\varphi A],
$$

$$
\text{with} \quad \rho_\varphi = \sum_{i=1}^{N} \alpha_i \mathbb{P}_i.
\tag{2.91}
$$

$\square$

(ii) **Proposition 2.47** *The state φ is faithful and normal.*

Proof The state φ is normal by Theorem 2.29, using the fact that ρ_φ is a density matrix and since we have

$$\langle \varphi; A \vee I \rangle = \mathrm{Tr}[\rho_\varphi A]. \tag{2.92}$$

Its faithfulness comes from Proposition 2.30 by use of the equivalence (2) $\Leftrightarrow$ (3), since $\mathfrak{A}_l \subset \mathcal{L}(\mathfrak{H})$ using Eq. (2.34) in the Remark 2.26, we have, with $\mathbb{P} = |\zeta_i\rangle\langle\zeta_i|$,

$$
\begin{aligned}
\langle \varphi; (A \vee I)^*(A \vee I) \rangle &:= \varphi\left(\{(A \vee I)^*(A \vee I)\}\right) \\
&= \mathrm{Tr}[\rho_\varphi A^* A] \qquad \text{[by (2.34)]} \\
&= \sum_{k=1}^{N} \langle \zeta_k | \rho_\varphi A^* A | \zeta_k \rangle \qquad \text{[by (2.67)]} \\
&= \sum_{k=1}^{N} \langle \zeta_k | \left\{ \sum_{i=1}^{N} \alpha_i |\zeta_i\rangle\langle\zeta_i| \right\} A^* A | \zeta_k \rangle \qquad \text{[by (2.91)]} \\
&= \sum_{k=1}^{N} \sum_{i=1}^{N} \alpha_i \langle \zeta_i | A^* A | \zeta_k \rangle \langle \zeta_k | \zeta_i \rangle \\
&= \sum_{i=1}^{N} \alpha_i \|A\zeta_i\|^2, \quad \alpha_i > 0, \quad A \in \mathcal{L}(\mathfrak{H}) \tag{2.93}
\end{aligned}
$$

where the $\{\zeta_i\}_{i=1}^{N}$ form an orthonormal basis set of $\mathfrak{H}$. Φ is separating for $\mathfrak{A}_l$, by use of Theorem 2.29, and the relation

$$
\begin{aligned}
\langle \varphi; (A \vee I)^*(A \vee I) \rangle = 0 &\Longleftrightarrow \sum_{i=1}^{N} \alpha_i \|A\zeta_i\|^2 = 0, \qquad \forall i = 1, 2, \ldots, N \\
&\Longleftrightarrow A \vee I = 0 \Longleftrightarrow A = 0. \tag{2.94}
\end{aligned}
$$

Thereby, $\langle \varphi; (A \vee I)^*(A \vee I) \rangle = 0$ if and only if $A \vee I = 0$. $\qquad\square$

(iii) **Proposition 2.48** *The vector Φ is cyclic and separating for $\mathfrak{A}_l$.*

Proof If $X \in \mathcal{B}_2(\mathfrak{H})$ is orthogonal to all $(A \vee I)\Phi$, $A \in \mathcal{L}(\mathfrak{H})$, then

$$\mathrm{Tr}[X^* A\Phi] = \sum_{i=1}^{N} \alpha_i^{\frac{1}{2}} \langle \zeta_i | X^* A \zeta_i \rangle = 0, \qquad \forall A \in \mathcal{L}(\mathfrak{H}). \tag{2.95}$$

Taking $A = X_{kl}$, it follows from the above equality, $\langle \zeta_l | X^* \zeta_k \rangle = 0$ and since

this holds for all k, l, we get $X = 0$. Indeed, let $\Phi = \sum_{i=1}^{N} \alpha_i^{\frac{1}{2}} \mathbb{P}_i = \sum_{i=1}^{N} \alpha_i^{\frac{1}{2}} X_{ii} =$

$\sum_{i=1}^{N} \alpha_i^{\frac{1}{2}} |\zeta_i\rangle \langle \zeta_i|$, then by definition, see Eq. (2.67),

$$
\begin{aligned}
\langle X | (A \vee I)\Phi \rangle_2 &= \mathrm{Tr}\left[X^*(A \vee I)\Phi \right] \\
&= \mathrm{Tr}\left[X^* A \Phi I^* \right] \qquad \text{[by (2.69)]} \\
&= \mathrm{Tr}\left[X^* A \Phi \right] \\
&= \sum_{k=1}^{N} \langle \zeta_k | X^* A \Phi | \zeta_k \rangle \qquad \text{[by (2.67)]} \\
&= \sum_{k=1}^{N} \langle \zeta_k | X^* A \left\{ \sum_{i=1}^{N} \alpha_i^{\frac{1}{2}} |\zeta_i\rangle \langle \zeta_i| \right\} | \zeta_k \rangle \qquad \text{[by (2.88)]} \\
&= \sum_{k=1}^{N} \sum_{i=1}^{N} \alpha_i^{\frac{1}{2}} \langle \zeta_k | X^* A \zeta_i \rangle \langle \zeta_i | \zeta_k \rangle \\
&= \sum_{i=1}^{N} \alpha_i^{\frac{1}{2}} \langle \zeta_i | X^* A \zeta_i \rangle
\end{aligned}
\tag{2.96}
$$

such that the orthogonality implies

$$
\langle X | (A \vee I)\Phi \rangle_2 = 0 \implies \sum_{i=1}^{N} \alpha_i^{\frac{1}{2}} \langle \zeta_i | X^* A \zeta_i \rangle = 0.
\tag{2.97}
$$

Now taking $A = X_{kl} = |\zeta_k\rangle \langle \zeta_l|$, it follows that

$$
\begin{aligned}
\sum_{i=1}^{N} \alpha_i^{\frac{1}{2}} \langle \zeta_i | X^* A \zeta_i \rangle &= \sum_{i=1}^{N} \alpha_i^{\frac{1}{2}} \langle \zeta_i | X^* \{ |\zeta_k\rangle \langle \zeta_l| \} | \zeta_i \rangle \\
&= \sum_{i=1}^{N} \alpha_i^{\frac{1}{2}} \langle \zeta_i | X^* \zeta_k \rangle \delta_{il} \\
&= \alpha_l^{\frac{1}{2}} \langle \zeta_l | X^* \zeta_k \rangle.
\end{aligned}
\tag{2.98}
$$

From (2.97) and (2.98) together, it follows

$$
\sum_{i=1}^{N} \alpha_i^{\frac{1}{2}} \langle \zeta_i | X^* A \zeta_i \rangle = 0 \implies \alpha_l^{\frac{1}{2}} \langle \zeta_l | X^* \zeta_k \rangle = 0, \quad \forall \alpha_l > 0.
\tag{2.99}
$$

Thereby,

$$\langle \zeta_l | X^* \zeta_k \rangle = 0, \quad \forall k, l \Longrightarrow X = 0. \tag{2.100}$$

Therefore, we have

$$\langle X | (A \vee I) \Phi \rangle_2 = 0 \Longrightarrow X = 0 \tag{2.101}$$

implying that the set $\{(A \vee I)\Phi, A \in \mathfrak{A}_l\}$ is dense in $\mathcal{B}_2(\mathfrak{H})$, proving from the Definition 2.19 that Φ is cyclic for $\mathfrak{A}_l$. $\qquad \square$

The fact that Φ is separating for $\mathfrak{A}_l$ is obtained through the relation

$$(A \vee I)\Phi = (B \vee I)\Phi \Longleftrightarrow A \vee I = B \vee I, \quad \forall A, B \in \mathfrak{A}_l. \tag{2.102}$$

Proof Let $A, B \in \mathfrak{A}_l$, such that $(A \vee I)\Phi = (B \vee I)\Phi$, and take $X \neq 0$, $X \in \mathcal{B}_2(\mathfrak{H})$. We have

$$
\begin{aligned}
&\langle X | \{(A \vee I) - (B \vee I)\} \Phi \rangle_2 \\
&= \mathrm{Tr}\left[X^* \{(A \vee I) - (B \vee I)\} \Phi \right] \\
&= \mathrm{Tr}\left[X^* (A - B) \Phi I^* \right] \qquad \text{[by (2.69)]} \\
&= \sum_{k=1}^{N} \langle \zeta_k | X^*(A - B) \left\{ \sum_{i=1}^{N} \alpha_i^{\frac{1}{2}} |\zeta_i\rangle\langle\zeta_i| \right\} |\zeta_k\rangle \qquad \text{[by (2.88)]} \\
&= \sum_{k=1}^{N} \sum_{i=1}^{N} \alpha_i^{\frac{1}{2}} \langle \zeta_k | X^*(A - B)\zeta_i \rangle \langle \zeta_i | \zeta_k \rangle \\
&= \sum_{i=1}^{N} \alpha_i^{\frac{1}{2}} \langle \zeta_i | X^*(A - B)\zeta_i \rangle \\
&= \sum_{i=1}^{N} \alpha_i^{\frac{1}{2}} \langle \zeta_i | X^*(A - B)\zeta_i \rangle. \qquad \text{[by (2.96) and (2.98)]} \tag{2.103}
\end{aligned}
$$

Taking $(A \vee I)\Phi = (B \vee I)\Phi$, the equality $\langle X | \{(A \vee I) - (B \vee I)\} \Phi \rangle_2 = 0$ leads to

$$
\langle X | \{(A \vee I) - (B \vee I)\} \Phi \rangle_2 = 0 \Longleftrightarrow \sum_{i=1}^{N} \alpha_i^{\frac{1}{2}} \langle \zeta_i | X^*(A - B)\zeta_i \rangle = 0, \quad \alpha_i > 0
$$
$$
\Longleftrightarrow A \vee I = B \vee I \tag{2.104}
$$

which completes the proof. $\qquad \square$

In the same way, Φ is also cyclic for $\mathfrak{A}_r = \{A_r = I \vee A | A \in \mathcal{L}(\mathfrak{H})\}$, which corresponds to the operators given with A in the right of the identity operator $I_{\mathfrak{H}}$ on $\mathfrak{H}$, hence separating for $\mathfrak{A}_r$, i.e. $(I \vee A)\Phi = (I \vee B)\Phi \Longleftrightarrow I \vee A = I \vee B$.

Then, starting to the above setup, to the pair $\{\mathfrak{A}, \varphi\}$ is associated:

- a *one parameter unitary group* $t \mapsto \Delta_\varphi^{-\frac{i}{t}\beta} \in \mathcal{L}(\mathfrak{H})$
- and a *conjugate-linear isometry* $J_\varphi : \mathfrak{H} \to \mathfrak{H}$ that:

$$\Delta_\varphi^{\frac{i}{t}\beta} \mathfrak{A} \Delta_\varphi^{-\frac{i}{t}\beta} = \mathfrak{A}, \quad t \in \mathbb{R}, \tag{2.105}$$

$$J_\varphi \mathfrak{A} J_\varphi = \mathfrak{A}', \tag{2.106}$$

$$J_\varphi \circ J_\varphi = I_{\mathfrak{H}}, \quad J_\varphi \circ \Delta_\varphi^{\frac{i}{t}\beta} = \Delta_\varphi^{-\frac{i}{t}\beta} \circ J_\varphi. \tag{2.107}$$

Denote the automorphisms by $\alpha_\varphi(t)$, and deal with operators $A \in \mathfrak{A}$ with $\mathfrak{A} \subset \mathcal{L}(\mathfrak{H})$. Then, taking into account the Definition 2.25 and the Remark 2.26, from the expression (2.85), the automorphisms, in this case, satisfy the following relation:

$$\alpha_\varphi(t)[A] = \Delta_\varphi^{\frac{i}{t}\beta} A \Delta_\varphi^{-\frac{i}{t}\beta}, \quad \forall A \in \mathfrak{A}. \tag{2.108}$$

The KMS condition with respect to the automorphism group $\alpha_\varphi(t), t \in \mathbb{R}$, is obtained for any two $A, B \in \mathfrak{A}$, such that the function

$$F_{A,B}(t) = \langle \varphi; A\alpha_\varphi(t)[B] \rangle \tag{2.109}$$

has an extension to the strip $\{z = x + iy | t \in \mathbb{R}, y \in [0, \beta]\} \subset \mathbb{C}$ such that $F_{A,B}(z)$ is analytic in the strip $(0, \beta)$ and continuous on its boundaries. In addition, it also satisfies the boundary condition, *at an inverse temperature β*

$$\langle \varphi; A\alpha_\varphi(t + i\beta)[B] \rangle = \langle \varphi; \alpha_\varphi(t)[B]A \rangle, \quad t \in \mathbb{R}. \tag{2.110}$$

Setting the generator of the one-parameter group by $\mathbf{H}_\varphi$, the operators $\Delta_\varphi^{-\frac{i}{t}\beta}$ verify the relation

$$\Delta_\varphi^{-\frac{i}{t}\beta} = e^{it\mathbf{H}_\varphi} \quad \text{and} \quad \Delta_\varphi = e^{-\beta\mathbf{H}_\varphi}. \tag{2.111}$$

2.3.3 von Neumann Algebras Generated by Unitary Operators

Before introducing the von Neumann algebra generated by the unitary operators, let us consider the following:

Definition 2.49 Consider the unitary operator $U(x, y)$ on $\mathfrak{H} = L^2(\mathbb{R})$ given by

$$(U(x, y)\Phi)(\xi) = e^{-ix(\xi - y/2)}\Phi(\xi - y), \tag{2.112}$$

$x, y, \xi \in \mathbb{R}$, with $U(x, y) = e^{-i(xQ + yP)}$, where Q, P are the usual position and momentum operators given on $\mathfrak{H} = L^2(\mathbb{R})$, with $[Q, P] = i\mathbb{I}_{\mathfrak{H}}$, and the Wigner transform, given by

$$\mathcal{W} : B_2(\mathfrak{H}) \rightarrow L^2(\mathbb{R}^2, dxdy)$$

$$(\mathcal{W}X)(x, y) = \frac{1}{(2\pi)^{1/2}} Tr[(U(x, y))^*X], \tag{2.113}$$

where $X \in B_2(\mathfrak{H})$, $x, y \in \mathbb{R}$. $\mathcal{W}$ is unitary.

Indeed, given $X_1, X_2 \in B_2(\mathfrak{H})$,

$$\int_{\mathbb{R}^2} \overline{(\mathcal{W}X_2(x, y))}(\mathcal{W}X_1(x, y))dxdy = \langle X_2|X_1\rangle_2 = \langle X_2|X_1\rangle_{B_2(\mathfrak{H})}. \tag{2.114}$$

On $\tilde{\mathfrak{H}} = L^2(\mathbb{R}^2, dxdy)$, $\forall (x, y) \in \mathbb{R}^2$, consider the operators

$$U_1(x, y) = \mathcal{W}[U(x, y) \vee I_{\mathfrak{H}}]\mathcal{W}^{-1},$$
$$U_2(x, y) = \mathcal{W}[I_{\mathfrak{H}} \vee U(x, y)^*]\mathcal{W}^{-1} \tag{2.115}$$

and let $\mathfrak{A}_i$, $i = 1, 2$, be the von Neumann algebra generated by the unitary operators [25] $\{U_i(x, y)/(x, y) \in \mathbb{R}^2\}$. Then, it follows that:

Proposition 2.50 ([1])

(i) *The algebra $\mathfrak{A}_1$ is the commutant of the algebra $\mathfrak{A}_2$ (i.e., each element of $\mathfrak{A}_1$ commutes with every element of $\mathfrak{A}_2$) and vice versa with a factor, i.e,*

$$\mathfrak{A}_1 \cap \mathfrak{A}_2 = \mathbb{C}I_{\tilde{\mathfrak{H}}}. \tag{2.116}$$

Considering the antiunitary operator J_β (i.e. $\langle J\phi|J\psi\rangle = \langle\psi|\phi\rangle$, $\forall\phi, \psi \in \mathfrak{H} = L^2(\mathbb{R})$) such that:

$$J_\beta\Psi_{nl} = \Psi_{ln}, \quad J_\beta^2 = I_{\mathfrak{H}}, \quad J_\beta\Phi_\beta = \Phi_\beta,$$

it comes

$$J_\beta\mathfrak{A}_1 J_\beta = \mathfrak{A}_2. \tag{2.117}$$

The relation (2.117) and the property (i) provide the modular structure of the triplet $\{\mathfrak{A}_1, \mathfrak{A}_2, J_\beta\}$.

(ii) The map [1]

$$S_\beta : \tilde{\mathfrak{H}} \to \tilde{\mathfrak{H}}, \quad S_\beta \left[U_1(x, y)\Phi_\beta \right] = U_1(x, y)^* \Phi_\beta,$$

is closable and has the polar decomposition

$$S_\beta = J_\beta \Delta_\beta^{\frac{1}{2}}, \tag{2.118}$$

where J_β is the antiunitary operator, with

$$J_\beta \Psi_{nl} = \Psi_{ln}, \quad J_\beta^2 = I_{\mathfrak{H}}, \quad J_\beta \Phi_\beta = \Phi_\beta, \tag{2.119}$$

$$J_\beta \mathfrak{A}_1 J_\beta = \mathfrak{A}_2. \tag{2.120}$$

Indeed, J_β is by definition an antiunitary operator. Then, it is self-adjoint, symmetric, and consequently closable. $\Delta_\beta^{\frac{1}{2}}$ also being self-adjoint by definition, is closable too. From (2.118), the map S_β given as the product of two closable operators is then closable.

Proof of (2.118) The proof is achieved as follows: The vectors Ψ_{jk}, $j, k = 0, 1, 2, \cdots, \infty$, form an orthonormal basis of $\tilde{\mathfrak{H}} = L^2(\mathbb{R}^2, dxdy)$. We have $\Phi_\beta = \sum_{i=0}^{\infty} \lambda_i^{\frac{1}{2}} \Psi_{ii}$. Applying $U_1(x, y)$ to both sides leads to

$$U_1(x, y)\Phi_\beta = \sum_{i=0}^{\infty} \lambda_i^{\frac{1}{2}} U_1(x, y)\Psi_{ii}.$$

Since $\sum_{j,k=0}^{\infty} |\Psi_{jk}\rangle\langle\Psi_{jk}| = I_{\tilde{\mathfrak{H}}}$, we get

$$U_1(x, y)\Phi_\beta = \sum_{i=0}^{\infty} \lambda_i^{\frac{1}{2}} U_1(x, y)\Psi_{ii} = \sum_{i,j,k=0}^{\infty} \lambda_i^{\frac{1}{2}} \langle \Psi_{jk} | U_1(x, y)\Psi_{ii} \rangle_{\tilde{\mathfrak{H}}} \Psi_{jk}.$$

From the relations

$$\phi_{nl} = |\phi_n\rangle\langle\phi_l| \quad \text{and} \quad \mathcal{W}\phi_{nl} = \Psi_{nl}, \quad n, l = 0, 1, 2, \cdots, \infty$$

we get

$$\mathcal{W}\phi_{jk} = \mathcal{W}(|\phi_j\rangle\langle\phi_k|) = \Psi_{jk}, \forall j, k.$$

Using the fact that ϕ_i, $i = 0, 1, 2, \cdots, \infty$, form a basis of $\mathfrak{H} = L^2(\mathbb{R})$, we have

$$\langle \Psi_{jk} | U_1(x, y) \Psi_{ii} \rangle_{\tilde{\mathfrak{H}}} = Tr \left[|\phi_k \rangle \langle \phi_j | U(x, y) | \phi_i \rangle_{\tilde{\mathfrak{H}}} \langle \phi_i | \right]$$

$$= \sum_{l=0}^{\infty} \langle \phi_l | \phi_k \rangle \langle \phi_j | U(x, y) | \phi_i \rangle_{\tilde{\mathfrak{H}}} \delta_{il}$$

$$= \delta_{ik} \langle \phi_j | U(x, y) | \phi_i \rangle_{\tilde{\mathfrak{H}}}$$

$$= \delta_{ik} \overline{\langle \phi_i | U(x, y) | \phi_j \rangle_{\tilde{\mathfrak{H}}}}$$

$$= (2\pi)^{\frac{1}{2}} \delta_{ik} \overline{\mathcal{W}(|\phi_j \rangle \langle \phi_i|)(x, y)}.$$

Thus $\langle \Psi_{jk} | U_1(x, y) \Psi_{ii} \rangle_{\tilde{\mathfrak{H}}} = (2\pi)^{\frac{1}{2}} \delta_{ik} \overline{\Psi_{ji}(x, y)}$. From (2.121), it comes

$$U_1(x, y) \Phi_\beta = (2\pi)^{\frac{1}{2}} \sum_{i,j,k=0}^{\infty} \lambda_i^{\frac{1}{2}} \overline{\Psi_{ji}(x, y)} \Psi_{jk} \delta_{ki} \tag{2.121}$$

i.e.,

$$U_1(x, y) \Phi_\beta = (2\pi)^{\frac{1}{2}} \sum_{i,j=0}^{\infty} \lambda_i^{\frac{1}{2}} \overline{\Psi_{ji}(x, y)} \Psi_{ji}. \tag{2.122}$$

Let us calculate $U_1(x, y)^* \Phi_\beta$. We have

$$U_1(x, y)^* \Phi_\beta = \sum_{j=0}^{\infty} \lambda_j^{\frac{1}{2}} U_1(x, y)^* \Psi_{jj} = \sum_{i,j,k=0}^{\infty} \lambda_j^{\frac{1}{2}} \langle \Psi_{ik} | U_1(x, y)^* \Psi_{jj} \rangle_{\tilde{\mathfrak{H}}} \Psi_{ik},$$

$$\tag{2.123}$$

where

$$\langle \Psi_{ik} | U_1(x, y)^* \Psi_{jj} \rangle_{\tilde{\mathfrak{H}}} = Tr \left[|\phi_k \rangle \langle \phi_i | U(x, y)^* | \phi_j \rangle_{\tilde{\mathfrak{H}}} \langle \phi_j | \right]$$

$$= \sum_{l=0}^{\infty} \langle \phi_l | \phi_k \rangle \langle \phi_i | U(x, y)^* | \phi_j \rangle_{\tilde{\mathfrak{H}}} \delta_{jl}$$

$$= \langle \phi_j | \phi_k \rangle \langle \phi_i | U(x, y)^* | \phi_j \rangle_{\tilde{\mathfrak{H}}}$$

$$= \delta_{jk} \left(\langle \phi_j | U(x, y) | \phi_i \rangle_{\tilde{\mathfrak{H}}} \right)^*$$

$$= (2\pi)^{\frac{1}{2}} \delta_{kj} \left(\mathcal{W}(|\phi_i \rangle \langle \phi_j|) \right)^* (x, y)$$

$$= (2\pi)^{\frac{1}{2}} \delta_{kj} \Psi_{ij}^*(x, y)$$

$$= (2\pi)^{\frac{1}{2}} \delta_{kj} \Psi_{ji}(x, y). \tag{2.124}$$

Putting (2.124) in (2.123) leads to

$$U_1(x, y)^* \Phi_\beta = (2\pi)^{\frac{1}{2}} \sum_{i,j,k=0}^{\infty} \lambda_j^{\frac{1}{2}} \Psi_{ji}(x, y) \Psi_{ik} \delta_{kj} = (2\pi)^{\frac{1}{2}} \sum_{i,j=0}^{\infty} \lambda_j^{\frac{1}{2}} \Psi_{ji}(x, y) \Psi_{ij}.$$

$$(2.125)$$

From (2.121), we have

$$U_1(x, y)\Phi_\beta = (2\pi)^{\frac{1}{2}} \sum_{i,j=0}^{\infty} \lambda_i^{\frac{1}{2}} \overline{\Psi_{ji}(x, y)} \Psi_{ji}.$$

Applying S_β to both sides of the equality gives

$$S_\beta U_1(x, y)\Phi_\beta = (2\pi)^{\frac{1}{2}} \sum_{i,j=0}^{\infty} \lambda_i^{\frac{1}{2}} S_\beta \overline{\Psi_{ji}(x, y)} S_\beta \Psi_{ji}.$$

Since $S_\beta \left[U_1(x, y)\Phi_\beta \right] = U_1(x, y)^* \Phi_\beta$, then $S_\beta \overline{\Psi_{ji}(x, y)} = \Psi_{ji}(x, y)$. Thus,

$$S_\beta U_1(x, y)\Phi_\beta = (2\pi)^{\frac{1}{2}} \sum_{i,j=0}^{\infty} \lambda_i^{\frac{1}{2}} \Psi_{ji} S_\beta \Psi_{ji},$$

which rewrites

$$S_\beta U_1(x, y)\Phi_\beta = U_1(x, y)^* \Phi_\beta = (2\pi)^{\frac{1}{2}} \sum_{i,j=0}^{\infty} \lambda_i^{\frac{1}{2}} \Psi_{ji}(x, y) S_\beta \Psi_{ji}. \quad (2.126)$$

From the relations (2.125) and (2.126) together, it follows that

$$\lambda_i^{\frac{1}{2}} S_\beta \Psi_{ji} = \lambda_j^{\frac{1}{2}} \Psi_{ij} \quad \text{i.e.} \quad S_\beta \Psi_{ji} = \left[\frac{\lambda_j}{\lambda_i} \right]^{\frac{1}{2}} \Psi_{ij}, \quad (2.127)$$

for all $\Psi_{ij} \in \tilde{\mathfrak{H}}$, $i, j = 0, 1, 2, \cdots, \infty$. $\qquad\qquad\qquad\qquad\qquad\qquad\Box$

Proof of (2.119) Consider the operator J_β with $J_\beta \Psi_{ji} = \Psi_{ij}$. We have

$$J_\beta^2 \Psi_{ji} = J_\beta \Psi_{ij} = \Psi_{ji}, \forall i, j \quad \text{i.e.} \quad J_\beta^2 = I_{\tilde{\mathfrak{H}}}.$$

Besides,

$$\Phi_\beta = \sum_{i=0}^{\infty} \lambda_i^{\frac{1}{2}} \Psi_{ii} \quad \text{i.e.} \quad J_\beta \Phi_\beta = \sum_{i=0}^{\infty} \lambda_i^{\frac{1}{2}} (J_\beta \Psi_{ii}) = \sum_{i=0}^{\infty} \lambda_i^{\frac{1}{2}} \Psi_{ii} = \Phi_\beta.$$

Thus $J_\beta \Phi_\beta = \Phi_\beta$. Therefore, J_β is such that

$$J_\beta \Psi_{nl} = \Psi_{ln}, \quad J_\beta^2 = I_{\tilde{\mathfrak{H}}}, \quad J_\beta \Phi_\beta = \Phi_\beta.$$

From (2.118) and

$$\Delta_\beta = \sum_{n,l=0}^{\infty} \frac{\lambda_n}{\lambda_l} |\Psi_{nl}\rangle\langle\Psi_{nl}| = e^{-\beta H}, \quad H = H_1 - H_2,$$

we get

$$S_\beta = J_\beta \Delta_\beta^{\frac{1}{2}} \quad \text{and} \quad \Delta_\beta = \sum_{n,l=0}^{\infty} \frac{\lambda_n}{\lambda_l} |\Psi_{nl}\rangle\langle\Psi_{nl}|.$$

Using this above relations yields

$$\forall i, j, \quad (J_\beta \Delta_\beta^{\frac{1}{2}})|\Psi_{ji}\rangle = J_\beta \left[\Delta_\beta^{\frac{1}{2}}|\Psi_{ji}\rangle \right] = J_\beta \left[\sum_{n,l=0}^{\infty} \left(\frac{\lambda_n}{\lambda_l}\right)^{\frac{1}{2}} |\Psi_{nl}\rangle\langle\Psi_{nl}|\Psi_{ji}\rangle \right]$$

$$= J_\beta \left[\sum_{n,l=0}^{\infty} \left(\frac{\lambda_n}{\lambda_l}\right)^{\frac{1}{2}} |\Psi_{nl}\rangle\delta_{nj}\delta_{il} \right] = J_\beta \left[\left(\frac{\lambda_j}{\lambda_i}\right)^{\frac{1}{2}} |\Psi_{ji}\rangle \right]$$

$$= \left(\frac{\lambda_j}{\lambda_i}\right)^{\frac{1}{2}} \left[J_\beta|\Psi_{ji}\rangle \right].$$

From (2.127),

$$S_\beta|\Psi_{ji}\rangle = \left(\frac{\lambda_j}{\lambda_i}\right)^{1/2} |\Psi_{ij}\rangle = (J_\beta \Delta_\beta^{\frac{1}{2}})|\Psi_{ji}\rangle, \quad \forall i, j \tag{2.128}$$

i.e.,

$$S_\beta = J_\beta \Delta_\beta^{\frac{1}{2}}. \tag{2.129}$$

$\square$

Proposition 2.51 ([1])

If $\{\lambda_n\}_{n=0}^{\infty}$ is a sequence of non-zero positive numbers such that $\sum_{n=0}^{\infty} \lambda_n = 1$, then the vector

$$\Phi = \sum_{n=0}^{\infty} \lambda_n^{\frac{1}{2}} \Psi_{nn} \tag{2.130}$$

is cyclic (that is the set of vectors $\{A\Phi/A \in \mathfrak{A}_1\}$ is dense in $\tilde{\mathfrak{H}}$) and separating (i.e., if $A\Phi = 0$, for all $A \in \mathfrak{A}_1$ then $A = 0$) for $\mathfrak{A}_1$.

Proof Let $X \in \mathcal{B}_2(\tilde{\mathfrak{H}})$ and consider the operator $\mathcal{W}[U(x, y) \vee I_{\tilde{\mathfrak{H}}}]\mathcal{W}^{-1} \in \mathfrak{A}_1$. Taking $U(x, y) \vee I_{\tilde{\mathfrak{H}}}$, we have, since $\mathcal{W}$ is unitary,

$$
\begin{aligned}
\langle X | \mathcal{W}[U(x, y) \vee I_{\tilde{\mathfrak{H}}}]\mathcal{W}^{-1}\Phi\rangle_{\tilde{\mathfrak{H}}} &= \langle X | (U(x, y) \vee I_{\tilde{\mathfrak{H}}})\Phi\rangle_{\mathcal{B}_2(\tilde{\mathfrak{H}})} \\
&= \mathrm{Tr}[X^*(U(x, y) \vee I_{\tilde{\mathfrak{H}}})\Phi] \\
&= \mathrm{Tr}[(X^*U(x, y))\Phi] \\
&= (2\pi)^{\frac{1}{2}}\overline{(\mathcal{W}X\Phi)(x, y)} \quad [\text{by (2.113)}]
\end{aligned}
$$
(2.131)

and the complex conjugate of (2.131) given by

$$
\overline{\langle X | (U(x, y) \vee I_{\tilde{\mathfrak{H}}})\Phi\rangle_{\mathcal{B}_2(\tilde{\mathfrak{H}})}} = \mathrm{Tr}[(X^*U(x, y))\Phi]^* = (2\pi)^{\frac{1}{2}}(\mathcal{W}X\Phi)(x, y).
$$
(2.132)

Then, integrating over $\mathbb{R}^2$ the modulus squared

$$
\begin{aligned}
&\langle X | (U(x, y) \vee I_{\tilde{\mathfrak{H}}})\Phi\rangle_{\mathcal{B}_2(\tilde{\mathfrak{H}})}\overline{\langle X | (U(x, y) \vee I_{\tilde{\mathfrak{H}}})\Phi\rangle_{\mathcal{B}_2(\tilde{\mathfrak{H}})}} \\
&= \left| \langle X | (U(x, y) \vee I_{\tilde{\mathfrak{H}}})\Phi\rangle_{\mathcal{B}_2(\tilde{\mathfrak{H}})} \right|^2
\end{aligned}
$$
(2.133)

with respect to x, y, we get

$$
\begin{aligned}
&\int_{\mathbb{R}^2} \overline{((\mathcal{W}X\Phi)(x, y))}((\mathcal{W}X\Phi)(x, y))dxdy \\
&= \int_{\mathbb{R}^2} \sum_{i,j=0}^{\infty} \sum_{l,k=0}^{\infty} \langle \Psi_{kk} | \lambda_i^{\frac{1}{2}} \lambda_k^{\frac{1}{2}} \overline{\mathcal{W}\phi_{jk}(x, y)} \mathcal{W}\phi_{li}(x, y)dxdy | \Psi_{ii}\rangle_{\tilde{\mathfrak{H}}} \\
&= \sum_{i,j=0}^{\infty} \sum_{l,k=0}^{\infty} \lambda_i^{\frac{1}{2}} \lambda_k^{\frac{1}{2}} \langle \Psi_{kk} | \left\{ \int_{\mathbb{R}^2} \overline{\mathcal{W}\phi_{jk}(x, y)} \mathcal{W}\phi_{li}(x, y)dxdy \right\} | \Psi_{ii}\rangle_{\tilde{\mathfrak{H}}} \\
&= \sum_{i,j=0}^{\infty} \sum_{l,k=0}^{\infty} \lambda_i^{\frac{1}{2}} \lambda_k^{\frac{1}{2}} \langle \Psi_{kk} | \left\{ \int_{\mathbb{R}^2} \overline{\mathcal{W}|\phi_j\rangle\langle\phi_k|(x, y)} \mathcal{W}|\phi_l\rangle\langle\phi_i|(x, y)dxdy \right\} | \Psi_{ii}\rangle_{\tilde{\mathfrak{H}}}
\end{aligned}
$$

$$= \sum_{i,j=0}^{\infty} \sum_{l,k=0}^{\infty} \lambda_i^{\frac{1}{2}} \lambda_k^{\frac{1}{2}} \langle \Psi_{kk}| \left\{ \int_{\mathbb{R}^2} \overline{\Psi_{jk}(x,y)} \Psi_{li}(x,y) dx dy \right\} |\Psi_{ii}\rangle_{\tilde{\mathfrak{H}}}$$

$$= \sum_{i,j=0}^{\infty} \sum_{l,k=0}^{\infty} \lambda_i^{\frac{1}{2}} \lambda_k^{\frac{1}{2}} \langle \Psi_{kk}| \left\{ \int_{\mathbb{R}^2} \Psi_{jk}(x,y)^* \Psi_{li}(x,y) dx dy \right\} |\Psi_{ii}\rangle_{\tilde{\mathfrak{H}}}$$

$$= \sum_{i,j=0}^{\infty} \sum_{l,k=0}^{\infty} \lambda_i^{\frac{1}{2}} \lambda_k^{\frac{1}{2}} \langle \Psi_{kk}|\Psi_{ii}\rangle_{\tilde{\mathfrak{H}}} \delta_{ij} \delta_{kl}$$

$$= \sum_{i,k=0}^{\infty} \lambda_i^{\frac{1}{2}} \lambda_k^{\frac{1}{2}} \delta_{ik}$$

$$= \sum_{i=0}^{\infty} \lambda_i = 1. \tag{2.134}$$

Since $\mathcal{W}$ is unitary, we may write

$$\int_{\mathbb{R}^2} \overline{((\mathcal{W}X\Phi)(x,y))}((\mathcal{W}X\Phi)(x,y)) dx dy = 0$$

$$\implies \left| \langle X|\mathcal{W}[U(x,y) \vee I_{\tilde{\mathfrak{H}}}]\mathcal{W}^{-1}\Phi \rangle_{\tilde{\mathfrak{H}}} \right|^2 = 0$$

$$\implies \left| \langle X|(U(x,y) \vee I_{\tilde{\mathfrak{H}}})\Phi \rangle_{\mathcal{B}_2(\tilde{\mathfrak{H}})} \right|^2 = 0$$

$$\implies X = 0. \tag{2.135}$$

This implies that the set $\left\{ \mathcal{W}[U(x,y) \vee I_{\tilde{\mathfrak{H}}}]\mathcal{W}^{-1}\Phi, \mathcal{W}[U(x,y) \vee I_{\tilde{\mathfrak{H}}}]\mathcal{W}^{-1} \in \mathfrak{A}_1 \right\}$ is dense in $\tilde{\mathfrak{H}}$, proving from the Definition 2.19 that Φ is cyclic for $\mathfrak{A}_1$. $\quad\square$

The fact that Φ is separating for $\mathfrak{A}_1$ is obtained through the relation

$$\mathcal{W}(U(x,y) \vee I_{\tilde{\mathfrak{H}}})\mathcal{W}^{-1}\Phi = \mathcal{W}(U'(x,y) \vee I_{\tilde{\mathfrak{H}}})\mathcal{W}^{-1}\Phi$$

$$\iff \mathcal{W}(U(x,y) \vee I_{\tilde{\mathfrak{H}}})\mathcal{W}^{-1} = \mathcal{W}(U'(x,y) \vee I_{\tilde{\mathfrak{H}}})\mathcal{W}^{-1}. \tag{2.136}$$

Proof Let $U(x,y), U'(x,y)$ such that $\mathcal{W}(U(x,y) \vee I_{\tilde{\mathfrak{H}}})\mathcal{W}^{-1}\Phi = \mathcal{W}(U'(x,y) \vee I_{\tilde{\mathfrak{H}}})\mathcal{W}^{-1}\Phi$. Take $X \neq 0$, $X \in \mathcal{B}_2(\tilde{\mathfrak{H}})$ and set $\Phi = \sum_{i=1}^{N} \lambda_i^{\frac{1}{2}} |\zeta_i\rangle\langle\zeta_i|$. We have

$$\langle X|\mathcal{W}\left\{ (U(x,y) \vee I_{\tilde{\mathfrak{H}}}) - (U'(x,y) \vee I_{\tilde{\mathfrak{H}}}) \right\} \mathcal{W}^{-1}\Phi \rangle_{\tilde{\mathfrak{H}}}$$

$$= \langle X| \left\{ (U(x,y) \vee I_{\tilde{\mathfrak{H}}}) - (U'(x,y) \vee I_{\tilde{\mathfrak{H}}}) \right\} \Phi \rangle_{\mathcal{B}_2(\tilde{\mathfrak{H}})}$$

$$= \mathrm{Tr}\left[X^* \left\{(U(x, y) \vee I_{\tilde{\mathfrak{H}}}) - (U'(x, y) \vee I_{\tilde{\mathfrak{H}}})\right\} \Phi\right]$$

$$= \sum_{k=1}^{N} \langle \zeta_k | X^* \left\{U(x, y) - U'(x, y)\right\} \left\{\sum_{i=1}^{N} \lambda_i^{\frac{1}{2}} |\zeta_i\rangle\langle\zeta_i|\right\} |\zeta_k\rangle$$

$$= \sum_{k=1}^{N}\sum_{i=1}^{N} \lambda_i^{\frac{1}{2}} \langle \zeta_k | X^* \left\{U(x, y) - U'(x, y)\right\} \zeta_i \rangle \langle \zeta_i | \zeta_k \rangle$$

$$= \sum_{k=1}^{N}\sum_{i=1}^{N} \lambda_i^{\frac{1}{2}} \langle \zeta_k | X^* \left\{U(x, y) - U'(x, y)\right\} \zeta_i \rangle \delta_{ik}$$

$$= \sum_{i=1}^{N} \lambda_i^{\frac{1}{2}} \langle \zeta_i | X^* \left\{U(x, y) - U'(x, y)\right\} \zeta_i \rangle. \tag{2.137}$$

Then, we have

$$\langle X | \mathcal{W} \left\{(U(x, y) \vee I_{\tilde{\mathfrak{H}}}) - (U'(x, y) \vee I_{\tilde{\mathfrak{H}}})\right\} \mathcal{W}^{-1} \Phi \rangle_{\tilde{\mathfrak{H}}}$$
$$= \langle X | \left\{(U(x, y) \vee I_{\tilde{\mathfrak{H}}}) - (U'(x, y) \vee I_{\tilde{\mathfrak{H}}})\right\} \Phi \rangle_{\mathcal{B}_2(\tilde{\mathfrak{H}})} = 0$$

$$\Longleftrightarrow \sum_{i=1}^{N} \lambda_i^{\frac{1}{2}} \langle \zeta_i | X^* \left\{U(x, y) - U'(x, y)\right\} \zeta_i \rangle = 0, \ \lambda_i > 0, \ \forall i$$

$$\Longleftrightarrow U(x, y) \vee I_{\tilde{\mathfrak{H}}} = U'(x, y) \vee I_{\tilde{\mathfrak{H}}} \tag{2.138}$$

which completes the proof. $\qquad\square$

2.4 Modular Theory-Thermal State

Here, we give two examples of thermal states as known from the literature. For more details, see [1, 2, 7, 8, 11, 24–26].

1. Let $\alpha_i, i = 1, 2, \ldots, N$ be a sequence of non-zero, positive numbers, satisfying $\sum_{i=1}^{N} \alpha_i = 1$. Then, the thermal state is defined as:

$$\Phi := \sum_{i=1}^{N} \alpha_i^{\frac{1}{2}} \mathbb{P}_i = \sum_{i=1}^{N} \alpha_i^{\frac{1}{2}} X_{ii} \in \mathcal{B}_2(\mathfrak{H}), \tag{2.139}$$

where $\mathbb{P}_i = X_{ii} = |\zeta_i\rangle\langle\zeta_i|$ is defined as in (2.88)–(2.90), with $\{X_{ij} = |\zeta_i\rangle\langle\zeta_j|, i, j = 1, 2, \ldots, N\}$ forming an orthonormal basis of $\mathcal{B}_2(\mathfrak{H})$.

2. The thermal equilibrium state Φ at inverse temperature β, corresponding to the harmonic oscillator Hamiltonian

$$H_{OSC} = \frac{1}{2}(P^2 + Q^2), \quad \text{with } H_{OSC}\phi_n = \omega(n + \frac{1}{2})\phi_n, n = 0, 1, 2, \ldots,$$

$$(2.140)$$

where the density matrix is

$$\rho_\beta = \frac{e^{-\beta H_{OSC}}}{Tr\left[e^{-\beta H_{OSC}}\right]} = (1 - e^{-\omega\beta})\sum_{n=0}^{\infty} e^{-n\omega\beta}|\phi_n\rangle\langle\phi_n|,$$

$$\mathrm{Tr}[e^{-\beta H_{OSC}}] = \frac{e^{-\frac{\beta\omega}{2}}}{1 - e^{-\beta\omega}}, \tag{2.141}$$

is

$$\Phi = \left[1 - e^{-\omega\beta}\right]^{\frac{1}{2}} \sum_{n=0}^{\infty} e^{-\frac{n}{2}\omega\beta}|\phi_n\rangle\langle\phi_n|. \tag{2.142}$$

3. Let the two von Neumann algebras be given by

$$\mathfrak{A}_l = \{A_l = A \vee I | A \in \mathcal{L}(\mathfrak{H})\}, \quad \mathfrak{A}_r = \{A_r = I \vee A | A \in \mathcal{L}(\mathfrak{H})\}$$

$$(2.143)$$

where $\mathfrak{A}_l$ corresponds to the operators given with A in the left, and $\mathfrak{A}_r$ corresponds to the operators given with A in the right of the identity operator $I_{\mathfrak{H}}$ on $\mathfrak{H}$, respectively. Φ defines a vector state φ, called *KMS* state, on the von Neumann algebra $\mathfrak{A}_l$. For any $A \vee I \in \mathfrak{A}_l$, one has the state φ on $\mathfrak{A}_l$ given by

$$\langle \varphi; A \vee I \rangle = \langle \Phi | (A \vee I)(\Phi) \rangle_2 = Tr[\Phi^* A \Phi] = Tr[\rho_\varphi A],$$

$$\text{with } \rho_\varphi = \sum_{n=1}^{N} \alpha_n \mathbb{P}_n \tag{2.144}$$

with $\mathbb{P}_n = |\phi_n\rangle\langle\phi_n|$, where

$$\rho_\varphi = \frac{e^{-\beta H_{OSC}}}{Tr[e^{-\beta H_\varphi}]} = (1 - e^{-\omega\beta})\sum_{n=0}^{\infty} e^{-n\omega\beta}|\phi_n\rangle\langle\phi_n| \tag{2.145}$$

and

$$H_\varphi = -\frac{1}{\beta}\sum_{n=0}^{\infty}(\ln \alpha_n)\mathbb{P}_n, \quad \alpha_n = (1 - e^{-\omega\beta})e^{-n\omega\beta}. \tag{2.146}$$

2.5 Coherent States Built from the Harmonic Oscillator Thermal State

Before dealing with the CS construction, we shall first extract few facts and notations about the modular structures emerging for von Neumann algebras in the study of an electron in a magnetic field as needed for the development of this paragraph. For details see [1] and references therein.

2.5.1 Electron in a Magnetic Field

Considering the quantum Hilbert space $\mathfrak{H} = L^2(\mathbb{R})$ of the Hamiltonian H_{osc} in (2.140), take $\mathcal{B}_2(\mathfrak{H}) \simeq \mathfrak{H} \otimes \bar{\mathfrak{H}}$ the space of Hilbert-Schmidt operators on $\mathfrak{H}$ with an orthonormal basis given by

$$\phi_{nl} := |\phi_n\rangle\langle\phi_l|, \quad n, l = 0, 1, 2, \ldots, \infty. \tag{2.147}$$

Taking the classical Hamiltonian describing an electron placed in the xy plane and subjected to a constant magnetic field [1],

$$H_{elec} = \frac{1}{2}(\vec{p} + \vec{A})^2 = \frac{1}{2}\left(p_x + \frac{y}{2}\right)^2 + \frac{1}{2}\left(p_y - \frac{x}{2}\right)^2 \tag{2.148}$$

let the following quantum Hamiltonians

$$H_1 = \frac{1}{2}(P_1^2 + Q_1^2), \quad [Q_1, P_1] = i I_{\tilde{\mathfrak{H}}} \tag{2.149}$$

with the magnetic field aligned along the negative z axis, $\vec{A} = \frac{1}{2}(y, -x, 0)$, where the quantized observables given on $\tilde{\mathfrak{H}} = L^2(\mathbb{R}^2, dxdy)$ by

$$p_x + \frac{y}{2} \rightarrow Q_1 = -i\frac{\partial}{\partial x} + \frac{y}{2}; \quad p_y - \frac{x}{2} \rightarrow P_1 = -i\frac{\partial}{\partial y} - \frac{x}{2} \tag{2.150}$$

and

$$H_2 = \frac{1}{2}(P_2^2 + Q_2^2), \quad [Q_2, P_2] = i I_{\tilde{\mathfrak{H}}} \tag{2.151}$$

with the magnetic field aligned along the positive z axis, $\vec{A} = \frac{1}{2}(-y, x, 0)$, with the quantized observables given on $\tilde{\mathfrak{H}} = L^2(\mathbb{R}^2, dxdy)$ by

$$p_y + \frac{x}{2} \rightarrow Q_2 = -i\frac{\partial}{\partial y} + \frac{x}{2}; \quad p_x - \frac{y}{2} \rightarrow P_2 = -i\frac{\partial}{\partial x} - \frac{y}{2}. \tag{2.152}$$

Since $[H_1, H_2] = 0$, the eigenvectors Ψ_{nl} of H_1 can be so chosen that they are also the eigenvectors of H_2 as follows:

$$H_1 \Psi_{nl} = \omega \left(n + \frac{1}{2} \right) \Psi_{nl}, \qquad H_2 \Psi_{nl} = \omega \left(l + \frac{1}{2} \right) \Psi_{nl} \qquad (2.153)$$

so that H_2 lifts the degeneracy of H_1 and vice versa. Next, from the Definition 2.49, it is established that [1]

$$\mathcal{W} \phi_{nl} = \mathcal{W}(|\phi_n\rangle \langle \phi_l|) = \Psi_{nl} \qquad (2.154)$$

where the ϕ_{nl} are the basis vectors given in (2.147) and the Ψ_{nl} the normalized eigenvectors in (2.153). Then, from (2.154), the vectors $\Psi_{nl}, n, l = 0, 1, 2, \ldots, \infty$ form a basis of $\tilde{\mathfrak{H}} = L^2(\mathbb{R}^2, dxdy)$.

Note that, in the sequel, the CS will be constructed from the thermal state Φ_β, identified with the vector Φ given in (2.142), denoted as a ket state $|\Phi_\beta\rangle$, the normalized eigenvectors (2.154) Ψ_{nl} also denoted $|\Psi_{nl}\rangle$ as proceeded in [1].

2.5.2 Coherent States built from the Thermal State

Take the cyclic vector Φ of the von Neumann algebra $\mathfrak{A}_1$ generated by the unitary operator (2.115)

$$U_1(x, y) = \mathcal{W}[U(x, y) \vee I_{\mathfrak{H}}] \mathcal{W}^{-1}, \qquad (2.155)$$

where $\mathcal{W}$ and $U(x, y) = e^{-i(xQ+yP)}$ are defined by (2.112) and (2.113), and consider Proposition 2.51 with the thermal state Φ_β, instead of Φ, such that

$$\Phi_\beta = \left[1 - e^{-\omega\beta} \right]^{\frac{1}{2}} \sum_{n=0}^{\infty} e^{-n\frac{\omega\beta}{2}} \Psi_{nn}, \text{ i.e., } \lambda_n = (1 - e^{-\omega\beta})e^{-n\omega\beta}. \quad (2.156)$$

The CS, denoted $|z, \bar{z}, \beta\rangle^{\text{KMS}}$, built from the thermal state in ket notation $|\Phi_\beta\rangle$ (see [1]), are given by

$$|z, \bar{z}, \beta\rangle^{\text{KMS}} = U_1(z)|\Phi_\beta\rangle := e^{zA_1^\dagger - \bar{z}A_1}|\Phi_\beta\rangle \qquad (2.157)$$

where the annihilation and creation operators, A_1 and $A_1^\dagger$ with [1]

$$A_1 = \frac{1}{\sqrt{2}}(Q_1 + i P_1), \qquad A_1^\dagger = \frac{1}{\sqrt{2}}(Q_1 - i P_1) \qquad (2.158)$$

act on $|\Psi_{n,l}\rangle$, eigenstates of the oscillator Hamiltonian in one dimension (2.149) with eigenvalues $E_n = \omega(n + \frac{1}{2})$, as follows:

$$A_1^\dagger |\Psi_{n,l}\rangle = \sqrt{n+1}|\Psi_{n+1,l}\rangle, \qquad A_1|\Psi_{n,l}\rangle = \sqrt{n}|\Psi_{n-1,l}\rangle \qquad (2.159)$$

where the states $|\Psi_{nl}\rangle$ are here denoted $|\Psi_{n,l}\rangle$ by commodity.

Proposition 2.52 ([1]) *The CS* $|z, \bar{z}, \beta\rangle^{KMS}$ *satisfy the resolution of the identity condition*

$$\frac{1}{2\pi} \int_{\mathbb{C}} |z, \bar{z}, \beta\rangle^{KMS\,KMS}\langle z, \bar{z}, \beta| dx dy = I_{\tilde{\mathfrak{H}}}, \quad \tilde{\mathfrak{H}} = L^2(\mathbb{R}^2, dx dy). \qquad (2.160)$$

Proof Consider the unitary operator $U(x, y) = e^{-i(xQ+yP)}$, $\forall (x, y) \in \mathbb{R}^2$. Let

$$U(z)\phi := \phi(x, y) = e^{-i(xQ+yP)}\phi \quad \forall \phi \in \mathfrak{H}. \qquad (2.161)$$

Show that, for any normalized state $\phi \in \mathfrak{H} = L^2(\mathbb{R})$:

$$\frac{1}{2\pi} \int_{\mathbb{R}^2} |\phi(x, y)\rangle\langle\phi(x, y)| dx dy = I_{\mathfrak{H}}. \qquad (2.162)$$

$$\forall q \in \mathbb{R}, \quad \text{set } \phi(x, y)(q) = (\pi)^{-\frac{1}{4}} e^{-i(\frac{x}{2}-q)y} e^{-\frac{(q-x)^2}{2}},$$

such that $\forall \psi, \xi \in \mathfrak{H}$, we have

$$\langle\xi|\phi(x, y)\rangle_{\mathfrak{H}} = (\pi)^{-\frac{1}{4}} e^{-i\frac{xy}{2}} \int_{\mathbb{R}} \overline{\xi(q)} e^{iyq} e^{-\frac{(q-x)^2}{2}} dq$$

$$\langle\phi(x, y)|\psi\rangle_{\mathfrak{H}} = (\pi)^{-\frac{1}{4}} e^{i\frac{xy}{2}} \int_{\mathbb{R}} e^{-iyq'} e^{-\frac{(q'-x)^2}{2}} \psi(q') dq'.$$

Then,

$$\frac{1}{2\pi} \int_{\mathbb{R}^2} \langle\xi|\phi(x, y)\rangle\langle\phi(x, y)|\psi\rangle dx dy$$

$$= \frac{1}{2\pi\sqrt{\pi}} \int_{\mathbb{R}^2} dx dy \int_{\mathbb{R}^2} dq dq' \overline{\xi(q)} e^{iy(q-q')} e^{\left[-\frac{(x-q)^2}{2} - \frac{(q'-x)^2}{2}\right]} \psi(q')$$

$$= \frac{1}{\sqrt{\pi}} \int_{\mathbb{R}} dx \int_{\mathbb{R}^2} dq dq' \overline{\xi(q)} \left[\frac{1}{2\pi} \int_{\mathbb{R}} e^{iy(q-q')} dy\right] e^{\left[\frac{-(x-q)^2}{2} - \frac{(q'-x)^2}{2}\right]} \psi(q').$$

$$(2.163)$$

Since $\frac{1}{2\pi} \int_{\mathbb{R}} e^{iy(q-q')} dy = \delta(q - q')$, we get

$$\frac{1}{2\pi} \int_{\mathbb{R}^2} \langle \xi | \phi(x, y) \rangle \langle \phi(x, y) | \psi \rangle dx dy$$

$$= \frac{1}{\sqrt{\pi}} \int_{\mathbb{R}} dx \int_{\mathbb{R}^2} dq dq' \overline{\xi(q)} \delta(q - q') e^{\left[-\frac{(x-q)^2}{2} - \frac{(q'-x)^2}{2} \right]} \psi(q')$$

$$= \frac{1}{\sqrt{\pi}} \int_{\mathbb{R}} dx \int_{\mathbb{R}} dq\, e^{-\frac{(x-q)^2}{2}} \overline{\xi(q)} \left[\int_{\mathbb{R}} \psi(q') \delta(q - q') e^{-\frac{(q'-x)^2}{2}} dq' \right]$$

$$= \frac{1}{\sqrt{\pi}} \int_{\mathbb{R}^2} e^{-(x-q)^2} \overline{\xi(q)} \psi(q) dx dq = \int_{\mathbb{R}} dq \left(\frac{1}{\sqrt{\pi}} \int_{\mathbb{R}} e^{-u^2} du \right) \overline{\xi(q)} \psi(q)$$

$$= \langle \xi | \psi \rangle.$$

Thus

$$\frac{1}{2\pi} \int_{\mathbb{R}^2} | \phi(x, y) \rangle \langle \phi(x, y) | dx dy = I_{\mathfrak{H}}$$

i.e. $$\frac{1}{2\pi} \int_{\mathbb{C}} | U(z)\phi \rangle \langle U(z)\phi | dx dy = I_{\mathfrak{H}}. \tag{2.164}$$

Using the isometry property of $\mathcal{W}$, the states $|z, \bar{z}, \beta\rangle^{\mathrm{KMS}}$ satisfy the following resolution of the identity

$$\frac{1}{2\pi} \int_{\mathbb{C}} |z, \bar{z}, \beta\rangle^{\mathrm{KMS\,KMS}} \langle z, \bar{z}, \beta| dx dy = I_{\tilde{\mathfrak{H}}}. \tag{2.165}$$

Proof By definition of the states $|z, \bar{z}, \beta\rangle^{\mathrm{KMS}}$ and using (2.122), we have:

$$|z, \bar{z}, \beta\rangle^{\mathrm{KMS}} = U_1(z) \Phi_\beta = (2\pi)^{\frac{1}{2}} \sum_{i,j=0}^{\infty} \lambda_i^{\frac{1}{2}} \overline{\Psi_{ji}(x, y)} |\Psi_{ji}\rangle,$$

$$^{\mathrm{KMS}} \langle z, \bar{z}, \beta| = U_1(z)^* \Phi_\beta = (2\pi)^{\frac{1}{2}} \sum_{l,k=0}^{\infty} \langle \Psi_{lk} | \lambda_k^{\frac{1}{2}} \Psi_{lk}(x, y). \tag{2.166}$$

Thereby

$$|z, \bar{z}, \beta\rangle^{\mathrm{KMS\,KMS}} \langle z, \bar{z}, \beta| = 2\pi \sum_{i,j=0}^{\infty} \sum_{l,k=0}^{\infty} \lambda_i^{\frac{1}{2}} \lambda_k^{\frac{1}{2}} \overline{\mathcal{W}\phi_{ji}(x, y)} \mathcal{W}\phi_{lk}(x, y) |\Psi_{ji}\rangle \langle \Psi_{lk}|.$$

$$\tag{2.167}$$

Integrating the two members of Eq. (2.167) over $\mathbb{R}^2$, and using the Wigner map $\mathcal{W}$, we get

$$\frac{1}{2\pi} \int_{\mathbb{R}^2} |z, \bar{z}, \beta\rangle^{\mathrm{KMS\,KMS}} \langle z, \bar{z}, \beta| dxdy$$

$$= \int_{\mathbb{R}^2} \sum_{i,j=0}^{\infty} \sum_{l,k=0}^{\infty} \lambda_i^{\frac{1}{2}} \lambda_k^{\frac{1}{2}} |\Psi_{ji}\rangle \langle \Psi_{lk}| \overline{\mathcal{W}\phi_{ji}(x,y)} \mathcal{W}\phi_{lk}(x,y) dxdy$$

$$= \sum_{i,j=0}^{\infty} \sum_{l,k=0}^{\infty} \lambda_i^{\frac{1}{2}} \lambda_k^{\frac{1}{2}} |\Psi_{ji}\rangle \langle \Psi_{lk}| \int_{\mathbb{R}^2} \overline{\mathcal{W}\phi_{ji}(x,y)} \mathcal{W}\phi_{lk}(x,y) dxdy$$

$$= \sum_{i,j=0}^{\infty} \sum_{l,k=0}^{\infty} \lambda_i^{\frac{1}{2}} \lambda_k^{\frac{1}{2}} |\Psi_{ji}\rangle \langle \Psi_{lk}| \delta_{lj} \delta_{ki}$$

$$= \sum_{i,j=0}^{\infty} |\Psi_{ji}\rangle \langle \Psi_{ji}|$$

$$= I_{\tilde{\mathfrak{H}}}. \tag{2.168}$$

□

From $(A_1^{\dagger})|\Psi_{0n}\rangle = \sqrt{n!}|\Psi_{nn}\rangle$ and $U_1(z)\left|\Psi_{0n}\right\rangle = e^{-\frac{|z|^2}{2}} \sum_{k=0}^{\infty} \frac{(zA_1^{\dagger})^k}{k!}\left|\Psi_{0n}\right\rangle$, it comes that:

$$U_1(z)|\Psi_{nn}\rangle = \frac{1}{\sqrt{n!}} \left(A_1^{\dagger} - \bar{z} I_{\tilde{\mathfrak{H}}}\right)^n U_1(z)|\Psi_{0n}\rangle = \frac{1}{\sqrt{n!}} \left(-\frac{\partial}{\partial z} - \frac{\bar{z}}{2} I_{\tilde{\mathfrak{H}}}\right)^n U_1(z)|\Psi_{0n}\rangle. \tag{2.169}$$

Proof of the First Equality of (2.169) The operators P and Q verify the following relations:

$$\left[Q, P^n\right] = n P^{n-1} [Q, P] = \imath \hbar n P^{n-1}, \quad \left[Q^n, P\right] = n Q^{n-1} [Q, P] = \imath \hbar n Q^{n-1}. \tag{2.170}$$

We establish that

$$e^{-\imath Pu} Q e^{\imath Pu} = Q - u, \qquad e^{-\imath Qu} P e^{\imath Qu} = P + u, \ \forall u \in \mathbb{R}. \tag{2.171}$$

Multiplying the first and second equalities of (2.171), by $\frac{1}{\sqrt{2}}$ and $\frac{-\imath}{\sqrt{2}}$, respectively, provides:

$$e^{-\imath Pu} \frac{Q}{\sqrt{2}} e^{\imath Pu} = \frac{Q}{\sqrt{2}} - \frac{u}{\sqrt{2}}, \qquad e^{-\imath Qu} \left(\frac{-\imath P}{\sqrt{2}}\right) e^{\imath Qu} = \frac{-\imath P}{\sqrt{2}} - \frac{\imath u}{\sqrt{2}}. \tag{2.172}$$

Setting $u = -x$ and $u = -y$, in the first and second relations of (2.172), respectively, gives with replacing P by P_1, and Q by Q_1, respectively:

$$e^{\iota P_1 x} \frac{Q_1}{\sqrt{2}} e^{-\iota P_1 x} = \frac{Q_1}{\sqrt{2}} + \frac{x}{\sqrt{2}}, \qquad e^{\iota Q_1 y} \left(-\frac{\iota P_1}{\sqrt{2}} \right) e^{-\iota Q_1 y} = \frac{-\iota P_1}{\sqrt{2}} + \frac{\iota y}{\sqrt{2}}.$$

(2.173)

From $A_1^\dagger = \frac{Q_1 - \iota P_1}{\sqrt{2}}$, set $z = \frac{-x + \iota y}{\sqrt{2}}$. Summing both equalities of (2.173) gives:

$$e^{\iota P_1 x} \frac{Q_1}{\sqrt{2}} e^{-\iota P_1 x} + e^{\iota Q_1 y} \left(-\frac{\iota P_1}{\sqrt{2}} \right) e^{-\iota Q_1 y} = A_1^\dagger - \bar{z} I_{\tilde{\mathfrak{H}}}.$$

(2.174)

Since $U_1(z) = e^{z A_1^\dagger - \bar{z} A_1}$, with $z A_1^\dagger - \bar{z} A_1 = \iota(P_1 x + Q_1 y)$, it follows: $U_1(z) = e^{\iota(P_1 x + Q_1 y)}$. This latter equality with (2.174) together leads to:

$$
\begin{aligned}
(A_1^\dagger - \bar{z} I_{\tilde{\mathfrak{H}}}) U_1(z) &= e^{\iota P_1 x} \frac{Q_1}{\sqrt{2}} e^{-\iota P_1 x} e^{\iota(P_1 x + Q_1 y)} + e^{\iota Q_1 y} \left(-\iota \frac{P_1}{\sqrt{2}} \right) e^{-\iota Q_1 y} e^{\iota(P_1 x + Q_1 y)} \\
&= e^{\iota(P_1 x + Q_1 y)} e^{-\iota \frac{xy}{2}} e^{\iota \frac{xy}{2}} \frac{Q_1}{\sqrt{2}} + e^{\iota(Q_1 y + P_1 x)} e^{\iota \frac{xy}{2}} e^{-\iota \frac{xy}{2}} \left(-\iota \frac{P_1}{\sqrt{2}} \right) \\
&= e^{z A_1^\dagger - \bar{z} A_1} \left(\frac{Q_1}{\sqrt{2}} - \iota \frac{P_1}{\sqrt{2}} \right) = U_1(z) A_1^\dagger.
\end{aligned}
$$

(2.175)

From (2.175), we have

$$U_1(z) |\Psi_{1n}\rangle = (A_1^\dagger - \bar{z} I_{\tilde{\mathfrak{H}}}) U_1(z) |\Psi_{0n}\rangle$$

(2.176)

such that, by recursion, we get

$$U_1(z) |\Psi_{nn}\rangle = \frac{1}{\sqrt{n!}} \left(A_1^\dagger - \bar{z} I_{\tilde{\mathfrak{H}}} \right)^n U_1(z) |\Psi_{0n}\rangle.$$

(2.177)

□

Proof of the Second Equality of (2.169) Considering $A_1^\dagger = \frac{1}{\sqrt{2}}(Q_1 - \iota P_1)$, where $z = \frac{1}{\sqrt{2}}(y - \iota x)$, with

$$\frac{\partial}{\partial x} = \frac{-\iota}{\sqrt{2}} \frac{\partial}{\partial z} + \frac{\iota}{\sqrt{2}} \frac{\partial}{\partial \bar{z}}, \quad \frac{\partial}{\partial y} = \frac{1}{\sqrt{2}} \frac{\partial}{\partial z} + \frac{1}{\sqrt{2}} \frac{\partial}{\partial \bar{z}}$$

(2.178)

provides

$$A_1^\dagger = -\frac{\partial}{\partial z} + \frac{\bar{z}}{2} I_{\tilde{\mathfrak{H}}}, \quad \text{i.e.,} \quad A_1^\dagger - \bar{z} I_{\tilde{\mathfrak{H}}} = -\frac{\partial}{\partial z} - \frac{\bar{z}}{2} I_{\tilde{\mathfrak{H}}}$$

(2.179)

which completes the proof.

□

Since

$$|z, \bar{z}, \beta\rangle^{\text{KMS}} = U_1(z)|\Phi_\beta\rangle = (1 - e^{-\omega\beta})^{\frac{1}{2}} \sum_{n=0}^{\infty} e^{-n\frac{\omega\beta}{2}} U_1(z)|\Psi_{nn}\rangle, \quad (2.180)$$

setting $|z; n\rangle = U_1(z)|\Psi_{0n}\rangle$ and from (2.169), it comes

$$|z, \bar{z}, \beta\rangle^{\text{KMS}} = (1 - e^{\omega\beta})^{\frac{1}{2}} \sum_{n=0}^{\infty} \frac{1}{\sqrt{n!}} e^{-n\frac{\omega\beta}{2}} \left(-\frac{\partial}{\partial z} - \frac{\bar{z}}{2} I_{\tilde{\mathfrak{H}}}\right)^n |z; n\rangle. \quad (2.181)$$

Set $|z; n\rangle = U_2(z)|\Psi_{n0}\rangle$ with $(A_2^\dagger)^n|\Psi_{n0}\rangle = \sqrt{n!}|\Psi_{nn}\rangle$ and

$$U_2(z)|\Psi_{n0}\rangle = e^{-\frac{|z|^2}{2}} \sum_{k=0}^{\infty} \frac{(zA_2^\dagger)^k}{k!} |\Psi_{n0}\rangle. \quad (2.182)$$

Taking $|z; 0\rangle = U_2(z)|\Psi_{00}\rangle$ leads to

$$(A_2^\dagger)^n|z; 0\rangle = \sqrt{n!}|z; n\rangle \quad \text{with} \quad U_2(z)\left[(A_2^\dagger)^n|\Psi_{00}\rangle\right] = \sqrt{n!}U_2(z)|\Psi_{n0}\rangle. \quad (2.183)$$

Then, $|z; n\rangle = \frac{1}{\sqrt{n!}}(A_2^\dagger)^n|z; 0\rangle$. In (2.181), we get

$$|z, \bar{z}, \beta\rangle^{\text{KMS}} = (1 - e^{-\omega\beta})^{\frac{1}{2}} \sum_{n=0}^{\infty} \frac{1}{n!} e^{-n\frac{\omega\beta}{2}} \left(-\frac{\partial}{\partial z} - \frac{\bar{z}}{2}\right)^n (A_2^\dagger)^n|z; 0\rangle. \quad (2.184)$$

Using the relation $U_1(x, y)^* = U_1(-x, -y)$, the CS $|-z, -\bar{z}, \beta\rangle^{\text{KMS}}$ are obtained as follows:

$$|-z, -\bar{z}, \beta\rangle^{\text{KMS}} = S_\beta|z, \bar{z}, \beta\rangle^{\text{KMS}}. \quad (2.185)$$

Indeed, by definition $|z, \bar{z}, \beta\rangle^{\text{KMS}} = U_1(z)|\Phi_\beta\rangle := U_1(x, y)|\Phi_\beta\rangle$ such that

$$S_\beta\left[U_1(x, y)|\Phi_\beta\rangle\right] = U_1(x, y)^*|\Phi_\beta\rangle = U_1(-x, -y)|\Phi_\beta\rangle,$$

$$\text{i.e., } U_1(-z)|\Phi_\beta\rangle = e^{-zA_1^\dagger + \bar{z}A_1}|\Phi_\beta\rangle \quad (2.186)$$

where $e^{-zA_1^\dagger + \bar{z}A_1}|\Phi_\beta\rangle := |-z, -\bar{z}', \beta\rangle^{\text{KMS}}$, leading to (2.185). The CS (2.185) satisfy a resolution of the identity analogue to (2.165), i.e.,

$$\frac{1}{2\pi} \int_{\mathbb{C}} |-z, -\bar{z}, \beta\rangle^{\text{KMS}}\,^{\text{KMS}}\langle-z, -\bar{z}, \beta|dxdy = I_{\tilde{\mathfrak{H}}}. \quad (2.187)$$

Proof Similar to the proof of (2.165). □

3 Noncommutative Quantum Harmonic Oscillator Hilbert Space

Without loss of generality, we restrict our developments to the noncommutative quantum mechanics formalism [6, 11, 15, 22] for the physical system of harmonic oscillator. We focus on the application of Hilbert-Schmidt operators, bounded operators on the noncommutative classical configuration space denoted by

$$\mathcal{H}_c = \text{span} \left\{ |n\rangle = \frac{1}{\sqrt{n!}} (a^\dagger)^n |0\rangle \right\}_{n=0}^\infty. \tag{3.1}$$

This space is isomorphic to the boson Fock space $\mathcal{F} = \{|n\rangle\}_{n=0}^\infty$, where the annihilation and creation operators $a, a^\dagger$ obey the Fock algebra $[a, a^\dagger] = 1$. The physical states of the system represented on $\mathcal{H}_q$ known as the set of Hilbert-Schmidt operators is equivalent to the Hilbert space of square integrable function, with

$$\mathcal{H}_q = \left\{ \psi(\hat{x}_1, \hat{x}_2) : \psi(\hat{x}_1, \hat{x}_2) \in \mathcal{B}(\mathcal{H}_c), \, tr_c(\psi(\hat{x}_1, \hat{x}_2)^\dagger, \psi(\hat{x}_1, \hat{x}_2)) < \infty \right\} \tag{3.2}$$

where $\mathcal{B}(\mathcal{H}_c)$ is the set of bounded operators on $\mathcal{H}_c$. $\mathcal{H}_q$ is defined as the set of bounded operators, with the form $|\cdot\rangle\langle\cdot|$, acting on the classical configuration space $\mathcal{H}_c$, with a general element of the quantum Hilbert space, in "bra-ket" notation given by

$$|\psi) = \sum_{n,m=0}^\infty c_{n,m} |n, m), \tag{3.3}$$

with $\{|n, m) := |n\rangle\langle m|\}_{n,m=0}^\infty$ a basis of $\mathcal{H}_q$ endowed with the inner product

$$(\tilde{n}, \tilde{m}|n, m) = tr_c[(|\tilde{n}\rangle\langle\tilde{m}|)^\ddagger |n\rangle\langle m|] = \delta_{\tilde{n},n} \delta_{\tilde{m},m}. \tag{3.4}$$

Considering the unitary *Wigner map* $\mathcal{W} : \mathcal{B}_2(\mathfrak{H}) \to L^2(\mathbb{R}^2, dxdy)$ let us discuss a correspondence between $L^2(\mathbb{R}^2, dxdy)$ and $\mathcal{B}_2(\mathfrak{H})$.

Proposition 3.1 *Given the Hilbert space* $\mathfrak{H} = L^2(\mathbb{R})$, *the inverse of the map* $\mathcal{W}$ *is defined on the dense set of vectors* $f \in L^2(\mathbb{R}^2, dxdy)$ *as follows:*

$$\mathcal{W}^{-1} : L^2(\mathbb{R}^2, dxdy) \to \mathfrak{H} \otimes \overline{\mathfrak{H}}$$

$$\mathcal{W}^{-1} f = \int_\mathbb{R} \int_\mathbb{R} U(x, y) \mathcal{W}(|\phi\rangle\langle\psi|)(x, y) dxdy, \tag{3.5}$$

where the integral is defined weakly, $|\phi\rangle\langle\psi|$ is an element of $\mathcal{B}_2(\mathfrak{H}) \simeq \mathfrak{H} \otimes \overline{\mathfrak{H}}$ and $f = \mathcal{W}(|\phi\rangle\langle\psi|)$.

Proof Let us derive the inverse of the map $\mathcal{W}$ on $L^2(\mathbb{R}^2, dxdy)$ where the group G and the Duflo-Moore operator C with domain $\mathcal{D}(C^{-1})$ given in [3] are identified here to $\mathbb{R}^2$, $I_{\mathfrak{H}}$, the identity operator on $\mathfrak{H} = L^2(\mathbb{R})$, respectively, with $\mathcal{D}(C^{-1}) = \mathfrak{H}$ and $\mathcal{D}(C^{-1})^\dagger = \overline{\mathfrak{H}}$. Consider an element in $\mathcal{B}_2(\mathfrak{H}) \simeq \mathfrak{H} \otimes \overline{\mathfrak{H}}$ of the type $|\phi\rangle\langle\psi|$, with $\phi, \psi \in \mathfrak{H}$ and let $f = \mathcal{W}(|\phi\rangle\langle\psi|)$. For $\phi', \psi' \in \mathfrak{H}$, we have from the definition of $\mathcal{W}$ in (2.113)

$$\int_{\mathbb{R}} \int_{\mathbb{R}} \langle\phi'|U(x, y)\psi'\rangle \mathcal{W}(|\phi\rangle\langle\psi|)(x, y)dxdy$$

$$= \int_{\mathbb{R}} \int_{\mathbb{R}} \langle\phi'|U(x, y)\psi'\rangle Tr(U(x, y)^*|\phi\rangle\langle\psi|)dxdy$$

$$= \int_{\mathbb{R}} \int_{\mathbb{R}} \overline{\langle\phi|U(x, y)\psi\rangle}\langle\phi'|U(x, y)\psi'\rangle dxdy. \tag{3.6}$$

By the orthogonality relation, we get

$$\int_{\mathbb{R}} \int_{\mathbb{R}} \langle\phi'|U(x, y)\psi'\rangle \mathcal{W}(|\phi\rangle\langle\psi|)(x, y)dxdy = \langle\phi'|\phi\rangle\langle\psi|\psi'\rangle. \tag{3.7}$$

The relation $|\langle\phi'|\phi\rangle\langle\psi|\psi'\rangle| \leqslant \|\phi'\|\|\psi'\|\|\phi\|\|\psi\|$ implies

$$\left|\int_{\mathbb{R}} \int_{\mathbb{R}} \langle\phi'|U(x, y)\psi'\rangle \mathcal{W}(|\phi\rangle\langle\psi|)(x, y)dxdy\right| \leqslant \|\phi'\|\|\psi'\|\|\phi\|\|\psi\|. \tag{3.8}$$

Then, (3.7) holds for all $\phi', \psi' \in \mathfrak{H}$. Then, we obtain

$$|\phi\rangle\langle\psi| = \int_{\mathbb{R}} \int_{\mathbb{R}} U(x, y)\mathcal{W}(|\phi\rangle\langle\psi|)(x, y)dxdy \tag{3.9}$$

which completes the proof. $\square$

4   Application

Consider the motion of an electron in the xy-plane, subjected to a constant magnetic field pointing along the positive z-direction, i.e., in the symmetric gauge $\mathbf{A}^\uparrow = \left(-\frac{B}{2}y, \frac{B}{2}x\right)$, in the presence of a harmonic potential, described by the following Hamiltonian [5]

$$H_\theta = \frac{1}{2M}\left(p_x - \frac{eB}{2c}y\right)^2 + \frac{1}{2M}\left(p_y + \frac{eB}{2c}x\right)^2 + \frac{M\omega_0^2}{2}(x^2 + y^2) \tag{4.1}$$

where $\omega_c = \frac{eB}{2c}$ is the cyclotron frequency, $\Omega^2 = \omega_0^2 + \frac{\omega_c^2}{4}$, with the commutation relations:

$$[x^i, x^j] = i\theta\epsilon^{ij}, \quad [x^i, p^j] = i\hbar\delta^{ij}, \quad [p^i, p^j] = 0,$$
$$i, j = 1, 2, \quad \epsilon^{12} = -\epsilon^{21} = 1. \tag{4.2}$$

Next, introduce the dimensionless complex variables, related to the chiral decomposition of the physical model, given by

$$z_+ = \frac{1}{\sqrt{2}}(x_+^1 - ix_+^2), \qquad z_- = \frac{1}{\sqrt{2}}(x_-^1 + ix_-^2) \tag{4.3}$$

such that they satisfy, with

$$\partial_{z_+} = \frac{1}{\sqrt{2}}[\partial_{x_+^1} + i\partial_{x_+^2}], \quad \partial_{z_-} = \frac{1}{\sqrt{2}}[\partial_{x_-^1} - i\partial_{x_-^2}], \tag{4.4}$$

the relations

$$[\partial_{z_\pm}, z_\pm] = 1 = [\partial_{\bar{z}_\pm}, \bar{z}_\pm], \qquad [\partial_{z_\pm}, z_\mp] = 0 = [\partial_{\bar{z}_\mp}, \bar{z}_\pm],$$
$$[z_+, z_-] = 0 = [\partial_{z_+}, \partial_{z_-}]. \tag{4.5}$$

Set

$$A_+ = \zeta\frac{\bar{z}_+}{2} + \frac{\iota}{\zeta\hbar}p_{z_+}, \quad A_+^\dagger = \zeta\frac{z_+}{2} - \frac{\iota}{\zeta\hbar}p_{\bar{z}_+}$$

$$A_- = \zeta\frac{z_-}{2} + \frac{\iota}{\zeta\hbar}p_{\bar{z}_-}, \quad A_-^\dagger = \zeta\frac{\bar{z}_-}{2} - \frac{\iota}{\zeta\hbar}p_{z_-}, \quad \zeta = \sqrt[4]{\frac{(M\Omega/\hbar)^2}{1 - \frac{M\omega_c}{2}\theta + \left(\frac{M\Omega}{4}\theta\right)^2}}$$

$$\tag{4.6}$$

satisfying the commutation relations

$$[A_-, A_+^\dagger] = 0 = [A_+, A_-^\dagger], \quad [A_\pm, A_\pm^\dagger] = 1. \tag{4.7}$$

Then, taking $\tilde{\Omega}_\pm = \tilde{\Omega} \pm \frac{\tilde{\omega}_c}{2}$, where [15]

$$\tilde{\Omega} = \Omega\sqrt{1 - \frac{M\omega_c}{2}\theta + \left(\frac{M\Omega}{4}\theta\right)^2} \qquad \tilde{\omega}_c = \omega_c\left(1 - \left(\frac{\omega_c}{4} + \frac{\omega_0^2}{\omega_c}\right)M\theta\right), \tag{4.8}$$

the Hamiltonian H_θ is obtained as follows:

$$H_\theta = \hbar\tilde\Omega_+ \left(N_+ + \frac{1}{2}\right) + \hbar\tilde\Omega_- \left(N_- + \frac{1}{2}\right), \quad N_\pm = A_\pm^\dagger A_\pm, \quad (4.9)$$

$N_\pm$ being the number operators such that H_θ writes

$$H_\theta = H_+ \otimes \mathbb{I}_{\mathcal{H}_{q,-}} + \mathbb{I}_{\mathcal{H}_{q,+}} \otimes H_-, \quad H_\pm = \hbar\tilde\Omega_\pm \left(N_\pm + \frac{1}{2}\right). \quad (4.10)$$

The Hilbert spaces $\mathcal{H}_{q,\pm}$ are given by $\mathcal{H}_{q,\pm} = span\{|n_\pm\rangle\langle m_\pm|\}_{n_\pm,m_\pm=0}^\infty$, with $\mathbb{I}_{\mathcal{H}_{q,\pm}}$ their corresponding identity operators. The system $\{A_\pm, A_\pm^\dagger\}$ forms an irreducible set of operators on the chiral boson Fock space $\mathcal{F} = \{|n_\pm\rangle\}_{n_\pm=0}^\infty$, and has the following realization on the states

$$|n_+, n_-; m_+, m_-) := |n_+\rangle\langle m_+| \otimes |n_-\rangle\langle m_-|, \quad n_\pm, m_\pm = 0, 1, 2, \ldots, \quad (4.11)$$

of the Hilbert space $\mathcal{H}_{q,+} \otimes \mathcal{H}_{q,-}$:

$$A_+|n_+, n_-; m_+, m_-) := \sqrt{n_+}|n_+ - 1, n_-; m_+, m_-)$$
$$A_+^\dagger|n_+, n_-; m_+, m_-) := \sqrt{n_+ + 1}|n_+ + 1, n_-; m_+, m_-), \quad (4.12)$$

$$A_-|n_+, n_-; m_+, m_-) := \sqrt{n_-}|n_+, n_- - 1; m_+, m_-)$$
$$A_-^\dagger|n_+, n_-; m_+, m_-) := \sqrt{n_- + 1}|n_+, n_- + 1; m_+, m_-). \quad (4.13)$$

The operators $A_\pm$ act on the right by $A_\pm^\dagger$ by conjugation of (4.12) and (4.13). We have

$$|n_+, n_-; 0, 0) = \frac{1}{\sqrt{n_+!n_-!}} \left(A_+^\dagger\right)^{n_+} \left(A_-^\dagger\right)^{n_-} |0, 0\rangle\langle 0, 0| \quad (4.14)$$

with $|0, 0\rangle\langle 0, 0|$ standing for the vacuum state of $\mathcal{H}_{q,+} \otimes \mathcal{H}_{q,-}$.

Then, the eigenvalues of the Hamiltonian H_θ are derived from the relation

$$H_\theta \left(|n_+\rangle\langle m_+| \otimes |n_-\rangle\langle m_-|\right) = E_{n_+,n_-} \left(|n_+\rangle\langle m_+| \otimes |n_-\rangle\langle m_-|\right) \quad (4.15)$$

as follows:

$$E_{n_+,n_-} = \hbar\tilde\Omega_+ \left(n_+ + \frac{1}{2}\right) + \hbar\tilde\Omega_- \left(n_- + \frac{1}{2}\right). \quad (4.16)$$

Given a state $|m\rangle\langle n|$ on $\mathcal{H}_q \simeq \mathfrak{H} \otimes \overline{\mathfrak{H}}$, the left a_L and right b_R annihilation operators act as follows:

$$a_L|m\rangle\langle n| = (a \otimes I_{\bar{\mathfrak{H}}})|m\rangle\langle n| = a|m\rangle\langle n|I_{\bar{\mathfrak{H}}}$$
$$= \sqrt{m}|m-1\rangle\langle n|$$
$$b_R|m\rangle\langle n| = (I_{\mathfrak{H}} \otimes b)|m\rangle\langle n| = I_{\mathfrak{H}}|m\rangle\langle n|b$$
$$= \sqrt{n+1}|m\rangle\langle n+1|, \qquad (4.17)$$

where

$$a_L := a \otimes I_{\bar{\mathfrak{H}}} = \frac{1}{\sqrt{2}}\left[(Q+iP) \otimes I_{\bar{\mathfrak{H}}}\right], \quad a_L^\dagger := a^\dagger \otimes I_{\bar{\mathfrak{H}}} = \frac{1}{\sqrt{2}}\left[(Q-iP) \otimes I_{\bar{\mathfrak{H}}}\right],$$

$$(4.18)$$

$$b_R := I_{\mathfrak{H}} \otimes b = \frac{1}{\sqrt{2}}\left[I_{\mathfrak{H}} \otimes (iQ - P)\right], \quad b_R^\dagger := I_{\mathfrak{H}} \otimes b^\dagger = \frac{1}{\sqrt{2}}\left[I_{\mathfrak{H}} \otimes (-iQ - P)\right].$$

$$(4.19)$$

Q, P are the usual position and momentum operators given on $\mathfrak{H} = L^2(\mathbb{R})$, with $[Q, P] = i\mathbb{I}_{\mathfrak{H}}$.

4.1 Coherent States Construction

Using the operators $\{A_\pm, A_\pm^\dagger\}$, the eigenstates $|z_\pm\rangle$ satisfy

$$A_\pm|z_\pm\rangle = z_\pm|z_\pm\rangle, \qquad \langle z_\pm|A_\pm^\dagger = \langle z_\pm|\bar{z}_\pm \qquad (4.20)$$

with the complex eigenvalues $z_\pm$,

$$|z_\pm\rangle = e^{-\frac{|z_\pm|^2}{2}} e^{\{z_\pm A_\pm^\dagger\}}|0\rangle, \qquad (4.21)$$

given in terms of the chiral Fock basis. Provided the Baker-Campbell-Hausdorff identity

$$e^{\{z_\pm A_\pm^\dagger - \bar{z}_\pm A_\pm\}} = e^{-\frac{|z_\pm|^2}{2}} e^{\{z_\pm A_\pm^\dagger\}} e^{\{-\bar{z}_\pm A_\pm\}}, \qquad (4.22)$$

the CS of the Hamiltonian (4.10) in the noncommutative plane, denoted by $|z_+, z_-\rangle$, are defined by

$$|z_+, z_-\rangle = e^{-(|z_+|^2 + |z_-|^2)} \sum_{n_+, m_+ = 0}^{\infty} \sum_{n_-, m_- = 0}^{\infty} \frac{z_+^{n_+} \bar{z}_+^{m_+} z_-^{n_-} \bar{z}_-^{m_-}}{\sqrt{n_+! m_+! n_-! m_-!}} |n_+\rangle \langle m_+| \otimes |n_-\rangle \langle m_-|$$

(4.23)

where we have used [6]:

$$|z_+\rangle \langle z_+| = D_R D_L \left(|0\rangle \langle 0|\right) = e^{-|z_+|^2} e^{z_+ A_+^{\dagger}} |0\rangle \langle 0| e^{\bar{z}_+ A_+}$$

(4.24)

with $D_R = e_R^{-z_+ A_+^{\dagger} + \bar{z}_+ A_+}$ and $D_L = e_L^{-\bar{z}_+ A_+ + z_+ A_+^{\dagger}}$. The lower indices R, L of the exponential operators refer to the right and left actions, respectively.

These CS satisfy the resolution of the identity [15]

$$\frac{1}{\pi^2} \int_{\mathbb{C}^2} |z_+, z_-\rangle \langle z_+, z_-| d^2 z_+ d^2 z_- = \mathbb{I}_{\mathcal{H}_{q,+} \otimes \mathcal{H}_{q,-}} := \mathbb{I}_q \otimes \mathbb{I}_q,$$

(4.25)

where $\mathbb{I}_q$ stands for the identity on $\mathcal{H}_q$ given by Ben Geloun and Scholtz [22]:

$$\mathbb{I}_q = \frac{1}{\pi} \int_{\mathbb{C}} dz d\bar{z} |z\rangle e^{\overleftarrow{\partial_{\bar{z}}} \overrightarrow{\partial_z}} \langle z|.$$

(4.26)

and the identity operator on $\mathcal{H}_{q,+} \otimes \mathcal{H}_{q,-}$ is:

$$\mathbb{I}_{\mathcal{H}_{q,+} \otimes \mathcal{H}_{q,-}} = \sum_{n_+, m_+ = 0}^{\infty} \sum_{n_-, m_- = 0}^{\infty} |n_+, n_-; m_+, m_-\rangle \langle n_+, n_-; m_+, m_-|.$$

(4.27)

Proof In order to provide an equivalence between (4.25) and (4.26), let us consider the following relations

$$\begin{aligned}
\mathbb{I}_q |\psi\rangle &= \frac{1}{\pi^2} \int_{\mathbb{C}^2} dz d\bar{z} dw d\bar{w} |z\rangle \langle w| \langle z|\psi|w\rangle \\
&= \frac{1}{\pi^2} \int_{\mathbb{C}^2} dz d\bar{z} du d\bar{u} |z\rangle \langle z + u| \langle z|\psi|z + u\rangle \\
&= \frac{1}{\pi} \int_{\mathbb{C}} dz d\bar{z} \frac{1}{\pi} \int_{\mathbb{C}} d^2 u e^{-|u|^2} |z\rangle \langle z| e^{\bar{u} \overleftarrow{\partial_{\bar{z}}} + u \overrightarrow{\partial_z}} \langle z|\psi|z\rangle
\end{aligned}$$

(4.28)

where $w = z + u$ with $d^2 w = d^2 u$, and $e^{u \partial_z} f(z) = f(z + u)$. Then, set

$$\frac{1}{\pi} \int_{\mathbb{C}} d^2 u e^{-|u|^2} |z\rangle \langle z| e^{\bar{u} \overleftarrow{\partial_{\bar{z}}} + u \overrightarrow{\partial_z}} \langle z|\psi|z\rangle = \frac{1}{\pi} \int_{\mathbb{C}} d^2 u e^{-|u|^2} |z\rangle \langle z| e^{\bar{u} \overleftarrow{\partial_{\bar{z}}}} e^{u \overrightarrow{\partial_z}} \langle z|\psi|z\rangle$$

(4.29)

and

$$I = |z\rangle \langle z| e^{\bar{u} \overleftarrow{\partial_{\bar{z}}}} e^{u \overrightarrow{\partial_z}} \langle z|\psi|z\rangle. \tag{4.30}$$

We have

$$I = \left[\sum_{n',m'=0}^{\infty} |n'\rangle \langle m'| e^{-\bar{z}z} \frac{\bar{z}^{m'}}{\sqrt{m'!}} \frac{z^{n'}}{\sqrt{n'!}} \right] e^{\bar{u} \overleftarrow{\partial_{\bar{z}}}} e^{u \overrightarrow{\partial_z}} \left[\sum_{n,m=0}^{\infty} \langle m|\psi|n\rangle e^{-\bar{z}z} \frac{\bar{z}^{n}}{\sqrt{n!}} \frac{z^{m}}{\sqrt{m!}} \right]$$

$$= \left[\sum_{n,m=0}^{\infty} \sum_{n',m'=0}^{\infty} \frac{z^{n'}}{\sqrt{n'!}} \frac{\bar{z}^{n}}{\sqrt{n!}} |n'\rangle \langle m'|\langle m|\psi|n\rangle \right] \left(e^{-\bar{z}z} \frac{\bar{z}^{m'}}{\sqrt{m'!}} \right) e^{\bar{u} \overleftarrow{\partial_{\bar{z}}}} e^{u \overrightarrow{\partial_z}} \left(e^{-\bar{z}z} \frac{z^{m}}{\sqrt{m!}} \right). $$

$$\tag{4.31}$$

Let

$$K(z) = \left(e^{-\bar{z}z} \frac{\bar{z}^{m'}}{\sqrt{m'!}} \right) e^{\bar{u} \overleftarrow{\partial_{\bar{z}}}} e^{u \overrightarrow{\partial_z}} \left(e^{-\bar{z}z} \frac{z^{m}}{\sqrt{m!}} \right). \tag{4.32}$$

We obtain

$$K(z) = \frac{1}{\sqrt{m'!}} \frac{1}{\sqrt{m!}} \sum_{k=0}^{\infty} \sum_{l=0}^{\infty} \frac{1}{k!} \left(\bar{u}^k \partial_{\bar{z}}^k [\bar{z}^{m'} e^{-\bar{z}z}] \right) \frac{1}{l!} \left(u^l \partial_z^l \left[z^m e^{-\bar{z}z} \right] \right) \tag{4.33}$$

which supplies, by performing a radial parametrization,

$$\frac{1}{\pi} \int_{\mathbb{C}} d^2 u e^{-|u|^2} K(z)$$

$$= \frac{1}{\sqrt{m'!}} \frac{1}{\sqrt{m!}} \sum_{k=0}^{\infty} \sum_{l=0}^{\infty} \frac{1}{k!} \frac{1}{l!} \frac{1}{\pi} \int_{\mathbb{C}} d^2 u e^{-|u|^2} \frac{\bar{u}^k}{k!} \frac{u^l}{l!} \partial_{\bar{z}}^k [\bar{z}^{m'} e^{-\bar{z}z}] \partial_z^l \left[z^m e^{-\bar{z}z} \right]$$

$$= \frac{1}{\sqrt{m'!}} \frac{1}{\sqrt{m!}} \sum_{k=0}^{\infty} \sum_{l=0}^{\infty} \frac{1}{\pi} \int_0^{\infty} r dr e^{-r^2} \frac{r^{k+l}}{k!l!} \int_0^{2\pi} e^{-i(l-k)\phi} d\phi$$

$$\times \partial_{\bar{z}}^k [\bar{z}^{m'} e^{-\bar{z}z}] \partial_z^l \left[z^m e^{-\bar{z}z} \right]$$

$$= \frac{1}{\sqrt{m'!}} \frac{1}{\sqrt{m!}} \sum_{k=0}^{\infty} \left[\frac{1}{k!} \int_0^{\infty} 2r^{2k+1} e^{-r^2} dr \right] \left[\frac{1}{k!} \partial_{\bar{z}}^k [\bar{z}^{m'} e^{-\bar{z}z}] \partial_z^k \left[z^m e^{-\bar{z}z} \right] \right]$$

$$= \frac{1}{\sqrt{m'!}} \frac{1}{\sqrt{m!}} \sum_{k=0}^{\infty} \left[\frac{1}{k!} \partial_{\bar{z}}^k [\bar{z}^{m'} e^{-\bar{z}z}] \partial_z^k \left[z^m e^{-\bar{z}z} \right] \right]. \tag{4.34}$$

Besides,

$$\left(e^{-\bar{z}z} \frac{\bar{z}^{m'}}{\sqrt{m'!}} \right) e^{\overleftarrow{\partial_{\bar{z}}} \overrightarrow{\partial_z}} \left(e^{-\bar{z}z} \frac{z^m}{\sqrt{m!}} \right) = \frac{1}{\sqrt{m'!}} \frac{1}{\sqrt{m!}} \sum_{k=0}^{\infty} \left[\frac{1}{k!} \partial_{\bar{z}}^k [\bar{z}^{m'} e^{-\bar{z}z}] \partial_z^k \left[z^m e^{-\bar{z}z} \right] \right]$$

(4.35)

implying

$$\frac{1}{\pi} \int_{\mathbb{C}} d^2 u e^{-|u|^2} K(z) = \left(e^{-\bar{z}z} \frac{\bar{z}^{m'}}{\sqrt{m'!}} \right) e^{\overleftarrow{\partial_{\bar{z}}} \overrightarrow{\partial_z}} \left(e^{-\bar{z}z} \frac{z^m}{\sqrt{m!}} \right). \qquad (4.36)$$

Then,

$$\frac{1}{\pi} \int_{\mathbb{C}} d^2 u e^{-|u|^2} |z\rangle \langle z| e^{\bar{u}\overleftarrow{\partial_{\bar{z}}} + u \overrightarrow{\partial_z}} \langle z|\psi|z\rangle$$

$$= \left[\sum_{n,m=0}^{\infty} \sum_{n',m'=0}^{\infty} \frac{z^{n'}}{\sqrt{n'!}} \frac{\bar{z}^n}{\sqrt{n!}} |n'\rangle \langle m'|\langle m|\psi|n\rangle \right] \left(e^{-\bar{z}z} \frac{\bar{z}^{m'}}{\sqrt{m'!}} \right) e^{\overleftarrow{\partial_{\bar{z}}} \overrightarrow{\partial_z}} \left(e^{-\bar{z}z} \frac{z^m}{\sqrt{m!}} \right)$$

$$= \left[\sum_{n',m'=0}^{\infty} |n'\rangle \langle m'| e^{-\bar{z}z} \frac{\bar{z}^{m'}}{\sqrt{m'!}} \frac{z^{n'}}{\sqrt{n'!}} \right] e^{\overleftarrow{\partial_{\bar{z}}} \overrightarrow{\partial_z}} \left[\sum_{n,m=0}^{\infty} \langle m|\psi|n\rangle e^{-\bar{z}z} \frac{\bar{z}^n}{\sqrt{n!}} \frac{z^m}{\sqrt{m!}} \right]$$

$$= |z\rangle e^{\overleftarrow{\partial_{\bar{z}}} \overrightarrow{\partial_z}} \langle z|\psi\rangle \qquad (4.37)$$

allowing to obtain (4.28) under the form:

$$\mathbb{I}_q |\psi\rangle = \frac{1}{\pi} \int_{\mathbb{C}} dz d\bar{z} \frac{1}{\pi} \int_{\mathbb{C}} d^2 u e^{-|u|^2} |z\rangle \langle z| e^{\bar{u}\overleftarrow{\partial_{\bar{z}}} + u \overrightarrow{\partial_z}} \langle z|\psi|z\rangle$$

$$= \frac{1}{\pi} \int_{\mathbb{C}} dz d\bar{z} |z\rangle e^{\overleftarrow{\partial_{\bar{z}}} \overrightarrow{\partial_z}} \langle z|\psi\rangle \qquad (4.38)$$

which completes the proof.   □

4.2 Density Matrix and Diagonal Elements

Considering that the quantum system obeys the canonical distribution [5, 9, 12], let us take the partition function $\mathcal{Z}$ as that of a composite system made of two independent systems such that it is the product of the partition functions of the components, i.e. $\mathcal{Z} = \mathcal{Z}_+ \mathcal{Z}_-$. The diagonal elements of the density operator $\hat{\rho} = \frac{1}{\mathcal{Z}} e^{-\beta H_\theta}$, where H_θ is given in (4.9) with eigenvalues E_{n_+,n_-} in (4.16), are then derived, in the CS $|z_+, z_-\rangle$ (4.23) representation, as

$$(z_+, z_- | \hat{\rho} | z_+, z_-)$$

$$= (z_+, z_- | \left\{ \sum_{n_+, m_+ = 0}^{\infty} \sum_{n_-, m_- = 0}^{\infty} \frac{e^{-\beta H_\theta}}{\mathcal{Z}} |n_+, n_-; m_+, m_-)(n_+, n_-; m_+, m_-| \right\} | z_+, z_-)$$

$$= \left\{ \frac{1}{\mathcal{Z}_+} e^{-\frac{\beta \hbar \tilde{\Omega}_+}{2}} e^{-|z_+|^2} \left[e^{e^{-\beta \hbar \tilde{\Omega}_+} |z_+|^2} \right] \right\} \left\{ \frac{1}{\mathcal{Z}_-} e^{-\frac{\beta \hbar \tilde{\Omega}_-}{2}} e^{-|z_-|^2} \left[e^{e^{-\beta \hbar \tilde{\Omega}_-} |z_-|^2} \right] \right\} \quad (4.39)$$

where $\mathcal{Z} = \mathcal{Z}_+ \mathcal{Z}_-$.

Since

$$\frac{1}{\mathcal{Z}_+} = \left[\frac{e^{-\frac{\beta \hbar \tilde{\Omega}_+}{2}}}{1 - e^{-\beta \hbar \tilde{\Omega}_+}} \right]^{-1}, \quad \text{and} \quad \frac{1}{\mathcal{Z}_-} = \left[\frac{e^{-\frac{\beta \hbar \tilde{\Omega}_-}{2}}}{1 - e^{-\beta \hbar \tilde{\Omega}_-}} \right]^{-1} \quad (4.40)$$

then,

$$(z_+, z_- | \hat{\rho} | z_+, z_-) = \left[\frac{e^{-\frac{\beta \hbar \tilde{\Omega}_+}{2}}}{1 - e^{-\beta \hbar \tilde{\Omega}_+}} \right]^{-1} e^{-\frac{\beta \hbar \tilde{\Omega}_+}{2}} e^{-|z_+|^2} \left[e^{e^{-\beta \hbar \tilde{\Omega}_+} |z_+|^2} \right]$$

$$\times \left[\frac{e^{-\frac{\beta \hbar \tilde{\Omega}_-}{2}}}{1 - e^{-\beta \hbar \tilde{\Omega}_-}} \right]^{-1} e^{-\frac{\beta \hbar \tilde{\Omega}_-}{2}} e^{-|z_-|^2} \left[e^{e^{-\beta \hbar \tilde{\Omega}_-} |z_-|^2} \right].$$

$$(4.41)$$

Thereby,

$$(z_+, z_- | \hat{\rho} | z_+, z_-)$$

$$= \left[1 - e^{-\beta \hbar \tilde{\Omega}_+} \right] e^{-(1 - e^{-\beta \hbar \tilde{\Omega}_+}) |z_+|^2} \times \left[1 - e^{-\beta \hbar \tilde{\Omega}_-} \right] e^{-(1 - e^{-\beta \hbar \tilde{\Omega}_-}) |z_-|^2}$$

$$= \frac{1}{\bar{n} + 1} e^{-\frac{1}{\bar{n} + 1} |z_+|^2} \times \frac{1}{\bar{n}^* + 1} e^{-\frac{1}{\bar{n}^* + 1} |z_-|^2}$$

$$= Q(|z_+|^2) Q(|z_-|^2). \quad (4.42)$$

$\bar{n} = \left[e^{\beta \hbar \tilde{\Omega}_+} - 1 \right]^{-1}$ and $\bar{n}^* = \left[e^{\beta \hbar \tilde{\Omega}_-} - 1 \right]^{-1}$ are the corresponding thermal expectation values of the number operator (i. e. the Bose-Einstein distribution functions for oscillators with angular frequencies $\tilde{\Omega}_+$ and $\tilde{\Omega}_-$, respectively), also called the thermal mean occupancy for harmonic oscillators with the angular frequencies $\tilde{\Omega}_+$ and $\tilde{\Omega}_-$, respectively.

Performing the variable changes $r_+ = \left[1 - e^{-\beta\hbar\tilde{\Omega}_+}\right]^{1/2} |z_+|$ and $r_- = \left[1 - e^{-\beta\hbar\tilde{\Omega}_-}\right]^{1/2} |z_-|$ with $\frac{d^2z}{\pi} = r\,dr\frac{d\varphi}{\pi}$, $r \in [0, \infty)$, $\varphi \in (0, 2\pi]$, we obtain

$$
\begin{aligned}
Tr\hat{\rho} &= \frac{1}{\pi^2} \int_{\mathbb{C}^2} d^2z_+ d^2z_- (z_+, z_- |\hat{\rho}| z_+, z_-) \\
&= \frac{1}{\pi^2} \int_{\mathbb{C}^2} d^2z_+ d^2z_- \left[1 - e^{-\beta\hbar\tilde{\Omega}_+}\right] e^{-(1-e^{-\beta\hbar\tilde{\Omega}_+})|z_+|^2} \left[1 - e^{-\beta\hbar\tilde{\Omega}_-}\right] e^{-(1-e^{-\beta\hbar\tilde{\Omega}_-})|z_-|^2} \\
&= \frac{1}{\pi^2} \int_0^\infty r_+ dr_+ \left[e^{-r_+^2}\right] \int_0^{2\pi} d\theta_+ \times \int_0^\infty r_- dr_- \left[e^{-r_-^2}\right] \int_0^{2\pi} d\theta_- \\
&= 1,
\end{aligned}
\tag{4.43}
$$

with here $n_\pm = 0$, where we have used the following integral

$$
\int_0^\infty \frac{1}{n_\pm!} 2r_\pm^{2n_\pm+1} dr_\pm e^{-r_\pm^2} = 1,
\tag{4.44}
$$

ensuring that the normalization condition of the density matrix is accomplished. The right-hand side of (4.42) corresponds to the product of two harmonic oscillators Husimi distributions [16].

4.3 Lowest Landau Levels and Reproducing Kernel

Let us make a relationship between the quantum numbers $n_\pm \in \mathbb{N}$ which label the energy levels per sector and the quantum numbers n, m where n labels the levels and m describes the degeneracy [4]. Fixing $n_- = 0$ (resp. $n = 0$), one obtains a state corresponding to the quantum number m in the lowest Landau level (LLL) given by

$$
\phi_{n=0,m}(z_+, \bar{z}_+) = \frac{1}{\sqrt{2\pi l_0^2 m!}} \left(\frac{z_+}{\sqrt{2}l_0}\right)^m e^{-|z_+|^2/4l_0^2}
\tag{4.45}
$$

where $l_0 = \sqrt{\frac{1}{eB}} \equiv 1$ (with $\hbar = 1$, $e = 1$) is the scale of lengths associated with the Landau problem.

Equivalently, fixing $n_+ = 0$ (resp. $m = 0$), one gets a state centered at the origin ($m = 0$) in the Landau level n given by

$$
\phi_{n,m=0}(z_-, \bar{z}_-) = \frac{1}{\sqrt{2\pi l_0^2 n!}} \left(\frac{\bar{z}_-}{\sqrt{2}l_0}\right)^n e^{-|z_-|^2/4l_0^2}.
\tag{4.46}
$$

Consider the projector onto the LLL given by

$$\mathbb{P}_0 = \sum_{m=0}^{\infty} |0, m)(0, m|. \tag{4.47}$$

In the LLL $|0, m)$, the state $|0, \bar{z}_+)$, where $z_+ = x_+^1 - ix_+^2$, is such that

$$(0, \bar{z}_+|0, m) = e^{-\frac{|\bar{z}_+|^2}{2}} \frac{\bar{z}_+^m}{\sqrt{m!}}, \qquad \bar{z}_+ = \frac{z_+}{l_0\sqrt{2}}. \tag{4.48}$$

and also

$$\overline{(0, \bar{z}_+|0, m)} = (0, m|0, \bar{z}_+) = e^{-\frac{|\bar{z}_+|^2}{2}} \frac{\bar{z}_+^m}{\sqrt{m!}}. \tag{4.49}$$

The matrix elements of the projector $\mathbb{P}_0$ are obtained as

$$(0, \bar{z}_+|\mathbb{P}_0|0, \bar{z}'_+) = e^{-\frac{1}{2}[|\bar{z}'_+|^2+|\bar{z}_+|^2-2\bar{z}_+\bar{z}'_+]}. \tag{4.50}$$

Let $|\psi) \in \mathcal{H}_q$, a state given on the LLL by

$$|\psi) = \sum_{m=0}^{\infty} a_m |0, m), \ a_m \in \mathbb{C}. \tag{4.51}$$

We obtain that $|\psi)$ is analytic up to the Landau gaussian factor $e^{-\frac{|\bar{z}_+|^2}{2}}$ as follows:

$$(0, \bar{z}_+|\psi) = e^{-\frac{|\bar{z}_+|^2}{2}} f(\bar{z}_+), \qquad f(\bar{z}_+) = \sum_{m=0}^{\infty} a_m \frac{\bar{z}_+^m}{\sqrt{m!}} \in L_{hol}^2(\mathbb{C}, dv(z, \bar{z})) \tag{4.52}$$

with $dv(z, \bar{z}) = \frac{e^{-|z|^2}}{2\pi} \frac{dz \wedge d\bar{z}}{i}$. Next, let us define the projection operator

$$\mathbb{P}_{hol} : L^2(\mathbb{C}, dv(z, \bar{z})) \longrightarrow L_{hol}^2(\mathbb{C}, dv(z, \bar{z})) \tag{4.53}$$

which is an integral operator with the reproducing kernel

$$K(\bar{z}_+, \bar{z}'_+) = e^{\frac{1}{2}[|\bar{z}'_+|^2+|\bar{z}_+|^2]}(0, \bar{z}_+|\mathbb{P}_0|0, \bar{z}'_+) = e^{\bar{z}_+ + \bar{z}'_+} \tag{4.54}$$

for $L_{hol}^2(\mathbb{C}, dv(z, \bar{z}))$ [3]. $L_{hol}^2(\mathbb{C}, dv(z, \bar{z}))$ is the subspace of the Hilbert space $L^2(\mathbb{C}, dv(z, \bar{z}))$ of dv-square integrable holomorphic functions on $\mathbb{C}$

in the variable z. Then, given an operator $\mathcal{O}$ on $L^2(\mathbb{C}, dv(z, \bar{z}))$ and $f \in L^2_{hol}(\mathbb{C}, dv(z, \bar{z}))$, we have

$$\frac{1}{\pi} \int_{\mathbb{C}} e^{\tilde{z} + \tilde{\bar{z}}'_+}(\mathcal{O}f)(\tilde{z}'_+) e^{-|\tilde{z}'_+|^2} d^2 \tilde{z}'_+ = \int_{\mathbb{C}} e^{\tilde{z} + \tilde{\bar{z}}'_+}(\mathcal{O}f)(\tilde{z}'_+) e^{-|\tilde{z}'_+|^2} \frac{d\tilde{z}'_+ \wedge d\tilde{\bar{z}}'_+}{2i\pi}$$
$$=: (\mathbb{P}_{hol}\mathcal{O}f)(\tilde{z}_+). \tag{4.55}$$

4.4 Statistical Properties

Let us consider the operators given on $\mathcal{H}_q \otimes \mathcal{H}_q$ by

$$\hat{P}_X = \frac{-i\hbar}{\sqrt{2\theta}}[a_R - a_R^\dagger, \, .] \qquad \hat{P}_Y = \frac{-\hbar}{\sqrt{2\theta}}[a_R + a_R^\dagger, \, .] \tag{4.56}$$

$$\hat{X} = \sqrt{\frac{\theta}{2}}[a_R + a_R^\dagger] \qquad \hat{Y} = i\sqrt{\frac{\theta}{2}}[a_R^\dagger - a_R]. \tag{4.57}$$

From (4.17), we obtain in a state $|\tilde{n}\rangle\langle\tilde{m}| \otimes |m\rangle\langle n| \in \mathcal{H}_q \otimes \mathcal{H}_q$

$$[a_R - a_R^\dagger, \, |\tilde{n}\rangle\langle\tilde{m}| \otimes |m\rangle\langle n|] = \sqrt{n+1}|\tilde{n}\rangle\langle\tilde{m}| \otimes |m\rangle\langle n+1| - \sqrt{n}|\tilde{n}\rangle\langle\tilde{m}| \otimes |m\rangle\langle n|. \tag{4.58}$$

We get the following expressions:

$$(\Delta\hat{X})^2 = \frac{\theta}{2}, \qquad (\Delta\hat{Y})^2 = \frac{\theta}{2} \tag{4.59}$$

$$(\Delta\hat{P}_X)^2 = \frac{\hbar^2}{\theta}, \qquad (\Delta\hat{P}_Y)^2 = \frac{\hbar^2}{\theta} \tag{4.60}$$

leading to the following uncertainties:

$$\begin{aligned}
[\Delta\hat{X}\Delta\hat{Y}]^2 &= \frac{\theta^2}{4} = \frac{1}{4}|\langle[\hat{X}, \hat{Y}]\rangle|^2, \\
[\Delta\hat{X}\Delta\hat{P}_X]^2 &= \frac{\hbar^2}{2} \geqslant \frac{1}{4}|\langle[\hat{X}, \hat{P}_X]\rangle|^2, \\
[\Delta\hat{Y}\Delta\hat{P}_Y]^2 &= \frac{\hbar^2}{2} \geqslant \frac{1}{4}|\langle[\hat{Y}, \hat{P}_Y]\rangle|^2, \\
[\Delta\hat{P}_X\Delta\hat{P}_Y]^2 &= \frac{\hbar^4}{4\theta^2} \geqslant \frac{1}{4}|\langle[\hat{P}_X, \hat{P}_Y]\rangle|^2 = 0. \tag{4.61}
\end{aligned}$$

5   Concluding Remarks

We have first dealt with some preliminaries about definitions, and remarkable properties on Hilbert-Schmidt operators and the Tomita-Takesaki modular theory. Then, the construction of CS built from the thermal state has been achieved and discussed, with the resolution of the identity. Besides, some detailed proofs have been provided in the study of the modular theory and Hilbert-Schmidt operators. The relation between the noncommutative quantum mechanics formalism and the modular theory, both using Hilbert-Schmidt operators, has been evidenced by the use of the Wigner map as an interplay between them. The formalism has been illustrated with the physical model of a charged particle on the flat plane xy in the presence of a constant magnetic field along the z-axis with a harmonic potential. CS have been constructed. Then, the density matrix, the projection onto the lowest Landau level (LLL), and main statistical properties have been discussed on the CS basis.

Acknowledgements This work is supported by TWAS Research Grant RGA No.17-542 RG/MATHS/AF/AC_G -FR3240300147. The ICMPA-UNESCO Chair is in partnership with Daniel Iagolnitzer Foundation (DIF), France, supporting the development of mathematical physics in Africa.

References

1. S.T. Ali, F. Bagarello, Some physical appearances of vector coherent states and coherent states related to degenerate Hamiltonians. J. Math. Phys. **46**, 053518 (2005)
2. S.T. Ali, F. Bagarello, G. Honnouvo, Modular structures on trace class operators and applications to Landau levels. J. Phys. A: Math. Theor. **43**, 105202 (2010); S.T. Ali, An interesting modular structure associated to Landau levels. J. Phys: Conf. Ser. **237**, 012001 (2010)
3. S.T. Ali, J.P. Antoine, J.P. Gazeau, *Coherent States, Wavelets and their Generalizations.* Theoretical and Mathematical Physics, 2nd edn. (Springer, New York, 2014)
4. I. Aremua, M.N. Hounkonnou, Coherent states for the exotic Landau model and related properties (unpublished work)
5. I. Aremua, M.N. Hounkonnou, E. Baloïtcha, Coherent states for Landau levels: algebraic and thermodynamical properties. Rep. Math. Phys. **76**(2), 247–269 (2015)
6. J. Ben Geloun, F.G. Scholtz, Coherent states in noncommutative quantum mechanics. J. Math. Phys. **50**, 043505 (2009)
7. P. Bertozzini, Non-commutative geometries via modular theory, in *RIMS International Conference on Noncommutative Geometry in Physics*, Kyoto (2010)
8. O. Bratelli, D.W. Robinson, *Operator Algebras and Quantum Statistical Mechanics*, vol. 1 (Springer, Berlin/Heidelberg, 2002); O. Bratelli, D.W. Robinson, *Operator Algebras and Quantum Statistical Mechanics*, vol. 2 (Springer, Berlin/Heidelberg, 2002)
9. K.E. Cahill, R.J. Glauber, Density operators and quasiprobability distributions. Phys. Rev. **177**, 1882 (1969)

10. A. Connes, Groupe modulaire d'une algèbre de von Neumann. C. R. Acad. Sci. Paris Série A **274**, 1923–1926 (1972); A. Connes, Caractérisation des algèbres de von Neumann comme espaces vectoriels ordonnés. Annales de l'Institut Fourier **24**, 121–155 (1974); A. Connes, E. Størmer, Entropy for automorphisms of finite von Neumann algebras. Acta Math. **134**, 289–306 (1975); A. Connes, M. Takesaki, The flow of weights on factors of type III. Tohoku Math. J. **29**, 473–575 (1977); A. Connes, Von Neumann algebras, *Proceedings of the International Congress of Mathematicians Helsinki*, pp. 97–109 (1978); A. Connes, C. Rovelli, Von Neumann algebra automorphisms and time-thermodynamics relation in generally covariant quantum theories. Class. Quantum Grav. **11**, 2899–2917 (1994)

11. A. Connes, Noncommutative differential geometry. Inst. Hautes Etudes Sci. Publ. Math. **62**, 257–360 (1985); A. Connes, *Noncommutative Geometry* (Academic Press, San Diego, CA, 1994); A. Connes, M. Douglas and A. Schwarz, JHEP **02**003; [e-print hep-th/9711162] (1998).

12. J.P. Gazeau, *Coherent States in Quantum Physics* (Wiley, Berlin, 2009)

13. M.O. Goerbig, P. Lederer, C.M. Smith, Competition between quantum-liquid and electron-solid phases in intermediate Landau levels. Phys. Rev. B **69**, 115327 (2004)

14. H. Grosse, P. Prešnajder, The construction on noncommutative manifolds using coherent states. Lett. Math. Phys. **28**, 239 (1993)

15. M.N. Houkonnou, I. Aremua, Landau Levels in a two-dimensional noncommutative space: matrix and quaternionic vector coherent states. J. Nonlinear Math. Phys. **19**, 1250033 (2012)

16. K. Husimi, Some formal properties of the density matrix. Proc. Phys. Math. Soc. Jpn. **22**, 264 (1940)

17. J.R. Klauder, B.S. Skagerstam, *Coherent States, Applications in Physics and Mathematical Physics* (World Scientific, Singapore, 1985)

18. L.D. Landau, E.M. Lifshitz, *Quantum Mechanics, Non-relativistic Theory* (Oxford, Pergamon, 1977)

19. F.J. Murray, J.v. Neumann, On rings of operators. Ann. Math. **37**, 116–229 (1936)

20. A.M. Perelomov, *Generalized Coherent States and Their Applications* (Springer, Berlin, 1986)

21. E. Prugovečki, *Quantum Mechanics in Hilbert Spaces*, 2nd edn. (Academic Press, New York, 1981)

22. F.G. Scholtz, L. Gouba, A. Hafver, C.M. Rohwer, Formulation, interpretation, and applications of non-commutative quantum mechanics. J. Phys. A: Math. Theor. **42**, 175303 (2009)

23. E. Schrödinger, Der stetige Übergang von der Mikro-zur Makromechanik. Naturwissenschaften **14**, 664 (1926)

24. S.J. Summers, Tomita-Takesaki Modular Theory (2005), arxiv: math-ph/0511034v1

25. M. Takesaki, *Tomita's Theory of Modular Hilbert Algebras and Its Applications* (Springer, New York, 1970); M. Takesaki, *Theory of Operator Algebras. I.* Encyclopaedia of Mathematical Sciences, vol. 124 (Springer, Berlin, 2002); Reprint of the first (1979) edition, *Operator Algebras and Non-commutative Geometry*, **5**; M. Takesaki, *Structure of Factors and Automorphism Groups*. CBMS Regional Conference Series in Mathematics, vol. 51 (American Mathematical Society, Providence, 1983); M. Takesaki, *Theory of Operator Algebras. II.* Encyclopaedia of Mathematical Sciences, vol. 125 (Springer, Berlin, 2003); *Operator Algebras and Non-commutative Geometry*, vol. 6; M. Takesaki, *Theory of Operator Algebras. III.* Encyclopaedia of Mathematical Sciences, vol. 127 (Springer, Berlin, 2003); *Operator Algebras and Non-commutative Geometry*, vol. 8

26. M. Tomita, Standard forms of von Neumann algebras, in *V-th Functional Analysis Symposium of the Mathematical Society of Japan*, Sendai (1967)

27. J. von Neumann, On rings of operators III. Ann. Math. **41**, 94–161 (1940)

Symplectic Affine Action and Momentum with Cocycle

Augustin Batubenge and Wallace Haziyu

Abstract Let G be a Lie group, $\mathfrak{g}$ its Lie algebra, and $\mathfrak{g}^*$ the dual of $\mathfrak{g}$. Let Φ be the symplectic action of G on a symplectic manifold (M, ω). If the momentum mapping $\mu : M \to \mathfrak{g}^*$ is not Ad^*-equivariant, it is a fact that one can modify the coadjoint action of G on $\mathfrak{g}^*$ in order to make the momentum mapping equivariant with respect to the new G-structure in $\mathfrak{g}^*$, and the orbit of the coadjoint action is a symplectic manifold. With the help of a two cocycle $\sum : \mathfrak{g} \times \mathfrak{g} \to \mathbf{R}$, $(\xi, \eta) \mapsto \sum(\xi, \eta) = d\hat{\sigma}_\eta(e) \cdot \xi$ associated with one cocycle $\sigma : G \to \mathfrak{g}^*$; $\sigma(g) = \mu(\phi_g(m)) - Ad_g^*\mu(m)$, we show that a symplectic structure can be defined on the orbit of the affine action $\Psi(g, \beta) := Ad_g^*\beta + \sigma(g)$ of G on $\mathfrak{g}^*$, the orbit of which is a symplectic manifold with the symplectic structure $\omega_\beta(\xi_{\mathfrak{g}^*}(v), \eta_{\mathfrak{g}^*}(v)) = -\beta([\xi, \eta]) + \sum(\eta, \xi)$.

Furthermore, we introduce a deformed Poisson bracket on (M, ω) with which some classical results of conservative mechanics still hold true in a new setting.

Keywords Symplectic action · Momentum mapping · Equivariance · Poisson bracket

1 Introduction

The study of coadjoint orbits was introduced by Kirillov in the 1960s (see [1]). Coadjoint orbits arise through the action of a Lie group G by means of a coadjoint representation Ad^* on the dual $\mathfrak{g}^*$ of the Lie algebra $\mathfrak{g}$ of G. The orbits so obtained by this action are called coadjoint orbits (see [2, 3]). Kostant and Souriau showed

A. Batubenge (✉)
Départment de Mathématiques et Statistique, Université de Montréal, Montréal, QC, Canada

University of Zambia,, Lusaka, Zambia

W. Haziyu
Department of Mathematics and Statistics, University of Zambia, Lusaka, Zambia
e-mail: whaziyu@unza.zm

© Springer Nature Switzerland AG 2018
T. Diagana, B. Toni (eds.), *Mathematical Structures and Applications*,
STEAM-H: Science, Technology, Engineering, Agriculture,
Mathematics & Health, https://doi.org/10.1007/978-3-319-97175-9_4

that there is (up to covering) an isomorphism between a symplectic manifold (M, ω), homogeneous under the action of a Lie group G and a coadjoint orbit [3]. We shall now show that if we substitute the coadjoint action of G on $\mathfrak{g}^*$ by another action defined through a one-cocycle σ, the orbits of the affine action of G on $\mathfrak{g}^*$ so obtained are symplectic manifolds. Notice that the cocycle map of a momentum mapping measures its lack of equivariance [2, p. 279], and as stated by Iglesias-Zemmour P., one co-cycles arise in different ways, see [4, p. 323].

The work is organized as follows.

In Sect. 2 we gather the basics on symplectic actions on a symplectic manifold. In Sect. 3 we recall the definition of Hamiltonian action, inducing the key concept of momentum mapping with one cocycle, which induces another action, making the momentum mapping be equivariant. To end the section we will construct a symplectic structure on coadjoint orbits of an affine action. In Sect. 4 we will provide a way forward on a deformation of the standard Poisson bracket on the algebra of smooth functions on a symplectic manifold for further investigations on this topic, which opens possible applications to theoretical physics.

Note that in this work, the mere topological assumptions are assumed. The contents are well detailed in our main reference literature (see [2]), as are most notations.

2 Lie Group Action

2.1 Preliminaries

Definition 2.1 A Lie group is a group G that is also a smooth manifold such that the group operations of multiplication $G \times G \to G$, defined by $(x, y) \mapsto xy$, and inversion $G \to G$ defined by $x \mapsto x^{-1}$, are compatible with the smooth structure.

The vector space $\mathfrak{g} = T_e G$ is called the Lie algebra of the corresponding Lie group G, where $e \in G$ is the identity element. We denote the dual of $LieG$ by $\mathfrak{g}^*$.

Definition 2.2 If G is a group and X a set, the map $\Phi : G \times X \to X$ is called an action of G on X if the following two conditions are satisfied:

(i) If e is the identity element of G, then $\Phi(e, x) = x$ for all $x \in X$.
(ii) If $g, h \in G$ then $\Phi(g, \Phi(h, x)) = \Phi(gh, x)$ for all $x \in X$.

Note that for each $g \in G$, $\Phi_g : X \to X$ defined by $\Phi_g(x) = \Phi(g, x)$ is a diffeomorphism. For more on Lie group actions, we refer the reader to [5].

Definition 2.3 Let M be a smooth manifold. A symplectic structure on M is a 2-form on M ($\omega \in \Omega^2(M)$) which satisfies the following two conditions:

(i) ω is closed. That is $d\omega = 0$.
(ii) ω is nondegenerate. That is, $\omega(X, Y) = 0$ for all $Y \in \mathfrak{X}(M)$ implies that $X = 0$. In other words, on each tangent space $T_m M$, $m \in M$, if $\omega_m(X_m, Y_m) = 0$ for all $Y_m \in T_m M$, then $X_m = 0$.

A manifold M is called a symplectic manifold if there is defined on M a closed 2-form ω which is nondegenerate.

Let $\Phi : G \times M \to M$, $(g, m) \mapsto \Phi_g(m) = g \cdot m$ be an action of a Lie group G on a symplectic manifold (M, ω). Then the action Φ is called symplectic if for each $g \in G$, the diffeomorphism $\Phi_g : M \to M$, $m \mapsto \Phi_g(m)$ is such that $\Phi_g^* \omega = \omega$. Details on these preliminaries can be found in [1–3].

Let G be a Lie group and let Φ be an action of G on a manifold M. Let $\mathfrak{g}$ be the Lie algebra of G. We define the infinitesimal generator of the action Φ corresponding to $X \in \mathfrak{g}$ to be

$$X_M(m) = \frac{d}{dt} \Phi_{\exp t X}(m) \mid_{t=0},$$

where $\exp : \mathfrak{g} \to G$ is the exponential map.

3 Momentum Mapping

Definition 3.1 Let $\Phi : G \times M \to M$, be a symplectic action of a Lie group G on a symplectic manifold (M, ω), and let X_M be the infinitesimal generator of the action corresponding to $X \in \mathfrak{g}$. Then the map

$$\mu : M \to \mathfrak{g}^*$$

is called the **momentum mapping** for the action if for every $X \in \mathfrak{g}$ there is a function $\hat{\mu}_X : M \to \mathbb{R}$ such that the relation $\hat{\mu}_X(m) = \mu(m) \cdot X$ holds, and where $d\hat{\mu}_X = i_{X_M}\omega$.

Definition 3.2 The space (M, ω, Φ, μ) is called a Hamiltonian G-space.

3.1 Coadjoint Cocycle

Definition 3.3 Let G be a Lie group, $\mathfrak{g}$ its Lie algebra, and $\mathfrak{g}^*$ the dual of its Lie algebra. The function $\sigma : G \to \mathfrak{g}^*$ defined by

$\sigma(g) = \mu(\Phi_g(m)) - Ad_g^*\mu(m)$ for all $m \in M$ is called a coadjoint cocycle on G or simply one-cocycle, where we define $< Ad_g^*\beta, \xi >=< \beta, Ad_{g^{-1}}\xi >$. See [6].

The map σ satisfies the cocycle identity

$$\sigma(gh) = \sigma(g) + Ad_g^*\sigma(h) \qquad (3.1)$$

for all $g, h \in G$.

To see this, from $\sigma(g) = \mu(\Phi_g(m)) - Ad_g^*\mu(m)$ we have

$$\begin{aligned}
\sigma(gh) &= \mu(\Phi_{gh}(m)) - Ad_{gh}^*\mu(m) \\
&= \mu((\Phi_g \circ \Phi_h)(m)) - (Ad_g^* \circ Ad_h^*)(\mu(m)) \\
&= \mu(\Phi_g(\Phi_h(m))) - Ad_g^*(Ad_h^*(\mu(m))) \\
&= \mu(\Phi_g(\Phi_h(m))) - Ad_g^*(\mu(\Phi_h(m))) \\
&\quad + Ad_g^*(\mu(\Phi_h(m))) - Ad_g^*(Ad_h^*\mu(m)) \\
&= \sigma(g) + Ad_g^*(\mu(\Phi_h(m)) - Ad_h^*\mu(m)) \\
&= \sigma(g) + Ad_g^*\sigma(h)
\end{aligned}$$

as required.

Proposition 3.4 *Let Φ be a symplectic action of a Lie group G on a symplectic manifold (M, ω) which admits a momentum mapping μ. Let σ be a one-cocycle. Define a map*

$$\Psi : G \times \mathfrak{g}^* \to \mathfrak{g}^*$$

by

$$\Psi(g, \alpha) = Ad_g^*\alpha + \sigma(g).$$

Then the map Ψ is an action and the momentum mapping is equivariant with respect to this action.

Proof First we need to check that the conditions of an action are satisfied. From the definition

$$\sigma(g) = \mu(\Phi_g(m)) - Ad_g^*\mu(m),$$

we have

$$\begin{aligned}
\sigma(e) &= \mu(\Phi_e(m)) - Ad_e^*\mu(m) \\
&= \mu(m) - \mu(m) = 0
\end{aligned}$$

since Ad_e^* is the identity map. Thus

$$\Psi(e, \alpha) = Ad_e^* \alpha + \sigma(e) = \alpha,$$

i.e.

$$\Psi(e, \alpha) = \alpha.$$

Using the cocycle identity (3.1) above we have

$$
\begin{aligned}
\Psi(gh, \alpha) &= Ad_{gh}^* \alpha + \sigma(gh) \\
&= Ad_g^*(Ad_h^* \alpha) + \sigma(g) + Ad_g^* \sigma(h) \\
&= Ad_g^*(Ad_h^* \alpha + \sigma(h)) + \sigma(g) \\
&= \sigma(g) + Ad_g^*(\Psi(h, \alpha)) \\
&= \Psi(g, \Psi(h, \sigma)).
\end{aligned}
$$

Hence Ψ is an action. To see that the momentum mapping is equivariant with respect to this action, we have

$$
\begin{aligned}
\mu(\Phi_g(m)) - \Psi(g, \mu(m)) &= \mu(\Phi_g(m)) - (Ad_g^* \mu(m) + \sigma(g)) \\
&= (\mu(\Phi_g(m)) - Ad_g^* \mu(m)) - \sigma(g) \\
&= \sigma(g) - \sigma(g) = 0.
\end{aligned}
$$

Thus,

$$\mu(\Phi_g(m)) = \Psi(g, \mu(m))$$

This concludes the proof of the proposition.   □

Note that the action

$$\Psi : G \times \mathfrak{g}^* \to \mathfrak{g}^*$$

is an affine action. That it is equivariant is illustrated by the following example.

Example 3.5 Consider the Lie group $G = \mathbf{R}^4$ with coordinates $(\eta_1, \eta_2, \eta_3, \eta_4)$ under addition. Let $M = \mathbf{R}^4$ be the symplectic manifold with coordinates (x_1, x_2, x_3, x_4) with symplectic structure $\omega = dx_1 \wedge dx_3 + dx_2 \wedge dx_4$. Let $\Phi : G \times M \to M$ be the action of G on M defined by

$$\Phi_{(\eta_1, \eta_2, \eta_3, \eta_4)}(x_1, x_2, x_3, x_4) = (x_1 + \eta_1, x_2 + \eta_2, x_3 + \eta_3, x_4 + \eta_4)$$

We shall first obtain the momentum mapping for this action. Let $\xi = (\xi_1, \xi_2, \xi_3, \xi_4) \in T_e G$, then the infinitesimal generator of the action is given by:

$$\xi_{\mathbf{R}^4}(x_1, x_2, x_3, x_4) = \frac{d}{dt} \Phi_{(\exp t\xi_1, \exp t\xi_2, \exp t\xi_3, \exp t\xi_4)}(x_1, x_2, x_3, x_4)\mid_{t=0}$$

$$= \frac{d}{dt}(x_1 + \exp t\xi_1, x_2 + \exp t\xi_2, x_3 + \exp t\xi_3, x_4$$

$$+ \exp t\xi_4)\mid_{t=0}$$

$$= (\xi_1, \xi_2, \xi_3, \xi_4).$$

From the relation $X_{\hat{\mu}_\xi} = \xi_M$, we get

$$X_{\hat{\mu}_\xi} = \xi_1 \frac{\partial}{\partial x_1} + \xi_2 \frac{\partial}{\partial x_2} + \xi_3 \frac{\partial}{\partial x_3} + \xi_4 \frac{\partial}{\partial x_4}.$$

Since $\omega = dx_1 \wedge dx_3 + dx_2 \wedge dx_4$, we have

$$i_{\xi_{\mathbf{R}^4}}\omega = i_{\xi_1 \frac{\partial}{\partial x_1} + \xi_2 \frac{\partial}{\partial x_2} + \xi_3 \frac{\partial}{\partial x_3} + \xi_4 \frac{\partial}{\partial x_4}}(dx_1 \wedge dx_3 + dx_2 \wedge dx_4)$$

$$= \xi_1 dx_3 - \xi_3 dx_1 + \xi_2 dx_4 - \xi_4 dx_2.$$

Thus

$$d\hat{\mu}_\xi(x_1, x_2, x_3, x_4) = \xi_1 dx_3 - \xi_3 dx_1 + \xi_2 dx_4 - \xi_4 dx_2.$$

Hence,

$$\hat{\mu}_\xi(x_1, x_2, x_3, x_4) = \xi_1 x_3 - \xi_3 x_1 + \xi_2 x_4 - \xi_4 x_2.$$

Therefore, the momentum mapping is

$$\mu(x_1, x_2, x_3, x_4) \cdot (\xi_1, \xi_2, \xi_3, \xi_4) = \xi_1 x_3 - \xi_3 x_1 + \xi_2 x_4 - \xi_4 x_2.$$

To obtain its one cocycle we use the definition

$$\sigma(g) = \mu(\Phi_g(m)) - Ad_g^* \mu(m).$$

But since $G = \mathbf{R}^4$ under addition is commutative, we have $Ad_g^* = id$ for all $g \in G$. It follows that

$$\hat{\sigma}_{\bar{\xi}}(\bar{\eta}) = \mu(\Phi_{\bar{\eta}}(\bar{x}) \cdot \bar{\xi} - \mu(\bar{x}) \cdot \bar{\xi}$$

$$= \xi_1 \eta_3 - \xi_3 \eta_1 + \xi_2 \eta_4 - \xi_4 \eta_2.$$

where $\bar{\xi} = (\xi_i)$, $i = 1, 2, 3, 4$, $\bar{\eta} = (\eta_i)$, $i = 1, 2, 3, 4$. Define a map $\Psi : G \times \mathfrak{g}^* \to \mathfrak{g}^* = (\mathbf{R}^4)^* = \mathbf{R}^4$ by

$$\Psi_{(\eta_1, \eta_2, \eta_3, \eta_4)}(\alpha_1, \alpha_2, \alpha_3, \alpha_4) = (\alpha_1 + \eta_3, \alpha_2 + \eta_4, \alpha_3 - \eta_1, \alpha_4 - \eta_2).$$

Note that Ψ is an action since

(i) $\Psi_{(0,0,0,0)}(\alpha_1, \alpha_2, \alpha_3, \alpha_4) = (\alpha_1, \alpha_2, \alpha_3, \alpha_4)$.

(ii)

$$\Psi_{(\eta_1+\gamma_1, \eta_2+\gamma_2, \eta_3+\gamma_3, \eta_4+\gamma_4)}(\alpha_1, \alpha_2, \alpha_3, \alpha_4)$$
$$= (\alpha_1 + (\eta_3 + \gamma_3), \alpha_2 + (\eta_4 + \gamma_4), \alpha_3 - (\eta_1 + \gamma_1), \alpha_4 - (\eta_2 + \gamma_2))$$
$$= [(\alpha_1 + \gamma_3) + \eta_3, (\alpha_2 + \gamma_4) + \eta_4, (\alpha_3 - \gamma_1) - \eta_1, (\alpha_4 - \gamma_2) - \eta_2]$$
$$= \Psi_{(\eta_1, \eta_2, \eta_3, \eta_4)}(\alpha_1 + \gamma_3, \alpha_2 + \gamma_4, \alpha_3 - \gamma_1, \alpha_4 - \gamma_2)$$
$$= \Psi_{(\eta_1, \eta_2, \eta_3, \eta_4)}\left(\Psi_{(\gamma_1, \gamma_2, \gamma_3, \gamma_4)}(\alpha_1, \alpha_2, \alpha_3, \alpha_4)\right).$$

showing Ψ is also a homomorphism.

We shall now show that the momentum mapping is equivariant with respect to the action Ψ of G on $\mathfrak{g}^*$ and the action Φ of G on M. That is, $\mu \circ \Phi = \Psi \circ \mu$.

Let $\mu(x_1, x_2, x_3, x_4) = (\alpha_1, \alpha_2, \alpha_3, \alpha_4) \in \mathfrak{g}^* = (\mathbf{R}^4)^*$, $\xi = (\xi_1, \xi_2, \xi_3, \xi_4) \in \mathfrak{g} = \mathbf{R}^4$, and take the standard inner product on $\mathbf{R}^4$.

Then

$$\mu(\Phi_{(\eta_1, \eta_2, \eta_3, \eta_4)}(x_1, x_2, x_3, x_4)) \cdot (\xi_1, \xi_2, \xi_3, \xi_4)$$
$$= \mu(x_1 + \eta_1, x_2 + \eta_2, x_3 + \eta_3, x_4 + \eta_4) \cdot (\xi_1, \xi_2, \xi_3, \xi_4)$$
$$= \xi_1(x_3 + \eta_3) - \xi_3(x_1 + \eta_1) + \xi_2(x_4 + \eta_4) - \xi_4(x_2 + \eta_2)$$
$$= \xi_1 x_3 - \xi_3 x_1 + \xi_2 x_4 - \xi_4 x_2 + \xi_1 \eta_3 - \xi_3 \eta_1 + \xi_2 \eta_4 - \xi_4 \eta_2$$
$$= \mu(x_1, x_2, x_3, x_4) \cdot (\xi_1, \xi_2, \xi_3, \xi_4) + \xi_1 \eta_3 - \xi_3 \eta_1 + \xi_2 \eta_4 - \xi_4 \eta_2$$
$$= (\alpha_1, \alpha_2, \alpha_3, \alpha_4) \cdot (\xi_1, \xi_2, \xi_3, \xi_4) + \xi_1 \eta_3 - \xi_3 \eta_1 + \xi_2 \eta_4 - \xi_4 \eta_2$$
$$= \xi_1 \alpha_1 + \xi_2 \alpha_2 + \xi_3 \alpha_3 + \xi_4 \alpha_4 + \xi_1 \eta_3 - \xi_3 \eta_1 + \xi_2 \eta_4 - \xi_4 \eta_2$$
$$= \xi_1(\alpha_1 + \eta_3) + \xi_2(\alpha_2 + \eta_4) + \xi_3(\alpha_3 - \eta_1) + \xi_4(\alpha_4 - \eta_2)$$
$$= (\alpha_1 + \eta_3, \alpha_2 + \eta_4, \alpha_3 - \eta_1, \alpha_4 - \eta_2) \cdot (\xi_1, \xi_2, \xi_3, \xi_4)$$
$$= \Psi_{(\eta_1, \eta_2, \eta_3, \eta_4)}(\alpha_1, \alpha_2, \alpha_3, \alpha_4) \cdot (\xi_1, \xi_2, \xi_3, \xi_4)$$
$$= \Psi_{(\eta_1, \eta_2, \eta_3, \eta_4)}\mu(x_1, x_2, x_3, x_4) \cdot (\xi_1, \xi_2, \xi_3, \xi_4).$$

Since $g = (\eta_1, \eta_2, \eta_3, \eta_4)$ and $x = (x_1, x_2, x_3, x_4)$ were arbitrary, we conclude that $\mu(\Phi_g(x)) = \Psi_g \mu(x)$ for all $g \in G, x \in M$.

Theorem 3.6 *Let* $\Phi : G \times M \to M$ *be a symplectic action of* G *on* (M, ω) *which admits a momentum mapping* $\mu : M \to \mathfrak{g}^*$ *and let* $\sigma : G \to \mathfrak{g}^*$ *be the cocycle of* μ. *Let the function* $\hat{\sigma}_\eta : G \to \mathbb{R}$ *be defined by* $\hat{\sigma}_\eta(g) = \sigma(g) \cdot \eta$.

Define also a function $\Sigma : \mathfrak{g} \times \mathfrak{g} \to \mathbb{R}$ *by* $\Sigma(\xi, \eta) = d\hat{\sigma}_\eta(e) \cdot \xi$ *for all* $\xi, \eta \in \mathfrak{g}$. *Then* Σ *is skew symmetric bilinear form on* $\mathfrak{g}$ *and satisfies the Jacobi's identity*
$$0 = \Sigma(\xi, [\eta, \xi]) + \Sigma(\eta, [\xi, \xi]) + \Sigma(\xi, [\xi, \eta]).$$

Proof We first obtain an expression for $\Sigma(\xi, \eta)$. From the expression

$$\hat{\sigma}_\eta(g) = \mu(\Phi_g(x)) \cdot \eta - Ad_g^* \mu(x) \cdot \eta$$
$$= \hat{\mu}_\eta(\Phi_g(x)) - \hat{\mu}_{Ad_{g^{-1}}\eta}(x).$$

Differentiating with respect to g at $g = e$ in the direction of $\xi \in \mathfrak{g}$ we get

$$d\hat{\sigma}_\eta(e) \cdot \xi = d(\hat{\mu}_\eta(\Phi_g(x)) \cdot \xi - \hat{\mu}_{Ad_{g^{-1}}\eta}(x) \cdot \xi)$$
$$= \frac{d}{dt}\hat{\mu}_\eta(\Phi_{\exp t\xi}(x))\,|_{t=0} - \frac{d}{dt}\hat{\mu}_{Ad_{\exp(-t\xi)}\eta}(x)\,|_{t=0}$$
$$= (i_{\eta_M}\omega)\frac{d}{dt}\Phi_{\exp t\xi}(x)\,|_{t=0} - \frac{d}{dt}\langle Ad_{\exp(-t\xi)}\eta, \mu(x)\rangle\,|_{t=0}$$
$$= (i_{\eta_M}\omega)(\xi_M(x)) - \langle\frac{d}{dt}Ad_{\exp(-t\xi)}\eta\,|_{t=0}, \mu(x)\rangle$$
$$= (i_{\xi_M}i_{\eta_M}\omega)(x) - \langle[\eta, \xi], \mu(x)\rangle$$
$$= -\{\hat{\mu}_\xi, \hat{\mu}_\eta\}(x) - \hat{\mu}_{[\eta,\xi]}(x)$$
$$= -\{\hat{\mu}_\xi, \hat{\mu}_\eta\}(x) + \hat{\mu}_{[\xi,\eta]}(x)$$

Thus,

$$\Sigma(\xi, \eta) = -\{\hat{\mu}_\xi, \hat{\mu}_\eta\} + \hat{\mu}_{[\xi,\eta]} \tag{3.2}$$

But both the Poisson bracket $\{\hat{\mu}_\xi, \hat{\mu}_\eta\}$ and the Lie bracket $[\xi, \eta]$ are skew symmetric bilinear. This implies that the right side of (3.1) is skew symmetric and bilinear. Therefore, $\Sigma(\xi, \eta)$ is skew symmetric and bilinear form on $\mathfrak{g}$. The right side also satisfies Jacobi's identity which implies that $\Sigma(\xi, \eta)$ also satisfies the Jacobi's identity. □

Example 3.7 We shall use the previous example to obtain an expression for $\Sigma(\xi, \eta)$. Let $\xi, \eta \in \mathfrak{g}$ where $\xi = (\xi_1, \xi_2, \xi_3, \xi_4)$ and $\eta = (\eta_1, \eta_2, \eta_3, \eta_4)$. We have already seen, for example, that

$$\xi_{\mathbf{R}^4}(x_1, x_2, x_3, x_4) = \xi_1\frac{\partial}{\partial x_1} + \xi_2\frac{\partial}{\partial x_2} + \xi_3\frac{\partial}{\partial x_3} + \xi_4\frac{\partial}{\partial x_4}.$$

From the commutation relation

$$\{\hat{\mu}_\xi, \hat{\mu}_\eta\} = \hat{\mu}_{[\xi,\eta]} - \Sigma(\xi, \eta).$$

Now, evaluating the left side

$$\{\hat{\mu}_\xi, \hat{\mu}_\eta\} = -i_{\xi_M} i_{\eta_M} \omega$$

$$= -i_{\xi_1 \frac{\partial}{\partial x_1} + \xi_2 \frac{\partial}{\partial x_2} + \xi_3 \frac{\partial}{\partial x_3} + \xi_4 \frac{\partial}{\partial x_4}} i_{\eta_1 \frac{\partial}{\partial x_1} + \eta_2 \frac{\partial}{\partial x_2} + \eta_3 \frac{\partial}{\partial x_3} + \eta_4 \frac{\partial}{\partial x_4}} (dx_1 \wedge dx_3 + dx_2 \wedge dx_4)$$

$$= -i_{\xi_1 \frac{\partial}{\partial x_1} + \xi_2 \frac{\partial}{\partial x_2} + \xi_3 \frac{\partial}{\partial x_3} + \xi_4 \frac{\partial}{\partial x_4}} (i_{\eta_1 \frac{\partial}{\partial x_1} + \eta_2 \frac{\partial}{\partial x_2} + \eta_3 \frac{\partial}{\partial x_3} + \eta_4 \frac{\partial}{\partial x_4}} (dx_1 \wedge dx_3 + dx_2 \wedge dx_4))$$

$$= -i_{\xi_1 \frac{\partial}{\partial x_1} + \xi_2 \frac{\partial}{\partial x_2} + \xi_3 \frac{\partial}{\partial x_3} + \xi_4 \frac{\partial}{\partial x_4}} (\eta_1 dx_3 - \eta_3 dx_1 + \eta_2 dx_4 - \eta_4 dx_2)$$

$$= -(\xi_3 \eta_1 - \xi_1 \eta_3 + \xi_4 \eta_2 - \xi_2 \eta_4).$$

For the right side, note that since $G = \mathbf{R}^4$ is commutative under addition $[\xi, \eta] = 0$. Therefore, we have

$$-(\xi_3 \eta_1 - \xi_1 \eta_3 + \xi_4 \eta_2 - \xi_2 \eta_4) = -\Sigma(\xi, \eta).$$

Thus

$$\Sigma(\xi, \eta) = \xi_3 \eta_1 - \xi_1 \eta_3 + \xi_4 \eta_2 - \xi_2 \eta_4.$$

3.2 The Orbits of an Affine Action

Proposition 3.8 *Let G be a Lie group, $\mathfrak{g}$ its Lie algebra, and $\mathfrak{g}^*$ the dual of its Lie algebra. Let $\Psi : G \times \mathfrak{g}^* \to \mathfrak{g}^*$ defined by $\Psi(g, \alpha) = Ad_g^* \alpha + \sigma(g)$ be the affine action of G on $\mathfrak{g}^*$. Then the orbit $G \cdot \beta = \{\Psi(g, \beta) : g \in G\}$ is a symplectic manifold with the 2-form given by*

$$\omega_\beta(\xi_{\mathfrak{g}^*}(v), \eta_{\mathfrak{g}^*}(v)) = -\beta[\xi, \eta] + \Sigma(\eta, \xi).$$

Proof We shall first show that the orbit $O_\beta = \{\Psi(g, \beta) : g \in G\}$ is a manifold. Thereafter we shall define a symplectic structure on it.

Let $O_\beta = \{\Psi(g, \beta) : g \in G\} \subset \mathfrak{g}^*$ be the orbit. The isotropy group of β is given by

$$G_\beta = \{g \in G : \Psi(g, \beta) = \beta\}.$$

This is a closed subgroup of G since if g_n is a sequence in G_β which converges to $g \in G$ then we have

$$\beta = \lim_{n \to \infty} \Psi(g_n, \beta)$$
$$= \Psi(\lim_{n \to \infty} g_n, \beta)$$
$$= \Psi(g, \beta).$$

The second equality holds because Ψ is an action and so it is smooth. This shows that $g \in G_\beta$.

Now we show that $O_\beta \cong G/G_\beta$. To this end, let us define a map

$$\varphi : O_\beta \to G/G_\beta$$

by

$$\varphi(\eta) = gG_\beta$$

for $\eta \in O_\beta$, where $\eta = \Psi(g, \beta)$ for some $g \in G$.

The map φ is well-defined since if $\varphi(\eta) = hG_\beta$ also, then we have

$$\Psi(g, \beta) = \Psi(h, \beta)$$

so that

$$\Psi(h^{-1}, \Psi(g, \beta)) = \beta.$$

Hence, $\Psi(h^{-1}g, \beta) = \beta$, which implies that $h^{-1}g \in G_\beta$ and consequently

$$gG_\beta = hG_\beta.$$

This map φ is injective. To see this let $\eta = \Psi(g, \beta)$, $\xi = \Psi(h, \beta)$ and $gG_\beta = hG_\beta$ for $h, g \in G$ and $\eta, \xi \in O_\beta$. Then $h^{-1}g \in G_\beta$ so that $\Psi(h^{-1}g, \beta) = \beta$. This implies that $\Psi(h^{-1}, \Psi(g, \beta)) = \beta$. It follows that $\Psi(g, \beta) = \Psi(h, \beta)$ so that $\eta = \xi$.

Furthermore, the map is surjective since if

$$gG_\beta \in G/G_\beta,$$

then

$$\eta = \Psi(g, \beta) \in O_\beta$$

gives

$$\varphi(\eta) = gG_\beta$$

by construction.

Thus, the map φ is an isomorphism.   □

Proposition 3.9 *Suppose that $\eta \in O_\beta$ so that $\eta = \Psi(h, \beta)$ for some $h \in G$, then the isotropy groups G_β and G_η are conjugates.*

Proof We shall change the notation a bit and write $\Psi_g(\beta)$ for $\Psi(g, \beta)$. We have already seen that Ψ is an action and so, it is a homomorphism

$$\Psi(gh, \beta) = \Psi(g, \Psi(h, \beta)).$$

Next, we define a map $\gamma : G/G_\beta \to G/G_\eta$ by

$$[g]_\beta \mapsto [hgh^{-1}]_\eta.$$

Then γ is a well-defined isomorphism. To see this, let $x \in G_\beta$ so that $\Psi_x(\beta) = \beta$. Since $\eta = \Psi_h(\beta)$, we have

$$\Psi_h \circ \Psi_x \circ \Psi_{h^{-1}}(\eta) = \Psi_h \circ \Psi_x \circ \Psi_{h^{-1}}(\Psi(h, \beta))$$

$$= \Psi_h \circ \Psi_x(\Psi(hh^{-1}, \beta))$$

$$= \Psi_h \circ \Psi_x(\beta)$$

$$= \Psi(h, \Psi(x, \beta))$$

$$= \Psi(h, \beta)$$

$$= \eta.$$

Since $x \in G_\beta$ was arbitrary, it follows that $\Psi_h G_\beta \Psi_{h^{-1}}$ is a subgroup of G_η. Taking $\beta = \Psi(h^{-1}, \eta)$ gives the reverse inclusion. Thus $G_\eta = \Psi_h G_\beta \Psi_{h^{-1}}$.

Hence γ is an isomorphism.

We now write the orbit of Ψ through β as

$$G \cdot \beta = G/G_\beta \cong O_\beta.$$

From the discussion above, it is clear that the orbit $G \cdot \beta$ does not depend on the choice of the element β in its orbit. We already have that G_β is a closed subgroup of G. Thus $G \cdot \beta = G/G_\beta$ is a manifold.

We shall now define a symplectic structure on the orbit of the action Ψ through β.

Let $\xi \in \mathfrak{g}$. We define the vector field on $\mathfrak{g}^*$, called the infinitesimal generator of the action to be:

$$\xi_{\mathfrak{g}^*}(\beta) = \frac{d}{dt}\Psi(\exp t\xi, \beta)\mid_{t=0}$$

$$= \frac{d}{dt}[Ad^*_{\exp t\xi}\beta + \sigma(\exp t\xi)]\mid_{t=0}$$

$$= \frac{d}{dt}Ad^*_{\exp t\xi}\beta\mid_{t=0} + \frac{d}{dt}\sigma(\exp t\xi)\mid_{t=0}$$

$$= \frac{d}{dt}Ad^*_{\exp t\xi}\beta\mid_{t=0} + d\sigma(e)\cdot\xi$$

$$= \frac{d}{dt}Ad^*_{\exp t\xi}\beta\mid_{t=0} + d\hat{\sigma}_\xi(e)$$

If now $\eta \in \mathfrak{g}$, then we have

$$(\xi_{\mathfrak{g}^*}(\beta))\eta = \frac{d}{dt}(Ad^*_{\exp t\xi}\beta)\eta\mid_{t=0} + d\hat{\sigma}_\xi(e)\cdot\eta$$

$$= \beta(\frac{d}{dt}Ad_{\exp(-t\zeta)}\eta)\mid_{t=0} + \sum(\eta, \xi)$$

$$= \beta(-[\xi, \eta]) + \sum(\eta, \xi).$$

To compute the tangent space to the orbit $G \cdot \beta$ at β, for $\xi \in \mathfrak{g}$ let $x(t) = \exp t\xi$ be a curve in G which is tangent to ξ at $t = 0$, then $\beta(t) = \Psi(x(t), \beta)$ is the curve in $G \cdot \beta$ such that $\beta(0) = \beta$ since $\sigma(e) = 0$.

If $\eta \in \mathfrak{g}$, then

$$\langle\beta(t), \eta\rangle = \langle\Psi(x(t), \beta), \eta\rangle$$

$$= \langle Ad^*_{x(t)}\beta + \sigma(x(t)), \eta\rangle$$

$$= \langle Ad^*_{\exp t\xi}\beta, \eta\rangle + \langle\sigma(\exp t\xi), \eta\rangle$$

where $\langle\cdot, \cdot\rangle$ is the natural pairing of $\mathfrak{g}$ and its dual $\mathfrak{g}^*$.

Differentiating with respect to t at $t = 0$ gives
$$\langle\beta'(0), \eta\rangle = \langle ad^*_\xi\beta, \eta\rangle + \sum(\eta, \xi).$$
This implies that

$$\beta'(0) = ad^*_\xi\beta + \sum(\cdot, \xi).$$

Therefore, the tangent space to $G \cdot \beta$ at β is given by

$$T_\beta G \cdot \beta = \{ad^*_\xi\beta + \sum(\cdot, \xi) : \xi \in \mathfrak{g}\}.$$

Consider now the function $\omega_\beta : \mathfrak{g} \times \mathfrak{g} \to \mathbb{R}$ defined by

$$\omega_\beta(\xi, \eta) = \beta(-[\xi, \eta]) + \sum(\eta, \xi).$$

Clearly ω_β is skew symmetric and bilinear on $\mathfrak{g}$ since both the Lie bracket and $\sum(\cdot,\cdot)$ are skew symmetric bilinear.

$$\ker\omega_\beta = \{\xi \in \mathfrak{g} : \omega_\beta(\xi,\eta) = 0, \forall \eta \in \mathfrak{g}\}$$

$$= \{\xi \in \mathfrak{g} : \beta(-[\xi,\eta]) + \sum(\eta,\xi) = 0, \forall \eta \in \mathfrak{g}\}$$

$$= LieG_\beta$$

Now, for $\xi \in \mathfrak{g}$ let $\tilde{\xi}$ denote the vector field on $\mathfrak{g}^*$ generated by ξ. That is,

$$\tilde{\xi}_\beta = \tilde{\xi}(\beta)$$
$$= \tfrac{d}{dt}\Psi(\exp t\xi, \beta)\,|_{t=0}\,.$$

Then for $\beta \in \mathfrak{g}^*$, define the function
$\Omega_\beta : T_\beta G \cdot \beta \times T_\beta G \cdot \beta \to \mathbb{R}$ by

$$\Omega_\beta(\tilde{\xi},\tilde{\eta}) = \omega_\beta(\xi,\eta)$$

for all $\xi, \eta \in \mathfrak{g}$.

Theorem 3.10 Ω_β *defined above is a well-defined 2-form on $G \cdot \beta$, the orbit of the affine action $\Psi : G \times \mathfrak{g}^* \to \mathfrak{g}^*$ through β.*

Proof First note that if $\zeta \in LieG_\beta$, then $\Omega_\beta(\tilde{\zeta},\tilde{\xi}) = \omega_\beta(\zeta,\xi) = 0$ for all $\xi \in \mathfrak{g}$. Now let $\xi, \eta \in \mathfrak{g}$. If $\zeta \in LieG_\beta$, then

$$\Omega_\beta(\tilde{\xi}+\tilde{\zeta},\tilde{\eta}) = \omega_\beta(\xi+\zeta,\eta)$$

$$= \omega_\beta(\xi,\eta) + \omega_\beta(\zeta,\eta) \text{ since } \omega_\beta \text{ is bilinear.}$$

$$= \omega_\beta(\xi,\eta) \text{ since } \omega_\beta(\zeta,\eta) = 0$$

$$= \Omega_\beta(\tilde{\xi},\tilde{\eta}).$$

Thus Ω_β does not depend on the choice of $\xi, \eta \in \mathfrak{g}$. Hence Ω_β is well-defined. Since locally ω_β is skew symmetric, bilinear on the tangent space T_eG, it follows that Ω_β is skew symmetric bilinear on the tangent space $T_\beta G \cdot \beta$. It remains to show that Ω_β is non-degenerate and closed on $G \cdot \beta$.

To prove non-degeneracy let $\xi \in \mathfrak{g}$ be such that $\xi \notin LieG_\beta$, we must show that there exists $\eta \in \mathfrak{g}$ such that $\Omega_\beta(\tilde{\xi},\tilde{\eta}) \neq 0$

But now if $\eta \in \mathfrak{g}$ and $\xi \notin LieG_\beta$ then

$$\Omega_\beta(\tilde{\xi},\tilde{\eta}) = \omega_\beta(\xi,\eta) \neq 0$$

if and only if

$$\xi \notin \ker \omega_\beta = Lie G_\beta.$$

This shows that if $\xi \notin Lie G_\beta$, there exists $\eta \in \mathfrak{g}$ such that $\Omega_\beta(\tilde{\xi}, \tilde{\eta}) \neq 0$. Hence Ω_β is non-degenerate. To show that Ω_β is closed we use the formula

$$d\omega(X, Y, Z) = (L_X\omega)(Y, Z) - (L_Y\omega)(X, Z) + (L_Z\omega)(X, Y)$$

$$+\omega(X, [Y, Z]) - \omega(Y, [X, Z]) + \omega(Z, [X, Y])$$

whose proof can be found in [3] on page 53.

Let $\xi, \eta, \zeta \in \mathfrak{g}$, then

$$d\Omega_\beta(\tilde{\xi}, \tilde{\eta}, \tilde{\zeta}) = d\omega_\beta(\xi, \eta, \zeta)$$

$$= (L_\xi \omega_\beta)(\eta, \zeta) - (L_\eta \omega_\beta)(\xi, \zeta) + (L_\zeta \omega_\beta)(\xi, \eta) + \omega_\beta(\xi, [\eta, \zeta])$$

$$- \omega_\beta(\eta, [\xi, \zeta]) + \omega_\beta(\zeta, [\xi, \eta]).$$

Repeated application of Jacobi identity then shows that $d\Omega_\beta = 0$ which means that Ω_β is closed.

We have therefore shown that if the affine action $\Psi : G \times \mathfrak{g}^* \to \mathfrak{g}^*$ defined by

$$\Psi(g, \alpha) = Ad_g^*\alpha + \sigma(g)$$

is used in place of the coadjoint action, then the orbit $G \cdot \alpha$ is a symplectic manifold with the 2-form given by

$$\omega_\alpha(\xi_{\mathfrak{g}^*}(v), \eta_{\mathfrak{g}^*}(v)) = -\alpha[\xi, \eta] + \sum(\eta, \xi).$$

4   Towards a Generalization

In Theorem 3.6 above and other related results on conservation laws which are not explicitly mentioned in this text, the function Σ or other identities make use of the canonical Poisson bracket. It would be worth extending them using a deformed Poisson bracket that looks more general and is defined as follows.

Let M be a C^∞ manifold, and let

$$\{f, g\}_p = p^r \left\{ \frac{\partial f}{\partial q} \frac{\partial g}{\partial p} - \frac{\partial f}{\partial p} \frac{\partial g}{\partial q} \right\},$$

where f and g lie in $C^\infty(M)$, q, p are canonical coordinates of points of M, r is a nonnegative integer. It is easily seen that the case $p = 1$ yields the usual Poisson bracket.

Proposition 4.1 *Let M be a C^∞ manifold. Then M endowed with $\{.,.\}_p$ on $C^\infty(M)$ is a Poisson manifold.*

Proof For all $f, g, h, f_1, f_2, g_1, g_2 \in C^\infty(M)$, for all $\alpha_i \in \mathbb{C}$, $i = 1, 2$,

(i) $\{.,.\}_p : C^\infty(M) \times C^\infty(M) \longrightarrow C^\infty(M)$ is a bilinear mapping. For, $\dfrac{\partial}{\partial t}$ being a linear operator, it is easily seen that

$$\{\alpha_1 f_1 + \alpha_2 f_2, g\}_p = \alpha_1\{f_1, g\}_p + \alpha_2\{f_2, g\}_p$$

and

$$\{f, \alpha_1 g_1 + \alpha_2 g_2\}_p = \alpha_1\{f, g_1\}_p + \alpha_2\{f, g_2\}_p$$

(ii) $\{.,.\}_p$ is skew-symmetric as it can be seen that $\{f, f\}_p = 0$.

(iii) $\{.,.\}_p$ satisfies the Jacobi identity, i.e.

$$\{f, \{g, h\}_p\}_p + \{g, \{h, f\}_p\}_p + \{h, \{f, g\}_p\}_p = 0$$

as by the straightforward calculation below:
    We have

$$
\begin{aligned}
\{f, \{g, h\}_p\}_p &= p^r \left[\frac{\partial f}{\partial q} \frac{\partial \{g, h\}_p}{\partial p} - \frac{\partial f}{\partial p} \frac{\partial \{g, h\}_p}{\partial q} \right] \\
&= p^r \left[\frac{\partial f}{\partial q} \frac{\partial}{\partial p} \left(p^r \left(\frac{\partial g}{\partial q} \frac{\partial h}{\partial p} - \frac{\partial g}{\partial p} \frac{\partial h}{\partial q} \right) \right) \right. \\
&\quad \left. - \frac{\partial f}{\partial p} \frac{\partial}{\partial q} \left(p^r \left(\frac{\partial g}{\partial q} \frac{\partial h}{\partial p} - \frac{\partial g}{\partial p} \frac{\partial h}{\partial q} \right) \right) \right] \\
&= p^r \left[\frac{\partial f}{\partial q} \left(r p^{r-1} \left(\frac{\partial g}{\partial q} \frac{\partial h}{\partial p} - \frac{\partial g}{\partial p} \frac{\partial h}{\partial q} \right) + p^r \left(\frac{\partial^2 g}{\partial p \partial q} \frac{\partial h}{\partial p} + \frac{\partial g}{\partial q} \frac{\partial^2 h}{\partial p^2} \right. \right. \right. \\
&\quad \left. \left. \left. - \left(\frac{\partial^2 g}{\partial p^2} \frac{\partial h}{\partial q} + \frac{\partial g}{\partial p} \frac{\partial^2 h}{\partial p \partial q} \right) \right) \right) \right. \\
&\quad \left. - \frac{\partial f}{\partial p} \left(p^r \left(\frac{\partial^2 g}{\partial q^2} \frac{\partial h}{\partial p} + \frac{\partial g}{\partial q} \frac{\partial^2 h}{\partial q \partial p} - \left(\frac{\partial^2 g}{\partial q \partial p} \frac{\partial h}{\partial q} + \frac{\partial g}{\partial p} \frac{\partial^2 h}{\partial q^2} \right) \right) \right) \right] \\
&= r p^{2r-1} \left(\frac{\partial f}{\partial q} \left(\frac{\partial g}{\partial q} \frac{\partial h}{\partial p} - \frac{\partial g}{\partial p} \frac{\partial h}{\partial q} \right) \right) + p^{2r} \left(\frac{\partial f}{\partial q} \frac{\partial^2 g}{\partial p \partial q} \frac{\partial h}{\partial p} + \frac{\partial f}{\partial q} \frac{\partial g}{\partial q} \frac{\partial^2 h}{\partial p^2} \right. \\
&\quad - \frac{\partial f}{\partial q} \frac{\partial^2 g}{\partial p^2} \frac{\partial h}{\partial q} - \frac{\partial f}{\partial q} \frac{\partial g}{\partial p} \frac{\partial^2 h}{\partial p \partial q} - \frac{\partial f}{\partial p} \frac{\partial^2 g}{\partial q^2} \frac{\partial h}{\partial p} \\
&\quad \left. - \frac{\partial f}{\partial p} \frac{\partial g}{\partial q} \frac{\partial^2 h}{\partial q \partial p} + \frac{\partial f}{\partial p} \frac{\partial^2 g}{\partial q \partial p} \frac{\partial h}{\partial q} + \frac{\partial f}{\partial p} \frac{\partial g}{\partial p} \frac{\partial^2 h}{\partial q^2} \right).
\end{aligned}
$$

$$(*)$$

Similarly,

$$\{g, \{h, f\}_p\}_p = rp^{2r-1} \left(\frac{\partial g}{\partial q} \left(\frac{\partial h}{\partial q} \frac{\partial f}{\partial p} - \frac{\partial h}{\partial p} \frac{\partial f}{\partial q} \right) \right) + p^{2r} \left(\frac{\partial g}{\partial q} \frac{\partial^2 h}{\partial p \partial q} \frac{\partial f}{\partial p} + \frac{\partial g}{\partial q} \frac{\partial h}{\partial q} \frac{\partial^2 f}{\partial p^2} \right.$$
$$- \frac{\partial g}{\partial q} \frac{\partial^2 h}{\partial p^2} \frac{\partial f}{\partial q} - \frac{\partial g}{\partial q} \frac{\partial h}{\partial p} \frac{\partial^2 f}{\partial p \partial q}$$
$$\left. - \frac{\partial g}{\partial p} \frac{\partial^2 h}{\partial q^2} \frac{\partial f}{\partial p} - \frac{\partial g}{\partial p} \frac{\partial h}{\partial q} \frac{\partial^2 f}{\partial q \partial p} + \frac{\partial g}{\partial p} \frac{\partial^2 h}{\partial q \partial p} \frac{\partial f}{\partial q} + \frac{\partial g}{\partial p} \frac{\partial h}{\partial p} \frac{\partial^2 f}{\partial q^2} \right).$$ $(**)$

and

$$\{h, \{f, g\}_p\}_p = rp^{2r-1} \left(\frac{\partial h}{\partial q} \left(\frac{\partial f}{\partial q} \frac{\partial g}{\partial p} - \frac{\partial f}{\partial p} \frac{\partial g}{\partial q} \right) \right) + p^{2r} \left(\frac{\partial h}{\partial q} \frac{\partial^2 f}{\partial p \partial q} \frac{\partial g}{\partial p} + \frac{\partial h}{\partial q} \frac{\partial f}{\partial q} \frac{\partial^2 g}{\partial p^2} \right.$$
$$- \frac{\partial h}{\partial q} \frac{\partial^2 f}{\partial p^2} \frac{\partial g}{\partial q} - \frac{\partial h}{\partial q} \frac{\partial f}{\partial p} \frac{\partial^2 g}{\partial p \partial q}$$
$$\left. - \frac{\partial h}{\partial p} \frac{\partial^2 f}{\partial q^2} \frac{\partial g}{\partial p} - \frac{\partial h}{\partial p} \frac{\partial f}{\partial q} \frac{\partial^2 g}{\partial q \partial p} + \frac{\partial h}{\partial p} \frac{\partial^2 f}{\partial q \partial p} \frac{\partial g}{\partial q} + \frac{\partial h}{\partial p} \frac{\partial f}{\partial p} \frac{\partial^2 g}{\partial q^2} \right).$$ $(**)$

Adding the three expressions (*), (**), and (***) taking note of the equality of mixed partials, the terms with coefficient p^{2r} cancel out. We remain with

$$\{f, \{g, h\}_p\}_p + \{g, \{h, f\}_p\}_p + \{h, \{f, g\}_p\}_p$$
$$= rp^{2r-1} \left[\frac{\partial f}{\partial q} \left(\frac{\partial g}{\partial q} \frac{\partial h}{\partial p} - \frac{\partial g}{\partial p} \frac{\partial h}{\partial q} \right) + \frac{\partial g}{\partial q} \left(\frac{\partial h}{\partial q} \frac{\partial f}{\partial p} - \frac{\partial h}{\partial p} \frac{\partial f}{\partial q} \right) \right.$$
$$\left. + \frac{\partial h}{\partial q} \left(\frac{\partial f}{\partial q} \frac{\partial g}{\partial p} - \frac{\partial f}{\partial p} \frac{\partial g}{\partial q} \right) \right]$$

Expanding and adding gives the desired results.

Thus, the Jacobi identity holds.

(iii) Finally, $\{., .\}$ is a derivation in each variable. That is,

$$\{fg, h\}_p = \{f, h\}_p g + f\{g, h\}_p$$

which ends the proof.   □

Proposition 4.2 *Let M be a Poisson manifold and $H \in C^\infty(M)$. There is a unique vector field X_H on M, so-called Hamiltonian vector field with Hamiltonian function H such that*

$$X_H[g]_p = \{g, H\}_p,$$

for all $g \in C^\infty(M)$. Thus,

$$\{f, g\}_p = X_g[f]_p = -X_f[g]_p.$$

Momentum mapping was introduced by Souriau for the purposes of reduction of mechanical systems with symmetries. Therefore, such a deformed Poisson bracket introduces a new formalism for Hamiltonian mechanics, a setting that goes beyond the scope of this paper.

Acknowledgements Augustin Batubenge extends his thanks to Professor François Lalonde for his financial support during a stay at the University of Montreal in Summer 2016, which allowed the writing-up of this paper; along with Professor Norbert Hounkonnou for suggesting the use of a generalized Poisson bracket.

Wallace Haziyu is thankful to Doctors I.D. Tembo, M. Lombe, and A. Ngwengwe for the encouragements they rendered towards this project.

References

1. A. Kirillov, *Introduction to Lie Groups and Lie Algebras*. Department of Mathematics (Suny at Stony Brook, New York, 2008)
2. R. Abraham, J.E. Marsden, *Foundations of Mechanics*, 2nd edn. (The Benjamin/Cummings Publishing Company, Reading, MA, 1978)
3. R. Berndt, *An Introduction to Symplectic Geometry*, vol. 26 (American Mathematical Society, Rhodes Island, 2001)
4. P. Iglesias-Zemmour, *Diffeology*. Mathematical Surveys and Monographs, vol. 185 (American Mathematical Society, Providence, RI, 2010)
5. F.W. Warner, *Foundations of Differentiable Manifolds and Lie Groups* (Springer, New York, 1983)
6. J. Śniatycki, *Differential Geometry of Singular Spaces and Reduction of Symmetry*. New Mathematical Monographs, vol. 23 (Cambridge University Press, Cambridge, 2013)

Some Difference Integral Inequalities

G. Bangerezako and J. P. Nuwacu

Dedicated to Prof. Hounkonnou M N, for his 60th birthday.

Abstract We establish difference versions of the classical integral inequalities of Hölder, Cauchy-Schwartz, Minkowski and integral inequalities of Grönwall, Bernoulli and Lyapunov based on the Lagrange method of linear difference equation of first order.

Keywords Hölder · Cauchy-Schwartz · Minkowski · Grönwall · Bernoulli and Lyapunov inequalities · Lagrange method

1   Introduction

Considering the most general divided difference derivative [5, 6],

$$\mathcal{D}f(t(s)) = \frac{f(t(s+\frac{1}{2}))-f(t(s-\frac{1}{2}))}{t(s+\frac{1}{2})-t(s-\frac{1}{2})}, \tag{1}$$

admitting the property that if $f(t) = P_n(t(s))$ is a polynomial of degree n in $t(s)$, then $\mathcal{D}f(t(s)) = \tilde{P}_{n-1}(t(s))$ is a polynomial in $t(s)$ of degree $n-1$, one is led to the following most important canonical forms for $t(s)$ in order of increasing complexity:

$$t(s) = t(0); \tag{2}$$

$$t(s) = s; \tag{3}$$

G. Bangerezako (✉) · J. P. Nuwacu
University of Burundi, Faculty of Sciences, Department of Mathematics, Bujumbura, Burundi
e-mail: gaspard.bangerezako@ub.edu.bi; jean-paul.nuwacu@ub.edu.bi

© Springer Nature Switzerland AG 2018
T. Diagana, B. Toni (eds.), *Mathematical Structures and Applications*,
STEAM-H: Science, Technology, Engineering, Agriculture,
Mathematics & Health, https://doi.org/10.1007/978-3-319-97175-9_5

$$t(s) = q^s;$$ (4)

$$t(s) = \frac{q^s + q^{-s}}{2}, \quad q \in \mathbf{C}, s \in \mathbb{Z}.$$ (5)

When the function $t(s)$ is given by (2)–(4), the divided difference derivative (1) leads to the ordinary differential derivative $Df(t) = \frac{d}{dt}f(t)$, finite difference derivative

$$\Delta f(s) = f(s+1) - f(s) = (e^{\frac{d}{ds}} - 1)f(s)$$ (6)

and q-difference derivative (or Jakson derivative [4])

$$D_q f(t) = \frac{f(qt) - f(t)}{qt - t} = \frac{q^{\frac{d}{dt}} - 1}{qt - t} f(t)$$ (7)

respectively. When $x(s)$ is given by (5), the corresponding derivative is usually referred to as the *Askey-Wilson* first order divided difference operator [1] that one can write:

$$\mathcal{D}f(x(z)) = \frac{f(x(q^{\frac{1}{2}}z)) - f(x(q^{-\frac{1}{2}}z))}{x(q^{\frac{1}{2}}z) - x(q^{-\frac{1}{2}}z)},$$ (8)

where $x(z) = \frac{z+z^{-1}}{2}$, having in mind that $z = q^s$.

The calculus related to the differential derivative, the continuous or differential calculus, is clearly the classical one. The one related to the derivatives (6)–(8) (difference, q-difference and q-nonuniform difference respectively) is referred to as the *discrete calculus*. Its interest is two folds: On the one hand it generalizes the continuous calculus, and on the other hand it uses discrete variable.

This work is concerned in the difference calculus. We particularly aim to establish difference versions of the well-known in differential calculus, integral inequalities of Hölder, Cauchy-Schwartz, Minkowski, Grönwall, Bernoulli, and Lyapunov. We will note that the raised inequalities were proved in [3] for a more general difference operator than (6), but one will remark that except classical recipes used for the inequalities of classical analysis (Hölder, Cauchy-Schwartz and Minkowski), our approach here is essentially different. It is essentially based on the Lagrange method and it is so that it can be extended to the more general derivative (7) or even (8) (see [2]), the latter being, at our best knowledge, the largest one having the mentioned property of sending a polynomial of degree n in a polynomial of degree $n - 1$.

In the following lines, we first introduce basic concepts of difference calculus and linear first order difference equations necessary for the sequel, and then study the mentionned integral inequalities.

2 Preliminaries

2.1 Difference Derivative and Integral

Consider again the difference derivative that is the derivative related to the grid in (2):

$$\Delta F(s) = F(s+1) - F(s) = f(s) \tag{1}$$

Basing on this derivative, one defines the integration that is the inverse of the differentiation operation as follows:

$$\int_{s_0}^{s} f(s) \overset{def}{=} \sum_{i=s_0}^{s-1} f(i). \tag{2}$$

The defined integral admits the following properties:

Fundamental Principle of Analysis One easily verifies that

$$(i) \ \Delta \left(\int_{s_0}^{s} f(s) d_\Delta s \right) = \Delta \left(\sum_{s_0}^{s-1} f(i) \right) = f(s), \tag{3}$$

$$(ii) \ \int_{s_0}^{s} (\Delta F(s)) \, d_\Delta s = \sum_{s_0}^{s-1} \Delta F(i) = F(s) - F(s_0). \tag{4}$$

Integration by Parts Integrating the two members of the equality

$$f(s)\Delta g(s) = \Delta(f(s)g(s)) - g(s+1)\Delta f(s) \tag{5}$$

and applying (4), one gets

$$\int_{s_0}^{s} f(s)\Delta g(s) d_\Delta s = [f(s)g(s)]_{s_0}^{s} - \int_{s_0}^{s} g(s+1)\Delta f(s) d_\Delta s, \tag{6}$$

which is the integration by parts formula.

Positivity of the Integral We finally remark that when f(s) is positive, the integral in (2) is clearly positive, which gives the following property and its corollary useful for the sequel.

Property 2.1 *If* $f(s) \geqslant 0$ *and* $s_1 < s_2$, *then*

$$\int_{s_1}^{s_2} f(s) d_\Delta s \geqslant o. \tag{7}$$

Corollary 2.1 *If* $f(s) \geqslant g(s)$ *and* $s_1 < s_2$, *then*

$$\int_{s_1}^{s_2} f(s) d_\Delta s \geqslant \int_{s_1}^{s_2} g(s) d_\Delta s. \tag{8}$$

2.2 Linear Difference Equations of First Order

A linear difference equation of first order can be written as

$$\Delta y(s) = a(s)y(s+1) + b(s) \tag{9}$$

or

$$\Delta y(s) = a(s)y(s) + b(s). \tag{10}$$

Consider first the homogenous equation corresponding to (9):

$$\Delta y(s) = a(s)y(s+1). \tag{11}$$

Equation (11) gives

$$y(s+1) = \left(\frac{1}{1-a(s)}\right)y(s), \tag{12}$$

which by recursion leads to

$$y(s) = E_a(n_0, n)y(n_0), \tag{13}$$

where

$$E_a(n_0, n) =^{def} \begin{cases} \prod_{i=s_0}^{s-1} \frac{1}{1-a(i)}, & s > s_0 \\ 1, & s \leqslant s_0 \end{cases} \tag{14}$$

is a difference version of the exponential function (since Eq. (11) is a difference version of the differential equation, $y'(x) = a(x)y(x)$). Consider now the homogenous equation corresponding to (10):

$$\Delta y(s) = a(s)y(s). \tag{15}$$

Equation (15) gives

$$y(s+1) = (1 + a(s))y(s), \tag{16}$$

which by recursion leads to

$$y(s) = e_a(n_0, n)y(n_0), \tag{17}$$

where

$$e_a(n_0, n) \overset{def}{=} \begin{cases} \prod_{i=s_0}^{s-1}(1 + a(i)), & s > s_0 \\ 1, & s \leqslant s_0 \end{cases} \tag{18}$$

is another difference version of the exponential function. Clearly, we have

Theorem 2.1

$$e_a(n_0, n).E_{-a}(n_0, n) = e_{-a}(n_0, n).E_a(n_0, n) = 1. \tag{19}$$

More generally, we have

Theorem 2.2 *If*

$$\Delta y(s) = a(s)y(s + 1),$$
$$\Delta z(s) = -a(s)z(s), \tag{20}$$

with

$$y(s_0)z(s_0) = 1, \tag{21}$$

then

$$y(s)z(s) = 1. \tag{22}$$

Prof. $\Delta(y(s)z(s)) = y(s + 1)\Delta z(s) + z(s)\Delta y(s) = y(s + 1)(-a(s))z(s) + z(s)a(s)y(s+1) = 0$. This implies that $y(s)z(s) = const.$, which by (21) gives (22), and the theorem is proved. $\square$

Nonhomogenous Cases Consider first the equation

$$\Delta y(s) = a(s)y(s + 1) + b(s). \tag{23}$$

Solving (23) by the method of variation of constants or method of Lagrange, we suppose that

$$\Delta y_0(s) = a(s)y_0(s + 1) \tag{24}$$

and search the solution of (23) as

$$y(s) = c(s)y_0(s) \tag{25}$$

where $c(s)$ is to be determined. Placing (25) in (23) and using (24), we get

$$y_0(s)\Delta c(s) = b(s), \tag{26}$$

or

$$c(s) = c + \sum_{i=s_0}^{s-1} y_0^{-1}(i)b(i). \tag{27}$$

Placing this in (25), we get

$$y(s) = y_0(s)c + y_0(s) \sum_{i=s_0}^{s-1} y_0^{-1}(i)b(i), \tag{28}$$

with $c = y_0^{-1}(s_0)y_0(s_0)$ (we suppose that $\sum_{i=s_1}^{s_2} h(i) = 0$, if $s_1 > s_2$), or equivalently

$$y(s) = \phi(s, s_0)\left[y(s_0) + \sum_{i=s_0}^{s-1} \phi(s_0, i)b(i)\right], \tag{29}$$

where $\phi(a, b) = y_0(a)y_0^{-1}(b)$.

Consider now the nonhomogenous equation

$$\Delta y(s) = a(s)y(s) + b(s). \tag{30}$$

Here also, solving the equation by the method of Lagrange, we get

$$y(s) = y_0(s)c + y_0(s) \sum_{i=s_0}^{s-1} y_0^{-1}(i+1)b(i), \tag{31}$$

where

$$\Delta y_0(s) = a(s)y_0(s) \tag{32}$$

and $c = y_0^{-1}(s_0)y(s_0)$, or equivalently,

$$y(s) = \phi(s, s_0)\left[y(s_0) + \sum_{i=s_0}^{s-1} \phi(s_0, i+1)b(i)\right]. \tag{33}$$

3   Difference Integral Inequalities

In this section, we deal with the main content of the work, that is we establish the mentioned integral inequalities. In the first two subsections, where we prove the Hölder, Cauchy-Schwartz, and Minkowski inequalities, we refer to classical recipes currently used in differential situations. In the last three sections, where we prove the Grönwall, Bernoulli, and Lyapunov inequalities, we mainly rely on the method of variation of constants of Lagrange.

3.1 Hölder and Cauchy-Schwartz Inequalities

Theorem 3.1 (Hölder Inequality) *Let $a, b \in \mathbb{Z}$. For all functions $f, g : [a, b] \cap \mathbb{Z} \longrightarrow \mathbb{R}$, we have*

$$\int_a^b |f(s)g(s)| d_\Delta s \leqslant \left(\int_a^b |f(s)|^\alpha d_\Delta s \right)^{\frac{1}{\alpha}} \left(\int_a^b |g(s)|^\beta d_\Delta s \right)^{\frac{1}{\beta}}, \qquad (1)$$

with $\frac{1}{\alpha} + \frac{1}{\beta} = 1$.

Proof For $A, B \in [0, \infty[$, by the concavity of the logarithm function, we have

$$log \left(\frac{A^{\frac{1}{\alpha}}}{\alpha} + \frac{B^{\frac{1}{\beta}}}{\beta} \right) \geqslant \frac{log(A^\alpha)}{\alpha} + \frac{log(B^\beta)}{\beta} = log(AB). \qquad (2)$$

which leads to

$$A^{\frac{1}{\alpha}} B^{\frac{1}{\beta}} \leqslant \frac{A}{\alpha} + \frac{B}{\beta}. \qquad (3)$$

Now let

$$A(s) = \frac{|f(s)|^\alpha}{\int_a^b |f(s)|^\alpha d_\Delta s}; \; B(s) = \frac{|g(s)|^\beta}{\int_a^b |g(s)|^\beta d_\Delta s} \qquad (4)$$

with

$$\left(\int_a^b |f(s)|^\alpha d_\Delta s \right) \left(\int_a^b |g(s)|^\beta d_\Delta s \right) \neq 0, \qquad (5)$$

since $\int_a^b |f(s)|^\alpha d_\Delta s = 0$ or $\int_a^b |g(s)|^\beta d_\Delta s = 0$ implies that $f(s) \equiv 0$ or $g(s) \equiv 0$ and (1) becomes an identity.

Next, substituting A and B in (3) and integrating from a to b, considering Corollary 2.1, one gets

$$\int_a^b \frac{|f(s)|}{\left(\int_a^b |f(s)|^\alpha d_\Delta s \right)^{\frac{1}{\alpha}}} \frac{|g(s)|}{\left(\int_a^b |g(s)|^\beta d_\Delta s \right)^{\frac{1}{\beta}}} d_\Delta s$$

$$\leqslant \int_a^b \left\{ \frac{1}{\alpha} \frac{|f(s)|^\alpha}{\int_a^b |f(s)|^\alpha d_\Delta s} + \frac{1}{\beta} \frac{|g(s)|^\beta}{\int_a^b |g(s)|^\beta d_\Delta s} \right\} d_\Delta s$$

$$= \frac{1}{\alpha} + \frac{1}{\beta} = 1, \qquad (6)$$

which gives directly the Hölder inequality and the theorem is proved. $\square$

If we set $\alpha = \beta = 2$ in the Hölder inequality (1), we get the Cauchy-Schwartz inequality.

Corollary 3.1 (Cauchy-Schwartz Inequality) *Let $a, b \in \mathbb{Z}$. For all functions $f, g : [a, b] \cap \mathbb{Z} \longrightarrow \mathbb{R}$, we have*

$$\int_a^b |f(s)g(s)| d_\Delta s \leqslant \sqrt{\left(\int_a^b |f(s)|^2 d_\Delta s \right) \left(\int_a^b |g(s)|^2 d_\Delta s \right)}. \qquad (7)$$

Next, we can use the Hölder inequality to prove the Minkowski one.

3.2 Minkowski Inequality

Theorem 3.2 (Minkowski Inequality) *Soient $a, b \in \mathbb{Z}$. For all functions $f, g : [a, b] \cap \mathbb{Z} \longrightarrow \mathbb{R}$, we have*

$$\left(\int_a^b |f(s) + g(s)| d_\Delta s \right)^{\frac{1}{\alpha}} \leqslant \left(\int_a^b |f(s)|^\alpha d_\Delta s \right)^{\frac{1}{\alpha}} + \left(\int_a^b |g(s)|^\alpha d_\Delta s \right)^{\frac{1}{\alpha}}. \qquad (8)$$

Proof We apply the Hölder inequality to obtain

$$\int_a^b |f(s) + g(s)|^\alpha d_\Delta s = \int_a^b |f(s) + g(s)|^{\alpha-1} |f(s) + g(s)| d_\Delta s$$

$$\leqslant \int_a^b |f(s) + g(s)|^{\alpha-1} |f(s)| d_\Delta s + \int_a^b |f(s) + g(s)|^{\alpha-1} |g(s)| d_\Delta s$$

$$\leqslant \left(\int_a^b |f(s) + g(s)|^{(\alpha-1)\beta} d_\Delta s \right)^{\frac{1}{\beta}} \left[\left(\int_a^b |f(s)|^\alpha d_\Delta s \right)^{\frac{1}{\alpha}} + \left(\int_a^b |g(s)|^\alpha d_\Delta s \right)^{\frac{1}{\alpha}} \right].$$

Dividing the two members of the inequality by $\left(\int_a^b |f(s) + g(s)|^{(\alpha-1)\beta} d_\Delta s \right)^{\frac{1}{\beta}}$, with $(\alpha - 1)\beta = \alpha$, we get

$$\left(\int_a^b |f(s) + g(s)|^\alpha d_\Delta s \right)^{1 - \frac{1}{\beta}} \leqslant \left[\left(\int_a^b |f(s)|^\alpha d_\Delta s \right)^{\frac{1}{\alpha}} + \left(\int_a^b |g(s)|^\alpha d_\Delta s \right)^{\frac{1}{\alpha}} \right],$$

which is the Minkowski inequality since $1 - \frac{1}{\beta} = \frac{1}{\alpha}$. $\square$

3.3 Grönwall Inequality

Let's prove first the following:

Lemma 3.1 *Given y, f, a real valued functions defined on $\mathbb{Z}$, with $a(s) \geqslant 0$. Suppose that $y_0(s)$ is the solution of $\Delta y_0(s) = a(s)y_0(s)$, such that $y_0(s_0) = 1$.*
In that case, if

$$\Delta y(s) \leqslant a(s)y(s) + f(s) \tag{9}$$

for all $s \in \mathbb{Z}$, then

$$y(s) \leqslant y_0(s)y(s_0) + y_0(s) \int_{s_0}^{s} y_0^{-1}(s+1) f(s) d_\Delta s. \tag{10}$$

Proof Let $y_0(s)$ be the solution of the homogenous equation

$$\Delta y_0(s) = a(s)y_0(s). \tag{11}$$

Searching the solution $y(s)$ of (9) verifying (10), by the method of variation of constants

$$y(s) = c(s)y_0(s), \tag{12}$$

where $c(s)$ is unknown, we place (12) in (9), considering (11) and get

$$y_0(s+1)\Delta c(s) \leqslant f(s). \tag{13}$$

Given the fact that $a(s) \geqslant 0$, we have that $y_0(s) > 0$ and the relation (13) simplifies in

$$\Delta c(s) \leqslant y_0^{-1}(s+1) f(s). \tag{14}$$

Integrating the two members of the inequality from s_0 to s, we get

$$c(s) - c(s_0) \leqslant \int_{s_0}^{s} y_0^{-1}(s+1) f(s) d_\Delta s. \tag{15}$$

Since $y_0(s_0) = 1$, (12) gives $c(s_0) = y(s_0)$, and (15) simplifies in

$$c(s) \leqslant y(s_0) + \int_{s_0}^{s} y_0^{-1}(s+1) f(s) d_\Delta s. \tag{16}$$

Hence

$$c(s)y_0(s) \leqslant y_0(s)\left[y(s_0) + \int_{s_0}^{s} y_0^{-1}(s+1)f(s)d_\Delta s\right], \tag{17}$$

which gives the expected result:

$$y(s) \leqslant y_0(s)y(s_0) + y_0(s)\int_{s_0}^{s} y_0^{-1}(s+1)f(s)d_\Delta s.$$

□

Considering the Theorem 2.1, we obtain the following

Corollary 3.2 *If the functions* y, f, a *verify the conditions of Lemma 3.1, then*

$$y(s) \leqslant y(s_0)e_a(s_0,s) + e_a(s_0,s)\int_{s_0}^{s} E_{-a}(s_0,s+1)f(s)d_\Delta s.$$

Lemma 3.2 *Given* y, f, a *a real valued functions defined on* $\mathbb{Z}$, *with* $a(s) \leqslant 0$.
Suppose that $y_0(s)$ *is the solution of* $\Delta y_0(s) = a(s)y_0(s+1)$, *such that* $y_0(s_0) = 1$.
In that case, if

$$\Delta y(s) \leqslant a(s)y(s+1) + f(s) \tag{18}$$

for all $s \in \mathbb{Z}$, *then*

$$y(s) \leqslant y_0(s)y(s_0) + y_0(s)\int_{s_0}^{s} y_0^{-1}(s)f(s)d_\Delta s \tag{19}$$

Proof Let $y_0(s)$ be the solution of the homogenous equation

$$\Delta y_0(s) = a(s)y_0(s+1). \tag{20}$$

Searching the solution $y(s)$ of (18) verifying (19), by the method of variation of constants

$$y(s) = c(s)y_0(s), \tag{21}$$

where $c(s)$ is unknown, we place (21) in (18), considering (20) and get

$$y_0(s)\Delta c(s) \leqslant f(s). \tag{22}$$

Given the fact that $a(s) \leqslant 0$, we have that $y_0(s) > 0$ and the relation (22) simplifies in

$$\Delta c(s) \leqslant y_0^{-1}(s) f(s).$$ (23)

Integrating the two members of the inequality from s_0 to s, we get

$$c(s) - c(s_0) \leqslant \int_{s_0}^{s} y_0^{-1}(s) f(s) d_\Delta s.$$ (24)

Since $y_0(s_0) = 1$, (21) gives $c(s_0) = y(s_0)$, and (24) simplifies in

$$c(s) \leqslant y(s_0) + \int_{s_0}^{s} y_0^{-1}(s) f(s) d_\Delta s.$$ (25)

Hence

$$c(s) y_0(s) \leqslant y_0(s) \left[y(s_0) + \int_{s_0}^{s} y_0^{-1}(s) f(s) d_\Delta s \right],$$ (26)

which gives the expected result:

$$y(s) \leqslant y_0(s) y(s_0) + y_0(s) \int_{s_0}^{s} y_0^{-1}(s) f(s) d_\Delta s.$$

□

For the same reasons as the Corollary 3.2, we obtain the following:

Corollary 3.3 *If the functions y, f, a verify the conditions of Lemma 3.2, then*

$$y(s) \leqslant y(s_0) E_a(s_0, s) + E_a(s_0, s) \int_{s_0}^{s} e_{-a}(s_0, s) f(s) d_\Delta s.$$

We can now prove the following:

Theorem 3.3 (Grönwall Inequality) *Let y, f, a be real valued functions defined on $\mathbb{Z}$, with $a(s) \geqslant 0$.*
Suppose that $y_0(s)$ is the solution of $\Delta y_0(s) = a(s) y_0(s)$, such that $y_0(s_0) = 1$. In that case if

$$y(s) \leqslant f(s) + \int_{s_0}^{s} y(s) a(s) d_\Delta s,$$ (27)

then

$$y(s) \leqslant f(s) + e_a(s_0, s) \int_{s_0}^{s} a(s) f(s) E_{-a}(s_0, s+1) d_\Delta s.$$ (28)

Proof Defining

$$v(s) = \int_{s_0}^{s} y(s)a(s)d_\Delta s, \tag{29}$$

(27) gives

$$y(s) \leqslant f(s) + v(s), \tag{30}$$

and

$$\Delta v(s) = y(s)a(s) \leqslant f(s)a(s) + a(s)v(s). \tag{31}$$

By the Corollary 3.2 of Lemma 3.1, the inequality (31) leads to

$$v(s) \leqslant v(s_0)e_a(s_0, s) + e_a(s_0, s) \int_{s_0}^{s} a(s)f(s)E_{-a}(s_0, s+1)d_\Delta s \tag{32}$$

Since $v(s_0) = 0$, (30) and (32) imply that

$$y(s) \leqslant f(s) + e_a(s_0, s) \int_{s_0}^{s} a(s)f(s)E_{-a}(s_0, s+1)d_\Delta s, \tag{33}$$

which is the expected Grönwall inequality. □

As direct consequences, we obtain the following results:

Corollary 3.4 *Let y, f, a be real valued functions defined on $\mathbb{Z}$, with $a(s) \geqslant 0$. If*

$$y(s) \leqslant \int_{s_0}^{s} y(s)a(s)d_\Delta s, \tag{34}$$

for all $s \in \mathbb{Z}$, then

$$y(s) \leqslant 0. \tag{35}$$

Proof This follows from the Theorem 3.3 with $f(s) \equiv 0$. □

Corollary 3.5 *Let $a(s) \geqslant 0$ and $\alpha \in \mathbb{R}$. If*

$$y(s) \leqslant \alpha + \int_{s_0}^{s} y(s)a(s)d_\Delta s, \tag{36}$$

for all $s \in \mathbb{Z}$, then

$$y(s) \leqslant \alpha e_a(s_0, s). \tag{37}$$

Proof From the Grönwall inequality with $f(s) = \alpha$, one gets

$$y(s) \leq \alpha + e_a(s_0, s) \int_{s_0}^{s} \alpha a(s) E_{-a}(s_0, s+1) d_\Delta s$$

$$= \alpha \left(1 - e_a(s_0, s) \int_{s_0}^{s} \Delta E_{-a}(s_0, s) d_\Delta s \right)$$

$$= \alpha \left(1 - e_a(s_0, s) [E_{-a}(s_0, s) - E_{-a}(s_0, s_0)] \right)$$

$$= \alpha - \alpha e_a(s_0, s) E_{-a}(s_0, s) + \alpha e_a(s_0, s)$$

$$= \alpha e_a(s_0, s),$$

$$(38)$$

which gives the expected inequality. $\square$

3.4   Bernoulli Inequality

Theorem 3.4 (Bernoulli Inequality) *Let $\alpha \in \mathbb{R}$. Then for all $s, s_0 \in \mathbb{Z}$, with $s > s_0$, we have*

$$e_a(s_0, s) \geq 1 + \alpha(s - s_0). \tag{39}$$

Proof Let $y(s) = \alpha(s - s_0), s > s_0$. Then $\Delta y(s) = \alpha$ and $\alpha y(s) + \alpha = \alpha^2(s - s_0) + \alpha \geq \alpha = \Delta y(s)$, which implies that $\Delta y(s) \leq \alpha y(s) + \alpha$.

By the Corollary 3.2 of Lemma 3.1, we obtain

$$y(s) \leq y(s_0) e_\alpha(s_0, s) + e_\alpha(s_0, s) \int_{s_0}^{s} \alpha E_{-\alpha}(s_0, s+1) d_\Delta s,$$

$$= -e_\alpha(s_0, s) \int_{s_0}^{s} \Delta E_{-\alpha}(s_0, s) d_\Delta s, \ (y(x_0) = 0)$$

$$= -e_\alpha(s_0, s) [E_{-\alpha}(s_0, s) - 1]$$

$$= -1 + e_\alpha(s_0, s).$$

$$(40)$$

Hence $e_\alpha(s_0, s) \geq 1 + \alpha(s - s_0)$, with $s > s_0$, as expected. $\square$

3.5 Lyapunov Inequality

Let $f : \mathbb{Z} \longrightarrow [0, \infty[$. Consider the Sturm-Liouville difference equation

$$\Delta^2 u(s) + f(s)u(s+1) = 0, \, s \in \mathbb{Z}. \tag{41}$$

Define the function F by

$$F(y) = \int_a^b \left[(\Delta y(s))^2 - f(s)y^2(s+1) \right] d_\Delta s. \tag{42}$$

We prove first the following lemmas:

Lemma 3.3 *Let $u(s)$ be a nontrivial solution of the Sturm-Liouville difference equation (41). In that case, for all y belonging to the domain of definition of F, the following equality is verified,*

$$F(y) - F(u) - F(y - u) = 2(y - u)(b)\Delta u(b) - 2(y - u)(a)\Delta u(a). \tag{43}$$

Proof We have

$$
\begin{aligned}
& F(y) - F(u) - F(y - u) \\
&= \int_a^b [(\Delta y(s))^2 - f(s)y^2(s+1) - (\Delta u(s))^2 \\
&\quad + f(s)u^2(s+1) - (\Delta(y-u)(s))^2 + f(s)(y-u)^2(s+1)]d_\Delta s \\
&= 2 \int_a^b [-(\Delta u(s))^2 + f(s)u^2(s+1) + \Delta y(s)\Delta u(s) \\
&\quad - f(s)y(s+1)u(s+1)]d_\Delta s \\
&= 2 \int_a^b [\Delta y(s)\Delta u(s) + y(s+1)\Delta^2 u(s) - (\Delta u(s))^2 \\
&\quad - \Delta^2 u(s)u(s+1)]d_\Delta s \\
&= 2 \int_a^b [\Delta[y(s)\Delta u(s)] - \Delta[u(s)\Delta u(s)]]d_\Delta s \\
&= 2 \int_a^b \Delta[(y(s) - u(s))\Delta u(s)]d_\Delta s \\
&= 2(y(b) - u(b))\Delta u(b) - 2(y(a) - u(a))\Delta u(a), \tag{44}
\end{aligned}
$$

which proves the lemma. $\square$

Lemma 3.4 *Let y be in the domain of definition of F. For all $c, d \in [a, b] \cap \mathbb{Z}$, $a, b \in \mathbb{Z}$ and $a \leqslant c \leqslant d \leqslant b$, we have*

$$\int_c^d (\Delta y(s)^2) d_\Delta s \geqslant \frac{(y(d) - y(c))^2}{d - c}. \tag{45}$$

Proof Let $u(s) = \frac{y(d) - y(c)}{d - c} s + \frac{dy(c) - cy(d)}{d - c}$. Then $\Delta u(s) = \frac{y(d) - y(c)}{d - c}$ and $\Delta^2 u(s) = 0$. This proves that $u(s)$ is a solution of (41) with $f(s) = 0$ for all $s \in \mathbb{Z}$ and $F(y) = \int_a^b (\Delta y(s))^2 d_\Delta s$, for all y from the domain of definition of F. By Lemma 3.3, we get $F(s) - F(u) - F(y - u) = 0$, and consequently $F(y) = F(u) + F(y - u) \geqslant F(u)$. This leads to the following result:

$$\int_c^d (\Delta y(s))^2 d_\Delta s \geqslant \int_c^d (\Delta u(s))^2 d_\Delta s$$

$$= \int_c^d \left(\frac{y(d) - y(c)}{d - c} \right)^2 d_\Delta s$$

$$= \frac{(y(d) - y(c))^2}{d - c}, \tag{46}$$

which proves the lemma. $\square$

Theorem 3.5 (Lyapunov Inequality) *Given $f : \mathbb{Z} \longrightarrow [0, \infty[$ and u a nontrivial solution of Eq. (41) with $u(a) = u(b) = 0$, $a, b \in \mathbb{Z}$ and $a < b$, then*

$$\int_a^b f(s) d_\Delta s \geqslant \frac{4}{b - a}. \tag{47}$$

Proof By the Lemma 3.3 with $y = 0$ and $u(a) = u(b) = 0$, one gets $F(0) - F(u) - F(-u) = -2u(b)\Delta u(b) + 2u(a)\Delta u(a)$. This gives $F(u) = 0$ since $F(0) = 0$ and $F(u) = F(-u)$. Thus

$$F(u) = \int_a^b \left[(\Delta u(s))^2 - f(s) u^2(s + 1) \right] d_\Delta s = 0. \tag{48}$$

Let $M = max \left[u^2(s); s \in [a, b] \cap \mathbb{Z} \right]$ and $c \in [a, b] \cap \mathbb{Z}$ such that $u^2(c) = M$. Then $M = u^2(c) \geqslant u^2(s+1)$ and using (48), Lemma 3.4 and the fact that $u(a) = u(b) = 0$, we get

$$M \int_a^b f(s) d_\Delta s \geqslant \int_a^b f(s) u^2(s + 1) d_\Delta s$$

$$= \int_a^b (\Delta(us))^2 d_\Delta s$$

$$= \int_a^c (\Delta(us))^2 d_\Delta s + \int_c^b (\Delta(us))^2 d_\Delta s$$

$$\geqslant \frac{(u(c) - u(a))^2}{c - a} + \frac{(u(b) - u(c))^2}{b - c}$$

$$= M \left[\frac{1}{c - a} + \frac{1}{b - c} \right] \geqslant M \frac{4}{b - a}.$$

(49)

which proves the Lyapunov inequality. $\square$

References

1. R. Askey, J. Wilson Some basic hypergeometric orthogonal polynomials that generalize the Jacobi polynomials. Mem. Am. Math. Soc. **54**, 1–55 (1985)
2. G. Bangerezako, J.P. Nuwacu, q-Non uniform difference calculus and classical integral inequalities. J. Inequal. Appl. (accepted for publication)
3. S. Elaydi, *An Introduction to Difference Equations* (Springer, New York, 2006)
4. H.F. Jackson, q-Difference equations. Am. J. Math. **32**, 305–314 (1910)
5. A.P. Magnus, *Associated Askey-Wilson polynomials as Laguerre-Hahn Orthogonal Polynomials*. Springer Lectures Notes in Mathematics, vol. 1329 (Springer, Berlin 1988), pp. 261–278
6. A.P. Magnus, Special nonuniform lattices (snul) orthogonal polynomials on discrete dense sets of points. J. Comput. Appl. Math. **65**, 253–265 (1995)

Theoretical and Numerical Comparisons of the Parameter Estimator of the Fractional Brownian Motion

Jean-Marc Bardet

This paper is dedicated to Norbert Hounkonnou who is not really a specialist of probability and statistics (...) but who often taught these topics and has mainly allowed them to significantly develop in Benin by his unwavering support to Masters and PhD of Statistics.

Abstract The fractional Brownian motion which has been defined by Kolmogorov (CR (Doklady) Acad Sci URSS (N.S.) 26:115–118) and numerous papers was devoted to its study since its study in Mandelbrot and Van Ness (SIAM Rev 10:422–437, 1968) [19] present it as a paradigm of self-similar processes. The self-similarity parameter, also called the Hurst parameter, commands the dynamic of this process and the accuracy of its estimation is often crucial. We present here the main and used methods of estimation, with the limit theorems satisfied by the estimators. A numerical comparison is also provided allowing to distinguish between the estimators.

Keywords Fractional Brownian motion · Long memory process · Parametric estimation

1 Introduction

The fractional Brownian motion (fBm for short) has been studied a lot since the seminal papers of Kolmogorov [16] and Mandelbrot and Van Ness [19]. A simple way to present this extension of the classical Wiener Brownian motion is to define it

J.-M. Bardet (✉)
University Paris 1 Panthéon-Sorbonne, Paris, France
e-mail: bardet@univ-paris1.fr

© Springer Nature Switzerland AG 2018
T. Diagana, B. Toni (eds.), *Mathematical Structures and Applications*,
STEAM-H: Science, Technology, Engineering, Agriculture,
Mathematics & Health, https://doi.org/10.1007/978-3-319-97175-9_6

153

from its both first moments. Hence, an fBm with parameter $(H, \sigma^2) \in (0, 1] \times (0, \infty)$ is a centered Gaussian process $X = \{X(t), t \in \mathbb{R}\}$ having stationary increments and such as

$$\mathrm{cov}\big(X(s), X(t)\big) = \sigma^2 R_H(s, t) = \frac{1}{2} \sigma^2 \left(|s|^{2H} + |t|^{2H} - |t - s|^{2H}\right) \qquad (s, t) \in \mathbb{R}^2.$$

As a consequence, $\mathrm{Var}(X(t)) = \sigma^2 |t|^{2H}$ for any $t \in \mathbb{R}$, which induces that X is the only H-self-similar centered Gaussian process having stationary increments. More detail on this process can be found in the monograph of Samorodnitsky and Taqqu [21].

We consider the following statistical problem. We suppose that a trajectory $(X(1), \cdots, X(N))$ of X is observed and we would like to estimate the parameters H and σ^2 which are unknown.

Remark 1.1 It is also possible to consider an observed path $(X(1/N), X(2/N), \cdots, X(1))$ of X. Using the self-similarity property of X, the distributions of $(X(1), \cdots, X(N))$ and $N^H(X(1/N), X(2/N), \cdots, X(1))$ are the same.

This statistical model is parametric and it is natural to estimate $\theta = (H, \sigma^2)$ using a parametric method. Hence, on the one hand, in the forthcoming Sect. 2, two possible parametric estimators are presented. On the other hand, several other used and famous semi-parametric estimators are studied in Sect. 3. For each estimator, its asymptotic behavior is stated and the main references recalled. Finally, Sect. 4 is devoted to a numerical study from Monte-Carlo experiments, allowing to obtain some definitive conclusions with respect to the estimators.

2   Two Classical Parametric Estimators

Since X is a Gaussian process for which the distribution is integrally defined when θ is known, a parametric method such as Maximum Likelihood Estimator (MLE) is a natural choice of estimation and it provides efficient estimators. As X is a process having stationary increments, it is appropriate to define Y the process of its increments, i.e. the fractional Gaussian noise,

$$Y = \{Y(t), t \in \mathbb{R}\} = \{X(t) - X(t - 1), t \in \mathbb{R}\},$$

with covariogram $r_Y(k) = \mathrm{cov}\big(Y(t), Y(t + k)\big)$ satisfying

$$r_Y(k) = \frac{1}{2} \sigma^2 \left(|k + 1|^{2H} + |k - 1|^{2H} - 2|k|^{2H}\right) \qquad k \in \mathbb{R}.$$

The spectral density f_Y of Y is defined for $\lambda \in [-\pi, \pi]$ by (see Sinaï [22] or Fox and Taqqu [11]):

$$f_H(\lambda) := \frac{1}{2\pi} \sum_{k \in \mathbb{Z}} r_Y(k) \, e^{ik\lambda}$$

$$= 2\sigma^2 \, \Gamma(2H + 1) \sin(\pi H) \, (1 - \cos \lambda) \sum_{k \in \mathbb{Z}} |\lambda + 2k\pi|^{-1-2H}. \quad (2.1)$$

Hence, since $(X(1), \cdots, X(N))$ is observed, $(Y(1), \cdots, Y(N))$ is observed (we assume $X(0) = 0$).

2.1 Maximum Likelihood Estimation

The likelihood $L_{(Y(1),\cdots,Y(N))}(y_1, \cdots, y_N)$ of $(Y(1), \cdots, Y(N))$ is the Gaussian probability density of $(Y(1), \cdots, Y(N))$, which can be written

$$L_{(Y(1),\cdots,Y(N))}(y_1, \cdots, y_N) = \frac{(2\pi)^{N/2}}{\det\left(\Sigma_N(H, \sigma^2)\right)} \times$$

$$\exp\left(-\frac{1}{2}(y_1, \cdots, y_N)\Sigma_N^{-1}(H, \sigma^2)(y_1, \cdots, y_N)'\right),$$

where the definite positive covariance matrix $\Sigma_N(H, \sigma^2)$ is such as

$$\Sigma_N(H, \sigma^2) = \left(r(|i - j|)\right)_{1 \leqslant i, j \leqslant N}.$$

Then the MLE $\widehat{\theta}_N = (\widehat{H}_N, \widehat{\sigma_N^2})$ of θ is defined as:

$$(\widehat{H}_N, \widehat{\sigma_N^2}) = \underset{(H,\sigma^2)\in(0,1)\times(0,\infty)}{\text{Arg max}} L_{(Y(1),\cdots,Y(N))}(Y(1), \cdots, Y(N)).$$

As it is generally done, it can be more convenient to minimize the contrast defined by $-2 \log\left(L_{(Y(1),\cdots,Y(N))}(Y(1), \cdots, Y(N))\right)$.

The asymptotic behavior of this estimator has been first obtained in Dahlhaus [9]. The main results are the following:

Theorem 2.1 *The estimator $(\widehat{H}_N, \widehat{\sigma_N^2})$ is asymptotically efficient and satisfies*

$$\begin{pmatrix} \widehat{H}_N \\ \widehat{\sigma_N^2} \end{pmatrix} \xrightarrow[N \to \infty]{a.s.} \begin{pmatrix} H \\ \sigma^2 \end{pmatrix} \quad (2.2)$$

and

$$\sqrt{N}\left(\begin{pmatrix}\widehat{H}_N\\\widehat{\sigma}_N^2\end{pmatrix} - \begin{pmatrix}H\\\sigma^2\end{pmatrix}\right) \xrightarrow[N\to\infty]{\mathscr{D}} \mathscr{N}_2\left(0,\,\Gamma_0^{-1}(H,\sigma^2)\right), \quad (2.3)$$

where $\Gamma_0^{-1}(H,\sigma^2)$ is the limit of $\frac{1}{N} I_N^{-1}(H,\sigma^2)$ and $I_N(H,\sigma^2)$ is the Fisher information matrix of $(Y(1),\cdots,Y(N))$. Moreover,

$$\Gamma_0(H,\sigma^2) = \frac{1}{4\pi} \int_{-\pi}^{\pi} \left(\frac{\partial}{\partial\theta}\log f_\theta(\lambda)\right)\left(\frac{\partial}{\partial\theta}\log f_\theta(\lambda)\right)' d\lambda, \quad (2.4)$$

and $f_\theta(\lambda) = \sigma^2 g_H(\lambda)$ where

$$g_H(\lambda) = 2\sin(\pi H)\Gamma(2H+1)(1-\cos\lambda)\sum_{j\in\mathbb{Z}}|\lambda+2\pi j|^{-2H-1}.$$

The asymptotic covariance $\Gamma_0(H,\sigma^2)$ cannot be really simplified, we just can obtain:

$$\Gamma_0^{-1}(H,\sigma^2) = \frac{1}{\frac{1}{2}a_H - b_H^2}\begin{pmatrix}\frac{1}{2} & -\sigma^2 b_H\\-\sigma^2 b_H & \sigma^4 a_H\end{pmatrix} \quad (2.5)$$

where $a_H = \frac{1}{4\pi}\int_{-\pi}^{\pi}\left(\frac{\partial}{\partial H}\log g_H(\lambda)\right)^2 d\lambda$ and $b_H = \frac{1}{4\pi}\int_{-\pi}^{\pi}\frac{\partial}{\partial H}\log g_H(\lambda)\,d\lambda$.

2.2 Whittle Estimation

The MLE is an asymptotically efficient estimator but it has two main drawbacks: first, it is a parametric estimator which can only be used, stricto sensu, to fBm and its use is numerically limited since the computation of the likelihood requires to inverse the matrix $\Sigma_N(H,\sigma^2)$ and this is extremely time consuming when $N \geqslant 5000$ and impossible when $N \geqslant 10000$ (with a 2014-software). In Whittle [28], a general approximation of the likelihood for Gaussian stationary processes Y has been first proposed. This consists in writing for Y depending on a parameter vector θ:

$$-\frac{1}{N}\log L_{(Y(1),\cdots,Y(N))}(Y(1),\cdots,Y(N)) - \frac{1}{2}\log(2\pi)$$

$$\xrightarrow[N\to\infty]{\mathscr{D}} \widehat{U}_N(\theta) = \frac{1}{4\pi}\int_{-\pi}^{\pi}\left(\log(f_\theta(\lambda)) + \frac{\widehat{I}_N(\lambda)}{f_\theta(\lambda)}\right)d\lambda$$

where $\widehat{I}_N(\lambda) = \frac{1}{2\pi N}\left|\sum_{k=1}^{N}Y(k)e^{-ik\lambda}\right|^2$ is the periodogram of Y.

Then $\widetilde{\theta}_N = (\widetilde{H}_W, \widetilde{\sigma}_N^2) = \mathrm{Argmin}_{\theta \in \Theta} \widehat{U}_N(\theta)$ is called the Whittle estimator of θ. In case of the fractional Gaussian noise, Dahlhaus [9] achieved the results of Fox and Taqqu [11] and proved the following limit theorem:

Theorem 2.2 *The estimator* $(\widetilde{H}_W, \widetilde{\sigma}_N^2)$ *is asymptotically efficient and satisfies*

$$
\begin{pmatrix} \widetilde{H}_W \\ \widetilde{\sigma}_N^2 \end{pmatrix} \xrightarrow[N \to \infty]{a.s.} \begin{pmatrix} H \\ \sigma^2 \end{pmatrix} \tag{2.6}
$$

and

$$
\sqrt{N}\left(\begin{pmatrix} \widetilde{H}_W \\ \widetilde{\sigma}_N^2 \end{pmatrix} - \begin{pmatrix} H \\ \sigma^2 \end{pmatrix} \right) \xrightarrow[N \to \infty]{\mathscr{D}} \mathscr{N}_2\left(0, \Gamma_0^{-1}(H, \sigma^2)\right), \tag{2.7}
$$

with $\Gamma_0(H, \sigma^2)$ *defined in (2.4).*

Hence, the Whittle estimator $(\widetilde{H}_N, \widetilde{\sigma}_N^2)$ has the same asymptotic behavior than the MLE while its numerical accuracy is clearly better: in case of fractional Gaussian noise, and therefore in case of the fBm, the Whittle estimator has to be preferred to the MLE.

3 Other Classical Semi-parametric Estimators

As we said previously, we present now some classical semi-parameteric methods frequently used for estimating the parameter H of an fBm, but also applied to other long-range dependent or self-similar processes.

3.1 R/S and Modified R/S Statistics

The first estimator which has been applied to an fBm, and more precisely to a fractional Gaussian noise has been the R/S estimator. This estimator defined by the hydrologist Hurst [14] was devoted to estimate the long-range dependent parameter (also called the Hurst parameter) of a long memory process. Lo [18] introduced the modified R/S statistic for a times series X which is defined as

$$
\widehat{Q}_N(q) = \frac{1}{\widehat{s}_{N,q}} \left(\max_{1 \leqslant k \leqslant N} \sum_{i=1}^{k} (X(i) - \overline{X}_N) - \min_{1 \leqslant k \leqslant N} \sum_{i=1}^{k} (X(i) - \overline{X}_N) \right) \tag{3.1}
$$

where $\overline{X}_N = \frac{1}{N}(X(1) + \cdots + X(N))$ is the sample mean and $\widehat{s}_{N,q}^2$ is an estimator of $\sigma^2 = \sum_{i \in \mathbb{Z}} \mathrm{cov}(X(0), X(j))$ defined by

$$\widehat{s}_{N,q}^2 = \frac{1}{N} \sum_{i=1}^{N} (X(i) - \overline{X}_N)^2 + 2 \sum_{i=1}^{q} \omega_i(q)\, \widehat{\gamma}_N(i) \qquad (3.2)$$

$$\text{with } \begin{cases} \omega_i(q) = 1 - i/(q+1) \\ \widehat{\gamma}_N(i) = \frac{1}{N} \sum_{j=1}^{N-i} (X(j) - \overline{X}_N)(X(j+i) - \overline{X}_N) \end{cases} \qquad (3.3)$$

The classical R/S statistic corresponds to $q = 0$. In such case, we have the following asymptotic behavior when X is an fGn:

Proposition 3.1 *If X is an fGn with parameter $H > 1/2$, then (see Li et al. [17])*

$$\frac{1}{N^H} E\big[\widehat{Q}_N(0)\big] \xrightarrow[N \to \infty]{} E\big[\max_{0 \leqslant t \leqslant 1} B_H(t) - \min_{0 \leqslant t \leqslant 1} B_H(t)\big], \qquad (3.4)$$

with B_H the fractional bridge with parameter H, i.e., $B_H(t) = X_H(t) - t X_H(1)$ for $t \in [0, 1]$ where X_H is a standardized fBm of parameter 1.

Using this asymptotic behavior of the expectation, an estimator of H has been defined. First the trajectory $(X(1), \cdots, X(N))$ is divided in K blocks of length N/K and $\widehat{Q}_{n_i}(0)$ is averaged for several values of n_i such as $n_i \xrightarrow[N \to \infty]{} \infty$. Then, a log-log regression of $(\widehat{Q}_{n_i}(0))$ onto (n_i) provides a slope $\widehat{H}_{RS}$ which is an estimator of H since $\log E\big[\widehat{Q}_{n_i}(0)\big] \simeq H \log(n_i) + C$ (see, for instance, Taqqu et al. [25]). Note that even in the "simple" case of the fGn, there still does not really exist a convincing asymptotic study of such an estimator.

Lo [18] and numerical experiments in Taqqu et al. [25] have shown that this estimator is not really accurate and Lo [18] proposed an extension of this R/S statistic: this is the modified R/S statistic. We have the following asymptotic behavior (see Giraitis et al. [12] and Li et al. [17]):

Proposition 3.2 *If $q \xrightarrow[N \to \infty]{} \infty$ and $q/N \xrightarrow[N \to \infty]{} 0$, and X is a fGn, then:*

- *if $H \leqslant 1/2$,* $\qquad N^{-1/2}\, \widehat{Q}_N(q) \xrightarrow[N \to \infty]{\mathscr{D}} U_{R/S}$ $\qquad\qquad$ (3.5)

- *if $H > 1/2$,* $\qquad q^{H-1/2} N^{-H}\, \widehat{Q}_N(q) \xrightarrow[N \to \infty]{\mathscr{D}} Z_{R/S}$ $\qquad\quad$ (3.6)

where $U_{R/S} = \max_{0 \leqslant t \leqslant 1} B_{1/2}(t) - \min_{0 \leqslant t \leqslant 1} B_{1/2}(t)$ and $Z_{R/S} = \max_{0 \leqslant t \leqslant 1} B_H(t) - \min_{0 \leqslant t \leqslant 1} B_H(t)$, with B_H the fractional bridge with parameter H.

Then, using several values of q, $(q_1, \cdots, q_m)$, a log-log regression of $\widehat{Q}_N(q_i)$ onto q_i provides an estimator $\widehat{H}_{RSM}$ of H (the slope of the regression line is $H - \frac{1}{2}$). But there does not exist more precise result about the convergence rate of such estimator of H in the literature. Moreover, in Teverovsky et al. [26], the difficulty of selecting a right range of values for q_i is highlighting.

As a conclusion, we can say that R/S or modified R/S statistics provides estimation of H but these estimators are not really accurate.

3.2 Second-Order Quadratic Variations

Contrary to the R/S method, the second-order quadratic variations can be directly applied to self-similar processes, and hence in particular to fBm.

For presenting this method, introduced by Guyon and Leòn [13] and Istas and Lang [15], first, for $a \in \mathbb{N}^*$, define the second-order quadratic variations of $X = (X(i))_{i \in \mathbb{Z}}$ by

$$V_t^{(a)} := (X(t + 2a)) - 2X(t + a) + X(t))^2 \text{ for } t \in \mathbb{N}^*. \tag{3.7}$$

The key-point of this method is the following property:

Property 3.3 *If X is a second moment order H-self-similar process having stationary increments, with $EX^2(1) = \sigma^2$, then for all $a \in \mathbb{N}^*$ and $t \in \mathbb{N}^*$*

$$E(V_t^{(a)}) = \sigma^2 \left(4 - 2^{2H}\right) a^{2H}. \tag{3.8}$$

Therefore $\log E(V_t^{(a)}) = C + 2H \log a$ for any (a, t). This provides an idea of estimating H: if $E(V_t^{(a_i)})$ can be estimated for several different scales a_i, then the slope of the log-log-regression of $(\widehat{E}(V_t^{(a_i)}))$ onto (a_i) is $2\widehat{H}_N$, which is an estimator of H.

The common choice for scales are $a_i = i$ and the estimator of $E(V_t^{(i)})$ is the empirical mean of $V_t^{(i)}$,

$$S_N(i) = \frac{1}{N - 2i} \sum_{k=1}^{N-2i} V_k^{(i)}. \tag{3.9}$$

Then a central limit theorem can be established for $i \in N^*$:

$$\sqrt{N}\left(S_N(i) - \sigma^2\left(4 - 2^{2H}\right)i^{2H}\right) \xrightarrow[N \to \infty]{\mathscr{D}} \mathscr{N}\left(0, \gamma(i)\right)$$

with $\gamma(i) = \frac{1}{2}\sigma^4 i^{2H+1}$

$$\sum_{\ell=-\infty}^{\infty} \left(|\ell + 2|^{2H} + |\ell - 2|^{2H} - 2|\ell + 1|^{2H} - 2|\ell - 1|^{2H} + 6|\ell + 2|^{2H}\right)^2.$$

Then we can define

$$\widehat{H}_N := \frac{1}{2} \frac{A}{A A^\mathsf{T}} \left(\log(S_N(i)) \right)^\mathsf{T}_{1 \leqslant i \leqslant p},$$

where $A := \left(\log i - \frac{1}{p} \sum_{j=1}^{p} \log j \right)_{1 \leqslant i \leqslant p} \in \mathbb{R}^p$ is a row vector, A^T its transposed vector (vector-column).

As a consequence, it can be shown (see Bardet [2] or [8]) that, with $(a_0, a_1, a_2) = (1, -2, 1)$,

$$\sqrt{N}\left(\widehat{H}_N - H\right) \qquad \xrightarrow[N \to \infty]{\mathscr{D}} \mathscr{N}\left(0, \Sigma(H)\right), \quad \text{where}$$

$$\Sigma(H): \qquad\qquad\qquad = \frac{A \Gamma(H) A^\mathsf{T}}{4(A^\mathsf{T} A)^2} \quad \text{and} \qquad\qquad (3.10)$$

$$\Gamma(H) := \left(\frac{2}{i^{2H} j^{2H}} \sum_{k \in \mathbb{Z}} \left[\frac{\sum_{k_1,k_2=0}^{q} a_{k_1}^{(1)} a_{k_2}^{(1)} |ik_1 - jk_2 + k|^{2H}}{\sum_{k_1,k_2=0}^{2} a_{k_1}^{(1)} a_{k_2}^{(1)} |k_1 - k_2|^{2H}} \right]^2 \right)_{1 \leqslant i,j \leqslant p} \quad (3.11)$$

This method has a lot of advantages: low time-consuming, convergence rate close to MLE convergence rate, etc.

However it is easy to slightly improve this estimator. First, the asymptotic covariance $\Gamma(H)$ of $(S_N(i))$ is a function of H; hence, using the estimator $\widehat{H}_N$, this asymptotic covariance can be estimated. Hence, a pseudo-generalized estimator of H can be defined. More precisely, define

$$\widehat{\Gamma}_N = \Gamma\left(\widehat{H}_N\right). \qquad\qquad (3.12)$$

Then the pseudo-generalized estimator $\widetilde{H}_N$ of H is defined by:

$$\widetilde{H}_N = \frac{(B_N L)^\mathsf{T} \widehat{\Gamma}_N^{-1} \left(\log(S_N(i)) \right)^\mathsf{T}_{1 \leqslant i \leqslant p}}{2 (B_N L)^\mathsf{T} \widehat{\Gamma}_N^{-1} (B_N L)} \qquad\qquad (3.13)$$

with $I_p = (1, 1, \cdots, 1)^\mathsf{T}$, $L = (\log i)_{1 \leqslant i \leqslant p}$ and $B_N = I - \dfrac{I_p I_p^\mathsf{T} \widehat{\Gamma}_N}{I_p^\mathsf{T} \widehat{\Gamma}_N^{-1} I_p}$.

Then, from Bardet [2],

Proposition 3.4 *If X is an fBm of parameter H, then with $B = I - \dfrac{I_p I_p^\mathsf{T} \Gamma}{I_p^\mathsf{T} \Gamma^{-1} I_p}$,*

$$\sqrt{N}(\widetilde{H}_N - H) \xrightarrow[N \to \infty]{\mathscr{D}} \mathscr{N}\left(0; \Sigma'(H)\right), \quad \text{with} \quad \Sigma'(H) = \frac{1}{4^t (BL) G^{-1} (BL)}. \quad (3.14)$$

From Gauss-Markov Theorem, the asymptotic variance $\Sigma'(H)$ is smaller or equal to $\Sigma(H)$ and thus the estimator $\widetilde{H}_N$ is more accurate than $\widehat{H}_N$.

Another improvement of this estimator consists in considering a number p of "'scales'" increasing with N: this is what we will use in simulations (theoretical results are not yet established, if they could be once).

3.3 Detrended Fluctuation Analysis (DFA)

The DFA method was introduced in Peng et al. [20] in a biological frame. The aim of this method is to highlight the self-similarity of a time series with a trend. Let $(Y(1), \ldots, Y(N))$ be a sample of a time series $(Y(n))_{n \in \mathbb{N}}$.

1. The first step of the DFA method is a division of $\{1, \ldots, N\}$ in $[N/n]$ windows of length n (for $x \in \mathbb{R}$, $[x]$ is the integer part of x). In each window, the least squares regression line is computed, which represents the linear trend of the process in the window. Then, we denote by $\widehat{Y}_n(k)$ for $k = 1, \ldots, N$ the process formed by this piecewise linear interpolation. Then the DFA function is the standard deviation of the residuals obtained from the difference between $Y(k)$ and $\widehat{Y}_n(k)$, therefore,

$$\widehat{F}(n) = \sqrt{\frac{1}{n \cdot [N/n]} \sum_{k=1}^{n \cdot [N/n]} \left(Y(k) - \widehat{Y}_n(k) \right)^2}$$

2. The second step consists in a repetition of the first step with different values $(n_1, \ldots, n_m)$ of the window's length. Then the graph of the $\log \widehat{F}(n_i)$ by $\log n_i$ is drawn. The slope of the least squares regression line of this graph provides an estimation of the self-similarity parameter of the process $(Y(k))_{k \in \mathbb{N}}$.

From the construction of the DFA method, it is interesting to define the restriction of the DFA function in a window. Thus, for $n \in \{1, \ldots, N\}$, one defines the partial DFA function computed in the j-th window, i.e.

$$F_j^2(n) = \frac{1}{n} \sum_{i=n(j-1)+1}^{nj} (X(i) - \widehat{X}_n(i))^2, \quad \text{for } j \in \{1, \ldots, [N/n]\}. \quad (3.15)$$

Then, it is obvious that

$$F^2(n) = \frac{1}{[N/n]} \sum_{j=1}^{[N/n]} F_j^2(n). \quad (3.16)$$

Let $\{X^H(t), t \geq 0\}$ be an FBM, built as a cumulated sum of stationary and centered fGn $\{Y^H(t), t \geq 0\}$. In Bardet and Kammoun [4], the following detailed asymptotic behavior of the DFA method is established. First some asymptotic properties of $F_1^2(n)$ can be established:

Proposition 3.5 *Let $\{X^H(t), t \geqslant 0\}$ be an fBm with parameters $0 < H < 1$ and $\sigma^2 > 0$. Then, for n and j large enough,*

1. $E(F_1^2(n)) \qquad = \sigma^2 f(H) n^{2H}\left(1 + O\left(\frac{1}{n}\right)\right),$

    $$\text{with } f(H) = \frac{(1-H)}{(2H+1)(H+1)(H+2)},$$

2. $Var\left(F_1^2(n)\right) \qquad = \sigma^4 g(H) n^{4H}\left(1 + O\left(\frac{1}{n}\right)\right),$

    with g depending only on H,

3. $cov(F_1^2(n), F_j^2(n)) = \sigma^4 h(H) n^{4H} j^{2H-3}\left(1 + O\left(\frac{1}{n}\right) + O\left(\frac{1}{j}\right)\right),$

    $$\text{with } h(H) = \frac{H^2(H-1)(2H-1)^2}{48(H+1)(2H+1)(2H+3)}.$$

In order to obtain a central limit theorem for the logarithm of the DFA function, we consider normalized DFA functions

$$\tilde{S}_j(n) = \frac{F_j^2(n)}{n^{2H}\sigma^2 f(H)} \quad \text{and} \quad \tilde{S}(n) = \frac{F^2(n)}{n^{2H}\sigma^2 f(H)} \tag{3.17}$$

for $n \in \{1, \ldots, N\}$ and $j \in \{1, \ldots, [N/n]\}$.

Under conditions on the asymptotic length n of the windows, one proves a central limit theorem satisfied by the logarithm of the empirical mean $\tilde{S}(n)$ of the random variables $(\tilde{S}_j(n))_{1 \leqslant j \leqslant [N/n]}$.

Proposition 3.6 *Under the previous assumptions and notations, let $n \in \{1, \ldots, N\}$ be such that $N/n \to \infty$ and $N/n^3 \to 0$ when $N \to \infty$. Then*

$$\sqrt{\left[\frac{N}{n}\right]} \cdot \log(\tilde{S}(n)) \xrightarrow[N \to \infty]{\mathscr{D}} \mathscr{N}(0, \gamma^2(H))),$$

where $\gamma^2(H) > 0$ depends only on H.

This result can be obtained for different lengths of windows satisfying the conditions $N/n \to \infty$ and $N/n^3 \to 0$. Let $(n_1, \ldots, n_m)$ be such different window lengths. Then, one can write for N and n_i large enough

$$\log(\tilde{S}(n_i)) \simeq \frac{1}{\sqrt{[N/n_i]}} \cdot \varepsilon_i \implies$$

$$\log(F(n_i)) \simeq H \cdot \log(n_i) + \frac{1}{2}\log(\sigma^2 f(H)) + \frac{1}{\sqrt{[N/n_i]}} \cdot \varepsilon_i,$$

with $\varepsilon_i \sim \mathcal{N}(0, \gamma^2(H))$. As a consequence, a linear regression of $\log(F(n_i))$ on $\log(n_i)$ provides an estimation $\widehat{H}_{DFA}$ of H. More precisely,

Proposition 3.7 *Under the previous assumptions and notations, let* $n \in \{1, \ldots, N\}$, $m \in \mathbb{N}^* \setminus \{1\}$, $r_i \in \{1, \ldots, [N/n]\}$ *for each* i *with* $r_1 < \cdots < r_m$ *and* $n_i = r_i n$ *be such that* $N/n \to \infty$ *and* $N/n^3 \to 0$ *when* $N \to \infty$. *Let* $\widehat{H}_{DFA}$ *be the estimator of* H, *defined as the slope of the linear regression of* $\log(F(r_i \cdot n))$ *on* $\log(r_i \cdot n)$, *i.e.*

$$\widehat{H}_{DFA} = \frac{\sum_{i=1}^{m}(\log(F(r_i \cdot n)) - \overline{\log(F)})(\log(r_i \cdot n) - \overline{\log(n)})}{\sum_{i=1}^{m}(\log(r_i \cdot n) - \overline{\log(n)})^2}.$$

Then $\widehat{H}_{DFA}$ *is a consistent estimator of* H *such that, with* $C(H, m, r_1, \ldots, r_m) > 0$,

$$E[(\widehat{H}_{DFA} - H)^2] \leqslant C(H, m, r_1, \ldots, r_m)\frac{1}{[N/n]}. \qquad (3.18)$$

Hence, this result shows that the convergence rate of $\widehat{H}_{DFA}$ is $\sqrt{N/n}$ that is a convergence rate $o(N^{1/3})$ from the condition $N/n^3 \to 0$. This is clearly less accurate than parametric estimators or even quadratic variations estimators. This estimator is devoted to trended long-range time series but even in such frame this estimator does not give satisfying results (see Bardet and Kammoun [4]).

3.4 Increment Ratio Statistic

The Increment Ratio (IR) statistic was first proposed in Surgailis et al. [24] in the frame of long-range dependent time series and extended to continuous time processes in Bardet and Surgailis [5]. For a time series $X = (X(k))_{k \in \mathbb{Z}}$ define the second-order variation as in (3.7) $D_t^{(a)} = X(t + 2a) - 2X(t + a) + X(t)$ for $t \in \mathbb{Z}$ and $a \in \mathbb{N}^*$. Assume that a trajectory $(X(0), X(1), \cdots, X(N))$ is observed. Define for $a \in \mathbb{N}^*$,

$$R_N^{(a)} := \frac{1}{N - 2a} \sum_{k=0}^{N-3a} \frac{|D_k^{(a)} + D_{k+1}^{(a)}|}{|D_k^{(a)}| + |D_{k+1}^{(a)}|}, \qquad (3.19)$$

with the convention $\frac{0}{0} := 1$. Note the ratio on the right-hand side of (3.19) is either 1 or less than 1 depending on whether the consecutive increments $D_k^{(a)}$ and $D_{k+1}^{(a)}$ have same signs or different signs; moreover, in the latter case, this ratio generally is small whenever the increments are similar in magnitude ("cancel each other").

If X is an fBm with parameter $H \in (0, 1)$, then it is established in Bardet and Surgailis [5] that for any $H \in (0, 1)$,

$$R_N^{(1)} \xrightarrow[N \to \infty]{a.s.} \Lambda_2(H) \tag{3.20}$$

$$\sqrt{N}\left(R_N^{(1)} - \Lambda_2(H)\right) \xrightarrow[N \to \infty]{\mathscr{D}} \mathscr{N}\left(0, \Sigma_2(H)\right). \tag{3.21}$$

The expressions of $\Lambda_2(H)$ and $\Sigma_2(H)$ are, respectively, given by:

$$\Lambda_2(H) := \lambda(\rho_2(H)), \tag{3.22}$$

$$\lambda(r) := \frac{1}{\pi} \arccos(-r) + \frac{1}{\pi}\sqrt{\frac{1+r}{1-r}} \log\left(\frac{2}{1+r}\right), \tag{3.23}$$

$$\rho_2(H) := \mathrm{corr}\left(D_0^{(1)}, D_1^{(1)}\right) = \frac{-3^{2H(t)} + 2^{2H(t)+2} - 7}{8 - 2^{2H(t)+1}} \tag{3.24}$$

$$\text{and} \quad \Sigma_2(H) := \sum_{j \in \mathbb{Z}} \mathrm{cov}\left(\frac{\left|D_0^{(1)} + D_1^{(1)}\right|}{\left|D_0^{(1)}\right| + \left|D_1^{(1)}\right|}, \frac{\left|D_j^{(1)} + D_{j+1}^{(1)}\right|}{\left|D_j^{(1)}\right| + \left|D_{j+1}^{(1)}\right|}\right). \tag{3.25}$$

and their graphs in Figs. 1 and 2.
The central limit (3.21) provides a way for estimating H: indeed, since $H \in (0, 1) \mapsto \Lambda_2(H)$ is an increasing $\mathscr{C}^1$ function, define

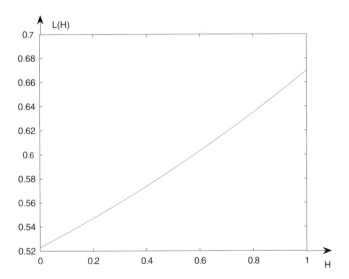

Fig. 1 The graph of $\Lambda_2(H)$

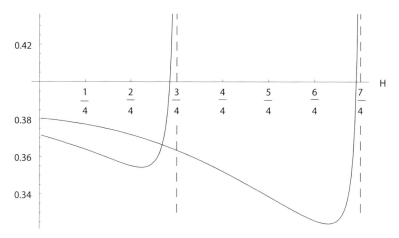

Fig. 2 The graphs of $\sqrt{\Sigma_p(H)}$, $p = 1$ (with a pole at 3/4) and $p = 2$ (with a pole at 7/4) (from Stoncelis and Vaičiulis [23]

$$\widehat{H}_N = \Lambda_2^{-1}\big(R_N^{(1)}\big).$$

From the Delta-method, we obtain the following central limit theorem for $\widehat{H}_N$:

Proposition 3.8 *For all $H \in (0, 1)$,*

$$\sqrt{N}\big(\widehat{H}_N - H\big) \xrightarrow[N \to \infty]{\mathscr{D}} \mathscr{N}\big(0, \gamma^2(H)\big)$$

with $\gamma^2(H) = \Sigma_2(H)\big[\big(\Lambda_2^{-1}\big)'(\Lambda_2(H))\big]^2$.

In Bardet and Surgailis [6], in a quite similar frame, an improvement of $\widehat{H}_N$ has been proposed. It consists first in obtaining a central limit theorem for the vector $(R_N^{(1)}, R_N^{(2)}, \cdots, R_N^{(m)})$ with $m \in \mathbb{N}^*$ and not only for $R_N^{(a)}$ with $a = 1$. Hence, we obtain the following multidimensional central limit theorem:

$$\sqrt{N}\big((R_N^{(i)})_{1 \leqslant i \leqslant m} - (\Lambda_2^{(i)})_{1 \leqslant i \leqslant m}\big) \xrightarrow[N \to \infty]{\mathscr{D}} \mathscr{N}\big(0, \Gamma_m(H)\big)$$

where $\Gamma_m(H) = (\gamma_{ij}(H))_{1 \leqslant i, j \leqslant m}$ and

$$\Lambda_2^{(i)}(H) = \lambda(\rho_2^{(i)}(H))$$

with $\rho_2^{(i)}(H) = \mathrm{Cor}(D_0^{(i)}, D_1^{(i)})$

$$= \frac{-|2i + 1|^{2H} - |2i - 1|^{2H} + 4|i + 1|^{2H} + 4|i - 1|^{2H} - 6}{8 - 2^{2H+1}}$$

and $\quad \gamma_{ij}(H) = \sum_{k \in \mathbb{Z}} \text{cov}\left(\dfrac{|D_0^{(i)} + D_1^{(i)}|}{|D_0^{(i)}| + |D_1^{(i)}|}, \dfrac{|D_k^{(j)} + D_{k+1}^{(j)}|}{|D_k^{(j)}| + |D_{k+1}^{(j)}|} \right).$

Then we define:

$$\widehat{H}_N^{(i)} = \left[\Lambda_2^{(i)}\right]^{-1}\left(R_N^{(i)}\right),$$

and using again the Delta-Method we obtain another multidimensional central limit theorem

$$\sqrt{N}\left((\widehat{H}_N^{(i)})_{1 \leqslant i \leqslant m} - H\, I_m\right) \xrightarrow[N \to \infty]{\mathscr{D}} \mathscr{N}\left(0,\, \Delta_m(H)\right)$$

and $\quad \Delta_m(H) := \left(\left[\dfrac{\partial}{\partial x}(\Lambda_2^{(i)})^{-1}(\Lambda_2^{(i)}(H))\right]\gamma_{ij}(H)\left[\dfrac{\partial}{\partial x}(\Lambda_2^{(j)})^{-1}(\Lambda_2^{(j)}(H))\right]\right)_{1 \leqslant i, j \leqslant p}.$

Finally a pseudo-generalized least squares estimator of H can be constructed (like for $\widehat{H}_{QV}$). Indeed $\Delta_m(H)$ can be estimated by $\widehat{\Delta}_m = \Delta_m(\widehat{H}_N^{(1)})$. Then we define

$$\widehat{H}_{IR} = \left(I_m^{\mathsf{T}}\, (\widehat{\Delta}_m\, I_m)\right)^{-1} I_m^{\mathsf{T}}\, (\widehat{\Delta}_m)^{-1} \left(\widehat{H}_N^{(i)}\right)_{1 \leqslant i \leqslant m} \tag{3.26}$$

and we obtain this proposition:

Proposition 3.9 *For all $H \in (0, 1)$ and $m \in \mathbb{N}^*$,*

$$\sqrt{N}\left(\widehat{H}_{IR} - H\right) \xrightarrow[N \to \infty]{\mathscr{D}} \mathscr{N}(0,\, s^2),$$

with $s^2 = \left(I_m^{\mathsf{T}}\left(\Delta_m(H)\right)^{-1} I_m\right)^{-1}.$

Then the convergence rate of $\widehat{H}_{IR}$ is $\sqrt{N}$, confidence intervals can also be easily computed.

3.5 Wavelet Based Estimator

This approach was introduced for fBm by Flandrin [10], and popularized by many authors to other self-similar or long-range dependent processes (see, for instance, Veitch and Abry [27], Abry et al. [1], or Bardet et al. [7]). Here we are going to follow Bardet [3], which is especially devoted to fBm as an extension of Flandrin [10].

First we define a (mother) wavelet function ψ such as $\psi : \mathbb{R} \to \mathbb{R}$ is a piecewise continuous and piecewise left (or right)-differentiable in $[0, 1]$, such that $|\psi_l'(t)|$ is

Riemann integrable in $[0, 1]$ with ψ'_l the left-derivative of ψ, with support included in $[0, 1]$ and Q first vanishing moments, i.e.

$$\int t^p \psi(t)dt = 0 \ for \ p = 0, 1, \cdots, Q-1 \qquad (3.27)$$

$$and \ \int t^Q \psi(t)dt \neq 0. \qquad (3.28)$$

For ease of writing, we have chosen a ψ supported in $[0, 1]$. But all the following results are still true, *mutatis mutandis*, if we work with any compactly supported wavelets. For instance, ψ can be any of the Daubechies wavelets.

Now we define the wavelet coefficients $d(a, i)$ of X where $a \in \mathbb{N}^*$ is called "scale" and $i \in \{1, 2, \cdots, [N/a] - 1\}$ is called "shift" by:

$$d(a, i) = \frac{1}{\sqrt{a}} \int_{-\infty}^{\infty} \psi(\frac{t}{a} - i)X(t)dt = \frac{1}{\sqrt{a}} \int_{0}^{a} \psi(\frac{t}{a})X(t + ai)dt. \qquad (3.29)$$

For each (a, i), $d(a, i)$ is a zero-mean Gaussian variable and its variance is a self-similar deterministic function in a, independent of the shift i, since for any $a \in \{1, 2, \cdots, [N/2]\}$ and $i \in \{1, 2, \cdots, [N/a] - 1\}$,

$$\mathrm{E}d^2(a, i) = a^{2H+1} C_\psi(H) \ where \ C_\psi(H) = -\frac{\sigma^2}{2} \int \int \psi(t)\psi(t')|t - t'|^{2H} dt dt'.$$

We assume now $C_\psi(H) > 0$ for all $H \in]0, 1[$. Now we consider an empirical variance $I_N(a)$ of $d(a, i)$ by

$$I_N(a) = \frac{1}{[N/a] - 1} \sum_{k=1}^{[N/a]-1} d^2(a, k). \qquad (3.30)$$

Using properties of ψ and particularly condition $\int t^p \psi(t)dt = 0$ for $p = 0, 1, \cdots, Q-1$, we can show that $\lim_{|i-j| \to \infty} |\mathrm{cov}(\tilde{d}(a, i), \tilde{d}(a, j))| = 0$ and limit theorems for $I_N(a)$. More precisely,

Proposition 3.10 *Under the previous assumptions, for $1 \leqslant a_1 < \cdots < a_m \in \mathbb{N}^*$, then*

$$\left[\sqrt{\frac{N}{a_i}} (\log I_N(a_i) - (2H+1) \log a_i - \log C_\psi(H)) \right]_{1 \leqslant i \leqslant m} \xrightarrow[N \to \infty]{\mathscr{D}} \mathscr{N}_m(0; F), \quad (3.31)$$

with $F = (f_{ij})_{1 \leqslant i, j \leqslant m}$ the matrix with $D_{ij} = GCD(a_i, a_j)$,

$$f_{ij} = \frac{\sigma^4 D_{ij}}{2C_\psi^2(H)a_i^{2H+1/2}a_j^{2H+1/2}}$$

$$\sum_{k=-\infty}^{\infty}\left(\int\int \psi(t)\psi(t')\left|kD_{ij}+a_it-a_jt'\right|^{2H}dtdt'\right)^2.$$

When a trajectory $(X(1),\cdots,X(N))$ is observed, d_X cannot be computed and an approximation has to be considered. Indeed, the wavelet coefficients $d(a,i)$, computed from a continuous process cannot be directly obtained and only approximated coefficients can be computed from a time series. It requires to choose large enough scales to fit well. Here, we will work with approximated coefficients $e(a,i)$ defined by:

$$e(a,i) = \frac{1}{\sqrt{a}}\sum_{k=-\infty}^{\infty}\psi\left(\frac{k}{a}-i\right)X(k) = \frac{1}{\sqrt{a}}\sum_{k=0}^{a}\psi\left(\frac{k}{a}\right)X(k+ai). \tag{3.32}$$

Denote also

$$J_N(a) = \frac{1}{[N/a]-1}\sum_{k=1}^{[N/a]-1}e^2(a,k). \tag{3.33}$$

The limit theorem of Proposition 3.10 can be rewritten with $\tilde{e}(a,i)$ instead of $\tilde{d}(a,i)$. The main difference is the use of scales $a_1(N),\cdots,a_m(N)$ satisfying $\lim_{N\to\infty} a_i(N) = \infty$. More precisely, the limit theorem is the following:

Proposition 3.11 *Let* $n_1 < \cdots < n_m$ *be integer numbers and let* $a_i(N) = n_i b(N)$ *for* $i = 1,\cdots,m$ *with* $b(N)$ *a sequence of integer numbers satisfying:* $[\frac{N}{b(N)}]\geqslant 2$, $\lim_{N\to\infty} b(N) = \infty$ *and* $\lim_{N\to\infty}\frac{N}{b^3(N)} = 0$. *Then, under previous assumptions,*

$$\sqrt{\frac{N}{b(N)}}\left(\log J_N(a_i(N))-(2H+1)\log a_i(N)-\log C_\psi(H)\right)_{1\leqslant i\leqslant m} \xrightarrow[N\to\infty]{\mathscr{D}} \mathscr{N}_m(0;G),$$

with $G = (g_{ij})_{1\leqslant i,j\leqslant m}$ *the matrix with* $D_{ij} = GCD(n_i,n_j)$,

$$g_{ij} = \frac{\sigma^4 D_{ij}}{2C_\psi^2(H)n_i^{2H+1/2}n_j^{2H+1/2}}$$

$$\sum_{k=-\infty}^{\infty}\left(\int\int\psi(t)\psi(t')\left|kD_{ij}+n_it-n_jt'\right|^{2H}dtdt'\right)^2.$$

From conditions on $b(N)$, the best convergence rate of this limit theorem is less than $N^{1/3}$ instead of $\sqrt{N}$ without the discretization problem. It is an important difference for the following estimation of H.

Indeed, Proposition 3.11 provides a method to estimate H from a linear regression. In fact, the central limit theorem of this proposition can be written :

$$\sqrt{\frac{N}{b(N)}}\,(Y_N - (2H+1)L - K\,I_m) \xrightarrow[N\to\infty]{\mathscr{D}} \mathscr{N}_m(0;\,G),$$

with

- $K = \log C_\psi(H)$, $I_m = (1, .., 1)'$, and $L = (\log n_i)_{1 \leqslant i \leqslant m}$,
- $Y_N = (\log J_N(a_i(N)) - \log b(N))_{1 \leqslant i \leqslant m}$ and $M = (L, I_m)$.

Under assumptions, there exists $\theta = {}^t(2H + 1, K)$, such that $Y_N = M\,\theta + \beta_N$, where β_N is a remainder which is asymptotically Gaussian. By the linearity of this model, one obtains an estimation $\widehat{\theta}_1(N)$ of θ by the regression of Y_N on M and ordinary least squares (O.L.S.).

But we can also identify the asymptotic covariance matrix of β_N. Indeed, the matrix G is a function $G_\psi(n_1, \cdots, n_m, H)$ and $\widehat{G}(N) = G_\psi(n_1, \cdots, n_m, \widehat{H}_1(N))$ converges in probability to G. So, it is possible to determine an estimation $\widehat{\theta}_2(N)$ of θ by generalized least squares (G.L.S.) of H by minimizing

$$\| Y_N - M\,\theta \|^2_{\widehat{G}(N)^{-1}} = (Y_N - M\,\theta)'\widehat{G}(N)^{-1}(Y_N - M\,\theta).$$

Thus, from the classical linear regression theory and the Gauss-Markov's Theorem:

Proposition 3.12 *Under previous assumptions,*

1. The O.L.S. estimator of H is $\widehat{H}_{OLS}$ such as $\widehat{H}_{OLS} = \left(\dfrac{1}{2}, 0\right)(M'\,M)^{-1}M'$

$$Y_N - \frac{1}{2} \text{ and } \sqrt{\frac{N}{b(N)}}(\widehat{H}_{OLS} - H) \xrightarrow[N\to\infty]{\mathscr{D}} \mathscr{N}(0,\,\sigma_1^2), \text{ with } \sigma_1^2$$

$$= \frac{1}{4}(M'\,M)^{-1}M\,G\,M'\,(M'\,M)^{-1}.$$

2. The G.L.S. estimator of H is $\widehat{H}_{Wave}$ such as $\widehat{H}_{Wave} = \left(\dfrac{1}{2}, 0\right)\left(M'\,\widehat{G}(N)^{-1}M\right)^{-1}$

$$M'\,\widehat{G}(N)^{-1}\,Y_N - \frac{1}{2} \text{ and } \sqrt{\frac{N}{b(N)}}(\widehat{H}_{Wave} - H) \xrightarrow[N\to\infty]{\mathscr{D}} \mathscr{N}(0,\,\sigma_2^2), \text{ with}$$

$$\sigma_2^2 = \frac{1}{4}\left(M'\,\widehat{G}(N)^{-1}M\right)^{-1} \leqslant \sigma_1^2.$$

Hence, as for other semi-parametric estimator of H (see the DFA estimator of H), the convergence rates of $\widehat{H}_{Wave}$ is $N^{1/3-\varepsilon}$, which is less accurate than the convergence rates of Whittle or generalized quadratic variations estimators.

4 Numerical Applications and Results of Simulations

4.1 Concrete Procedures of Estimation of H

In the previous section, we theoretically defined several estimators of H from a trajectory $(X(1), \cdots, X(N))$ of an fBm. Hereafter, we specify the concrete procedure of computation of these estimators:

- The Whittle estimator $\widehat{H}_W$ does not require to select auxiliary parameters or bandwidths. However we can notice that the integrals are replaced by Riemann sums computed for $\lambda = \pi k/n$, $k = 1, \cdots, N$.
- The classical R/S estimator $\widehat{H}_{RS}$ has been computed by averaging on uniformly distributed windows (see Taqqu et al. [25]).
- The modified R/S estimator $\widehat{H}_{RSM}$ has been computed using several uniformly distributed values of q around the optimal bandwidth $q = [N^{1/3}]$ as it is given by Lo [18]. More precisely we selected $q = \{[N^{0.3}], \cdots, [N^{0.5}]\}$.
- The second order quadratic variations estimator $\widehat{H}_{QV}$ requires the choice of the number of scales. After convincing auxiliary simulations, we selected $p = [3\log(N)]$.
- The DFA estimator $\widehat{H}_{DFA}$ requires the choice of windows. From the theoretical and numerical study in Bardet and Kammoun [4], we have chosen $n = \{[N^{0.3}], \cdots, [N^{0.5}]\}$.
- The Increment Ratio estimator $\widehat{H}_{IR}$ is computed with $M = 5$.
- The wavelet based estimator $\widehat{H}_{Wave}$ is computed with $b(N) = [N^{0}.3]$ and $m = [2 * log(N)]$.

4.2 Results of Simulations

We generated 1000 independent replications of trajectories $(X(1), \cdots, X(N))$ for $N = 500$ and $N = 5000$, with X an fBm of parameters $H = 0.1, 0.2, \cdots, 0.9$. We applied the estimators of H to these trajectories and compared the Mean Square Error (MSE) for each of them (Table 1).

Table 1 Values of the (empirical) MSE for the estimators of H when X is an fBm of parameter H and $N = 500$, $N = 5000$

H		0.1	0.2	0.3	0.4	0.5	0.6	0.7	0.8	0.9
$N = 500$	$\sqrt{MSE}$ for $\hat{H}_W$	**0.0280**	**0.0246**	**0.0255**	**0.0277**	**0.0294**	**0.0301**	**0.0310**	**0.0319**	**0.0305**
	$\sqrt{MSE}$ for $\hat{H}_{RS}$	0.2148	0.1918	0.1682	0.1420	0.1141	0.0847	0.0566	0.0431	0.0724
	$\sqrt{MSE}$ for $\hat{H}_{RSM}$	0.0642	0.0660	0.0675	0.0654	0.0651	0.0691	0.0831	0.1061	0.1437
	$\sqrt{MSE}$ for $\hat{H}_{QV}$	**0.0206**	0.0263	0.0292	0.0326	0.0339	0.0346	0.0356	0.0361	0.0354
	$\sqrt{MSE}$ for $\hat{H}_{DFA}$	0.0291	0.0424	0.0574	0.0723	0.0840	0.0994	0.1109	0.1195	0.1293
	$\sqrt{MSE}$ for $\hat{H}_{IR}$	0.0411	0.0483	0.0549	0.0561	0.0612	0.0624	0.0644	0.0603	0.0511
	$\sqrt{MSE}$ for $\hat{H}_{Wave}$	0.0970	0.1043	0.0791	0.0666	0.0608	0.0598	0.0606	0.0618	0.0637
$N = 5000$	$\sqrt{MSE}$ for $\hat{H}_W$	0.0246	**0.0094**	**0.0088**	**0.0085**	**0.0087**	**0.0090**	**0.0092**	**0.0094**	**0.0097**
	$\sqrt{MSE}$ for $\hat{H}_{RS}$	0.1640	0.1427	0.1191	0.0972	0.0751	0.0515	0.0261	0.0185	0.0568
	$\sqrt{MSE}$ for $\hat{H}_{RSM}$	0.0296	0.0300	0.0303	0.0293	0.0275	0.0309	0.0352	0.0497	0.0757
	$\sqrt{MSE}$ for $\hat{H}_{QV}$	**0.0059**	**0.0077**	0.0321	0.0491	0.0013	0.0104	0.0157	0.0107	0.0111
	$\sqrt{MSE}$ for $\hat{H}_{DFA}$	0.0152	0.0271	0.0362	0.0457	0.0510	0.0573	0.0632	0.0682	0.0710
	$\sqrt{MSE}$ for $\hat{H}_{IR}$	0.0119	0.0149	0.0159	0.0177	0.0183	0.0194	0.0192	0.0198	0.0196
	$\sqrt{MSE}$ for $\hat{H}_{Wave}$	0.0666	0.0359	0.0293	0.0281	0.0286	0.0312	0.0328	0.0428	0.0472

In bold, the minimal MSE is highlighted

5 Conclusion

We studied here several parametric and semi-parametric estimators of the Hurst parameter. In the part, we only consider the fBm. In such frame, we obtained:

1. **Theoretically** Only 3 estimators have a $\sqrt{N}$ convergence rate: $\widehat{H}_W$, $\widehat{H}_{QV}$, and $\widehat{H}_{IR}$. The first one is specific for fBm, both the other ones can also applied to other processes. The worst estimators are certainly R/S and modified R/S estimators. Wavelet based and DFA estimators can finally be written as generalized quadratic variations but the works with semi-parametric convergence rate $o(N^{1/2})$, and second-order quadratic variation estimator is clearly more accurate.

2. **Numerically** The ranking between the estimators is clear and follows the theoretical study: the Whittle estimator $\widehat{H}_W$ provides the best results, followed by the second-order quadratic variation estimator $\widehat{H}_{QV}$ and the IR estimator $\widehat{H}_{IR}$ which provides accurate estimations, followed by $\widehat{H}_{Wave}$ which is still efficient. The DFA and R/S estimators are not really interesting.

References

1. P. Abry, P. Flandrin, M.S. Taqqu, D. Veitch, Self-similarity and long-range dependence through the wavelet lens, in *Long-Range Dependence: Theory and Applications*, ed. by P. Doukhan, G. Oppenheim, M.S. Taqqu (Birkhäuser, New York, 2003)
2. J.-M. Bardet, Testing for the presence of self-similarity of Gaussian processes having stationary increments. J. Time Ser. Anal. **21**, 497–516 (2000)
3. J.-M. Bardet, Statistical study of the wavelet analysis of fractional Brownian motion. IEEE Trans. Inf. Theory **48**, 991–999 (2002)
4. J.-M. Bardet, I. Kammoun, Asymptotic properties of the detrended fluctuation analysis of long range dependent processes. IEEE Trans. Inf. Theory **54**, 1–13 (2008)
5. J.-M. Bardet, D. Surgailis, Measuring the roughness of random paths by increment ratios. Bernoulli **17**, 749–780 (2010)
6. J.-M. Bardet, D. Surgailis, A new nonparametric estimator of the local Hurst function of multifractional processes. Stoch. Process. Appl. **123**, 1004–1045 (2012)
7. J.-M. Bardet, G. Lang, E. Moulines, P. Soulier, Wavelet estimator of long range dependent processes. Stat. Infer. Stoch. Process **3**, 85–99 (2000)
8. J.-F. Coeurjolly, Identification of multifractional Brownian motion. Bernoulli **11**, 987–1008 (2005)
9. R. Dahlhaus, Efficient parameter estimation for self-similar processes. Ann. Stat. **17**, 1749–1766 (1989)
10. P. Flandrin, Wavelet analysis and synthesis of fractional Brownian motions. IEEE Trans. Inf. Theory **38**(2), 910–917 (1992)
11. R. Fox, M.S. Taqqu, Large-sample properties of parameter estimates for strongly dependent Gaussian time series. Ann. Stat. **14**, 517–532 (1986)
12. L. Giraitis, P. Kokoszka, R. Leipus, G. Teyssire, Rescaled variance and related tests for long memory in volatility and levels. J. Econ. **112**, 265–294 (2003)
13. X. Guyon, J. Leon, Convergence en loi des h-variations d'un processus gaussien stationnaire. Ann. Inst. Poincaré **25**, 265–282 (1989)

14. H. Hurst, Long term storage capacity of reservoirs. Trans. Am. Soc. Civ. Eng. **116**, 770–799 (1951)
15. J. Istas, G. Lang, Quadratic variations and estimation of the local Hölder index of a Gaussian process. Ann. Inst. Poincaré **33**, 407–436 (1997)
16. A.N. Kolmogorov, Wienersche Spiralen und einige andere interessante Kurven im Hilbertschen Raum. C.R. (Doklady) Acad. Sci. URSS (N.S.) **26**, 115–118 (1940)
17. W. Li, C. Yua, A. Carriquiry, W. Kliemann, The asymptotic behavior of the R/S statistic for fractional Brownian motion. Statist. Probab. Lett. **81**, 83–91 (2011)
18. A.W. Lo, Long-term memory in stock market prices. Econometrica **59**, 1279–1313 (1991)
19. B. Mandelbrot, J. Van Ness, Fractional Brownian motion, fractional noises and applications. SIAM Rev. **10**, 422–437 (1968)
20. C.K. Peng, S. Havlin, H.E. Stanley, A.L. Goldberger, Quantification of scaling exponents and crossover phenomena in nonstationary heartbeat time series. Chaos **5**, 82 (1995)
21. G. Samorodnitsky, M.S. Taqqu, *Stable Non-gaussian Random Processes*. Stochastic Modeling (Chapman and Hall, New York, 1994)
22. Y.G. Sinaï, Self-similar probability distributions. Theory Probab. Appl. **21**, 64–80 (1976)
23. M. Stoncelis, M. Vaičiulis, Numerical approximation of some infinite gaussian series and integrals. Nonlinear Anal. Modell. Control **13**, 397–415 (2008)
24. D. Surgailis, G. Teyssière, M. Vaičiulis, The increment ratio statistic. J. Multivar. Anal. **99**, 510–541 (2008)
25. M.S. Taqqu, V. Teverovsky, W. Willinger, Estimators for long-range dependance: an empirical study. Fractals **3**, 785–798 (1995)
26. V. Teverovsky, M.S. Taqqu, W. Willinger, A critical look at Lo's modified R/S statistic. J. Stat. Plann. Inference **80**, 211–227 (1999)
27. D. Veitch, P. Abry, A wavelet-based joint estimator of the parameters of long-range dependence. IEEE Trans. Inf. Theory **45**, 878–897 (1999)
28. P. Whittle, Gaussian estimation in stationary time series. Bull. Int. Stat. Inst. **39**, 105–129 (1962)

Minimal Lethal Disturbance for Finite Dimensional Linear Systems

Abdes Samed Bernoussi, Mina Amharref, and Mustapha Ouardouz

Abstract In this work we consider the problem of robust viability and viability radius for finite dimensional disturbed linear systems. The problem consists in the determination of the smallest disturbance f (in some disturbance set $\mathcal{F}$), for which a given viable state z_0 does not remains viable. We also consider the problem of the determination of the smallest disturbance f for which the viability set $Viab_{\mathcal{K}}^{f}$ becomes empty; the smallest disturbance that makes all the $\mathcal{K}$-viable states non viable, which we call the Minimal Lethal Disturbance (MLD). We give some characterizations of the viability radius and an illustration through some examples and connection with toxicity in biology.

Keywords Viability · Viability radius · Minimal lethal disturbance · Robustness

1 Introduction

The viability notion has been introduced by Aubin in [1–3] and has been developed by many authors. The principle consists in saying that a given state z_0 is $\mathcal{K}$ viable during a time interval $I = [0, T]$, where $\mathcal{K}$ is a given subset of the state space, if the state $z(t, z_0)$ remains in $\mathcal{K}$ for all $t \in]0, T[$. Since the introduction of the viability notion, it was used to study many applications related to environment, economic, political development phenomena, etc. [1].

Combining the concepts of control and viability, a more general definition was given by Aubin [1]: it is the viability kernel. A state $z(t = 0)$ is in the $\mathcal{K}$ viability kernel, for a controlled system during a time interval $I = [0, T]$, if there exists at least one control u such that the controlled state $z_u(t, z_0)$ remains in $\mathcal{K}$ for all

A. S. Bernoussi (✉) · M. Amharref
GAT, Tangier, Morocco

M. Ouardouz
MMC Team, Faculty of Sciences and Techniques, Tangier, Morocco

© Springer Nature Switzerland AG 2018
T. Diagana, B. Toni (eds.), *Mathematical Structures and Applications*,
STEAM-H: Science, Technology, Engineering, Agriculture,
Mathematics & Health, https://doi.org/10.1007/978-3-319-97175-9_7

$t \in]0, T[$. Some characterizations of the viability kernel, known as the viability theorems, are given in [2, 3] using the contingent concepts introduced by Bouligand in the years 30 to extend the tangent concept to the multivogue applications. The approximation of the viability kernel, by a numerical algorithms, has been considered by many authors and a wide literature is devoted to this problem.

In this work we consider the viability problem for a disturbed finite dimensional linear systems. Indeed consider, for a dynamical system and an initial state z_0 which is $\mathcal{K}$-viable for the autonomous system, the following problems: Can the initial state z_0 remain $\mathcal{K}$-viable in the presence of the disturbance f? If yes, for each disturbance? For that we introduce the so-called "viability radius" which is the smallest disturbance f for which the given viable state z_0 does not remain viable for the disturbed system. The problem is connected to the robust control problems but with some particularities due to the viability notion. In [4], I. Alvarez and S. Martin have considered the problem of the viability robustness from a geometric point of view. In this work we consider the problem in relation with the dynamic of the system, the disturbance, and control operators. Also we consider for a given subset $\mathcal{K}$ of the state space the problem of the determination of the smallest disturbance f for which the set $Viab_{\mathcal{K}}^{f}$ becomes empty; the smallest disturbance (in some disturbance set $\mathcal{F}$) that makes all the $\mathcal{K}$-viable states nonviable. This work is motivated by some problems in biology and particularly in toxicology where the determination of the so-called "Minimal Lethal Dose (MLD)" is very important: It consists to determinate the "smallest lethal dose" of a given poison. The problem has many applications as for example the vulnerability and risk management of groundwater pollution problem. For such problem the risk is linked to three parameters which are probability, vulnerability, and gravity. The gravity depends on the characteristics of the pollutant and particularity on the Minimal Lethal Dose (MLD) of the pollutant which characterize its proper gravity [5, 6, 9]. The problem will be detailed in the section application. In this paper, we recall in the second section the definition of the viability and we introduce the "viability radius." In the third section we consider the problem of the determination of a control which maximizes the viability radius. Some examples are presented to illustrate our approaches.

2 Viability Radius and Minimal Lethal Disturbance

2.1 Viability Concept: Definition and Characterization

In this section we recall the viability definition as it was introduced by Aubin in [1]. Consider the linear system given by the following state equation:

$$\begin{cases} \dot{z}(t) = Az(t) \; ; \quad t \in]0, T[\\ z(0) = z_0 \end{cases} \tag{1}$$

where $A \in \mathcal{M}_n(\mathbb{R})$, $z(.) \in \mathcal{C}([0, T], \mathbb{R}^n)$ and $z_0 \in \mathbb{R}^n$.

Let $\mathcal{K}$ be a non empty subset of $\mathbb{R}^n$. We recall the following definition [1].

Definition 2.1 *We say that a state $z_0 \in Z = \mathbb{R}^n$ is $\mathcal{K}$-viable or viable in $\mathcal{K}$ on $I = [0, T]$ if:*

(i) $z_0 \in \mathcal{K}$;
(ii) $z(t, z_0) \in \mathcal{K}$ for all $t \in [0, T]$.

where $z(t, z_0)$ is the solution of (1) with $z(0) = z_0$.

The set of all $\mathcal{K}$-viable states is called $\mathcal{K}$-viability set and denoted by:

$$Viab_{\mathcal{K}} = \{z_0 \in Z : z_0 \text{ is } \mathcal{K} - \text{viable}\}$$

In [1] some characterizations and examples of viability sets are given and we have the definition [1]:

Definition 2.2 *We say that*

- *a subset $\mathcal{K}$ is viable if: $\mathcal{K} = Viab_{\mathcal{K}}$*
- *$\mathcal{K}$ is a repeller if $Viab_{\mathcal{K}} = \emptyset$.*

Let $\mathcal{K}$ be a non empty bounded and connected subset of $\mathbb{R}^n$ with a smooth regular boundary $\partial_{\mathcal{K}}$ and consider the signed distance $\phi_{\mathcal{K}}$ defined by:

$$\phi_{\mathcal{K}}(z(t, z_0)) = \begin{cases} - d[z(t, z_0), \partial_{\mathcal{K}}] & \text{if} \quad z(t, z_0) \in \mathcal{K} \\[2mm] + d[z(t, z_0), \partial_{\mathcal{K}}] & \text{else.} \end{cases} \tag{2}$$

where $d[z(t, z_0), \partial_{\mathcal{K}}] = \inf\limits_{\{y \in \partial_{\mathcal{K}}\}} \|z(t, z_0) - y\|_{\mathbb{R}^n}$.

$\phi_{\mathcal{K}}$ is the signed distance as defined and used by J. Sethian in [7] to characterize some level sets.

We define

$$\rho_{\mathcal{K}}(z_0) = \sup\limits_{t \in I} \phi_K(z(t, z_0))$$

Then we have the characterization:

Proposition 2.3 *Let $\mathcal{K}$ be a nonempty bounded and connected subset of $\mathbb{R}^n$ with a smooth regular boundary $\partial_{\mathcal{K}}$. Then if $\rho_{\mathcal{K}}(z_0) < 0$ (respectively $\rho_{\mathcal{K}}(z_0) > 0$), then the initial state z_0 is $\mathcal{K}$-viable (respectively is not $\mathcal{K}$-viable) (Fig. 1).*

Remark 2.4 *In the case where $\rho_{\mathcal{K}}(z_0) = 0$, if $\mathcal{K}$ is closed then z_0 is $\mathcal{K}$-viable but in the general cases we can't decide.*

For more details about viability definitions and characterizations, we refer to [1–3] and the references therein.

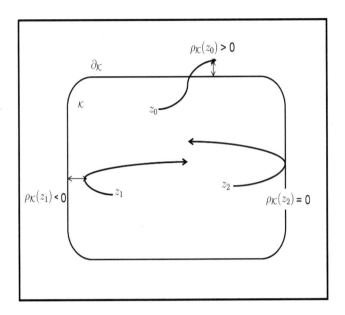

Fig. 1 Viable state and nonviable one

2.2 Viability Radius

2.2.1 Definitions and Examples

Consider now the disturbed system given by the following state equation:

$$\begin{cases} \dot{z}(t) = Az(t) + Gf(t) & 0 < t < T \\ \\ z(0) = z_0 \end{cases} \tag{3}$$

where $G \in \mathcal{M}_{n,p}(\mathbb{R})$ and $f \in \mathcal{F}$. $\mathcal{F}$ is the disturbance space which is in this paper a subspace of $\mathcal{C}(0, T; \mathbb{R}^p)$.

The solution of (3) is given by:

$$z_f(t, z_0) = z_f(t) = S(t)z_0 + \int_0^t S(t-s)Gf(s)ds$$

where $S(t) = e^{At}$.

Denote

$$H_t f = \int_0^t S(t-s)Gf(s)ds$$

then $z_f(t) = S(t)z_0 + H_t f$.

For a $\mathcal{K}$-viable state z_0 ($f = 0$ for the autonomous system) when the system is disturbed by a given disturbance $f \in \mathcal{F}$, the state z_0 can remain viable, i.e. $z_f(t, z_0) \in \mathcal{K}$ for all $t \in I$ or no. In this case we define the set of all $\mathcal{K}$-viable states which remain viable for the disturbed system as follows:

Definition 2.5 *For a given disturbance f we define the set of all $\mathcal{K}$-viable states which remain $\mathcal{K}$-viable for the disturbed system by f as:*

$$Viab_{\mathcal{K}}^{f} = \{z_0 \in Viab_{\mathcal{K}} : z_f(t, z_0) \in \mathcal{K},$$

$$\forall t \in [0, T] \}$$

Remark 2.6 *1. In the case where $f = 0$, we have $Viab_{\mathcal{K}}^{0} = Viab_{\mathcal{K}}$ and for each f, $Viab_{\mathcal{K}}^{f} \subset Viab_{\mathcal{K}}$*
2. We can define another set of all initial states which are viable under f which can contain some states which are not viable for the autonomous system. Such subset will be defined like the viability kernel for the controlled system but as here the system is disturbed (not controlled) we consider the set $Viab_{\mathcal{K}}^{f}$ as defined in Definition 2.5.

Definition 2.7 *We say that*

- *a subset $\mathcal{K}$ is f-viable if: $\mathcal{K} = Viab_{\mathcal{K}}^{f}$*
- *$\mathcal{K}$ is a f-repeller if $Viab_{\mathcal{K}}^{f} = \emptyset$.*

Remark 2.8 *If a subset $\mathcal{K}$ is f-viable, then it is viable. The converse is not true as it will be shown in the next sections.*
Consequently if $\mathcal{K}$ is a repeller then it is f-repeller for each $f \in \mathcal{F}$. The converse is not true.

Now for a given $\mathcal{K}$-viable state, can we determinate the smallest disturbance, which makes the state nonviable? For that we introduce the so-called viability radius. We have the following definition.

Definition 2.9 *Let z_0 be an initial $\mathcal{K}$-viable state. We define the $\mathcal{K}$-viability radius of z_0 as:*

$$R_{viab}^{\mathcal{K}}(z_0) = \sup\{r > 0 \,\forall\, f \in \mathcal{F} \text{ if } ||f||_{\infty} < r \text{ then } z_f(t, z_0) \in \mathcal{K}, \forall\, t \in I\}$$

where $||f||_{\infty} = \sup_{t \in I} ||f(t)||_{\mathbb{R}^p}$

Remark 2.10 *The viability radius is the "smallest disturbance" that can make z_0 nonviable: For all $f \in \mathcal{C}(0, T; \mathbb{R}^p)$ if $||f|| < R_{viab}^{\mathcal{K}}(z_0)$, then z_0 remains viable in presence of f and*

$$\forall \varepsilon > 0, \exists f \in \mathcal{C}(0, T; \mathbb{R}^p)$$

$$: R_{viab}^{\mathcal{K}}(z_0) < ||f|| < R_{viab}^{\mathcal{K}}(z_0) + \varepsilon \text{ such that } z_0 \text{ is not viable under } f$$

To illustrate the definitions and the remark, let us consider the following example in one-dimensional case for clarity.

Example 2.11 *Consider the linear system given by the following state equation:*

$$\begin{cases} \dot{z}(t) = 2z(t) + 4f(t) & 0 < t \leqslant T = 3 \\ \\ z(0) = z_0 \end{cases} \tag{4}$$

and $f \in \mathcal{C}([0, T]; \mathbb{R})$.
  The solution of Eq. (4) is given by:

$$z_f(t) = z_0 e^{2t} + \int_0^t 4e^{2(t-s)} f(s) ds$$

which gives:

$$z_f(t) = z_0 e^{2t} + 4e^{2t} \int_0^t e^{-2s} f(s) ds$$

Case 1. *Consider the subset* $\mathcal{K} =]0, 40[$.
  We have for $f = 0$, *the nondisturbed system:*

$$Viab_{\mathcal{K}} = \{z_0 \in]0, 40[\text{ such that } z_0 e^{2t} \in]0, 40[$$

$$\text{for all } t \in [0, 3]\}$$

which gives

$$Viab_{\mathcal{K}} =]0, \ 40e^{-6}[$$

Consider now the disturbed system ($f \neq 0$ *). We consider, for example, the initial state* $z_0 = 20e^{-6}$.
  z_0 *is* $\mathcal{K}$-*viable in the autonomous case and in the disturbed case it remains viable if and only if:*

$$20e^{-6}e^{2t} + 4e^{2t} \int_0^t e^{-2s} f(s) ds \in]0, \ 40[\ ; \ \forall \, t \in [0, 3]$$

which is equivalent to

$$-5e^{-6} < \int_0^t e^{-2s} f(s) ds < 10e^{-2t} - 5e^{-6} \ ; \ \forall \, t \in [0, 3] \tag{5}$$

Consider the case where $\mathcal{F} = \mathbb{R}$, i.e. $f(t) = f \in \mathbb{R}$, then Eq. (5) gives:

$$- 5e^{-6} < (\frac{1}{2} - \frac{1}{2}e^{-2t})f < 10e^{-2t} - 5e^{-6} ; \quad \forall t \in [0, 3] \qquad (6)$$

which is equivalent, for $0 < t < 3$ to

$$\frac{-10e^{-6}}{1 - e^{-2t}} < f < \frac{20e^{-2t} - 10e^{-6}}{1 - e^{-2t}} ; \quad \forall t \in]0, 3] \qquad (7)$$

and for $t = 0$, we obtain : $-5e^{-6} < 0f < 10 - 5e^{-6}$.
We remark also that:

$$\lim_{t \to 0^+} \frac{-10e^{-6}}{1 - e^{-2t}} = -\infty \quad and \quad \lim_{t \to 0^+} \frac{20e^{-2t} - 10e^{-6}}{1 - e^{-2t}} = +\infty$$

So, z_0 remains $\mathcal{K}$-viable for all f such that:

$$\frac{-10e^{-6}}{1 - e^{-6}} < f < \frac{10e^{-6}}{1 - e^{-6}} \qquad (8)$$

and finally we obtain:

$$R^{\mathcal{K}}_{viab}(z_0) = \frac{10e^{-6}}{1 - e^{-6}} \qquad (9)$$

Case 2. Consider the subset $\mathcal{K} = [0, 40]$.
We have for $f = 0$, the nondisturbed system:

$$Viab_{\mathcal{K}} = \{z_0 \in [0, 40] \text{ such that } z_0 e^{2t} \in [0, 40]$$

$$for \text{ all } t \in [0, 3]\}$$

which gives

$$Viab_{\mathcal{K}} = [0, 40e^{-6}]$$

Consider the state $z_0 = 0$ which is $\mathcal{K}$-viable and we have

$$R^{\mathcal{K}}_{viab}(z_0 = 0) = 0 \qquad (10)$$

Because for all $\varepsilon > 0$ and $f = -\varepsilon$,

$$z_f(t, z_0 = 0) = -\varepsilon \int_0^t 4e^{2(t-s)}ds \notin \mathcal{K} = [0, 40]$$

This example proof that we can have, for a given $\mathcal{K}$-viable state, $R^{\mathcal{K}}_{viab}(z_0) = 0$.

We remark that for $f = \varepsilon > 0$, the state z_0 remains viable while $R^{\mathcal{K}}_{viab}(z_0) = 0$ (Remark 2.10).

Case 3. Let us consider now another case where

$$\mathcal{F}_1 = \{\alpha e^t \; ; \; \alpha \in \mathbb{R} \text{ and } t \in [0, 3]\} \text{ and } \mathcal{K} =]0, 40[$$

Then we obtain: the $\mathcal{K}$-viable state $z_0 = 20e^{-6}$ remains $\mathcal{K}$-viable for the disturbed system if and only if:

$$z_f(t) = z_0 e^{2t} + 4\alpha e^{2t} \int_0^t e^{-2s} e^s \, ds \in]0, 40[$$

which gives

$$\frac{-20e^{-6}e^{2t}}{4(e^{2t} - e^t)} < \alpha < \frac{40 - 20e^{-6}e^{2t}}{4(e^{2t} - e^t)} < \forall t \in]0, 3[$$

which gives

$$\frac{-5e^{-6}}{e^t - 1}e^t < f(t) = \alpha e^t < \frac{40e^{-t} - 20e^{-6}e^t}{4(e^t - 1)} \tag{11}$$

and consequently:

$$R^{\mathcal{K}}_{viab}(z_0) = \frac{5e^{-6}}{1 - e^{-3}}$$

We remark that the $\mathcal{K}$-viability radius depends on the choice of the disturbances set and we will show in the section applications in biology that this property is very important.

2.2.2 Characterization

We have the following characterization:

Theorem 2.12 *Assume that $\mathcal{K}$ is a nonempty bounded connected and closed set of R^n with a smooth regular boundary $\partial_{\mathcal{K}}$ and $Im H_T = R^n$, then for a given $\mathcal{K}$-viable state z_0, the viability radius satisfies:*

$$R^{\mathcal{K}}_{viab}(z_0) \geqslant \inf_{t \in I} \frac{|\phi_{\mathcal{K}}(z(t, z_0))|}{\|H_t\|} \tag{12}$$

Proof of the Theorem 2.12. Consider the solution of the disturbed system

$$z_f(t, z_0) = S(t)z_0 + \int_0^t S(t - s)Gf(s)ds$$

So the state z_0 remains $\mathcal{K}$-viable if and only if

$$z_f(t, z_0) \in \mathcal{K}, \quad \forall t \in I$$

We have for $t \in I$,

$$\phi_\mathcal{K}(z_f(t, z_0)) = \begin{cases} -d[z_f(t, z_0), \partial_\mathcal{K}] & \text{if } z_f(t, z_0) \in \mathcal{K}, \\ +d[z_f(t, z_0), \partial_\mathcal{K}] & \text{else.} \end{cases} \tag{13}$$

Denote

$$\rho_{\mathcal{K}, f}(z_0) = \sup_{t \in I} \phi_K(z_f(t, z_0))$$

So if $\rho_{\mathcal{K}, f}(z_0) < 0$, then the initial state z_0 remains $\mathcal{K}$-viable and if $\rho_{\mathcal{K}, f}(z_0) > 0$, then the initial state z_0 does not remain $\mathcal{K}$-viable.

We have:

$$d[z_f(t, z_0), \partial_\mathcal{K}] = \inf_{\{y \in \partial_\mathcal{K}\}} \|z_f(t, z_0) - y\|$$

If

$$\|f\|_\infty < \inf_{t \in I} \frac{|\phi_K(z(t, z_0))|}{\|H_t\|}$$

we obtain

$$\|f\|_\infty < \frac{|\phi_K(z(t, z_0))|}{\|H_t\|} ; \quad \forall t \in I$$

As

$$\|H_t f\| \leqslant \|H_t\| \|f\|_\infty$$

then

$$\|H_t f\| < |\phi_K(z(t, z_0))\| ; \quad \forall t \in I \tag{14}$$

then for all $y \in \partial_{\mathcal{K}}$ and $t \in I$,

$$\|S(t)z_0 + H_t f - y\| \neq 0$$

because if there exists $t \in I$ and $y \in \partial_{\mathcal{K}}$, such that $\|S(t)z_0 + H_t f - y\| = 0$, then:

$$H_t f = -S(t)z_0 + y$$

which is in contradiction with the fact that $\|H_t f\| < |\phi_K(z(t, z_0))| \ \forall \ t \in I$.

Then

$$\phi_K(z_f(t, z_0)) < 0 \ ; \ \forall \ t \in [0, T]$$

and then $\rho_{\mathcal{K}, f}(z_0) \leqslant 0$ and consequently z_0 remains $\mathcal{K}$-viable because $\mathcal{K}$ is closed.

So we conclude that

$$R^{\mathcal{K}}_{viab}(z_0) \geqslant \inf_{t \in I} \frac{|\phi_K(z(t, z_0))|}{\|H_t\|} \tag{15}$$

Remark 2.13 *We can have equality in some situations as in the following example:*

Example 2.14 *Consider the example given by Eq. (4) and let us consider the two cases:*

Case 1. $\mathcal{K} = [0, 40]$, $z_0 = 20e^{-6}$ *and* $f(t) = f \in R$ *for all* $t \in [0, T]$.
In this case we have:

$$|\phi_{\mathcal{K}}(z(t, z_0)| = 20e^{-6}e^t$$

and

$$\|H_t\| = 2(e^{2t} - 1)$$

So we obtain the same result as in (9) and we have:

$$R^{\mathcal{K}}_{viab}(z_0) = \frac{10e^{-6}}{1 - e^{-6}} = \inf_{t \in I} \frac{|\phi_K(z(t, z_0))|}{\|H_t\|}$$

Case 2. $\mathcal{K} = [0, 40]$ *and* $z_0 = 0$.
In this case we remark that $R^{\mathcal{K}}_{viab}(z_0 = 0) = 0$ *(the same result as in (10)) because* $\phi_K(z(t, z_0 = 0)) = 0$

2.3 Minimal Lethal Disturbance (MLD)

In this subsection we consider the problem of the determination of the smallest disturbance f, for which the set $Viab^f_{\mathcal{K}}$ is empty (for which the subset $\mathcal{K}$ becomes

f-repeller); the smallest disturbance that makes all the $\mathcal{K}$-viable states nonviable which we call the Minimal Lethal Disturbance (MLD). For that we have the following definition:

Definition 2.15

- *We define the $\mathcal{K}$-viability radius as :*

$$R_{viab}^{\mathcal{K}} = \sup\{r > 0 \,:\, \forall\, f \,\in \mathcal{F}\,,\ \|f\|_{\infty} < r,\, Viab_K^f \neq \emptyset\}$$

- *If there exists a disturbance $f \in \mathcal{F}$ such that $R_{viab}^{\mathcal{K}} = \| f \|_{\infty}$ and $Viab_K^f = \emptyset$, we say that f is the $\mathcal{K}$-Minimal Lethal Disturbance (MLD).*

Remark 2.16 *We remark that for each disturbance f such that $\|f\|_{\infty} < R_{viab}^{\mathcal{K}}$, then there exists at least one viable state z_0 which remains f-viable. But if $\|f\|_{\infty} > R_{viab}^{\mathcal{K}}$ we do not have necessary $Viab_K^f = \emptyset$, but there exists another disturbance h such that $\|f\|_{\infty} = \|h\|_{\infty}$ and for which $Viab_K^h = \emptyset$.*

The $\mathcal{K}$-viability radius is the smallest disturbance f for which $\mathcal{K}$ becomes f-repeller. We have used the terminology Minimal Lethal Disturbance because in biology an equivalent term is used: Minimal Lethal Dose.

In some applications, the determination of the Minimal Lethal Disturbance is a very difficult problem as it will be shown in the section application where we consider an application in biology and particularly in toxicology and irradiation.

From Definition 2.15 we have the result:

Proposition 2.17 *We have*

$$R_{viab}^{\mathcal{K}} = \sup\{R_{viab}^{\mathcal{K}}(z) \,:\, z \,\in\, Viab_{\mathcal{K}}\}$$

Proof Let z be a $\mathcal{K}$-viable state, then

$$\forall\, f \,\in\, \mathcal{F} \text{ such that } \|f\| \geqslant R_{viab}^{\mathcal{K}}$$

there exists a disturbance h such that $\|h\| = \|f\|$ for which z is not viable for the system excited by h.

Then

$$R_{viab}^{\mathcal{K}}(z) \leqslant R_{viab}^{\mathcal{K}}$$

So we have

$$R_{viab}^{\mathcal{K}} \geqslant \sup\{R_{viab}^{\mathcal{K}}(z) \,:\, z \,\in\, Viab_{\mathcal{K}}\}$$

and consequently the result.

Example 2.18 *Consider the linear system given by the following state equation:*

$$\begin{cases} \dot{z}(t) = 2z(t) + 4f(t) & 0 < t \leqslant T = 3 \\ \\ z(0) = z_0 \end{cases} \tag{16}$$

and $\forall t \in I$, $f(t) = f \in \mathbb{R}$.
  The solution of Eq. (16) is given by:

$$z_f(t) = z_0 e^{2t} + 4e^{2t} \int_0^t e^{-2s} f(s) ds$$

Consider the subset $\mathcal{K} = [0, 40]$.
  We have for $f = 0$, *the nondisturbed system:*

$$Viab_{\mathcal{K}} = [0, 40e^{-6}]$$

For each state $z_0 \in Viab_{\mathcal{K}} =]0, 40e^{-6}[$, *the* $\mathcal{K}$-*viability radius is given by:*

$$R_{viab}^{\mathcal{K}}(z_0) = \inf_{t \in [0,3]} \frac{|\phi_K(z(t, z_0))|}{\|H_t\|}$$

For each $z_0 \in]0, 40e^{-6}[$, *we have :*

$$|\phi_K(z(t, z_0))| = \inf\{|z_0 e^{2t}|, |40e^{-6} - z_0 e^{2t}|\}$$

which gives:

$$|\phi_K(z(t, z_0))| = \begin{cases} z_0 e^{2t} & if \ z_0 \in [0, 20e^{-6}[\\ \\ 40 - z_0 e^{2t} & if \ z_0 \in]20e^{-6}, 40e^{-6}] \end{cases}$$

We have $\|H_t\| = 2(e^{2t} - 1)$ *and so we obtain:*

$$R_{viab}^{\mathcal{K}}(z_0) = \begin{cases} \inf_{t \in]0,3]} \frac{z_0 e^{2t}}{2(e^{2t}-1)} & if \ z_0 \in]0, 20e^{-6}[\\ \\ \inf_{t \in]0,3]} \frac{40 - z_0 e^{2t}}{2(e^{2t}-1)} & if \ z_0 \in]20e^{-6}, 40e^{-6}[\end{cases}$$

and finally we obtain

$$R_{viab}^{\mathcal{K}} = \frac{10e^{-6}}{1 - e^{-6}}$$

We remark that $R_{viab}^{\mathcal{K}} = R_{viab}^{\mathcal{K}}(z_0 = 20e^{-6})$ and it is normal because z_0 is in the center of $Viab_{\mathcal{K}}$ and consequently it is the farthest viable state from ∂_K.

3 Controlled System

3.1 Viability Kernel and Viability Radius

Consider a controlled system given by:

$$\begin{cases} \dot{z}(t) = Az(t) + Bu(t) \quad 0 < t < T \\ \\ z(0) = z_0 \end{cases} \tag{1}$$

where $u(.) \in \mathcal{U} = \mathcal{C}([0, T], \mathbb{R}^p)$.

Definition 3.1 *1. The viability kernel of $\mathcal{K}$, noted $Viab_{(A,B)}(\mathcal{K})$, is the subset of initial states $z_0 \in \mathcal{K}$ such that there exists at least one solution of (1) starting from z_0 and $\mathcal{K}$-viable during the time interval $I = [0, T]$. i. e. $z_u(x, t) \in \mathcal{K}$ for all $t \in [0, T]$, where $z_u(x, t)$ is the solution of (1).*
We say that $\mathcal{K}$ is a repeller if $Viab_{(A,B)}(\mathcal{K}) = \emptyset$.

Consider now the disturbed controlled system given by the following equation:

$$\begin{cases} \dot{z}(t) = Az(t) + Bu(t) + Gf \quad 0 < t < T \\ \\ z(0) = z_0 \end{cases} \tag{2}$$

where $f \in \mathcal{F}$.
We define the viability radius for each state in the viability kernel as follows:

Definition 3.2 *Let $\mathcal{K}$ be a nonempty subset of Z.*

(i) We define for each $z_0 \in Viab_{(A,B,G)}(\mathcal{K})$, the $\mathcal{K}$-viability radius, of z_0, as:

$$R_{Viab_{(A,B,G)}(\mathcal{K})}(z_0)$$
$$= \sup\{r > 0 \,\forall\, f \in \mathcal{F} \,:\, \|f\|_\infty < r : \exists\, u \in \mathcal{U} \,::\, z_{f,u}(t, z_0) \in \mathcal{K} \,;\, \forall\, t \in I\}$$

(ii) We define the $\mathcal{K}$-viability radius of the system as:

$$R_{Viab_{(A,B,G)}(\mathcal{K})}$$
$$= \sup_{r > 0}\{\|f\|_\infty < r : \forall z_0 \in Viab_{(A,B,G)}(\mathcal{K}) \exists\, u \in U \,::\, z_{f,u}(t, z_0) \in \mathcal{K} \,\forall\, t \in I\}$$

Remark 3.3

- *For each f such that $\|f\|_\infty > R_{Viab_F(\mathcal{K})}(z_0)$ there exists a function h (not necessary equal to f) such that $\|h\|_\infty = \|f\|_\infty$ and z_0 does not remain viable for the system excited by h.*

- We have a similar remark about $R_{Viab_F}(\mathcal{K})$. For each f such that $||f||_\infty > R_{Viab_F(\mathcal{K})}(z_0)$ there exists, for each viable state z_0, a function h (not necessarily equal to f) such that $||h||_\infty = ||f||_\infty$ and

$$\forall u \in U : \text{there exists } t \in I : z_{f,u}(t, z_0) \notin \mathcal{K}$$

Example 3.4 *Consider the disturbed and controlled system given by the state equation:*

$$\begin{cases} \dot{z}(t) = 2z(t) + 3u(t) + 4f(t) \quad 0 < t < T = 3 \\ \\ z(0) = z_0 \end{cases} \tag{3}$$

Let us consider the same set $\mathcal{K} = [0, 40]$.

In the autonomous case (nondisturbed noncontrolled), the $\mathcal{K}$-viability set is given by:

$$Viab_{\mathcal{K}} = [0, 40e^{-6}].$$

In the controlled and nondisturbed case, the viability kernel of $\mathcal{K}$ under F is given by:

$$Viab_F(\mathcal{K}) = \mathcal{K} = [0, 40]$$

In the disturbed noncontrolled case, the viability radius is given by:

$$R_{viab}^{\mathcal{K}} = \frac{10e^{-6}}{1 - e^{-6}}$$

and in the disturbed and controlled case, the $\mathcal{K}$-viability radius under F of the system is given by:

$$R_{Viab_F(\mathcal{K})} = \mathcal{K}$$

3.2 Robust Viability and Viability Radius for Linear Finite Dimensional System

3.2.1 Problem Statement

Consider the linear system given by the following state equation:

$$\begin{cases} \dot{z}(t) = Az(t) + Bu(t) + Gf(t) ; \ 0 < t < T \\ z(0) = z_0 \end{cases} \tag{4}$$

where $A \in \mathcal{M}_n(\mathbb{R})$, $z(.) \in \mathcal{C}([o, T], \mathbb{R}^n)$, $B \in \mathcal{M}_{n,p}(\mathbb{R})$ and $G \in \mathcal{M}_{q,n}(\mathbb{R})$.

The solution of the disturbed and controlled system is given by:

$$z_{u,f}(t, z_0) = e^t z_0 + \tilde{H}_t u + H_t f \tag{5}$$

where

$$\tilde{H}_t u = \int_0^t e^{A(t-s)} Bu(s) ds$$

and

$$H_t u = \int_0^t e^{A(t-s)} Gf(s) ds$$

In this subsection we consider the following problem:

$$(P) \begin{cases} \text{For a given state } z_0 \in Viab_F(\mathcal{K}) \\[2mm] \text{Determinate a control u} \\[2mm] \text{which maximize the viability radius of } z_0 \end{cases}$$

If the initial state z_0 is $\mathcal{K}$-viable, then $z_0 \in Viab_F(\mathcal{K})$, so we consider the problem for each state z_0 in the viability kernel.

Remark 3.5 *We have for all $t \in I$,*

$$\phi_{\mathcal{K}}(z_{u,0}(t, z_0)) = \begin{cases} -d[z_{u,0}(t, z_0), \partial_{\mathcal{K}}] & \text{if } z_{u,0}(t, z_0) \in \mathcal{K}, \\[2mm] +d[z_{u,0}(t, z_0), \partial_{\mathcal{K}}] & \text{else.} \end{cases} \tag{6}$$

Denote

$$\rho_{\mathcal{K}}^{u,0}(z_0) = \sup_{t \in I} \phi_{\mathcal{K}}(z_{u,0}(t, z_0))$$

then for all u such that $z_{u,0}(t, z_0) \in \mathcal{K} \ \forall \ t \in I$ we have:

$$-d(z_0, \partial_{\mathcal{K}}) \leqslant \rho_{\mathcal{K}}^{u,0}(z_0) \leqslant 0$$

So $|\rho_{\mathcal{K}}^{u,0}(z_0)| \leqslant d(z_0, \partial_{\mathcal{K}})$.

Let z_0 be an initial state in the viability kernel $Viab_F(\mathcal{K})$ so there exists at least one control u such that:

$$z_{u,0}(t, z_0) \in \mathcal{K} ; \ \forall t \in I$$

Consider the subset C_α of $\mathcal{K}$ defined by:

$$C_\alpha = \{x \in \mathcal{K} \; ; \quad \phi_{\mathcal{K}}(x) \leqslant \alpha\} \tag{7}$$

where $-d(z_0, \partial_{\mathcal{K}}) \leqslant \alpha \leqslant 0$, then

$$d_H(C_\alpha, \mathcal{K}) = |\alpha| \leqslant d(z_0, \partial_{\mathcal{K}})$$

where for two subset A and B, $d_H(A, B)$ is the Hausdorf distance defined by:

$$d_H(A, B) = max(\rho(A, B), \rho(B, A))$$

where $\rho(A, B) = \sup_{x \in A} d(x, B)$.

Then C_α is the largest subset of $\mathcal{K}$ such that

$$d(C_\alpha, \partial_{\mathcal{K}}) = |\alpha|$$

So we consider a relaxed problem:

$$(P1) \begin{cases} \text{For a given state } z_0 \in Viab_F(\mathcal{K}) \\\\ \text{Determinate a control u} \\\\ \text{which maximize } d(\mathcal{C}, \partial_K) \end{cases}$$

where $\mathcal{C}$ is the farthest subset of $\mathcal{K}$ from $\partial_{\mathcal{K}}$ where the state z_0 is $\mathcal{K}$-viable.

Remark 3.6 *The effect of such control is to make the state $z(., z_0)$ in $\mathcal{K}$ but as far as possible from ∂_K.*

3.2.2   Problem Approach

To solve the problem $(P1)$ we consider an approach based on the Dichotomy algorithm.

   Algorithm 1: Consider the largest subset C_0 of $\mathcal{K}$ such that

$$d(C_0, \partial_{\mathcal{K}}) = d(z_0, \partial_{\mathcal{K}})$$

then:

(1) if z_0 is in $Viab_F(C_0)$ (the viability kernel of C_0), stop because in this case

$$d(C_0, \partial_{\mathcal{K}}) = d(z_0, \partial_{\mathcal{K}})$$

(2) else define C_1 the largest subset of K such that

$$d(C_1, \partial_K) = \frac{d(z_0, \partial_K)}{2}$$

and

(2*) if z_0 is in $Viab_F(C_1)$ (the viability kernel of C_1), we define C_2 the largest subset of K such that

$$d(C_2, \partial_K) = \frac{d(z_0, \partial_K)}{4}$$

and we go to (2*)

else

we define C_2 the largest subset of K such that

$$d(C_2, \partial_K) = \frac{d(z_0, \partial_K)}{2} + \frac{d(z_0, \partial_K)}{4}$$

and we go to (2*).

We obtain the following algorithm:

If $z_0 \in Viab_F(C_i)$, we define C_{i+1} as the largest subset of K such that

$$d(C_{i+1}, \partial_K) = \frac{d(C_i, \partial_K)}{2}$$

Else, ($z_0 \notin Viab_F(C_i)$), we define C_{i+1} as the largest subset of K such that

$$d(C_{i+1}, \partial_K) = d(C_i, \partial_K) + \frac{d(C_i, \partial_K)}{2}$$

Denoting $r_m = d(C_m, \partial_K)$, we have the result:

Theorem 3.7 *We assume that K is a bounded nonempty and closed subset of $Z = \mathbb{R}^n$ with a smooth regular boundary ∂_K, then:*

(i) the sequence (r_m) converges to a real r^*;

(ii) there exists a control u^* solution of problem $(P1)$ and the subset C^* given by (7) with $\alpha = r^*$ is the farthest subset of K from ∂_K where z_0 is viable.

Proof (i) The sequence (r_m) converges to a real r^* due to the dichotomy principle. Indeed, we have

$$|d(C_{i+1}, \partial_K) - d(C_i, \partial_K)| = \frac{d(C_i, \partial_K)}{2}$$

and

$$d_H(C_{i+1}, C_i) = \frac{d_H(C_i, C_{i-1})}{2} \quad \text{for } i \geqslant 1$$

For (ii) denote by

$$\mathcal{C} = \{C_j \subset \mathcal{K} \ : \ z_0 \in Viab_F(C_j) \ ; \ j \geqslant 0\}$$

We have two cases:

1. If $\mathcal{C} = \emptyset$, then $r^* = 0$ and in this case $C^* = \mathcal{K}$.
   As $z_0 \in Viab_F(\mathcal{K})$, then there exists a control u such that $z_u(t, z_0 \in \mathcal{K}$ for all $t \in I$ and for such control u we have

$$\rho_{\mathcal{K}}^u(z_0) = 0$$

   and consequently u is a solution of problem $(P1)$.
2. $\mathcal{C} \neq \emptyset$, we consider the subsequence (r_j) of (r_m) defined for j such that $C_j \in \mathcal{C}$.

   The subsequence (r_j) is convergent to r^* and $z_0 \in Viab_F(C^*)$. Indeed, if $z_0 \notin Viab_F(C^*)$, then for all control u, there exists a time $t \in I$, such that $z_u(t, z_0) \notin C^*$. So as C^* is closed subset of $\mathcal{K}$, then for all u, we have

$$\rho_{C^*}^u(z_0) > 0$$

which is in contradiction with the fact that

$$\lim_{j \to \infty} d_H(C_j, C^*) = 0$$

So there exists a control u^* such that $z_{u^*}(t, z_0) \in C^*$ for all $t \in I$ and

$$\rho_{C^*}^{u^*}(z_0) = r^*$$

and as

$$r^* = d(C^*, \partial_{\mathcal{K}})$$

we obtain the result.

So the determination of the robust control which maximizes the viability radius consists to solve a sequence of viability kernel problems (steep (2^*)) and for that there exist many methods based on the viability theorems [1–4]. Also for the numerical approaches (which is not the aim of this paper) many algorithms are developed and can be used.

3.2.3 Example

Example 3.8 *To illustrate our approach we consider in this subsection an example of disturbed controlled system in* $\mathbb{R}^2$.
  Consider the system given by the following state equation:

$$\begin{cases} \dot{z}(t) = Az(t) + Bu(t) + Gf(t) \ \ 0 < t < T \\ z(0) = z_0 \end{cases} \tag{8}$$

where

$$A = \begin{pmatrix} 2 & 0 \\ 0 & 1 \end{pmatrix} \ ; \ B = \begin{pmatrix} 3 \\ b \end{pmatrix} \ and \ G = \begin{pmatrix} 2 \\ g \end{pmatrix}$$

where $b, g \in \mathbb{R}$ *and* $T = 3$.
  The solution of the system (8) is given by:

$$z_{u,f}(t) = e^t z_0 + \int_0^t e^{A(t-s)} Bu(s)ds + \int_0^t e^{A(t-s)} Gf(s)ds \tag{9}$$

Let $\mathcal{K} = [0, 40] \times [0, 10] \subset \mathbb{R}^2$.
  We consider three cases: Autonomous system, disturbed noncontrolled system, and disturbed controlled system.
  Case 1. Autonomous system. *In this case we determinate the* $\mathcal{K}$-*viability set,* $Viab_\mathcal{K}$.
  We have

$$Viab_\mathcal{K} = [0, 40e^{-6}] \times [0, e^{-3}] \tag{10}$$

Indeed, $z_0 = (z_{01}, z_{0,2})$ *is* $\mathcal{K}$-*viable if and only if* $e^t z_0 \in \mathcal{K}$ *which is equivalent to*

$$\begin{cases} 0 \leqslant z_{01} e^{2t} \leqslant 40 \ ; \ \forall t \in [0, 3[] \\ 0 \leqslant z_{02} e^t \leqslant 10 \ ; \ \forall t \in [0, 3] \end{cases}$$

So we obtain the result (10).
  Case 2. Disturbed and noncontrolled system *In this case we determinate the* $\mathcal{K}$-*viability radius for a given* $\mathcal{K}$-*viable state and the Minimal Lethal Disturbance of the set* $\mathcal{K}$.

- $\mathcal{K}$-**viability radius.** *To simplify we consider the case where* $f(t) = f \in \mathbb{R}$.
    Let $z_0 = (20e^{-6}, \frac{1}{2}e^{-3})$ *an initial* $\mathcal{K}$-*viable state. We have*

$$z(t, z_0) = (20e^{-6}e^{2t}, \frac{1}{2}e^{-3}e^t) \ ; \ \forall t \in [0, 3]$$

So we have

$$S(t)z_0 + H_t f = (20e^{-6}e^{2t} + 2f(e^{2t} - 1), \frac{1}{2}e^{-3}e^t + f2(e^t - 1))$$

So z_0 remains viable if

$$\begin{cases} \frac{-20e^{-6}e^{2t}}{2(e^{2t}-1)} \leqslant f \leqslant \frac{40-20e^{-6}e^{2t}}{2(e^{2t}-1)} \\ \\ \frac{-\frac{1}{2}e^{-3}e^t}{2(e^t-1)} \leqslant f \leqslant \frac{10-\frac{1}{2}e^{-3}e^t}{2(e^t-1)} \quad ; \; \forall t \in [0,3] \end{cases}$$

So

$$R^{\mathcal{K}}_{viab}(z_0) = inf(\frac{10e^{-6}}{1-e^{-6}}, \frac{e^{-3}}{4(1-e^{-3})})$$

and finally

$$R^{\mathcal{K}}_{viab}(z_0) = \frac{e^{-3}}{4(1-e^{-3})}$$

- **Minimal Lethal Disturbance.** *To assess the viability radius of $\mathcal{K}$ we use Proposition 2.17. We obtain*

$$R^{\mathcal{K}}_{viab} = R^{\mathcal{K}}_{viab}(z_0) = \frac{e^{-3}}{4(1-e^{-3})}$$

And it is natural because z_0 is in the "center" of the viability set.

Case 3. Disturbed and controlled system. *In this case we consider the state z_0 considered in the case 1 and we will determinate a control solution of problem $(P1)$. For this we consider two cases: the case where the system is controllable and the case which the system is not controllable.*

- ***The system is controllable.***
   The system is controllable if and only if $b \neq 0$. As an example we take $b = 1$. So we have

$$z_{u,0}(t, z_0) = (20e^{-6}e^{2t} + 3e^t \int_0^t e^{-2s} u(s)ds, \frac{1}{2}e^{-3}e^t + e^t \int_0^t e^{-s} u(s)ds)$$

for all $t \in [0,3]$.
   For $z_0 = (20e^{-6}, \frac{1}{2}e^{-3})$ we have for

$$C_0 = [20e^{-6}, 40 - \frac{1}{2}e^{-3}] \times [\frac{1}{2}e^{-3}, 10 - \frac{1}{2}e^{-3}]$$

we have $d(C_0, \partial_K) = d(z_0, \partial_K)$ and z_0 is C_0-viable. So the control $u = 0$ is a solution of problem $(P1)$.

Consider now the case where z_0 is not $\mathcal{K}$-viable but in the viability kernel. As an example we consider $z_0 = (20, 5)$ which is not $\mathcal{K}$-viable because

$$z_{0,0}(t, z_0) = (20e^{2t}, 5e^t)$$

and for $t = 2$, for example, $z_{0,0}(t, z_0) \notin \mathcal{K}$, but z_0 is in the viability kernel, because

$$z_{u,0}(t, z_0) = (20e^{2t} + 3e^t \int_0^t e^{-2s} u(s) ds,$$

$$5e^t + e^t \int_0^t e^{-s} u(s) ds)$$

and there exists a control u such that:

$$z_{u,0}(t, z_0) \in \mathcal{K}, \quad \forall t \in [0, 3]$$

- **The system is not controllable.** This is the case where $b = 0$. So the solution is given by

$$z_{u,0}(t, z_0 = (z_{01}, z_{02})$$

$$= (z_{01}e^{2t} + 3e^t \int_0^t e^{-2s} u(s) ds, z_{02}e^t)$$

$$; \forall t \in]0, 3[$$

In this case the state $z_0 = (20, 5)$ is not in the viability kernel.

Let us consider another state $z_1 = (30, (9.7)e^{-3})$ which is in the viability kernel. To determinate the robust control we have to find the smallest subset C for which z_1 is in the viability kernel using the Algorithm 1. As $z_1 = (30, (9.7)e^{-3})$ is in the viability kernel so $R^{\mathcal{K}}_{viab}(z_0) \geqslant 0$. Consider

$$C_0 =](9.7)e^{-3}, 40 - (9.7)e^{-3}[\times](9.7)e^{-3}, 10 - (9.7)e^{-3}[$$

By explicit calculus we proof that z_1 is not in the viability kernel of C_0. So

$$0 \leqslant r^* < (9.7)e^{-3}$$

we consider

$$\mathcal{C}_1 =$$

$$]\frac{1}{2}(9.7)e^{-3}, 40 - \frac{1}{2}(9.7)e^{-3}[\times]\frac{1}{2}(9.7)e^{-3}, 10 - \frac{1}{2}(9.7)e^{-3}[$$

In this case z_1 is in the viability kernel of $\mathcal{C}_1$, so we have

$$\frac{1}{2}(9.7)e^{-3} \leqslant r^* < (9.7)e^{-3}$$

and we consider

$$\mathcal{C}_2 =$$

$$](\frac{1}{2} + \frac{1}{4})(9.7)e^{-3}, 40 - (\frac{1}{2} + \frac{1}{4})(9.7)e^{-3}[\times$$

$$](\frac{1}{2}9.7)e^{-3} + \frac{1}{4}, 10 - (\frac{1}{2} + \frac{1}{4})(9.7)e^{-3}[$$

and in this case z_1 is in the viability kernel of $\mathcal{C}_2$, so we have

$$\frac{1}{2}(9.7)e^{-3} \leqslant c^* < (\frac{1}{2} + \frac{1}{4})(9.7)e^{-3}$$

which gives

$$0.2240 \leqslant r^* \leqslant 0.3360$$

And we continue the algorithm.

Remark 3.9 *We have presented in this paper, as examples, the one-dimensional case just for clarity, however for the case where the dimension is greater or equal to 2,the approch is the same but we have to do numerical computations for all the considered parameters.*

4   Connection with Minimal Lethal Dose in Toxicity

The viability radius depends on the disturbance set; $\mathcal{F}$. This property has been exploited in biology, and particularly in toxicology, to assess the toxicity of each poison. For that the concept of Minimal Lethal Dose was introduced. We recall the following definition [10, 11]:

Definition 4.1 *The Minimal Lethal Dose (MLD) is the lowest dose of a substance that can kill an animal by administering a slow intravenous drug. Death is assessed by cardiac arrest.*

We remark that in this definition one can take into account the manner of administering the substance: this is the equivalent operator H_t in our case. We recall that the dose is the ratio between the weight of the absorbed substance and the weight of the body absorbs. It was cited in Paracelsus *all substances are toxic, it is only a question of quantity, a question of dose* [9].

In practice, the determination of the Minimal Lethal Dose (MLD) for each poison is a very difficult problem and for that the biologists estimate the MLD in the laboratories. As in each measure in laboratories there are some errors, J.W. Trevan has introduced in 1927, the so-called *Lethal Dose 50* (LD50). We recall the definition:

Definition 4.2 *The LD50 value is a statistical estimate of a dose that can kill 50% of the population of animals from the same species. It is expressed in mg of substance per kg of body weight.*

The median lethal dose is a quantitative indicator of the toxicity of a given substance. Why 50%? This is for statistical representation reasons. In general we uses the value 50%, rather than 0, 5, 95, or 100%. In fact, Gaussian curve is "flat" to 50%, making a sample is more representative when a threshold is exceeded by 50% [8, 11].

The LD50 concept was introduced by J. W Trevan in 1927 and it permits the classification of all products by their dangerousness.

5 Conclusion

In this work we have considered, for finite dimensional linear systems, the viability radius and the robust viability control problems. Some characterizations are given. We have presented some examples to illustrate the definitions and approaches. We notice that this work was motivated by the concept of minimal lethal dose introduced for toxicity problems. It will be interesting to extend the results to Distributed Parameters Systems and consider the relationships between vulnerability and protector control concepts introduced by Bernoussi for such systems, and the robust viability control. Also as for Distributed Parameters Systems the measures function and sensors play an important role, it will be interesting to consider the problems through measures functions. Such problems are under investigation.

Acknowledgements This work, presented in Honor of Prof. Norbert Hounkonnou for his 60th Birthday, is supported by MESRCFC and CNRST, Morocco under the projectPPR2/2016/79, OGI-Env.

References

1. J.P. Aubin, *Viability Theory* (Birkhäuser, Boston, 1991)
2. J.P. Aubin, A viability approach to the inverse set-valued theorem. J. Evol. Equ. **6**, 419–432 (2006)

3. J.P. Aubin, Viability kernels and capture basins of sets under differential inclusions. SIAM J. Control Optim. **40**(3), 853–881 (2001)
4. I. Alvarez, S. Martin, Geometric robustness of viability kernels and resilience basins, in *Viability and Resilience of Complex Systems* (Springer, Berlin, Heidelberg) , pp. 193–218
5. M. Amharref, S. Aassine, A. Bernoussi, B. Haddouchi, Carthographie de la vulnérabilité à la pollution des eaux souterraines: Application à la plaine du Gharb. Revue des Sciences de l'Eau **20**(2), 185–199 (2007)
6. M. Amharref, A. Bernoussi, Vulnérabilité et risque de pollution des eaux souterraines. Actes des JSIRAUF (Journées Sientifiques Inter Réseaux de l'Agence Universitaire de la Francophonie), Hanoi, 6–9 Novembre 2007 (2007)
7. J.A. Sethian, *Level Set Methods and Fast Marching Methods: Evolving Interfaces in Computational Geometry, Fluid Mechanics, Computer Vision, and Materials Science*, vol. 3 (Cambridge University Press, Cambridge, 1999)
8. S. Adler, G. Bicker, H. Bigalke, C. Bishop, J. Blümel, D. Dressler et al., The current scientific and legal status of alternative methods to the LD50 test for botulinum neurotoxin potency testing, in *The Report and Recommendations of a ZEBET Expert Meeting*. ATLA, vol. 38(4) (August 2010), pp. 315–330
9. M. Amharref, A. Bernoussi, Groundwater pollution risk, in *Systems Theory: Modeling, Analysis and Control, Fes-2009*, ed. by A. El Jai, L. Afifi, E. Zerrik (PUP, Perpignan, 2009), pp. 499–506. ISBN 978-2-35412-043-6
10. T.S.S. Dikshith, P.V. Diwan, *Industrial Guide to Chemical and Drug Safety* (Wiley, New York, 2003)
11. E. Schlede, U. Mischke, R. Roll, D. Kayser, A national validation study of the acute-toxic-class method-an alternative to the LD50 test. Arch. Toxicol. **66**(7), 455–470 (1992)

Walker Osserman Metric of Signature (3, 3)

Abdoul Salam Diallo, Mouhamadou Hassirou, and Ousmane Toudou Issa

Dedicated to Prof. Mahouton Norbert Hounkonnou on the occasion of his 60th Birthday

Abstract A Walker m-manifold is a pseudo-Riemannian manifold, which admits a field of parallel null r-planes, with $r \leqslant \frac{m}{2}$. The Riemann extension is an important method to produce Walker metric on the cotangent bundle T^*M of any affine manifold (M, ∇). In this paper, we investigate the torsion-free affine manifold (M, ∇) and their Riemann extension $(T^*M, \bar{g})$ as concerns heredity of the Osserman condition.

Keywords Affine Jacobi operator · Affine Osserman connections · Osserman manifolds · Rieman extension · Walker metric

2010 Mathematics Subject Classification: 53B05, 53B20, 53B30, 53B50

A. S. Diallo (✉)
Université Alioune Diop de Bambey, UFR SATIC, Département de Mathématiques, Bambey, Senegal
e-mail: abdoulsalam.diallo@uadb.edu.sn

M. Hassirou
Département de Mathématiques et Informatique, Faculté des Sciences et Techniques, Université Abdou Moumouni, Niamey, Niger
e-mail: hassirou@refer.ne

O. T. Issa
Département de l'Environnement, Université de Tillaberi, Tillaberi, Niger
e-mail: outoud26@yahoo.fr

© Springer Nature Switzerland AG 2018
T. Diagana, B. Toni (eds.), *Mathematical Structures and Applications*,
STEAM-H: Science, Technology, Engineering, Agriculture,
Mathematics & Health, https://doi.org/10.1007/978-3-319-97175-9_8

1   Introduction

The pseudo-Riemannian geometry is the study of the Levi–Civita connection, which is the unique torsion-free connection compatible with the metric structure. The theory of affine connections is a classical topic in differential geometry, it was initially developed to solve pure geometrical problems. It provides an extremely important tool to study geometrical structures on manifolds and, as such, has been applied with great sources in many different setting. For affine connections, a survey of the development of the theory can be found in [19] and references therein. In [13], García-Rio et al. introduced the notion of the affine Osserman connections. Affine Osserman connections are well-understood in dimension two. For instance, in [6] and [13], the authors proved in a different way that an affine connection is Osserman if and only if its Ricci tensor is skew-symmetric. The situation is however more involved in higher dimensions where the skew-symmetry of the Ricci tensor is a necessary (but not a sufficient) condition for an affine connection to be Osserman. The concept of an affine Osserman connection has become a very active research subject (See [7–9] for more details.)

In this paper, we associate a pseudo-Riemannian structure of neutral signature to certain affine connections and use this correspondence to study both geometries. We examine affine Osserman connections, Riemann extensions, and Walker structures. Our paper is organized as follows. Section 1 introduces this topic. Section 2 contains some definitions and basic results we shall need. In Sect. 3, we study the Osserman condition on a family of affine connection (cf. Proposition 3.3). Finally in Sect. 4, we construct an example of pseudo-Riemannian Walker Osserman metric of signature (3, 3), using the Riemann extensions. The Riemann extension provides a link between affine and pseudo-Riemannian geometries. It plays an important role in various questions involving the spectral geometry of the curvature operator. (See, for example, [1–3, 7, 13] for more details.)

2   Preliminaries

2.1   Affine Manifolds

Let M be an m-dimensional smooth manifold and ∇ be an affine connection on M. Let us consider a system of coordinates $(u_1, \cdots, u_m)$ in a neighborhood $\mathcal{U}$ of a point p in M. In $\mathcal{U}$, the connection is given by

$$\nabla_{\partial_i} \partial_j = f_{ij}^k \partial_k, \tag{2.1}$$

where $\{\partial_i = \frac{\partial}{\partial u_i}\}_{1 \leqslant i \leqslant m}$ is a basis of the tangent space $T_p M$ and the functions $f_{ij}^k (i, j, k = 1, \cdots, m)$ are called the coefficients of the affine connection. The pair (M, ∇) shall be called *affine manifold*.

We define a few tensor fields associated to a given affine connection ∇. The *torsion tensor field* T^∇, which is of type (1, 2), is defined by

$$T^\nabla(X, Y) = \nabla_X Y - \nabla_Y X - [X, Y],$$

for any vector fields X, and Y on M. The components of the torsion tensor T^∇ in local coordinates are

$$T_{ij}^k = f_{ij}^k - f_{ji}^k.$$

If the torsion tensor of a given affine connection ∇ vanishes, we say that ∇ is torsion-free.

The *curvature tensor field* $\mathcal{R}^\nabla$, which is of type (1, 3), is defined by

$$\mathcal{R}^\nabla(X, Y)Z := \nabla_X \nabla_Y Z - \nabla_Y \nabla_X Z - \nabla_{[X,Y]}Z,$$

for any vector fields X, Y and Z on M. The components in local coordinates are

$$\mathcal{R}^\nabla(\partial_k, \partial_l)\partial_j = \sum_i R_{jkl}^i \partial_i.$$

We shall assume that ∇ is torsion-free. If $\mathcal{R}^\nabla = 0$ on M, we say that ∇ is *flat affine connection*. It is known that ∇ is flat if and only if around a point there exists a local coordinates system such that $f_{ij}^k = 0$ for all i, j, and k.

We define the *Ricci tensor* Ric^∇, of type (0, 2) by

$$Ric^\nabla(Y, Z) = \text{trace}\{X \mapsto \mathcal{R}^\nabla(X, Y)Z\}.$$

The components in local coordinates are given by

$$Ric^\nabla(\partial_j, \partial_k) = \sum_i R_{kij}^i.$$

It is known in Riemannian geometry that the Levi–Civita connection of a Riemannian metric has symmetric Ricci tensor, that is, $Ric(Y, Z) = Ric(Z, Y)$. But this property is not true for an arbitrary affine connection which is torsion-free. In fact, the property is closely related to the concept of parallel volume element (cf. [19] for more details).

In a 2-dimensional manifold, the curvature tensor $\mathcal{R}^\nabla$ and the Ricci tensor Ric^∇ are related by

$$\mathcal{R}^\nabla(X, Y)Z = Ric^\nabla(Y, Z)X - Ric^\nabla(X, Z)Y. \tag{2.2}$$

For $X \in \Gamma(T_p M)$, we define the *affine Jacobi operator* $J_{\mathcal{R}^\nabla}$ with respect to X by $J_{\mathcal{R}^\nabla}(X) : T_p M \longrightarrow T_p M$ such that

$$J_{\mathcal{R}^\nabla}(X)Y := \mathcal{R}^\nabla(Y, X)X. \tag{2.3}$$

for any vector field Y. The affine Jacobi operator satisfies $J_{\mathcal{R}^\nabla}(X)X = 0$ and $J_{\mathcal{R}^\nabla}(\alpha X) = \alpha^2 J_{\mathcal{R}^\nabla}(X)Y$, for $\alpha \in \mathbb{R} - \{0\}$ and $X \in T_pM$. Let (M, ∇) be a three-dimensional affine manifold and let $X = \sum_{i=1}^{3} \alpha_i \partial_i$ be a non-null vector on M, where $\{\partial_i\}$ denotes the coordinate basis and $\alpha_i \in \mathbb{R}^*$. Then the affine Jacobi operator is given by

$$\begin{aligned}
J_{\mathcal{R}^\nabla}(X) = &\; \alpha_1^2 \mathcal{R}^\nabla(\cdot, \partial_1)\partial_1 + \alpha_1\alpha_2 \mathcal{R}^\nabla(\cdot, \partial_1)\partial_2 + \alpha_1\alpha_3 \mathcal{R}^\nabla(\cdot, \partial_1)\partial_3 \\
&+ \alpha_1\alpha_2 \mathcal{R}^\nabla(\cdot, \partial_2)\partial_1 + \alpha_2^2 \mathcal{R}^\nabla(\cdot, \partial_2)\partial_2 + \alpha_2\alpha_3 \mathcal{R}^\nabla(\cdot, \partial_2)\partial_3 \\
&+ \alpha_1\alpha_3 \mathcal{R}^\nabla(\cdot, \partial_3)\partial_1 + \alpha_2\alpha_3 \mathcal{R}^\nabla(\cdot, \partial_3)\partial_2 + \alpha_3^2 \mathcal{R}^\nabla(\cdot, \partial_3)\partial_3.
\end{aligned}$$

2.2 Affine Osserman Manifolds

Let (M, ∇) be an m-dimensional affine manifold, i.e., ∇ is a torsion free connection on the tangent bundle of a smooth manifold M of dimension m. Let $\mathcal{R}^\nabla(X, Y)$ be the curvature operator and $J_{\mathcal{R}^\nabla}(X)$ the Jacobi operator with respect to a vector $X \in T_pM$ associated.

Definition 2.1 ([14]) *One says that an affine manifold (M, ∇) is affine Osserman at $p \in M$ if the characteristic polynomial of $J_{\mathcal{R}^\nabla}(X)$ is independent of $X \in T_pM$. Also (M, ∇) is called affine Osserman if (M, ∇) is affine Osserman at each $p \in M$.*

Theorem 2.2 ([14]) *Let (M, ∇) be an m-dimensional affine manifold. Then (M, ∇) is called affine Osserman at $p \in M$ if and only if the characteristic polynomial of $J_{\mathcal{R}^\nabla}(X)$ is*

$$P_\lambda[J_{\mathcal{R}^\nabla}(X)] = \lambda^m$$

for every $X \in T_pM$.

Corollary 2.3 *We say that (M, ∇) is affine Osserman if $\operatorname{Spect}\{J_{\mathcal{R}^\nabla}(X)\} = \{0\}$ for any vector X*

Corollary 2.4 *If (M, ∇) is affine Osserman at $p \in M$, then the Ricci tensor is skew-symmetric at $p \in M$.*

The affine Osserman connections are of interest not only in the affine geometry, but also in the study of the pseudo-Riemannian Osserman metrics since they provide some nice examples of Osserman manifolds whose Jacobi operators have non-trivial Jordan normal form and which are not nilpotent. It has long been a task in this field to build examples of Osserman manifolds were not nilpotent and which exhibited non-trivial Jordan normal form. We will refer [1, 3] for more information.

2.3   The Riemann Extension Construction

Let $N := T^*M$ be the cotangent bundle of an m-dimensional manifold and let $\pi : T^*M \to M$ be the natural projection. A point ξ of the cotangent bundle is represented by an ordered pair (ω, p), where $p = \pi(\xi)$ is a point on M and ω is an 1-form on $T_p M$. If $u = (u_1, \cdots, u_m)$ are local coordinates on M, let $u' = (u_{1'}, \cdots, u_{m'})$ be the associated dual coordinates on the fiber where we expand an 1-form ω as $\omega = u_{i'} du_i$ $(i = 1, \cdots, m; i' = i + m)$; we shall adopt the Einstein convention and sum over repeated indices henceforth.

For each vector field $X = X^i \partial_i$ on M, the evaluation map $\iota X(p, \omega) = \omega(X_p)$ defines on function on N which in local coordinates is given by

$$\iota X(u_i, u_{i'}) = u_{i'} X^i.$$

Vector fields on N are characterized by their action on function ιX; the complete lift X^C of a vector field X on M to N is characterized by the identity

$$X^C(\iota Z) = \iota[X, Z], \quad \text{for all} \quad Z \in C^\infty(TM).$$

Moreover, since a $(0, s)$-tensor field on M is characterized by its evaluation on complete lifts of vector fields on M, for each tensor field T of type $(1, 1)$ on M, we define a 1-form ιT on N which is characterized by the identity

$$\iota T(X^C) = \iota(TX).$$

Definition 2.5 *Let (M, ∇) be an affine manifold of dimension m. The Riemann extension $\bar{g}$ of (M, ∇) is the pseudo-Riemannian metric of neutral signature (m, m) on the cotangent bundle T^*M, which is characterized by the identity*

$$\bar{g}(X^C, Y^C) = -\iota(\nabla_X Y + \nabla_Y X).$$

In the system of induced coordinates $(u_i, u_{i'})$ on TM, the Riemann extension takes the form:

$$\bar{g} = \begin{pmatrix} -2u_{k'} \Gamma_{ij}^k & \mathrm{Id}_m \\ \mathrm{Id}_m & 0 \end{pmatrix},$$

with respect to $\{\partial_{u_1}, \ldots, \partial_{u_m}, \partial_{u_1'}, \ldots, \partial_{u_m'}\}$; here, the indices i and j range from $1, \ldots, m$, $i' = i + m$, and Γ_{ij}^k are the Christoffel symbols of the connection ∇ with respect to the coordinates (u_i) on M. More explicitly:

$$\bar{g}(\partial_{u_i}, \partial_{u_j}) = -2u_{k'} \Gamma_{ij}^k, \quad \bar{g}(\partial_{u_i}, \partial_{u_j'}) = \delta_i^j, \quad \bar{g}(\partial_{u_i'}, \partial_{u_j'}) = 0.$$

Let (M, g) be a pseudo-Riemannian manifold. The Riemann extension of the Levi–Civita connection inherits many of the properties of the base manifold. For instance, (M, g) has constant sectional curvature if and only if $(TM, \bar{g})$ is locally conformally flat. However, the main applications of the Riemann extensions appear when considering affine connections are not the Levi–Civita connection of any metric. We have the following result:

Theorem 2.6 ([13]) *Let $(T^*M, \bar{g})$ be the cotangent bundle of an affine manifold (M, ∇) equipped with the Riemann extension of the torsion free connection ∇. Then $(T^*M, \bar{g})$ is a pseudo-Riemannian globally Osserman manifold if and only if (M, ∇) is an affine Osserman manifold.*

3 Example of Affine Osserman Connections

In the following M denotes a three-dimensional manifold and ∇ a smooth torsion-free affine connection. Choose a system (u_1, u_2, u_3) of local coordinates in a domain $\mathcal{U} \subset M$ such that the affine connection ∇ is uniquely determined by six functions $f_1, \ldots, f_6$ given by the formulas

$$\begin{cases} \nabla_{\partial_1} \partial_1 = f_1(u_1, u_2, u_3) \partial_2; \\ \nabla_{\partial_1} \partial_2 = f_2(u_1, u_2, u_3) \partial_2; \\ \nabla_{\partial_1} \partial_3 = f_3(u_1, u_2, u_3) \partial_2; \\ \nabla_{\partial_2} \partial_2 = f_4(u_1, u_2, u_3) \partial_2; \\ \nabla_{\partial_2} \partial_3 = f_5(u_1, u_2, u_3) \partial_2; \\ \nabla_{\partial_3} \partial_3 = f_6(u_1, u_2, u_3) \partial_2. \end{cases} \tag{3.1}$$

One can easily show that the non-zero components of the Ricci tensor are given by

$$\begin{cases} Ric(\partial_1, \partial_1) = \partial_2 f_1 - \partial_1 f_2 + f_1 f_4 - f_2^2 \\ Ric(\partial_1, \partial_2) = \partial_2 f_2 - \partial_1 f_4 \\ Ric(\partial_1, \partial_3) = \partial_2 f_3 - \partial_1 f_5 + f_3 f_4 - f_2 f_5 \\ Ric(\partial_3, \partial_1) = \partial_2 f_3 - \partial_3 f_2 + f_3 f_4 - f_2 f_5 \\ Ric(\partial_3, \partial_2) = \partial_2 f_5 - \partial_3 f_4 \\ Ric(\partial_3, \partial_3) = \partial_2 f_6 - \partial_3 f_5 + f_4 f_6 - f_5^2. \end{cases} \tag{3.2}$$

The skew-symmetry of Ricci tensor means that, in any local coordinates, we have:

$$\begin{cases} Ric(\partial_1, \partial_1) = Ric(\partial_2, \partial_2) = Ric(\partial_3, \partial_3) = 0 \\ Ric(\partial_1, \partial_2) + Ric(\partial_2, \partial_1) \qquad\qquad = 0 \\ Ric(\partial_1, \partial_3) + Ric(\partial_3, \partial_1) \qquad\qquad = 0 \\ Ric(\partial_2, \partial_3) + Ric(\partial_3, \partial_2) \qquad\qquad = 0. \end{cases} \tag{3.3}$$

According (3.1) and (3.3), we have the following

Proposition 3.1 *The affine connection ∇ defined in (3.1) is skew-symmetric if the functions $f_i, i = 1, \ldots, 6$ satisfy the following partial differential equations:*

$$\partial_2 f_2 - \partial_1 f_4 = 0; \quad \partial_2 f_5 - \partial_3 f_4 = 0$$
$$\partial_2 f_1 - \partial_1 f_2 + f_1 f_4 - f_2^2 = 0$$
$$\partial_2 f_6 - \partial_3 f_5 + f_4 f_6 - f_5^2 = 0$$
$$2\partial_2 f_3 - \partial_1 f_5 - \partial_3 f_2 + 2 f_3 f_4 - 2 f_2 f_5 = 0. \tag{3.4}$$

Proof It follows from (3.1) and (3.3).

Corollary 3.2 ([8]) *Let ∇ be as (3.1). Assume that $f_2 = f_3 = f_5 = 0$, then the affine connection (3.1) is skew-symmetric if and only if the coefficients of the connection (3.1) satisfy*

$$f_4(u_1, u_2, u_3) = f_1(u_2), \quad \partial_2 f_1 + f_1 f_4 = 0, \quad and \quad \partial_2 f_6 + f_4 f_6 = 0. \tag{3.5}$$

We have the following result:

Proposition 3.3 *Let (M, ∇) be a 3-dimensional affine manifold with torsion free connection given by (3.1). Then (M, ∇) is affine Osserman if and only if the Ricci tensor is skew-symmetric.*

Proof Since the Ricci tensor of any affine Osserman connection is skew-symmetric, it follows from previous expression that we have the following necessary conditions for the affine connections (3.1) to be Osserman

$$\partial_2 f_2 - \partial_1 f_4 = 0; \quad \partial_2 f_5 - \partial_3 f_4 = 0$$
$$\partial_2 f_1 - \partial_1 f_2 + f_1 f_4 - f_2^2 = 0$$
$$\partial_2 f_6 - \partial_3 f_5 + f_4 f_6 - f_5^2 = 0$$

and

$$2\partial_2 f_3 - \partial_1 f_5 - \partial_3 f_2 + 2 f_3 f_4 - 2 f_2 f_5 = 0.$$

Then, the associated affine Jacobi operator can be expressed, with respect to the coordinate basis, as

$$(J_{\mathcal{R}^\nabla}(X)) = \begin{pmatrix} 0 & 0 & 0 \\ a & 0 & c \\ 0 & 0 & 0 \end{pmatrix},$$

with

$$a = \alpha_1\alpha_3\left(\partial_1 f_3 - \partial_3 f_1 + f_2 f_3 - f_1 f_4\right)$$
$$+\alpha_2\alpha_3\left(2\partial_1 f_5 - \partial_2 f_3 - \partial_3 f_2 + f_2 f_5 - f_3 f_4\right)$$
$$+\alpha_3^2\left(\partial_1 f_6 - \partial_3 f_3 + f_2 f_6 - f_3 f_5\right);$$
$$c = -\alpha_1^2\left(\partial_1 f_3 - \partial_3 f_1 + f_2 f_3 - f_1 f_5\right)$$
$$-\alpha_1\alpha_2\left(\partial_1 f_5 - \partial_2 f_3 - 2\partial_3 f_2 + f_3 f_4 - f_2 f_5\right)$$
$$-\alpha_1\alpha_3\left(\partial_1 f_6 - \partial_3 f_3 + f_2 f_6 - f_3 f_5\right).$$

The characteristic polynomial of the affine Jacobi operator is now seen to be:

$$P_\lambda[J_{\mathcal{R}^\nabla}(X)] = -\lambda^3$$

which has zero eigenvalues.   □

Example 3.4 *Following Corollary 3.2, one can construct examples of affine Osserman connections. The following connection on $\mathbb{R}^3$ whose non-zero coefficients of the cofficients are given by*

$$\nabla_{\partial_1}\partial_1 = u_1 u_3 \partial_2 \quad and \quad \nabla_{\partial_3}\partial_3 = (u_2 + u_3)\partial_2 \tag{3.6}$$

is nonflat affine Osserman.

The concept of an affine Osserman connection has become a very active research subject. In [10], the authors give examples of affine Osserman connections which are locally symmetric but not flat on 3-dimensional manifolds. In [11], affine Osserman connections which are Ricci flat but not flat on 3-dimensional manifolds are given. In [12], examples of affine Osserman connections which are Ricci flat and which are not Ricci flat on 3-dimensional manifolds are exhibited.

4   Example of Walker Osserman Metric

Let M be a pseudo-Riemannian manifold of signature (p, q). We suppose given a splitting of the tangent bundle in the form $TM = V_1 \oplus V_2$ where V_1 and V_2 are smooth subbundles which are called distribution. This defines two complementary projection π_1 and π_2 of TM onto V_1 and V_2. We say that V_1 is a parallel distribution if $\nabla \pi_1 = 0$. Equivalently this means that if X_1 is any smooth vector field taking values in V_1, then ∇X_1 again takes values in V_1. If M is Riemannian, we can take

$V_2 = V_1^{\perp}$ to be the orthogonal complement of V_1 and in that case V_2 is again parallel. In the pseudo-Riemannian setting, of course, $V_1 \cap V_2$ need not be trivial. We say that V_1 is a null parallel distribution if V_1 is parallel and if the metric restricted to V_1 vanishes identically. Manifolds which admit null parallel distribution are called Walker manifolds. More precisely, a Walker manifold is a triple $(M, g, \mathcal{D})$ where M is an m-dimensional manifold, g an indefinite metric, and $\mathcal{D}$ an r-dimensional parallel null distribution. Of special interest are those manifolds admitting a field of null planes of maximum dimension $r = \frac{m}{2}$. In this particular case, it is convenient to use special coordinate systems associated with any Walker metric.

Let (u_1, u_2, u_3) be the local coordinates on a 3-dimensional affine manifold (M, ∇). We expand $\nabla_{\partial_i} \partial_j = \sum_k f_{ij}^k \partial_k$ for $i, j, k = 1, 2, 3$ to define the Christoffel symbols of ∇. Let $\omega = u_4 du_1 + u_5 du_2 + u_6 du_3 \in T^*M : (u_4, u_5, u_6)$ be the dual fiber coordinates. The *Riemann extension* is the pseudo-Riemannian metric $\bar{g}$ on the cotangent bundle T^*M of neutral signature $(3, 3)$ defined by setting

$$\bar{g}(\partial_1, \partial_4) = \bar{g}(\partial_2, \partial_5) = \bar{g}(\partial_3, \partial_6) = 1,$$
$$\bar{g}(\partial_1, \partial_1) = -2u_4 f_{11}^1 - 2u_5 f_{11}^2 - 2u_6 f_{11}^3,$$
$$\bar{g}(\partial_1, \partial_2) = -2u_4 f_{12}^1 - 2u_5 f_{12}^2 - 2u_6 f_{12}^3,$$
$$\bar{g}(\partial_1, \partial_3) = -2u_4 f_{13}^1 - 2u_5 f_{13}^2 - 2u_6 f_{13}^3,$$
$$\bar{g}(\partial_2, \partial_2) = -2u_4 f_{22}^1 - 2u_5 f_{22}^2 - 2u_6 f_{22}^3,$$
$$\bar{g}(\partial_2, \partial_3) = -2u_4 f_{23}^1 - 2u_5 f_{23}^2 - 2u_6 f_{23}^3,$$
$$\bar{g}(\partial_3, \partial_3) = -2u_4 f_{33}^1 - 2u_5 f_{33}^2 - 2u_6 f_{33}^3.$$

Let us consider the affine Osserman connection given in (3.6). The Riemann extension $\bar{g}$ on $\mathbb{R}^6$ of the connection (3.6) has the form

$$\bar{g} = \begin{pmatrix} -2u_5 u_1 u_3 & 0 & 0 & 1 & 0 & 0 \\ 0 & 0 & 0 & 0 & 1 & 0 \\ 0 & 0 & -2u_5(u_1 + u_3) & 0 & 0 & 1 \\ 1 & 0 & 0 & 0 & 0 & 0 \\ 0 & 1 & 0 & 0 & 0 & 0 \\ 0 & 0 & 1 & 0 & 0 & 0 \end{pmatrix}.$$

The nonvanishing covariant derivatives of $\bar{g}$ are given by

$$\bar{\nabla}_{\partial_1} \partial_1 = u_1 u_3 \partial_2 - u_3 u_5 \partial_4 + u_1 u_5 \partial_6, \quad \bar{\nabla}_{\partial_1} \partial_3 = -u_1 u_5 \partial_4 - u_5 \partial_6,$$
$$\bar{\nabla}_{\partial_1} \partial_5 = -u_1 u_3 \partial_4, \quad \bar{\nabla}_{\partial_3} \partial_3 = (u_1 + u_3) \partial_2 + u_5 \partial_4 - u_5 \partial_6,$$
$$\bar{\nabla}_{\partial_3} \partial_5 = -(u_1 + u_3) \partial_6.$$

The nonvanishing components of the curvature tensor of $(\mathbb{R}^6, \bar{g})$ are given by

$$R(\partial_1, \partial_3)\partial_1 = -u_1\partial_2; \quad R(\partial_1, \partial_3)\partial_3 = \partial_2; \quad R(\partial_1, \partial_3)\partial_5 = u_1\partial_4 - \partial_6;$$

$$R(\partial_1, \partial_5)\partial_1 = -u_1\partial_6; \quad R(\partial_1, \partial_5)\partial_3 = u_1\partial_4; \quad R(\partial_3, \partial_5)\partial_1 = \partial_6;$$

$$R(\partial_3, \partial_5)\partial_3 = -\partial_4.$$

Now, if $X = \sum_{i=1}^{6} \alpha_i \partial_i$ is a vector field on $\mathbb{R}^6$, then the matrix associated with the Jacobi operator $J_{\mathcal{R}}(X) = \mathcal{R}(\cdot, X)X$ is given by

$$(J_{\mathcal{R}}(X)) = \begin{pmatrix} A & 0 \\ B & A^t \end{pmatrix},$$

where A is the 3×3 matrix given by

$$A = \begin{pmatrix} 0 & 0 & 0 \\ 1 - u_1 & 0 & u_1 - 1 \\ 0 & 0 & 0 \end{pmatrix};$$

and B is the 3×3 matrix given by

$$B = \begin{pmatrix} 2u_1 & 0 & -u_1 \\ 0 & 0 & 0 \\ -1 - u_1 & 0 & 1 \end{pmatrix}.$$

Then we have the following

Proposition 4.1 $(\mathbb{R}^6, \bar{g})$ *is a Walker Osserman metric of signature* $(3, 3)$.

Walker geometry is intimately related to many questions in mathematical physics. Note that the Riemann extension is necessarily a Walker metric. It is a remarkable fact that Walker metrics satisfying some natural curvature conditions are locally Riemann extensions, thus leading the corresponding classification problem to a task in affine geometry as shown in [2].

Chaichi et al. [4] have studied conditions for a Walker metric to be Einstein, Osserman, or locally conformally flat and obtained thereby exact solutions to the Einstein equations for a restricted Walker manifold.

Appendix 1: Components of the Curvature Tensor

The non-zero components of the curvature tensor of the affine connection (3.1) are given by

$$R(\partial_1, \partial_2)\partial_1 = (\partial_1 f_2 - \partial_2 f_1 + f_2^2 - f_1 f_4)\partial_2$$

$$R(\partial_1, \partial_2)\partial_2 = (\partial_1 f_4 - \partial_2 f_2)\partial_2$$

$$R(\partial_1, \partial_2)\partial_3 = (\partial_1 f_5 - \partial_2 f_3 + f_2 f_5 - f_3 f_4)\partial_2$$

$$R(\partial_1, \partial_3)\partial_1 = (\partial_1 f_3 - \partial_3 f_1 + f_2 f_3 - f_1 f_5)\partial_2$$

$$R(\partial_1, \partial_3)\partial_2 = (\partial_1 f_5 - \partial_3 f_2)\partial_2$$

$$R(\partial_1, \partial_3)\partial_3 = (\partial_1 f_6 - \partial_3 f_3 + f_2 f_6 - f_3 f_5)\partial_2$$

$$R(\partial_2, \partial_3)\partial_1 = (\partial_2 f_3 - \partial_3 f_2 + f_3 f_4 - f_2 f_5)\partial_2$$

$$R(\partial_2, \partial_3)\partial_2 = (\partial_2 f_5 - \partial_3 f_4)\partial_2$$

$$R(\partial_2, \partial_3)\partial_3 = (\partial_2 f_6 - \partial_3 f_5 + f_4 f_6 - f_5^2)\partial_2.$$

Appendix 2: Osserman Geometry

Let R be the curvature operator of a Riemannian manifold (M, g) of dimension m. The Jacobi operator $\mathcal{J}(x) : y \mapsto R(y, x)x$ is the self-adjoint endomorphism of the tangent bundle. Following the seminal work of Osserman [20], one says that (M, g) is *Osserman* if the eigenvalues of $\mathcal{J}$ are constant on the unit sphere bundle

$$S(M, g) := \{X \in TM : g(X, X) = 1\}.$$

Work of Chi [5], of Gilkey et al. [15], and of Nikolayevsky [16, 17] show that any complete and simply connected Osserman manifold of dimension $m \neq 16$ is a rank-one symmetric space; the 16-dimensional setting is exceptional and the situation is still not clear in that setting although there are some partial result due, again, to Nikolayevsky [18].

Suppose (M, g) is a pseudo-Riemannian manifold of signature (p, g) for $p > 0$ and $q > 0$. The pseudo-sphere bundles are defined by setting

$$S^{\pm}(M, g) := \{X \in TM : g(X, X) = \pm 1\}.$$

One says that (M, g) is spacelike (resp. timelike) Osserman if the eigenvalues of $\mathcal{J}$ are constant on $S^+(M, g)$ (resp. $S^-(M, g)$). The situation is rather different here as the Jacobi operator is no longer diagonalizable and can have nontrivial Jordan normal form as shown by Garcá-Ró et al. [13]. We refer to [14] for more information on Osserman manifolds.

References

1. E. Calviño-Louzao, E. García-Rio, P. Gilkey, R. Vázquez-Lorenzo, The geometry of modified Riemannian extensions. Proc. R. Soc. A **465**, 2023–2040 (2009)
2. E. Calviño-Louzao, E. García-Río, and R. Vázquez-Lorenzo, Riemann extensions of torsion-free connections with degenerate Ricci tensor. Can. J. Math. **62**(5), 1037–1057 (2010)

3. E. Calviño-Louzao, E. García-Río, P. Gilkey, R. Vázquez-Lorenzo, Higher-dimensional Osserman metrics with non-nilpotent Jacobi operators. Geom. Dedicata **156**, 151–163 (2012)
4. M. Chaichi, E. García-Río, Y. Matsushita, Curvature properties of four-dimensional Walker metrics. Classical Quantum Gravity **22**, 559–577 (2005)
5. Q.S. Chi, A curvature characterization of certain locally rank-one symmetric spaces. J. Differ. Geom. **28**, 187–202 (1988)
6. A.S. Diallo, Affine Osserman connections on 2-dimensional manifolds. Afr. Dispora J. Math. **11**(1), 103–109 (2011)
7. A.S. Diallo, The Riemann extension of an affine Osserman connection on 3-dimensional manifold. Glob. J. Adv. Res. Class. Mod. Geom. **2**(2), 69–75 (2013)
8. A.S. Diallo, M. Hassirou, Examples of Osserman metrics of (3, 3)-signature. J. Math. Sci. Adv. Appl. **7**(2), 95–103 (2011)
9. A.S. Diallo, M. Hassirou, Two families of affine Osserman connections on 3-dimensional manifolds. Afr. Diaspora J. Math. **14**(2), 178–186 (2012)
10. A.S. Diallo, P.G. Kenmogne, M. Hassirou, Affine Osserman connections which are locally symmetrics. Glob. J. Adv. Res. Class. Mod. Geom. **3**(1), 1–6 (2014)
11. A.S. Diallo, M. Hassirou, I. Katambé, Affine Osserman connections which are Ricci flat but not flat. Int. J. Pure Appl. Math. **91**(3), 305–312 (2014)
12. A.S. Diallo, M. Hassirou, I. Katambé, Examples of affine Osserman 3-manifolds. Far East J. Math. Sci. **94**(1), 1–11 (2014)
13. E. García-Rio, D.N. Kupeli, M.E. Vázquez-Abal, R. Vázquez-Lorenzo, Affine Osserman connections and their Riemannian extensions. Differential Geom. Appl. **11**, 145–153 (1999)
14. E. García-Rio, D.N. Kupeli, R. Vázquez-Lorenzo, *Osserman Manifolds in Semi-Riemannian Geometry*. Lectures Notes in Mathematics, vol. 1777 (Springer, Berlin, 2002)
15. P. Gilkey, A. Swann, L. Vanhecke, Isoparametric geodesic spheres and a conjecture of Osserman concerning the Jacobi operator. Q. J. Math. Oxford **46**, 299–320 (1995)
16. Y. Nikolayevsky, Osserman manifolds of dimension 8. Manuscr. Math. **115**, 31–53 (2004)
17. Y. Nikolayevsky, Osserman conjecture in dimension $\neq$ 8, 16. Math. Ann. **331**, 505–522 (2005)
18. Y. Nikolayevsky, On Osserman manifolds of dimension 16, in *Contemporary Geometry and Related Topics* (University of Belgrade, Belgrade, 2006), pp. 379–398
19. K. Nomizu, T. Sasaki, *Affine Differential Geometry*, vol. 111 (Cambridge University Press, Cambridge, 2008)
20. R. Osserman, Curvature in the eighties. Am. Math. Mon. **97**, 731–756 (1990)

Conformal Symmetry Transformations and Nonlinear Maxwell Equations

Gerald A. Goldin, Vladimir M. Shtelen, and Steven Duplij

It is a pleasure to dedicate this article to M. Norbert Hounkonnou on the occasion of his 60th birthday

Abstract We make use of the conformal compactification of Minkowski spacetime $M^{\#}$ to explore a way of describing general, nonlinear Maxwell fields with conformal symmetry. We distinguish the inverse Minkowski spacetime $[M^{\#}]^{-1}$ obtained via conformal inversion, so as to discuss a doubled compactified spacetime on which Maxwell fields may be defined. Identifying $M^{\#}$ with the projective light cone in $(4 + 2)$-dimensional spacetime, we write two independent conformal-invariant functionals of the 6-dimensional Maxwellian field strength tensors—one bilinear, the other trilinear in the field strengths—which are to enter general nonlinear constitutive equations. We also make some remarks regarding the dimensional reduction procedure as we consider its generalization from linear to general nonlinear theories.

Keywords Conformal symmetry · Inverse Minkowski spacetime · Maxwell fields · Nonlinear constitutive equations

G. A. Goldin (✉)
Department of Mathematics, Rutgers University, New Brunswick, NJ, USA

Department of Physics, Rutgers University, New Brunswick, NJ, USA
e-mail: geraldgoldin@dimacs.rutgers.edu

V. M. Shtelen
Department of Mathematics, Rutgers University, New Brunswick, NJ, USA
e-mail: shtelen@math.rutgers.edu

S. Duplij
University of Münster, Münster, Germany
e-mail: duplijs@math.uni-muenster.de

© Springer Nature Switzerland AG 2018
T. Diagana, B. Toni (eds.), *Mathematical Structures and Applications*,
STEAM-H: Science, Technology, Engineering, Agriculture,
Mathematics & Health, https://doi.org/10.1007/978-3-319-97175-9_9

211

1 Introduction

It is well-known that in $(3 + 1)$-dimensional spacetime (Minkowski space), denoted $M^{(4)}$, Maxwell's equations respect not only Poincaré symmetry, but also conformal symmetry. But the physical meaning of this conformal symmetry is still not entirely clear. A historical review is provided by Kastrup [6].

In our ongoing work, we have been investigating the characterization of general, nonlinear conformal-invariant Maxwell theories [2]. Our strategy is to make use of the identification of the conformal compactification $M^\#$ of Minkowski space with the projective light cone in $(4 + 2)$-dimensional spacetime $Y^{(6)}$ [1]. Poincaré transformations, dilations, and special conformal transformations act by rotations and boosts in $Y^{(6)}$. Nikolov and Petrov [9] consider a linear Maxwell theory in $Y^{(6)}$, and carry out a ray reduction and dimensional reduction procedure to obtain conformal-invariant theories in $M^{(4)}$. The result is a description of some additional fields that might survive in $M^{(4)}$. To handle nonlinear Maxwell theories, we allow the constitutive equations to depend explicitly on conformal-invariant functionals of the field strength tensors (with the goal of carrying out a similar dimensional reduction). This parallels, in a certain way, the approach taken by two of us in earlier articles describing general (Lagrangian and non-Lagrangian) nonlinear Maxwell and Yang–Mills theories with Lorentz symmetry in $M^{(4)}$ [3, 4].

This contribution surveys some of the key ideas underlying our investigation. A major tool is to focus on the behavior of the fields and the coordinates under conformal inversion. We introduce here the resulting "inverse Minkowski space" obtained via conformal inversion, and consider the possibility of defining Maxwellian fields independently on the inverse space. We also write two independent conformal-invariant functionals of the Maxwell field strength tensors in $Y^{(6)}$—one bilinear, the other trilinear in the field strengths. These are the functionals which are to enter general nonlinear constitutive equations in the $(4 + 2)$-dimensional theory. We also make some remarks regarding the dimensional reduction procedure from six to four dimensions, as we consider its generalization from linear to general nonlinear theories.

2 Conformal Transformations and Compactification

2.1 Conformal Transformations in Minkowski Space

The full conformal group for $(3 + 1)$-dimensional Minkowski spacetime $M^{(4)}$, as usually defined, includes the following transformations. For $x = (x^\mu) \in M^{(4)}$, $\mu = 0, 1, 2, 3$, we have:
translations:

$$x'^\mu = (T_b x)^\mu = x^\mu - b^\mu \, ; \qquad (2.1)$$

spatial rotations and Lorentz boosts, for example:

$$x'^0 = \gamma(x^0 - \beta x^1), \quad x'^1 = \gamma(x^1 - \beta x^0), \quad -1 < \beta = \frac{v}{c} < 1, \quad \gamma = (1 - \beta^2)^{-\frac{1}{2}};$$
$$(2.2)$$

or more generally,

$$x'^\mu = (\Lambda x)^\mu = \Lambda^\mu_\nu x^\nu \quad \text{(Einstein summation convention)};$$
$$(2.3)$$

and dilations:

$$x'^\mu = (D_\lambda x)^\mu = \lambda x^\mu, \quad \lambda > 0;$$
$$(2.4)$$

all of which are *causal* in $M^{(4)}$. Let us consider *conformal inversion* R, which acts singularly on $M^{(4)}$, and breaks causality:

$$x'^\mu = (R x)^\mu = x^\mu / x_\nu x^\nu, \quad \text{where}$$
$$(2.5)$$

$$x_\nu x^\nu = g_{\mu\nu} x_\mu x^\nu, \quad g_{\mu\nu} = \text{diag}\,[1, -1, -1, -1].$$
$$(2.6)$$

Evidently $R^2 = I$. That is, neglecting singular points, conformal inversion is like a reflection operator: inverting twice yields the identity operation. Conformal inversion preserves the set of light-like submanifolds (the "light rays"), but not the causal structure. Locally, we have:

$$g_{\mu\nu} dx'^\mu dx'^\nu = \frac{1}{(x_\sigma x^\sigma)^2} g_{\mu\nu} dx^\mu dx^\nu.$$
$$(2.7)$$

Combining inversion with translations, and inverting again, gives us the *special conformal transformations* C_b, which act as follows:

$$x'^\mu = (C_b x)^\mu = (RT_b R x)^\mu = (x^\mu - b^\mu x_\nu x^\nu)/(1 - 2b_\nu x^\nu + b_\nu b^\nu x_\sigma x^\sigma).$$
$$(2.8)$$

The operators C_b belong to the conformal group, and can be continuously connected to the identity.

2.2 Conformal Compactification

We can describe Minkowski space $M^{(4)}$ using light cone coordinates. Choose a particular (spatial) direction in $\mathbf{R}^3$. Such a direction is specified by a unit vector $\hat{u}$, labeled (for example) by an appropriate choice of angles in spherical coordinates. A vector $\mathbf{x} \in \mathbf{R}^3$ is then labeled by angles and by the coordinate u, with $-\infty < u <$

∞, and $\mathbf{x} \cdot \mathbf{x} = u^2$. With respect to this direction, we introduce the usual light cone coordinates,

$$u^{\pm} = \frac{1}{\sqrt{2}}(x^0 \pm u) \,. \tag{2.9}$$

Then for $x = (x^0, \mathbf{x})$, we have $x_\mu x^\mu = 2u^+ u^-$; so under conformal inversion (with obvious notation),

$$u'^+ = \frac{1}{2u^-}, \quad u'^- = \frac{1}{2u^+} \,. \tag{2.10}$$

To obtain the conformal compactification $M^\#$ of the $(3 + 1)$-dimensional Minkowski space $M^{(4)}$, we formally adjoin to it the set $\mathcal{J}$ of necessary "points at infinity." These are the images under inversion of the light cone $L^{(4)} \subset M^{(4)}$ (defined by either $u^+ = 0$ or $u^- = 0$), together with the formal limit points of $L^{(4)}$ itself at infinity (which form an invariant submanifold of $\mathcal{J}$). Here $\mathcal{J}$ is the well-known "extended light cone at infinity." The resulting space $M^\# = M^{(4)} \cup \mathcal{J}$ has the topology of $S^3 \times S^1 / Z_2$.

In the above, we understand the operators T_a, Λ_ν^μ, D_λ, R, and C_b as transformations of $M^\#$. Including these operators but leaving out R, we have what is often referred to as the "conformal group," all of whose elements are continuously connected to the identity. There are many different ways to coordinatize $M^\#$ and to visualize its structure, which we shall not discuss here.

3   Inverse Minkowski Space

3.1   Motivation and Definition

In the preceding construction, which is quite standard, there is a small problem with the units. We glossed over (as do nearly all authors) the fact that x^μ has the dimension of *length*, while the expression for $(R\,x)^\mu$ has the dimension of *inverse length*. Thus we cannot actually consider R as a transformation on Minkowski space (or on compactified Minkowski space) without arbitrarily fixing a unit of length!

Furthermore, regarding the formula for $(C_b\,x)^\mu$, it is clear that b must have the dimension of inverse length; but in the expression for $(T_b\,x)^\mu$, it has the dimension of length.

Kastrup [5] suggested introducing a Lorentz-invariant "standard of length" κ at every point, having the dimension of inverse length, and working with the dimensionless coordinates $\eta^\mu = \kappa x^\mu$ together with κ. This leads into a discussion of geometrical gauge properties of Minkowski space.

Let us consider instead the idea of introducing a separate "inverse Minkowski space" $[M^{(4)}]^{-1}$, whose points z have dimension of inverse length. Then we can let $z^\mu = (\hat{R}x)^\mu = x^\mu/x_\nu x^\nu$ belong to $[M^{(4)}]^{-1}$. As before, in order to define $\hat{R}$ on the light cone, we shall need to compactify: first, to compactify $[M^{(4)}]^{-1}$ so as to include the image points of $\hat{R}$ acting on the light cone in $M^{(4)}$, and then to compactify $[M^{(4)}]$, obtaining $[M^{\#}]^{-1}$ and $M^{\#}$. The two spaces are, of course, topologically and geometrically the same, with $\hat{R} : M^{\#} \to [M^{\#}]^{-1}$, and its inverse $\hat{R}^{-1} : [M^{\#}]^{-1} \to M^{\#}$, given by the same formula: $x^\mu = z^\mu/z_\nu z^\nu$.

Just as we have T_a, Λ_ν^μ, D_λ, and C_b acting in $M^{\#}$ (allowing a to have the dimension of length, and b to have the dimension of inverse length), we now define corresponding transformations, $\tilde{T}_b$, $\tilde{\Lambda}_\nu^\mu$, $\tilde{D}_\lambda$, and $\tilde{C}_a$ acting in $[M^{\#}]^{-1}$, using the same formulas as before, but with z replacing x. Thus, $(\tilde{T}_b z)^\mu = z^\mu - b^\mu$, and so forth. Now,

$$C_b = \hat{R}^{-1}\tilde{T}_b\hat{R}, \quad D_\lambda = \hat{R}^{-1}\tilde{D}_{1/\lambda}\hat{R}, \quad \Lambda = \hat{R}^{-1}\tilde{\Lambda}\hat{R}, \quad T_a = \hat{R}^{-1}\tilde{C}_b\hat{R}.$$
(3.1)

3.2 *Conformal Lie Algebra*

The well-known Lie algebra of the conformal group has 15 generators, as follows:

$$[P_\mu, P_\nu] = 0, \quad [K_\mu, K_\nu] = 0, \quad [P_\mu, \mathrm{d}] = P_\mu, \quad [K_\mu, \mathrm{d}] = -K_\mu,$$

$$[P_\mu, J_{\alpha\beta}] = g_{\mu\alpha}P_\beta - g_{\mu\beta}P_\alpha, \quad [K_\mu, J_{\alpha\beta}] = g_{\mu\alpha}K_\beta - g_{\mu\beta}K_\alpha, \quad (3.2)$$

$$[J_{\mu\nu}, J_{\alpha\beta}] = (\text{usual Lorentz algebra}), \quad [P_\mu, K_\nu] = 2(g_{\mu\nu}\mathrm{d} - J_{\mu\nu}),$$

where the P_μ generate translations, the K_μ generate special conformal transformations, the $J_{\alpha\beta}$ generate Lorentz rotations and boosts, and d generates dilations.

Evidently the exchange $P_\mu \to K_\mu$, $K_\mu \to P_\mu$, $\mathrm{d} \to -\mathrm{d}$ leaves the Lie algebra invariant. This fact is now easily understood, if we think of it as conjugating the operators in M with the operator $\hat{R}$ to obtain the generators of transformations in $[M^{\#}]^{-1}$:

$$\tilde{P}_\mu = \hat{R}K_\mu\hat{R}^{-1}, \quad \tilde{K}_\mu = \hat{R}P_\mu\hat{R}^{-1}, \quad \tilde{\mathrm{d}} = \hat{R}(-\mathrm{d})\hat{R}^{-1}, \quad \tilde{J} = \hat{R}J\hat{R}^{-1}. \quad (3.3)$$

3.3 *Some Comments*

To relate the original conformal inversion R to $\hat{R}$, we may introduce an arbitrary constant $A > 0$, having the dimension of area. Let $\hat{A} : [M^{(4)}]^{-1} \to M^{(4)}$ be the operator $x^\mu = Az^\mu$. Then define $x'^\mu = (R_A x)^\mu = (\hat{A}\hat{R}x)^\mu = Ax^\mu/x_\nu x^\nu$, for $A > 0$. Note that $R_A^2 = I$, independent of the value of A.

Now (letting b have units of length), we have $(R_A T_b R_A x)^\mu = (C_{b/A} x)^\mu$, and we can work consistently in the original Minkowski space and its compactification. The introduction of the constant A parallels Kastrup's introduction of the length parameter κ.

However, it is also interesting not to follow this path, but to consider the *doubled, compactified Minkowski space* $M^\# \cup [M^\#]^{-1}$; i.e., the disjoint union of $M^\#$ and its inverse space. It is possible to define Maxwell fields on the doubled space, making use of the conformal inversion.

Finally we remark that a similar construction of an "inverse spacetime" can be carried out for the Schrödinger group introduced by Niederer [8]. The Schrödinger group consists of the Galilei group, dilation of space and time given by $D_\lambda(t, \mathbf{x}) = (\lambda^2 t, \lambda \mathbf{x})$, and additional transformations that can be considered as analogues of special conformal transformations. The latter transformations can be obtained as the result of an inversion, followed by time translation, and then inversion again. Here the inversion is defined by $R : (t, \mathbf{x}) \to (-1/t, \mathbf{x}/t)$, with $R^2 : (t, \mathbf{x}) \to (t, -\mathbf{x})$. Note that for the Schrödinger group, there is only a one-parameter family of transformations obtained this way, in contrast to the four-parameter family of special conformal transformations; the Schrödinger group is only 12-dimensional, while the conformal group is 15-dimensional.

Under inversion, the dimensions again change. Here, they change from time and space to inverse time and velocity, respectively. Again one compactifies, and again we have the option to introduce a "doubled spacetime," where now it is a compactified Galilean spacetime which has been doubled.

4 Nonlinear Electrodynamics: General Approach

4.1 Motivation and Framework for Nonlinear Maxwell Fields

Let us write Maxwell's equations as usual (in SI units), in terms of the four fields $\mathbf{E}, \mathbf{B}, \mathbf{D}$, and $\mathbf{H}$:

$$\nabla \times \mathbf{E} = -\frac{\partial \mathbf{B}}{\partial t}, \quad \nabla \cdot \mathbf{B} = 0, \quad \nabla \times \mathbf{H} = \frac{\partial \mathbf{D}}{\partial t} + \mathbf{j}, \quad \nabla \cdot \mathbf{D} = \rho. \tag{4.1}$$

The *constitutive equations*, relating the pair $(\mathbf{E}, \mathbf{B})$ to the pair $(\mathbf{D}, \mathbf{H})$, may be linear or nonlinear. Our strategy is to introduce general constitutive equations respecting the desired symmetry at the "last possible moment."

Now the general nonlinear theory with Lorentz symmetry has constitutive equations of the form

$$\mathbf{D} = M\mathbf{B} + \frac{1}{c^2} N\mathbf{E}, \quad \mathbf{H} = N\mathbf{B} - M\mathbf{E}, \tag{4.2}$$

where M and N may depend on the field strengths via the two Lorentz invariants

$$I_1 = \mathbf{B}^2 - \frac{1}{c^2}\mathbf{E}^2, \quad I_2 = \mathbf{B} \cdot \mathbf{E}. \tag{4.3}$$

In the linear case, M and N are constants.

Our initial motivation for studying nonlinear Maxwell theories with symmetry was to explore the existence of a Galilean limit [3]. It is well known that taking a Galilean limit $c \to \infty$ in the linear case requires losing one of the time-derivative terms in Maxwell's equations, as described carefully by Le Bellac and Lévy–Leblond [7]. But in the general nonlinear case (allowing non-Lagrangian as well as Lagrangian theories), we showed that all four Maxwell equations can survive intact. Here I_1 and I_2 survive, and can yield nontrivial theories in the $c \to \infty$ limit.

We remark here that introducing conformal symmetry in this context further restricts the invariants, leaving only the ratio I_2/I_1 as an invariant.

In covariant form, Maxwell's equations are written (in familiar notation):

$$\partial_\alpha \tilde{F}^{\alpha\beta} = 0, \quad \partial_\alpha G^{\alpha\beta} = j^\beta, \tag{4.4}$$

where

$$\tilde{F}^{\alpha\beta} = \frac{1}{2}\epsilon^{\alpha\beta\mu\nu}F_{\mu\nu} \quad \text{and} \quad F_{\mu\nu} = \partial_\mu A_\nu - \partial_\nu A_\mu. \tag{4.5}$$

Here the constitutive equations relate G to F and $\tilde{F}$. With Lorentz symmetry, they take the general form

$$G^{\mu\nu} = NF^{\mu\nu} + cM\tilde{F}^{\mu\nu} \equiv M_1\frac{\partial I_1}{\partial F_{\mu\nu}} + M_2\frac{\partial I_2}{\partial F_{\mu\nu}}, \tag{4.6}$$

where M and N (or, equivalently, M_1 and M_2) are functions of the Lorentz invariants I_1 and I_2:

$$I_1 = \frac{1}{2}F_{\mu\nu}F^{\mu\nu}, \quad I_2 = -\frac{c}{4}F_{\mu\nu}\tilde{F}^{\mu\nu}. \tag{4.7}$$

4.2 Transformations Under Conformal Inversion

Under conformal inversion, we have the following symmetry transformations of the electromagnetic potential, and of spacetime derivatives:

$$A'_\mu(x') = x^2 A_\mu(x) - 2x_\mu(x^\alpha A_\alpha) \tag{4.8}$$

$$\partial'_\mu := \frac{\partial}{\partial x'} = x^2\partial_\mu - 2x_\mu(x \cdot \partial) \tag{4.9}$$

where we here abbreviate $x^2 = x_\mu x^\mu$ and $(x \cdot \partial) = x^\alpha \partial_\alpha$. Then with $F_{\mu\nu} = \partial_\mu A_\nu - \partial_\nu A_\mu$, we have:

$$F'_{\mu\nu}(x') = (x^2)^2 F_{\mu\nu}(x) - 2x^2 x^\alpha (x_\mu F_{\alpha\nu} + x_\nu F_{\mu\alpha}) \tag{4.10}$$

and

$$\Box' = (x^2)^2 \Box - 4x^2(x \cdot \partial) , \tag{4.11}$$

where the d'Alembertian $\Box = \partial_\mu \partial^\mu$. Additionally, the 4-current j_μ transforms by

$$j'_\mu(x') = (x^2)^3 j_\mu(x) - 2(x^2)^2 x_\mu (x^\alpha j_\alpha(x)) . \tag{4.12}$$

These transformations define a symmetry of the (linear) Maxwell equations,

$$\Box A_\nu - \partial_\nu (\partial^\alpha A_\alpha) = j_\nu . \tag{4.13}$$

That is, if $A(x)$ and $j(x)$ satisfy (4.13), then $A'(x')$ and $j'(x')$ satisfy the same equation with $\Box'$ and ∂' in place of $\Box$ and ∂, respectively. Combining this symmetry with that of the Poincaré transformations and dilations, we have the symmetry with respect to the usual conformal group.

But note that the symmetry under conformal inversion can be interpreted to suggest not only a relation among solutions to Maxwell's equations in $M^\#$, but also the definition of new Maxwell fields on the inverse compactified Minkowski space $[M^\#]^{-1}$.

4.3 Steps Toward General Nonlinear Conformal-Invariant Electrodynamics

We see the remaining steps in constructing general, nonlinear conformal invariant Maxwell theories (both Lagrangian and non-Lagrangian) as the following. Identifying $M^\#$ with the projective light cone in the $(4 + 2)$-dimensional space $Y^{(6)}$, we write Maxwell fields in $Y^{(6)}$, and constitutive equations in $Y^{(6)}$. The constitutive equations depend only on conformal-invariant functionals of the Maxwell fields in $Y^{(6)}$, which we identify. To restrict the theory to the projective light cone, we then carry out a dimensional reduction procedure, as discussed by Nikolov and Petrov [9]. In doing this we make use of the "hexaspherical space" $Q^{(6)}$—transforming all the expressions to hexaspherical coordinates, and proceeding from there.

5   Related (4+2)-Dimensional Spaces

In this section we review the $(4+2)$-dimensional spaces $Y^{(6)}$ and $Q^{(6)}$, highlighting how conformal inversion acts in these spaces.

5.1   The Space $Y^{(6)}$

For $y \in \mathbf{R}^6$, write $y = (y^m)$, $m = 0, 1, \ldots, 5$, define the flat metric tensor $\eta_{mn} = \mathrm{diag}[1, -1, -1, -1; -1, 1]$, so that

$$y_m y^m = \eta_{mn} y^m y^n = (y^0)^2 - (y^1)^2 - (y^2)^2 - (y^3)^2 - (y^4)^2 + (y^5)^2. \qquad (5.1)$$

This is the space we call $Y^{(6)}$. The light cone $L^{(6)}$ is then the submanifold specified by the condition,

$$y_m y^m = 0, \quad \text{or} \quad (y^1)^2 + (y^2)^2 + (y^3)^2 + (y^4)^2 = (y^0)^2 + (y^5)^2. \qquad (5.2)$$

To define the projective space $PY^{(6)}$ and the projective light cone $PL^{(6)}$, consider $y = (y^m) \in Y^{(6)}$, and define the projective equivalence relation,

$$(y^m) \sim (\lambda y^m) \quad \text{for } \lambda \in \mathbf{R}, \ \lambda \neq 0. \qquad (5.3)$$

The equivalence classes $[y]$ are just the *rays* in $Y^{(6)}$; and $PY^{(6)}$ is this space of rays.

To describe the projective light cone $PL^{(6)}$, we may choose one point in each ray in $L^{(6)}$. Referring back to Eq. (5.2), if we consider

$$(y^1)^2 + (y^2)^2 + (y^3)^2 + (y^4)^2 = (y^0)^2 + (y^5)^2 = 1, \qquad (5.4)$$

we see that we have $S^3 \times S^1$. But evidently the above condition selects two points in each ray; so $PY^{(6)}$ can in this way be identified with (and has the topology of) $S^3 \times S^1/Z_2$.

Furthermore, $PL^{(6)}$ can be identified with $M^\#$. When $y^4 + y^5 \neq 0$, the corresponding element of $M^\#$ belongs to $M^{(4)}$ (finite Minkowski space), and is given by

$$x^\mu = \frac{y^\mu}{y^4 + y^5}, \quad \mu = 0, 1, 2, 3. \qquad (5.5)$$

The "light cone at infinity" corresponds to the submanifold $y^4 + y^5 = 0$ in $PL^{(6)}$.

5.2 Conformal Transformations in $Y^{(6)}$

The 15 conformal group generators act via rotations in the $(4+2)$-dimensional space $Y^{(6)}$, so as to leave $PL^{(6)}$ invariant. Setting

$$X_{mn} = y_m \partial_n - y_n \partial_m \quad (m < n), \tag{5.6}$$

one has the 6 rotation and boost generators

$$M_{mn} = X_{mn} \quad (0 \le m < n \le 3), \tag{5.7}$$

the 4 translation generators

$$P_m = X_{m5} - X_{m4} \quad (0 \le m \le 3), \tag{5.8}$$

the 1 dilation generator

$$D = -X_{45}, \tag{5.9}$$

and the 4 special conformal generators,

$$K_m = -X_{m5} - X_{m4}, \quad (0 \le m < n \le 3). \tag{5.10}$$

But of course, from these infinitesimal transformations we can only construct the *special* conformal transformations, which act like (proper) rotations and boosts. Conformal inversion acts in $Y^{(6)}$ by *reflection* of the y^5 axis, which makes it easy to explore in other coordinate systems too:

$$y'^{\,m} = y^m \, (m = 0, 1, 2, 3, 4), \quad y'^5 = -y^5, \tag{5.11}$$

or more succinctly, $y'^{\,m} = K^m_n y^n$, where $K^m_n = \text{diag}\,[1, 1, 1, 1, 1, -1]$.

5.3 The Hexaspherical Space $Q^{(6)}$

This space is a different $(4 + 2)$-dimensional space, defined conveniently for dimensional reduction. For $q \in \mathbf{R}^6$, write $q = (q^a)$ with the index $a = 0, 1, 2, 3, +, -$. Then define, for $y \in Y^{(6)}$, with $y^4 + y^5 \ne 0$,

$$q^a = \frac{y^a}{y^4 + y^5} \ (a = 0, 1, 2, 3); \quad q^+ = y^4 + y^5; \quad q^- = \frac{y_m y^m}{(y^4 + y^5)^2}. \tag{5.12}$$

In $Q^{(6)}$ the metric tensor is no longer flat:

$$g_{ab}(q) = \begin{pmatrix} (q^+)^2 g_{\mu\nu} & 0 & 0 \\ 0 & q^- & \frac{q^+}{2} \\ 0 & \frac{q^+}{2} & 0 \end{pmatrix} \tag{5.13}$$

The projective equivalence is simply

$$(q^0, q^1, q^2, q^3, q^+, q^-) \sim (q^0, q^1, q^2, q^3, \lambda q^+, q^-), \quad \lambda \neq 0. \tag{5.14}$$

We comment, however, that with this metric tensor, the map from contravariant to covariant vectors in $Q^{(6)}$ is actually two-to-one; hence, it is not invertible. This suggests that one can improve on the hexaspherical coordinatization, a discussion we shall not pursue here.

When we take q^- to zero, we have the light cone in $Q^{(6)}$, while fixing the value $q^+ = 1$ is one way to select a representative vector in each ray. Another comment, however, is that fixing q^+ actually *breaks the conformal symmetry*. This is a subtle point that does not cause practical difficulty, but seems to have been unnoticed previously.

Convenient formulas for the transformation of q-coordinates under conformal inversion may be found in [2].

6   Maxwell Theory with Nonlinear Constitutive Equations in (4 + 2)-Dimensional Spacetime

6.1   Nonlinear Maxwell Equations in $Y^{(6)}$

Next we introduce 6-component fields A_m in $Y^{(6)}$, and write

$$F_{mn} = \partial_m A_n - \partial_n A_m, \tag{6.1}$$

so that

$$\frac{\partial F_{mn}}{\partial y^k} + \frac{\partial F_{nk}}{\partial y^m} + \frac{\partial F_{km}}{\partial y^n} = 0. \tag{6.2}$$

While this is not really the *most* general possible "electromagnetism" in 4 space and 2 time dimensions, it is the theory most commonly discussed in the linear case, and the one we wish to generalize. As before, we defer writing the constitutive relations, and we have:

$$\frac{\partial G^{mn}}{\partial y^m} = J^n, \tag{6.3}$$

where J^n is the 6-current.

For the nonlinear theory, we next need general conformal-invariant nonlinear constitutive equations relating G^{mn} to F_{mn}. But *conformal invariance now means rotational invariance* in $Y^{(6)}$. Thus we write

$$G^{mn} = R^{mnk\ell} F_{n\ell} + P^{mnk\ell rs} F_{k\ell} F_{rs},\tag{6.4}$$

where the tensors R and P take the general form,

$$R^{mnk\ell} = r(\cdots)(\eta^{mk}\eta^{\ell n} - \eta^{nk}\eta^{\ell m}), \text{ and } P^{mnk\ell rs} = p(\cdots)\epsilon^{mnk\ell rs}.\tag{6.5}$$

Here r and s must be functions of *rotational invariants*, which we next write down.

6.2 Invariants for the General Nonlinear Maxwell Theory with Conformal Symmetry

We can now write two rotation-invariant functionals of the field strength tensor in $Y^{(6)}$. The first invariant is, as expected,

$$I_1 = \frac{1}{2}F_{mn}F^{mn}.\tag{6.6}$$

But unlike in the $(3+1)$-dimensional case, the second rotational invariant is *trilinear* in the field strengths:

$$I_2 = \frac{1}{2}\epsilon^{mnk\ell rs} F_{mn} F_{k\ell} F_{rs}.\tag{6.7}$$

This is a new pattern. Then, in Eq. (6.3), we have

$$r = r(I_1, I_2), \quad p = p(I_1, I_2),\tag{6.8}$$

with I_1 and I_2 as above.

In $Q^{(6)}$, the invariants take the form,

$$I_1(q) = \frac{1}{2}F_{ab}(q)F^{ab}(q) = \frac{1}{2}g^{ac}g^{bd}F_{ab}(q)F_{cd}(q),\tag{6.9}$$

$$I_2(q) = \frac{1}{(q^+)^5}\epsilon^{abcdeg}F_{ab}(q)F_{cd}(q)F_{eg}(q)\tag{6.10}$$

$$= \frac{1}{2}(\det \bar{J})\epsilon^{abcdeg}F_{ab}(q)F_{cd}(q)F_{eg}(q),$$

where $\bar{J}$ is a Jacobian matrix for transforming between y and q-coordinates. Note that in the above, ϵ is the Levi–Civita *symbol*. The Levi–Civita *tensor* with raised indices is defined generally as $(1/\sqrt{|g|})\epsilon$, where $g = \det[g_{ab}]$. Here this becomes $(\det \bar{J}) \epsilon^{abcdeg}$.

The explicit presence of q^+ in the expression for I_2 explains why the condition $q^+ = 1$ does not respect the conformal symmetry: the value of q^+ can change under conformal transformations.

7 Dimensional Reduction to (3 + 1) Dimensions

The final steps are to carry out a ray reduction and dimensional reduction of the $(4 + 2)$-dimensional Maxwell theory with conformal symmetry.

A *prolongation condition* states that the Maxwell fields respect the ray equivalence in $Y^{(6)}$):

$$y^k \partial_k A_n \propto A_n . \tag{7.1}$$

A *splitting relation* allows the characterization of components tangential to $LC^{(6)}$:

$$\frac{\partial A_n}{\partial y^n} = 0 \quad \text{(a gauge condition)} . \tag{7.2}$$

One then expresses everything in $Q^{(6)}$ (hexaspherical coordinates), and restricts to the light cone by taking $q^- \to 0$, to obtain (as in the linear case) a general conformal nonlinear electromagnetism in $(3 + 1)$ dimensions, with some additional fields surviving the dimensional reduction.

In this article we have highlighted some new features suggested by the conformal symmetry of nonlinear Maxwell fields, including the idea of doubling the compactified Minkowski spacetime, and the trilinear form of one of the conformal invariant functionals in $Y^{(6)}$. For some additional details, see also [2].

Acknowledgements The first author (GG) wishes especially to thank Professor Hounkonnou, who has been the inspiration behind his many visits to Benin as a visiting lecturer, and as a participant in the international "Contemporary Problems in Mathematical Physics" (Copromaph) conference series and school series. Acknowledgment is due to the International Centre for Theoretical Physics in Trieste, Italy, for partial support of these visits.

References

1. P.A.M. Dirac, Wave equations in conformal space. Ann. Math. **37** (2), 429–442 (1935)
2. S. Duplij, G.A. Goldin, V.M. Shtelen, Conformal inversion and Maxwell field invariants in four- and six-dimensional spacetimes, in *Geometric Methods in Physics: XXXII Workshop, Bialowieza, 2013*, ed. by P. Kielanowski et al. (Springer, Basel, 2014)

3. G.A. Goldin, V.M. Shtelen, On Galilean invariance and nonlinearity in electrodynamics and quantum mechanics. Phys. Lett. A **279**, 321–326 (2001)
4. G.A. Goldin, V.M. Shtelen, Generalization of Yang-Mills theory with nonlinear constitutive equations. J. Phys. A Math. Gen. **37**, 10711–10718 (2004)
5. H.A. Kastrup, Conformal group in space-time. Phys. Rev. **142**(4), 1060–1071 (1966)
6. H.A. Kastrup, On the advancements of conformal transformations and their associated symmetries in geometry and theoretical physics. Ann. Phys. **17**(9–10), 631–690 (2008)
7. M. Le Bellac, J.-M. Lévy-Leblond, Galilean electromagnetism. Nuovo Cimento B **14**(2), 217–234 (1973)
8. U. Niederer, The maximal kinematical invariance group of the free Schrödinger equation. Helv. Phys. Acta **45**, 802–810 (1972)
9. P.A. Nikolov, N.P. Petrov, A local approach to dimensional reduction II: conformal invariance in Minkowski space. J. Geom. Phys. **44**(4), 539–554 (2003)

The Yukawa Model in One Space - One Time Dimensions

Laure Gouba

Abstract The Yukawa Model is revisited in one space - one time dimensions in an approach completely different to those available in the literature. We show that at the classical level it is a constrained system. We apply the Dirac method of quantization of constrained systems. Then by means of the bosonization procedure we uniformize the Hamiltonian at the quantum level in terms of a pseudo-scalar field and the chiral components of a real scalar field.

Keywords Constrained system · Dirac method of quantization · Bosonization procedure

In 1934 the Japanese physicist, Hideki Yukawa, predicted a new particle which later became known as the pi meson, or the pion for short [1]. He considered these pions as the carriers of the force exchanged between two nucleons. The Yukawa coupling is the coupling between nucleons and pion that has been generalized as any coupling between scalars and fermions. In particle physics, Yukawa's interaction or Yukawa coupling, is an interaction between a scalar field ϕ and a Dirac field ψ of the type $V = g\bar{\psi}\phi\psi$ for a scalar field or $V = g\bar{\psi}i\gamma_5\phi\psi$ for a pseudoscalar field, g is called a Yukawa coupling constant. Recently the scalar Yukawa model has been introduced, where the Dirac field is replaced by a complex scalar field [2–4]. The Yukawa interaction is also used in the Standard Model to describe the coupling between the Higgs field and massless quark and lepton fields (i.e., the fundamental fermion particles). Several papers about Yukawa models can be found in the literature [5–8]. We are interested in the Yukawa Model in one space - one time dimensions that we consider as a good testing ground of nonperturbative studies in Yukawa models. We start by considering the model at the classical level, then we apply the Dirac method

L. Gouba (✉)
The Abdus Salam International Centre for Theoretical Physics (ICTP), Trieste, Italy
e-mail: lgouba@ictp.it

© Springer Nature Switzerland AG 2018
T. Diagana, B. Toni (eds.), *Mathematical Structures and Applications*,
STEAM-H: Science, Technology, Engineering, Agriculture,
Mathematics & Health, https://doi.org/10.1007/978-3-319-97175-9_10

of quantization of constrained systems. By means of a bosonization procedure we
reformulate the model into a quantum model of scalar fields.

We consider in a Minkowski space-time, the Yukawa model where the coupling
is between a pseudo-scalar field and a Dirac field. The dynamics of this system is
given by the Lagrangian density,

$$\mathcal{L} = \frac{1}{2}\partial_\mu\phi\partial^\mu\phi + \frac{i}{2}\bar{\psi}\gamma^\mu\partial_\mu\psi - \frac{i}{2}\partial_\mu\bar{\psi}\gamma^\mu\psi - g\bar{\psi}i\gamma_5\phi\psi - V(\phi), \tag{1}$$

where ϕ is a pseudo-scalar field, ψ is the Dirac massless field, g is the coupling
constant, V is the potential that for the moment is left arbitrary. We consider the
model in $1+1$ dimensional space-time where the spacetime coordinates indices are
$\mu = 0, 1$ while spacetime metric is $\eta_{\mu\nu} = \text{diag}(+, -)$. An implicit choice of units is
such that $\hbar = c = 1$. The matrices γ^μ define the Clifford Dirac algebra associated
with the two-dimensional Minkowski space-time whose representation is given by
the Pauli matrices

$$\sigma^1 = \begin{pmatrix} 0 & +1 \\ +1 & 0 \end{pmatrix}, \quad \sigma^2 = \begin{pmatrix} 0 & -i \\ i & 0 \end{pmatrix}, \quad \sigma^3 = \begin{pmatrix} +1 & 0 \\ 0 & -1 \end{pmatrix}, \tag{2}$$

and

$$\gamma^0 = \sigma^1, \quad \gamma^1 = i\sigma^2, \quad \gamma_5 = \gamma^0\gamma^1 = -\sigma^3. \tag{3}$$

Then in this representation of Dirac, the spinor ψ is split into two components as
follows:

$$\psi = \begin{pmatrix} \psi_+ \\ \psi_- \end{pmatrix}, \quad \psi^\dagger = (\psi_+^\dagger \ \psi_-^\dagger), \quad \text{with} \quad \gamma_5\psi_\pm = \mp\psi_\pm, \quad \bar{\psi} = \psi^\dagger\gamma^0 \tag{4}$$

where ψ_+ is the left chirality spinor and ψ_- the right chirality spinor.

We consider a space-time topology τ that is cylindrical by compactifying the
spatial real line into a circle S^1 of circumference $L = 2\pi R$, where R is the radius
of the circle. We formally set

$$\tau = \mathbb{R} \times S^1, \tag{5}$$

with the above topology, it is necessary to define the periodic boundary conditions.
We choose the following boundary conditions

$$\psi_\pm(t, x+L) = -e^{2i\pi\alpha_\pm}\psi_\pm(t, x), \quad t \in \mathbb{R}, \ x \in S^1, \tag{6}$$

with $\alpha_\pm$ being real constants.

The dimensions of the fields follow from the corresponding kinetic energy terms
in the Lagrangian. In two-dimensional spacetime ($D = 2$), the dimensionality of
the fields is set

$$[\phi] : \frac{D-2}{2} = 0; \quad [\psi] : \frac{D-1}{2} = \frac{1}{2}. \tag{7}$$

The physical dimension of the coupling constant g is determined by using ordinary dimensional analysis as follows:

$$[g] : D - [\phi] - 2[\psi] = \frac{-D + 4}{2} = 1 .$$

(8)

Let's first determine the equation of motion for each variable. Given the Lagrangian density $\mathcal{L}$ and a degree of freedom X, the Euler Lagrange equations are determined by

$$\partial_\mu \frac{\partial \mathcal{L}}{\partial (\partial_\mu X)} = \frac{\partial \mathcal{L}}{\partial X} .$$

(9)

The explicit expression of the Lagrangian density in Eq. (1) is given by

$$\mathcal{L} = \frac{1}{2}(\partial_0 \phi)^2 - \frac{1}{2}(\partial_1 \phi)^2 + \frac{i}{2}\psi^\dagger \partial_0 \psi - \frac{i}{2}\partial_0 \psi^\dagger \psi + \frac{i}{2}\psi^\dagger \gamma_5 \partial_1 \psi$$

$$- \frac{i}{2}\partial_1 \psi^\dagger \gamma_5 \psi - ig\phi\psi^\dagger \gamma^1 \psi - V(\phi).$$

(10)

The fundamental degrees of freedom are the following:

$$\phi, \ \psi, \ \psi^\dagger.$$

(11)

The equation of motion for the variable ϕ is the following:

$$\partial_0 \frac{\partial \mathcal{L}}{\partial (\partial_0 \phi)} + \partial_1 \frac{\partial \mathcal{L}}{\partial (\partial_1 \phi)} = \frac{\partial \mathcal{L}}{\partial \phi} ,$$

(12)

that is

$$(\partial_0^2 - \partial_1^2)\phi + ig\psi^\dagger \gamma^1 \psi + \frac{\partial V}{\partial \phi} = 0 .$$

(13)

For the variable ψ, the equation of motion is

$$\partial_0 \frac{\partial \mathcal{L}}{\partial (\partial_0 \psi)} + \partial_1 \frac{\partial \mathcal{L}}{\partial (\partial_1 \psi)} = \frac{\partial \mathcal{L}}{\partial \psi} ,$$

(14)

that is

$$\partial_0 \psi^\dagger + \partial_1 \psi^\dagger \gamma_5 + 2g\phi\psi^\dagger \gamma^1 = 0.$$

(15)

The equation of motion for the variable $\psi^\dagger$ is determined by

$$\partial_0 \frac{\partial \mathcal{L}}{\partial (\partial_0 \psi^\dagger)} + \partial_1 \frac{\partial \mathcal{L}}{\partial (\partial_1 \psi^\dagger)} = \frac{\partial \mathcal{L}}{\partial \psi^\dagger} ,$$

(16)

that is

$$\partial_0 \psi + \partial_1 \gamma_5 \psi - 2g\phi\gamma^1\psi = 0 . \tag{17}$$

This model is characterized by the existence of constraints that appear naturally from the expressions of the conjugate momenta of the degrees of freedom of the system. The literature about constrained systems is wide, for more details, one can read, for instance, in [9]. The momenta variables associated with ϕ, ψ, $\psi^\dagger$ are, respectively,

$$\pi_\phi = \frac{\partial \mathcal{L}}{\partial(\partial_0\phi)} = \partial_0\phi; \quad \pi_\psi = \frac{\partial \mathcal{L}}{\partial(\partial_0\psi)} = -\frac{i}{2}\psi^\dagger; \quad \pi_{\psi^\dagger} = \frac{\partial \mathcal{L}}{\partial(\partial_0\psi^\dagger)} = -\frac{i}{2}\psi , \tag{18}$$

where the left derivation convention has been performed for the fermionic variables ψ and $\psi^\dagger$ that are Grassmann odd variables. The phase space is then characterized by the pairs

$$\{(\phi(t, x), \pi_\phi(t, x)), \ (\psi(t, x), \pi_\psi(t, x)), \ (\psi^\dagger(t, x), \pi_{\psi^\dagger}(t, x))\} . \tag{19}$$

By definition, these pairs are canonically conjugated. In other words, their elementary Poisson brackets at equal time are given by

$$\{\phi(t, x), \pi_\phi(t, y)\} = \delta(x - y) = -\{\pi_\phi(t, y), \phi(t, x)\}; \tag{20}$$

$$\{\psi(t, x), \pi_\psi(t, y)\} = -\delta(x - y) = \{\pi_\psi(t, y), \psi(t, x)\}; \tag{21}$$

$$\{\psi^\dagger(t, x); \pi_{\psi^\dagger}(t, y)\} = -\delta(x - y) = \{\pi_{\psi^\dagger}(t, y), \psi^\dagger(t, x)\} . \tag{22}$$

Without any confusion and ambiguity we choose to omit in the rest of the paper the variables (t, x). Now we apply the canonical formalism for quantizing theories with constraints (Dirac formalism). This formalism has been successfully used in [10] and widely in the literature, for instance in [9, 11, 12].

The conjugate momenta π_ψ and $\pi_{\psi^\dagger}$ induce the following constraints:

$$\sigma_1 = \pi_\psi + \frac{i}{2}\psi^\dagger; \quad \sigma_2 = \pi_{\psi^\dagger} + \frac{i}{2}\psi . \tag{23}$$

These constraints are space-time classical configurations that are in terms of the degrees of freedom of the system. These constraints are called the primary constraints. Since the dynamics of the system depends on the primary constraints, it is compulsory to study the dynamical evolution of the system. Given the Lagrangian density in (10), the canonical Hamiltonian density follows as

$$\mathcal{H}_0 = \partial_0\phi\pi_\phi + \partial_0\psi\pi_\psi + \partial_0\psi^\dagger\pi_{\psi^\dagger} - \mathcal{L} , \tag{24}$$

that is after substitution of $\mathcal{L}$ by its expression in (10)

$$\mathcal{H}_0 = \frac{1}{2}\pi_\phi^2 + \frac{1}{2}(\partial_1\phi)^2 - \frac{i}{2}\psi^\dagger\gamma_5\partial_1\psi + \frac{i}{2}\partial_1\psi^\dagger\gamma_5\psi + ig\phi\psi^\dagger\gamma^1\psi + V(\phi) \quad (25)$$

and the canonical Hamiltonian is

$$H = \int dx\,\mathcal{H}_0. \quad (26)$$

Once we have specified the phase space in Eq. (19), the fundamental Poisson brackets in Eqs. (20)–(22), and the canonical Hamiltonian density in Eq. (25), we can study now the evolution of the constraints in order to check if they generate other constraints and proceed to their classification according to the Dirac formalism.

The primary Hamiltonian is given by the summation of the canonical Hamiltonian and a linear combination of the primary constraints as follows:

$$H_1 = H_0 + \int dx(u_{(1)}\sigma_1 + u_{(2)}\sigma_2). \quad (27)$$

In order to check whether the constraints σ_i, $i = 1, 2$ generate other constraints, we solve the equations $\{\sigma_i, H_1\} = 0$, $i = 1, 2$.

$$\{\sigma_1, H_1\} = u_{(2)} \int dx(-\frac{i}{2}\delta(x - y)); \quad (28)$$

$$\{\sigma_2, H_1\} = u_{(1)} \int dx(-\frac{i}{2})\delta(x - y).$$

Solving $\{\sigma_i, H_1\} = 0$ implies some choices for $u_{(i)}$, $i = 1, 2$ and that means that the constraints σ_i, $i = 1, 2$ do not generate other constraints. The total number of constraints for this model is then equal to 2. An algebra of the constraints is as follows:

$$\{\sigma_1, \sigma_1\} = 0; \quad \{\sigma_1, \sigma_2\} = -\frac{i}{2}\delta(x - y); \quad (29)$$

$$\{\sigma_2, \sigma_1\} = -\frac{i}{2}\delta(x - y); \quad \{\sigma_2, \sigma_2\} = 0.$$

With the algebra in (29), we conclude that all the constraints are of second class. Let's call Δ the matrix of the Poisson brackets of the second class constraints. According to the Dirac formalism of quantization, we should define the algebra of the Dirac brackets by using the general formula

$$\{f, g\}_D = \{f, g\} - \sum_{s,s'}\{f, \sigma_s\}C^{ss'}\{\sigma_{s'}, g\}, \quad (30)$$

where f and g are two degrees of freedom, σ_s, $\sigma_{s'}$ the constraints and C the inverse matrix of the matrix Δ. Then it follows for our system the Dirac brackets

$$\{\phi, \pi_\phi\}_D = \delta(x - y) = -\{\pi_\phi, \phi\}_D; \tag{31}$$

$$\{\psi_\pm, \psi_\pm^\dagger\}_D = -i\delta(x - y) = \{\psi_\pm^\dagger, \psi_\pm\}_D. \tag{32}$$

The fundamental Hamiltonian formulation of the system is then given after the complete analysis of the system by the degrees of freedom $\phi(t, x)$, $\pi_\phi(t, x)$, $\psi(t, x)$, $\pi_\psi(t, x)$, $\psi^\dagger(t, x)$, $\pi_{\psi^\dagger}(t, x)$, the fundamental symplectic structure is given by the Dirac brackets, that appear now implicitly since we omit the index D as

$$\{\phi(t, x), \pi_\phi(t, y)\} = \delta(x - y) = -\{\pi_\phi(t, x), \phi(t, y)\}; \tag{33}$$

$$\{\psi_\pm(t, x), \psi_\pm^\dagger(t, y)\} = -i\delta(x - y) = \{\psi_\pm^\dagger(t, x), \psi_\pm(t, y)\}, \tag{34}$$

and the fundamental Hamiltonian density

$$\mathcal{H} = \frac{1}{2}\pi_\phi^2 + \frac{1}{2}(\partial_1\phi)^2 - \frac{i}{2}\psi^\dagger\gamma_5\partial_1\psi + \frac{i}{2}\partial_1\psi^\dagger\gamma_5\psi + ig\phi\psi^\dagger\gamma^1\psi + V(\phi). \tag{35}$$

Now we proceed by canonical quantization, that is the correspondence principle that states that to each of the classical structures should correspond a similar structure for the quantum system. Then at the quantum level, the phase space is the abstract Hilbert space whose elements are called quantum states. To the classical variables of the phase space correspond linear operators acting on the Hilbert space. To the Poisson brackets correspond now an algebraic structure of commutation relations for the quantum system. We still consider that $\hbar = 1 = c$. We have then the fundamental bosonic and fermionic operators:

$$\hat{\phi}(t, x), \ \hat{\pi}_\phi(t, x), \ \hat{\psi}_\pm(t, x), \ \hat{\psi}_\pm^\dagger(t, x), \tag{36}$$

that satisfy the fundamental commutation and anticommutation relations, respectively, for the bosonic operators and the fermionic operators:

$$[\hat{\phi}(t, x), \ \hat{\pi}_\phi(t, y)] = i\delta(x - y) = -[\hat{\pi}_\phi(t, x), \ \hat{\phi}(t, y)] ; \tag{37}$$

$$\{\hat{\psi}_\pm(t, x), \ \hat{\psi}_\pm^\dagger(t, y)\} = \delta(x - y) = \{\hat{\psi}_\pm^\dagger(t, x), \ \hat{\psi}_\pm(t, y)\} . \tag{38}$$

The quantum Hamiltonian density is given by

$$\hat{\mathcal{H}} = \frac{1}{2}\hat{\pi}_\phi^2 + \frac{1}{2}(\partial_1\hat{\phi})^2 - \frac{i}{2}\hat{\psi}^\dagger\gamma_5\partial_1\hat{\psi} + \frac{i}{2}\partial_1\hat{\psi}^\dagger\gamma_5\hat{\psi} + ig\hat{\phi}\hat{\psi}^\dagger\gamma^1\hat{\psi} + V(\hat{\phi}). \tag{39}$$

The Hamiltonian (39) can also be written in terms of their chiral components of the fermionic operators as

$$\hat{\mathcal{H}} = \frac{1}{2}\hat{\pi}_\phi^2 + \frac{1}{2}(\partial_1\hat{\phi})^2 + \frac{i}{2}\left(\hat{\psi}_+^\dagger \partial_1\hat{\psi}_+ - \partial_1\hat{\psi}_+^\dagger \hat{\psi}_+\right) - \frac{i}{2}\left(\hat{\psi}_-^\dagger \partial_1\hat{\psi}_- - \partial_1\hat{\psi}_-^\dagger \hat{\psi}_-\right)$$
$$+ig\hat{\phi}\left(\hat{\psi}_+^\dagger\hat{\psi}_- - \hat{\psi}_-^\dagger\hat{\psi}_+\right) + V(\hat{\phi}). \tag{40}$$

It is well known that in $1+1$ spacetime dimensions, the fermionic operators can be expressed in terms of bosonic operators by means of vertex operators and the Klein factors. This procedure is called bosonization. The inverse procedure called fermionization also exists but is less used in the literature [13]. The aim of this paper is to uniformize the quantum representation of the system, we choose to bosonize the fermionic operators in order to uniformize the Hamiltonian in terms of only bosonic operators. We consider the Schrödinger picture where the time variable is fixed and that we choose equal to zero. The notation : : expresses the normal ordering in which the creation operators should be placed at the left of the annihilation operators. We bosonize then the chiral fermionic operators in terms of chiral bosonic operators. This procedure has been already well done in [14, 15]. Referring then to the results in [14] and [15], the chiral components of the fermionic operators in (40) are bosonized as follows:

$$\hat{\psi}_\pm(x) = \frac{1}{L}e^{\pm\frac{i\pi}{L}x}e^{i\rho_\pm\frac{\pi}{2}\hat{p}_\mp} : e^{\pm i\lambda\hat{\varphi}_\pm(x)} : ; \tag{41}$$

$$\hat{\psi}_\pm^\dagger(x) = \frac{1}{L}e^{\pm\frac{i\pi}{L}x}e^{-i\rho_\pm\frac{\pi}{2}\hat{p}_\mp} : e^{\mp i\lambda\hat{\varphi}_\pm(x)} : , \tag{42}$$

where

$$\hat{\varphi}_\pm(x) = \hat{q}_\pm \pm \frac{2\pi}{L}\hat{p}_\pm\, x + \sum_{n=1}^{+\infty}\frac{1}{\sqrt{n}}\left(a_{\pm,n}^\dagger e^{\pm\frac{2i\pi}{L}nx} + a_{\pm,n}e^{\mp\frac{2i\pi}{L}nx}\right), \tag{43}$$

are the chiral components of a real scalar bosonic field $\hat{\varphi} = \hat{\varphi}_+ + \hat{\varphi}_-$. The parameters λ and $\rho_\pm$ are such that $\lambda = \pm 1$ and $\rho_\pm^2 = 1 = \lambda^2$. The Klein factor is given by $e^{i\rho_\pm\frac{\pi}{2}\hat{p}_\mp}$. We have

$$\partial_1\hat{\varphi}_\pm(x) = \pm\frac{2i\pi}{L}\left(p_\pm + i\sum_{n=1}^{+\infty}\sqrt{n}\left(a_{\pm,n}^\dagger e^{\pm\frac{2i\pi}{L}nx} - a_{\pm,n}e^{\mp\frac{2i\pi}{L}nx}\right)\right), \tag{44}$$

and the algebra of the bosonic representation is

$$\left[\hat{q}_\pm,\ \hat{p}_\pm\right] = i; \quad \left[a_{\pm,n},\ a_{\pm,m}^\dagger\right] = \delta_{nm}, \quad n,\ m \geq 1, \tag{45}$$

$$\left[\hat{\varphi}_\pm(x),\ \partial_1\hat{\varphi}_\pm(y)\right] = \pm 2i\pi\delta(x-y). \tag{46}$$

For the normal ordering, the convention is that the operators $\begin{pmatrix} \hat{q}_\pm \\ a^\dagger_{\pm,n} \end{pmatrix}$ should be placed at the left of the operators $\begin{pmatrix} \hat{p}_\pm \\ a_{\pm,n} \end{pmatrix}$. Using the procedure of point splitting, that is necessary for the well definition of the composites fermionic operators, and the Baker-Campbell-Hausdorf formula, we show that

$$\hat{\psi}^\dagger_+ \partial_1 \hat{\psi}_+ - \partial_1 \hat{\psi}^\dagger_+ \hat{\psi}_+ = (-i) \left(\frac{1}{\pi}(\partial_1\hat{\varphi}_+)^2 - \frac{\pi}{L^2} \right); \tag{47}$$

$$\hat{\psi}^\dagger_- \partial_1 \hat{\psi}_- - \partial_1 \hat{\psi}^\dagger_- \hat{\psi}_- = (i) \left(\frac{1}{\pi}(\partial_1\hat{\varphi}_-)^2 - \frac{\pi}{L^2} \right); \tag{48}$$

$$\hat{\psi}^\dagger_+ \hat{\psi}_- - \hat{\psi}^\dagger_- \hat{\psi}_+ = -\frac{2i}{L} : \sin[\frac{\pi}{2}(\rho_+\hat{p}_- - \rho_-\hat{p}_+) + \lambda(\hat{\varphi}_+ + \hat{\varphi}_-] : . \tag{49}$$

We set now $\lambda = 1$ and the uniformized quantum Hamiltonian density of the 1+1 Yukawa model is given by

$$\hat{\mathcal{H}} = : \frac{1}{2}\hat{\pi}^2_\phi + \frac{1}{2}(\partial_1\hat{\phi})^2 + \frac{1}{2\pi}(\partial_1\hat{\varphi}_+)^2 + \frac{1}{2\pi}(\partial_1\hat{\varphi}_-)^2$$
$$+ \frac{2}{L} g\,\hat{\phi}\,\sin\left(\frac{\pi}{2}(\rho_+\hat{p}_- - \rho_-\hat{p}_+) + \hat{\varphi}_+ + \hat{\varphi}_-\right) + V(\hat{\phi}) : \tag{50}$$
$$- \frac{\pi}{L^2},$$

where $\hat{\phi}$ is the quantum pseudo-scalar field in (39) and $\hat{\varphi}_\pm$ the chiral components of the real scalar field $\hat{\varphi}$ in Eqs. (41)–(43). The quantity $\frac{\pi}{L^2}$ can be interpreted as the Casimir Energy.

As concluding remarks, we can first notice the absence of first class constraints at the classical level, means there are no gauge symmetry generators, that makes the model more simple. We did not discuss about the symmetries and conserved charges. The potential is left arbitrary, a nice choice would be the Higgs potential. Some extensions of this work can be performed starting with Eq. (50). For instance, the coupling constant g has the dimension of mass, thus setting a mass scale. It would be interesting to understand how this mass scale, g, determines finally the mass spectrum of the different (pseudo) scalar fields in Eq. (50).

Acknowledgements L. Gouba is supported by the Abdus Salam International Centre for Theoretical Physics (ICTP), Trieste, Italy.

References

1. H. Yukawa, Proc. Phys. Math. Soc. Jpn. **17**, 48 (1935)
2. L.M. Abreu, E.S. Nery, A.P.C. Malbouisson, Size effects on the thermodynamic behavior of a simplified generalized scalar Yukawa model. Phys. Rev. D **91**, 087701 (2015)

3. L.M. Abreu, A.P.C. Malbouisson, E.S. Nery, Phase structure of the scalar Yukawa model with compactified spatial dimensions. Mod. Phys. Lett. A **31**(20), 1650121 (2016)

4. V.E. Rochev, Hermitian vs PT-Symmetric Scalar Yukawa model. J. Mod. Phys. **7**, 899–907 (2016)

5. P. Federbush, B. Gildas, Renormalization of the one-space dimensional Yukawa model by unitary transformations. Ann. Phys. **68**, 98–101 (1971)

6. A. Hague, S.D. Joglekar, Causality in 1+1 dimensional Yukawa model II. Pramana **81**, 569–578 (2013). arxiv: 1004.2344 v2 [hep-th]

7. Y. Li, V.A. Karmanov, P. Maris, J.P. Vary, Ab initio approach to the non-perturbative scalar Yukawa model. Phys. Lett. B **748**, 278–283 (2015)

8. O. Akerlund, P. de Forcrand, Higgs-Yukawa model with higher dimension operators via extended mean field theory. Phys. Rev. D **93**, 035015 (2016)

9. J. Govaerts, *Hamiltonian Quantisation and Constrained Dynamics* (Leuven University Press, Leuven, 1991)

10. L. Gouba, Théories de jauge abéliennes scalaire et spinorielle à 1 + 1 dimensions: une étude non perturbative, Thèse de Doctorat (IMSP), Novembre 2005

11. J. Barcelos-Neto, A. Das, W. Scherer, Canonical quantization of constrained systems. Acta Phys. Polon. B **18**(4), 269 (1987)

12. J.R. Klauder, Quantization of constrained systems, in *Proceedings of the 39th Schladming Winter School on "Methods of Quantization"*, Schladming, 26 February–4 March, 2000

13. L. Gouba, G.Y.H. Avossevou, J. Govaerts, M.N. Hounkonnou, Fermionization of a two-dimensional free massless complex scalar field, in *The Proceedings of the Third International Workshop on Contemporary Problems in Mathematical Physics*, ed. by J. Govaerts, M.N. Hounkonnou, A.Z. Msezane (World Scientific, Singapore, 2004), pp. 233–243. e-print, arXiv: hep-th/0408024 (August 2004)

14. G.Y.H. Avossevou, J. Govaerts, The Schwinger model and the physical projector: a nonperturbative quantization without gauge fixing, in *The Proceedings of the Second International Workshop on Contemporary Problems in Mathematical Physics*, ed. by J. Govaerts, M.N. Hounkonnou, A.Z. Mzezane (World Scientific, Singapore, 2002), pp. 374–394. e-print: arxiv: hep-th/0207277 (July 2002)

15. G.Y.H. Avossevou, Théories de jauge et états physiques à 0+1 et 1+1 dimensions, Thèse de Doctorat (IMSP), Avril 2002

Towards the Quantum Geometry of Saturated Quantum Uncertainty Relations: The Case of the (Q, P) Heisenberg Observables

Jan Govaerts

It is a pleasure to dedicate this article to M. Norbert
Hounkonnou on the occasion of his sixtieth birthday

Abstract This contribution to the present Workshop Proceedings outlines a general programme for identifying geometric structures—out of which to possibly recover quantum dynamics as well—associated with the manifold in Hilbert space of the quantum states that saturate the Schrödinger–Robertson uncertainty relation associated with a specific set of quantum observables which characterise a given quantum system and its dynamics. The first step in such an exploration is addressed herein in the case of the observables Q and P of the Heisenberg algebra for a single degree of freedom system. The corresponding saturating states are the well-known general squeezed states, whose properties are reviewed and discussed in detail together with some original results, in preparation of a study deferred to a separated analysis of their quantum geometry and of the corresponding path integral representation over such states.

Keywords Uncertainty relations · Coherent states · Squeezed states · Quantum symplectic geometry · Quantum Riemannian geometry · Quantum geometry

---

This author "Jan Govaerts" is Fellow of the Stellenbosch Institute for Advanced Study (STIAS), Stellenbosch, Republic of South Africa.

J. Govaerts (✉)
Centre for Cosmology, Particle Physics and Phenomenology (CP3), Institut de Recherche en Mathématique et Physique (IRMP), Université catholique de Louvain (U.C.L.),
Louvain-la-Neuve, Belgium

National Institute for Theoretical Physics (NITheP), Stellenbosch, Republic of South Africa

International Chair in Mathematical Physics and Applications (ICMPA–UNESCO Chair),
University of Abomey-Calavi, Cotonou, Republic of Benin
e-mail: Jan.Govaerts@uclouvain.be

© Springer Nature Switzerland AG 2018
T. Diagana, B. Toni (eds.), *Mathematical Structures and Applications*,
STEAM-H: Science, Technology, Engineering, Agriculture,
Mathematics & Health, https://doi.org/10.1007/978-3-319-97175-9_11

235

1 Introduction

Historically, Heisenberg's uncertainty principle [1] has proved to be pivotal in the emergence of quantum mechanics as the conceptual paradigm for physics at the smallest distance scales. To date the uncertainty principle remains a reliable guide in the exploration and the understanding of the physical consequences of the foundational principles of quantum dynamics.

In its original formulation, Heisenberg suggested that measurements of a quantum particle's (configuration space) coordinate, q, and (conjugate) momentum, p, are intrinsically limited in their precision in a way such that

$$\Delta q \, \Delta p \gtrsim h, \qquad h = 2\pi \hbar, \qquad \hbar = \frac{h}{2\pi} \simeq 6.626 \times 10^{-34} \,\text{J s}, \tag{1}$$

$\hbar$ being the reduced Planck constant. Soon thereafter, Schrödinger [2] and Robertson [3] made this statement both more precise and more general for any given pair of self-adjoint, or at least hermitian quantum observables A and B, in the form of the Schrödinger–Robertson[1] uncertainty relation (SR-UR),

$$(\Delta A)^2 \, (\Delta B)^2 \geq \frac{1}{4}\langle(-i)[A, B]\rangle^2 + \frac{1}{4}\langle\{A - \langle A\rangle, B - \langle B\rangle\}\rangle^2, \tag{2}$$

where as usual $(\Delta A)^2 = \langle(A - \langle A\rangle)^2\rangle$ and $(\Delta B)^2 = \langle(B - \langle B\rangle)^2\rangle$, while $\langle\mathcal{O}\rangle$ denotes the normalised expectation value of any quantum operator $\mathcal{O}$ given an arbitrary (normalisable) quantum state (see Appendix 1 for notations and a derivation of the SR-UR). As a by-product one thus also obtains the less tight (but better known, and generalised) Heisenberg uncertainty relation (H-UR),

$$(\Delta A) \, (\Delta B) \geq \frac{1}{2}|\langle(-i)[A, B]\rangle|. \tag{3}$$

In the case of the Heisenberg algebra, namely $[Q, P] = i\hbar\,\mathbb{I}$, indeed this becomes $\Delta q \, \Delta p \geq \hbar/2$.

In the classical limit $\hbar \to 0$, both terms of these inequalities vanish and the latter turn into strict equalities. The physical world, however, is not classical since Planck's constant, albeit small as measured in our macroscopic units, definitely has a finite and non-vanishing value. Yet, in certain regimes of their Hilbert spaces dynamical quantum systems must display classical behaviour as we experience it through quantum observables some of which are of a macroscopic character. Indeed, any quantum system is specified through a set of quantum observables of which the algebra of commutation relations is represented by the Hilbert space

---

[1]Robertson extended this statement to an arbitrary number of observables in terms of the determinant of their covariance matrix of bi-correlations.

which describes that quantum system and its quantum states. Given a particular choice of quantum observables and through measurements of the latter, experiments give access to the quantum states of such a system and enable their manipulation. If certain regimes of a quantum system display the hallmarks of a classical-like behaviour, certainly these regimes must correspond to quantum states which are as close as possible to being classical given a set ensemble of quantum observables characterising that system. In other words classical-like regimes of a quantum system which is characterised by a collection of quantum observables need to correspond to quantum states which saturate as exact equalities the generalised Schrödinger–Robertson uncertainty relation related to that ensemble of quantum observables. Indeed saturated uncertainty relations leave the least room possible for a genuine quantum dynamical behaviour which would otherwise potentially lead to large differences in the values taken by the two terms involved in the inequalities expressing such uncertainty relations.

For reasons recalled in Appendix 1, in the case of two observables the SR-UR is saturated by quantum states $|\psi_0\rangle$ which are such that

$$[(A - \langle A \rangle) - \lambda_0 (B - \langle B \rangle)] |\psi_0\rangle = 0, \tag{4}$$

$$[A - \lambda_0 B] |\psi_0\rangle = [\langle A \rangle - \lambda_0 \langle B \rangle] |\psi_0\rangle,$$

where the complex parameter λ_0 is given by the following combination of expectation values for the state $|\psi_0\rangle$,

$$\lambda_0 = \frac{\langle (B - \langle B \rangle)(A - \langle A \rangle) \rangle}{(\Delta B)^2} = \frac{(\Delta A)^2}{\langle (A - \langle A \rangle)(B - \langle B \rangle) \rangle}. \tag{5}$$

Such saturating quantum states are parametrised by collections of continuous parameters, if only for the expectation values $\langle A \rangle$ and $\langle B \rangle$ as well as the ratio $\Delta A / \Delta B$, for instance. Indeed, especially when considered in the form of the second relation in (4), such states determine classes of quantum coherent-like states (see [4, 5] and references therein), which share many of the remarkable properties of the well-known Schrödinger canonical coherent states for the Heisenberg algebra. In particular in order that their expectation values $\langle A \rangle$ and $\langle B \rangle$ retain finite non-vanishing classical values as $\hbar \to 0$, it is necessary that the saturating states $|\psi_0\rangle$ meeting the conditions (4) involve all possible linearly independent quantum states spanning the full Hilbert space of the system. Furthermore, usually the linear span of such coherent states encompasses the full Hilbert space, since they obey a specific overcompleteness relation or resolution of the unit operator, thereby providing a self-reproducing kernel representation of that Hilbert space [4, 6].

In other words, given a set of quantum observables such saturating quantum states for the corresponding collection of uncertainty relations determine a specific differentiable submanifold of Hilbert space, out of which the full Hilbert space of the quantum system may a priori be reconstructed (provided a sufficient number of quantum observables is considered). In particular quantum amplitudes may then be given a functional path integral representation over that manifold of coherent states, which involves specific geometrical structures of that manifold [6, 7]. Indeed, very naturally that manifold comes equipped then not only with a (quantum) symplectic structure[2] but also with a (quantum) Riemannian metric structure[3] [7], both of these geometric structures being compatible with one another (and dependent, generally, on Planck's constant). A quantum geometric representation of the quantum system thus arises out of its Hilbert space given a choice of its quantum observables and through the associated uncertainty relation. It may even be that, for instance through the corresponding path integral, the quantum system itself may be reconstructed out of these geometric structures (provided the original choice of quantum observables be large enough).

Such an approach connects directly with, and expands on Klauder's general programme of "Enhanced Quantisation" having been proposed for many years now (see [6] and references therein), as a path towards a geometrical formulation of genuine quantum dynamics which shares a number of similarities with other proposals for such geometrical formulations [8, 9]. For that same reason, the programme as briefly outlined above provides a possible avenue towards a further understanding of the underpinnings of the AdS/CFT correspondence and the holographic principle, for instance along lines similar to those having been explored already in [10] and H.J.R. van Zyl (2015; Constructing dualities from quantum state manifolds. Unpublished).

While the general programme outlined above, based on saturated uncertainty relations and the geometry of the associated coherent-like quantum states, is offered here as a project of possible interest to Professor Norbert Hounkonnou in celebration as well of his sixtieth birthday and on the occasion of this COPROMAPH Workshop organised in his honour, the present paper only deals with the construction of the quantum states which saturate the Schrödinger–Robertson uncertainty relation in the case of the Heisenberg algebra for a single quantum degree of freedom, leaving for separate work a discussion of the ensuing geometric structures. Besides some results which presumably are original, most of those being presented herein certainly are available in the literature (see [11–13] and references therein) even though in a scattered form.[4] However this author did not find them discussed along the lines addressed here, nor could he find them all brought together in one single place, as made available in the present contribution with the purpose of providing a basis towards a pursuit of the projected programme aiming at a better understanding of the geometric structures inherent to quantum systems and their dynamics.

Section 2 particularises the discussion to the Heisenberg algebra and identifies the saturating states for the SR-UR in the configuration space representation of that algebra. A construction in terms of Fock algebras and their canonical coherent states is then initiated in Sect. 3, beginning with a reference Fock algebra related to an intrinsic physical scale. Section 4 then presents the complete parametrised set of

---

[2]Because of the sesquilinear properties of the inner product defined over Hilbert space.

[3]Because of the hermitian and positive definite properties of the inner product defined over Hilbert space.

[4]For this reason no attempt is being made towards a complete list of references to the original literature which relates to many different fields of quantum physics.

saturating quantum states, leading to the general class of the well-known squeezed coherent states. Further specific results of interest for these states are then presented in Sects. 5 and 6, to conclude with some additional comments in the Conclusions. Complementary material of a more pedagogical character as befits the Proceedings of the present COPROMAPH Workshop is included in two Appendices.

2 The Uncertainty Relation for the Heisenberg Algebra

Given a single degree of freedom system whose configuration space has the topology of the real line, $q \in \mathbb{R}$, let us consider the corresponding Heisenberg algebra with its conjugate quantum observables, Q and P, such that

$$[Q, P] = i\hbar\, \mathbb{I}, \qquad Q^\dagger = Q, \qquad P^\dagger = P. \tag{6}$$

The configuration and momentum space representations of this algebra are well known, based on the corresponding eigenstate bases, $Q|q\rangle = q|q\rangle$ and $P|p\rangle = p|p\rangle$, with $q, p \in \mathbb{R}$. Our choices of normalisations and phase conventions for these bases states are such that[5]

$$\langle q|q'\rangle = \delta(q-q'), \quad \langle p|p'\rangle = \delta(p-p'), \quad \int_{-\infty}^{+\infty} dq\, |q\rangle\langle q| = \mathbb{I} = \int_{-\infty}^{+\infty} dp\, |p\rangle\langle p|, \tag{7}$$

$$\langle q|p\rangle = \frac{1}{\sqrt{2\pi\hbar}} e^{\frac{i}{\hbar}qp}, \qquad \langle p|q\rangle = \frac{1}{\sqrt{2\pi\hbar}} e^{-\frac{i}{\hbar}qp}. \tag{8}$$

Consider an arbitrary (normalisable) quantum state $|\psi_0\rangle$, which we assume also to have been normalised, $\langle\psi_0|\psi_0\rangle = 1$. In configuration space this state is represented by its wave function, $\psi_0(q) = \langle q|\psi_0\rangle \in \mathbb{C}$. Let q_0 and p_0 be its real valued expectation values for the Heisenberg observables,

$$q_0 = \langle\psi_0|Q|\psi_0\rangle, \qquad p_0 = \langle\psi_0|P|\psi_0\rangle, \qquad q_0, p_0 \in \mathbb{R}, \tag{9}$$

and introduce the shifted or displaced operators

$$\bar{Q} = Q - q_0, \qquad \bar{P} = P - p_0, \tag{10}$$

which again define a Heisenberg algebra of hermitian (ideally self-adjoint) quantum observables, $[\bar{Q}, \bar{P}] = i\hbar\, \mathbb{I}$, $\bar{Q}^\dagger = \bar{Q}$, $\bar{P}^\dagger = \bar{P}$. One also has $(\Delta Q)^2 = \langle\psi_0|\bar{Q}^2|\psi_0\rangle$ and $(\Delta P)^2 = \langle\psi_0|\bar{P}^2|\psi_0\rangle$.

---

[5]Hence the states $|q\rangle$, say, are determined up to a q-independent overall global phase factor which remains unspecified, relative to which all other phase factors are then identified accordingly.

The Schrödinger–Robertson uncertainty relation (SR-UR) then reads (see Appendix 1),

$$(\Delta Q)^2 \, (\Delta P)^2 \geq \frac{1}{4}\hbar^2 + \frac{1}{4}\langle\{\bar{Q}, \bar{P}\}\rangle^2, \tag{11}$$

$\{\bar{Q}, \bar{P}\}$ being the anticommutator of $\bar{Q}$ and $\bar{P}$. As a corollary note that one then also has the looser Heisenberg uncertainty relation (H-UR),

$$\Delta Q \, \Delta P \geq \frac{1}{2}\hbar. \tag{12}$$

However according to the general programme outlined in the Introduction, we are interested in identifying all quantum states that saturate the SR-UR, but not necessarily the H-UR. Quantum states that saturate the H-UR are certainly such that $\langle\{\bar{Q}, \bar{P}\}\rangle = 0$, namely they cannot possess any (Q, P) quantum correlation. The ensemble of states that saturate the SR-UR is thus certainly larger than that which saturates the H-UR. What distinguishes these two sets of states will be made explicit later on.

For reasons recalled in Appendix 1, those states which saturate the SR-UR are such that

$$\left[\bar{Q} - \lambda_0 \bar{P}\right]|\psi_0\rangle = 0, \qquad [(Q - q_0) - \lambda_0 \, (P - p_0)]\,|\psi_0\rangle = 0, \tag{13}$$

where the complex parameter λ_0 takes the value,

$$\lambda_0 = \frac{1}{(\Delta P)^2}\left(\frac{1}{2}\langle\{\bar{Q}, \bar{P}\}\rangle - \frac{1}{2}i\hbar\right) = (\Delta Q)^2 \frac{1}{\frac{1}{2}\langle\{\bar{Q}, \bar{P}\}\rangle + \frac{1}{2}i\hbar}. \tag{14}$$

The defining Eq. (13) of saturating states for the (Q, P) observables of the Heisenberg algebra is best solved by working in a wave function representation, say in configuration space. The above condition then reads

$$\left[(q - q_0) - \lambda_0\left(-i\hbar\frac{d}{dq} - p_0\right)\right]\psi_0(q) = 0. \tag{15}$$

Clearly its solution is

$$\psi_0(q) = N_0(q_0, p_0, \lambda_0) \, e^{\frac{i}{\hbar}qp_0} \, e^{\frac{i}{2\lambda_0\hbar}(q-q_0)^2}, \tag{16}$$

$$\psi_0^*(q) = N_0^*(q_0, p_0, \lambda_0) \, e^{-\frac{i}{\hbar}qp_0} \, e^{-\frac{i}{2\lambda_0^*\hbar}(q-q_0)^2},$$

where $N_0(q_0, p_0, \lambda_0)$ is a complex valued normalisation factor still to be determined. Requiring the state $|\psi_0\rangle$ to be normalised to unity implies the following value for the norm of $N_0(q_0, p_0, \lambda_0)$,

$$|N_0(q_0, p_0, \lambda_0)| = \left(2\pi \, (\Delta Q)^2\right)^{-1/4}. \tag{17}$$

Its overall phase, however, will be determined later on, once further phase conventions will have been specified. Note well that all quantum states saturating the SR-UR are of this simple form, specified in terms of four independent real parameters, namely q_0, p_0, $\Delta Q > 0$ (say) and $\langle\{\bar{Q}, \bar{P}\}\rangle$ (in terms of which $\Delta P > 0$ is then also determined since $(\Delta Q)^2 \, (\Delta P)^2 = (\hbar^2 + \langle\{\bar{Q}, \bar{P}\}\rangle^2)/4$). In the remainder of this paper, we endeavour to understand the structure of these saturating quantum states from the point of view of coherent states, as indeed the defining Eq. (13) invites us to do.

To conclude, let us also remark that for those saturating states such that in addition $\langle\{\bar{Q}, \bar{P}\}\rangle = 0$, in this particular case which thus saturates the H-UR rather than the SR-UR we have the following results (with a choice of phase factor for the wave function which complies with the specifications to be addressed later on),

$$\langle\{\bar{Q}, \bar{P}\}\rangle = 0 : \qquad \lambda_0 = -\frac{i\hbar}{2 \, (\Delta P)^2} = -2i \frac{(\Delta Q)^2}{\hbar}, \qquad \frac{1}{\lambda_0} = \frac{i\hbar}{2 \, (\Delta Q)^2}, \tag{18}$$

$$\psi_0(q) = \frac{1}{\left(2\pi \, (\Delta Q)^2\right)^{1/4}} \, e^{\frac{i}{\hbar} q p_0} \, e^{-\frac{1}{4(\Delta Q)^2}(q-q_0)^2}, \qquad \Delta Q \, \Delta P = \frac{1}{2}\hbar. \tag{19}$$

Note well, however, that even in this case the value of $\Delta P/\Delta Q = \hbar/(2(\Delta Q)^2)$ is still left as a free real and positive parameter. In the case of the ordinary Schrödinger canonical coherent states, which indeed saturate the H-UR, this latter ratio is implicitly set to a specific value in terms of physical parameters of the system under consideration.

3   A Reference Fock Algebra

A priori the quantum observables Q and P possess specific physical dimensions, of which the product has the physical dimension of $\hbar$. For the sake of the construction hereafter, let us denote by ℓ_0 an intrinsic physical scale which has the same physical dimension as Q, so that the physical dimension of P is that of $\hbar/\ell_0$. For instance, we may think of Q as a configuration space coordinate measured in a unit of length, in which case ℓ_0 has the dimension of length, hence the notation. However, note that the physical dimension of ℓ_0 could be anything, as may be relevant given the physical system under consideration. Furthermore ℓ_0 need not correspond to some fundamental physical scale or constant. The scale ℓ_0 may well be expressed in terms of fundamental physical constants in combination with other physical parameters related to the system under consideration. In particular ℓ_0 may involve Planck's constant itself, $\hbar$, and thus change value in the classical limit $\hbar \to 0$ (as is the case for the ordinary harmonic oscillator of mass m and angular frequency ω, with then the natural choice $\ell_0 = \sqrt{\hbar/(m\omega)}$. The purpose of the intrinsic physical scale ℓ_0 is to introduce a reference quantum Fock algebra, hence the corresponding reference canonical coherent states, in order to address the quantum content characterised by the defining Eq. (13) of the quantum states saturating the SR-UR of the Heisenberg algebra, which is indeed a condition characteristic of quantum coherent states.

Given the intrinsic physical scale ℓ_0, let us thus introduce the following reference Fock operators,

$$ a = \frac{1}{\sqrt{2}} \left(\frac{Q}{\ell_0} + i\frac{\ell_0}{\hbar} P \right), \qquad a^\dagger = \frac{1}{\sqrt{2}} \left(\frac{Q}{\ell_0} - i\frac{\ell_0}{\hbar} P \right), \tag{20} $$

with the inverse relations for the Heisenberg observables,

$$ Q = \frac{1}{\sqrt{2}} \ell_0 \left(a + a^\dagger \right), \qquad P = -\frac{i\hbar}{\ell_0 \sqrt{2}} \left(a - a^\dagger \right), \tag{21} $$

which indeed generate the corresponding Fock and Heisenberg algebras, respectively,

$$ \left[a, a^\dagger \right] = \mathbb{I}, \qquad [Q, P] = i\hbar\,\mathbb{I}. \tag{22} $$

The associated normalised reference Fock vacuum, $|\Omega_0\rangle$, such that

$$ a|\Omega_0\rangle = 0, \qquad \langle\Omega_0|\Omega_0\rangle = 1, \tag{23} $$

is chosen with a phase relative to the overall phase implicitly chosen for the position eigenstates $|q\rangle$ such that

$$ \langle q|\Omega_0\rangle = \left(\pi\ell_0^2 \right)^{-1/4} e^{-\frac{1}{2\ell_0^2} q^2}. \tag{24} $$

On account of the condition $a|\Omega_0\rangle = 0$ to be compared to the defining Eq. (13), it is clear that the reference Fock vacuum $|\Omega_0\rangle$ saturates not only the SR-UR but also the H-UR with vanishing expectation values for q_0, for p_0 and for the (Q, P) correlator $\langle\{Q, P\}\rangle$, while the values for ΔQ and ΔP given by

$$ (\Delta Q)^2 = \frac{1}{2}\ell_0^2, \quad (\Delta P)^2 = \frac{1}{2}\frac{\hbar^2}{\ell_0^2}, \quad \left(\frac{\Delta Q}{\ell_0}\right)^2 = \frac{1}{2}, \quad \left(\frac{\ell_0}{\hbar}\Delta P\right)^2 = \frac{1}{2}, \tag{25} $$

are such that

$$ (\Delta Q)\,(\Delta P) = \frac{1}{2}\hbar, \quad \left(\frac{\Delta Q}{\ell_0}\right)\left(\frac{\ell_0}{\hbar}\Delta P\right) = \frac{1}{2}, \quad \left(\frac{\Delta Q}{\ell_0}\right)^2 + \left(\frac{\ell_0}{\hbar}\Delta P\right)^2 = 1. \tag{26} $$

with in particular thus even the ratio $\Delta P / \Delta Q$ taking a predetermined value, $\Delta P / \Delta Q = \hbar / \ell_0^2$. As it turns out, all states saturating the SR-UR will be constructed out of this reference Fock vacuum (thereby also determining the overall phase of the wave function of these states, $\psi_0(q)$, left unspecified in (16) and (17) of Sect. 2).

In order to deal with the shifted or displaced observables $\bar{Q}$ and $\bar{P}$ which involve the expectation values q_0 and p_0, given the reference Fock algebra (20) let us introduce the following complex quantity,

$$u_0 = \frac{1}{\sqrt{2}} \left(\frac{q_0}{\ell_0} + i \frac{\ell_0}{\hbar} p_0 \right), \qquad \bar{u}_0 = u_0^* = \frac{1}{\sqrt{2}} \left(\frac{q_0}{\ell_0} - i \frac{\ell_0}{\hbar} p_0 \right), \qquad (27)$$

with the inverse relations,

$$q_0 = \frac{1}{\sqrt{2}} \ell_0 \left(u_0 + \bar{u}_0 \right), \qquad p_0 = -\frac{i\hbar}{\ell_0 \sqrt{2}} \left(u_0 - \bar{u}_0 \right). \qquad (28)$$

Correspondingly we have the Fock algebra of the associated shifted or displaced Fock generators,

$$b(u_0) = a - u_0, \qquad b^\dagger(u_0) = a^\dagger - \bar{u}_0, \qquad \left[b(u_0), b^\dagger(u_0) \right] = \mathbb{I}, \qquad (29)$$

which are such that

$$b(u_0) = \frac{1}{\sqrt{2}} \left(\frac{\bar{Q}}{\ell_0} + i \frac{\ell_0}{\hbar} \bar{P} \right), \qquad b^\dagger(u_0) = \frac{1}{\sqrt{2}} \left(\frac{\bar{Q}}{\ell_0} - i \frac{\ell_0}{\hbar} \bar{P} \right), \qquad (30)$$

as well as

$$\bar{Q} = \frac{1}{\sqrt{2}} \ell_0 \left(b(u_0) + b^\dagger(u_0) \right), \qquad \bar{P} = -\frac{i\hbar}{\ell_0 \sqrt{2}} \left(b(u_0) - b^\dagger(u_0) \right), \qquad (31)$$

$$[\bar{Q}, \bar{P}] = i\hbar \, \mathbb{I}.$$

The correspondence between the displaced Fock algebra and the reference one is best understood by considering the displacement operator [4] defined[6] in terms of the parameters u_0 or (q_0, p_0),

$$D(q_0, p_0) \equiv D(u_0) \equiv e^{u_0 a^\dagger - \bar{u}_0 a} = e^{-\frac{1}{2}|u_0|^2} e^{u_0 a^\dagger} e^{-\bar{u}_0 a}, \qquad (32)$$

$$D(u_0) \equiv D(q_0, p_0) \equiv e^{-\frac{i}{\hbar} q_0 P + \frac{i}{\hbar} p_0 Q} = e^{\frac{i}{2\hbar} q_0 p_0} e^{-\frac{i}{\hbar} q_0 P} e^{\frac{i}{\hbar} p_0 Q} \qquad (33)$$

$$= e^{-\frac{i}{2\hbar} q_0 p_0} e^{\frac{i}{\hbar} p_0 Q} e^{-\frac{i}{\hbar} q_0 P},$$

---

[6]All Baker–Campbell–Hausdorff (BCH) formulae necessary for this paper are discussed in Appendix 2.

which is a unitary operator defined over Hilbert space,

$$D^\dagger(u_0) = D^{-1}(u_0) = D(-u_0). \tag{34}$$

Indeed the following identities readily follow, which make explicit the displacement action of the displacement operator $D(u_0)$ on the different quantities being involved,

$$b(u_0) = D(u_0)\, a\, D^\dagger(u_0) = a - u_0, \quad b^\dagger(u_0) = D(u_0)\, a^\dagger\, D^\dagger(u_0) = a^\dagger - \bar{u}_0, \tag{35}$$

$$\bar{Q} = D(u_0)\, Q\, D^\dagger(u_0) = Q - q_0, \qquad \bar{P} = D(u_0)\, P\, D^\dagger(u_0) = P - p_0, \tag{36}$$

$$D(u_0)\, |q\rangle = e^{\frac{i}{2\hbar} q_0 p_0}\, e^{\frac{i}{\hbar} q p_0}\, |q + q_0\rangle, \qquad D(u_0)\, |p\rangle = e^{-\frac{i}{2\hbar} q_0 p_0}\, e^{-\frac{i}{\hbar} q_0 p}\, |p + p_0\rangle. \tag{37}$$

Consequently the normalised Fock vacuum, $|\Omega_0(u_0)\rangle$, of the displaced Fock algebra, such that $b(u_0)|\Omega_0(u_0)\rangle = 0$ and $\langle\Omega_0(u_0)|\Omega_0(u_0)\rangle = 1$, is obtained as being simply the displaced reference Fock vacuum since $b(u_0)D(u_0) = D(u_0)a$,

$$|\Omega_0(u_0)\rangle = D(u_0)\, |\Omega_0\rangle, \quad \langle\Omega_0(u_0)|\Omega_0(u_0)\rangle = 1, \tag{38}$$

$$b(u_0)|\Omega_0(u_0)\rangle = 0, \qquad (a - u_0)|\Omega_0(u_0)\rangle = 0, \qquad a|\Omega_0(u_0)\rangle = u_0|\Omega_0(u_0)\rangle. \tag{39}$$

In other words, the Fock vacuum $|\Omega_0(u_0)\rangle$ of the displaced Fock algebra is a canonical coherent state of the reference Fock vacuum $|\Omega_0\rangle$. This also implies that all such states $|\Omega_0(u_0)\rangle$ again saturate not only the SR-UR but also the H-UR with still a vanishing expectation value for the (Q, P) correlator, $\langle\{\bar{Q}, \bar{P}\}\rangle$, but this time with non-vanishing expectation values for Q and P which are specified by the choice for u_0,

$$\langle\Omega_0(u_0)|Q|\Omega_0(u_0)\rangle = q_0, \quad \langle\Omega_0(u_0)|P|\Omega_0(u_0)\rangle = p_0, \tag{40}$$

$$\langle\Omega_0(u_0)|\{\bar{Q}, \bar{P}\}|\Omega_0(u_0)\rangle = 0,$$

while the values for ΔQ and ΔP remain those of the reference Fock vacuum $|\Omega_0\rangle$,

$$(\Delta Q)^2 = \frac{1}{2}\ell_0^2, \quad (\Delta P)^2 = \frac{1}{2}\frac{\hbar^2}{\ell_0^2}, \quad (\Delta Q)^2\,(\Delta P)^2 = \frac{1}{4}\hbar^2. \tag{41}$$

As is well known, the coherent states $|\Omega_0(u_0)\rangle$ possess some remarkable properties [4, 6] of which two are worth to be emphasised in our discussion. Even though these states are not linearly independent among themselves as is made explicit by their non-vanishing overlap matrix elements, none of which is vanishing,

$$\langle \Omega_0(u_2)|\Omega_0(u_1)\rangle = e^{-\frac{1}{2}|u_2|^2 - \frac{1}{2}|u_1|^2 + \bar{u}_2 u_1} = e^{-\frac{1}{2}(u_2\bar{u}_1 - \bar{u}_2 u_1)} e^{-\frac{1}{2}|u_2 - u_1|^2}$$

$$= e^{\frac{i}{2\hbar}(q_2 p_1 - q_1 p_2)} e^{-\frac{1}{4\ell_0^2}(q_2 - q_1)^2 - \frac{1}{4}\left(\frac{\ell_0}{\hbar}\right)^2 (p_2 - p_1)^2}, \tag{42}$$

their linear span over all possible values of the parameter $u_0 \in \mathbb{C}$ encompasses the complete Hilbert space of the system. As a matter of fact this latter result remains valid whatever the choice of normalised reference quantum state on which the displacement operator acts. Thus given an arbitrary state $|\chi_0\rangle$ normalised to unity, $\langle \chi_0|\chi_0\rangle = 1$, consider the states obtained from the action on it of $D(u_0)$ for all possible values of $u_0 \in \mathbb{C}$,

$$|u_0, \chi_0\rangle \equiv D(u_0)|\chi_0\rangle. \tag{43}$$

One then has the following overcompleteness relation in Hilbert space,[7]

$$\int_{\mathbb{C}} \frac{du_0\, d\bar{u}_0}{\pi} |u_0, \chi_0\rangle \langle u_0, \chi_0| = \int_{\mathbb{R}^2} \frac{dq_0 dp_0}{2\pi\hbar} |u_0, \chi_0\rangle \langle u_0, \chi_0| = \mathbb{I}, \tag{44}$$

a result which may readily be established by computing the matrix elements of both terms of this equality in the Q eigenstate basis, for instance. In particular, by choosing for $|\chi_0\rangle$ the reference Fock vacuum $|\Omega_0\rangle$, one obtains the overcompleteness relation for the displaced Fock vacua $|\Omega_0(u_0)\rangle$,

$$\int_{\mathbb{C}} \frac{du_0 d\bar{u}_0}{\pi} |\Omega_0(u_0)\rangle \langle \Omega_0(u_0)| = \int_{\mathbb{R}^2} \frac{dq_0 dp_0}{2\pi\hbar} |\Omega_0(u_0)\rangle \langle \Omega_0(u_0)| = \mathbb{I}. \tag{45}$$

This specific result will be shown to extend to all saturating states of the SR-UR.

Another remarkable property of the states $|\Omega_0(u_0)\rangle$ which extends to all saturating states of the SR-UR is the following. Any finite order polynomial in the Heisenberg observables Q and P possesses a diagonal kernel integral representation in terms of the states $|\Omega_0(u_0)\rangle$, a result which extends the above overcompleteness relation valid specifically for the unit operator. Let us point out, however, that this property applies specifically for the states $|\Omega_0(u_0)\rangle$ constructed out of the reference Fock vacuum $|\Omega_0\rangle$. Generically, it does not apply[8] for other choices of reference state $|\chi_0\rangle$.

To establish such a result, first consider a general finite order polynomial in the operators Q and P. Such a composite operator may always be brought into the form of a finite sum of normal ordered monomials relative to the reference Fock algebra $(a, a^\dagger)$. A generating function of such normal ordered monomials

---

[7]With $du_0 d\bar{u}_0 \equiv d\mathrm{Re}\, u_0\, d\mathrm{Im}\, u_0$.

[8]Unless of course, one considers a reference state which itself is again the Fock vacuum for some other Fock algebra constructed out of the Heisenberg algebra of the observables Q and P, as is the case with the squeezed quantum states to be identified in Sect. 4.

is provided by the operator $\exp(\alpha a^\dagger)\exp(-\bar\alpha a)$ with α and $(-\bar\alpha = -\alpha^*)$ as independent generating parameters. Using the above overcompleteness relation and the fact that $a|\Omega_0(u_0)\rangle = u_0|\Omega_0(u_0)\rangle$, this generating operator may be given the following integral representation (see also (175) in Appendix 2),

$$
\begin{aligned}
e^{\alpha a^\dagger} e^{-\bar\alpha a} &= e^{|\alpha|^2} e^{-\bar\alpha a} e^{\alpha a^\dagger}\\
&= \int_\mathbb{C} \frac{du_0 d\bar u_0}{\pi} e^{|\alpha|^2} e^{-\bar\alpha a}|\Omega_0(u_0)\rangle\langle\Omega_0(u_0)|e^{\alpha a^\dagger}\\
&= \int_\mathbb{C} \frac{du_0 d\bar u_0}{\pi}|\Omega_0(u_0)\rangle\left(e^{|\alpha|^2}e^{-\bar\alpha u_0}e^{\alpha\bar u_0}\right)\langle\Omega_0(u_0)|. \quad (46)
\end{aligned}
$$

However the product of exponential factors appearing inside this integral is directly related to the diagonal matrix elements of the same operator for the coherent states $|\Omega_0(u_0)\rangle$,

$$
\langle\Omega_0(u_0)|e^{\alpha a^\dagger} e^{-\bar\alpha a}|\Omega_0(u_0)\rangle = e^{\alpha\bar u_0} e^{-\bar\alpha u_0}, \quad e^{-\partial_{u_0}\partial_{\bar u_0}} e^{\alpha\bar u_0} e^{-\bar\alpha u_0} = e^{\bar\alpha\alpha} e^{\alpha\bar u_0} e^{-\bar\alpha u_0}, \quad (47)
$$

leading to the final diagonal kernel integral representation of the generating operator,

$$
e^{\alpha a^\dagger} e^{-\bar\alpha a} = \int_\mathbb{C} \frac{du_0 d\bar u_0}{\pi}|\Omega_0(u_0)\rangle\left(e^{-\partial_{u_0}\partial_{\bar u_0}}\langle\Omega_0(u_0)|e^{\alpha a^\dagger} e^{-\bar\alpha a}|\Omega_0(u_0)\rangle\right)\langle\Omega_0(u_0)|. \quad (48)
$$

Therefore any composite operator, A, which is a finite order polynomial in the observables Q and P possesses the following diagonal kernel integral representation over the states $|\Omega_0(u_0)\rangle$,

$$
A = \int_\mathbb{C} \frac{du_0 d\bar u_0}{\pi}|\Omega_0(u_0)\rangle\, a(u_0,\bar u_0)\,\langle\Omega_0(u_0)|, \quad (49)
$$

where the diagonal kernel $a(u_0,\bar u_0)$ is constructed as follows out of the diagonal matrix elements of A in the states $|\Omega_0(u_0)\rangle$,

$$
a(u_0,\bar u_0) = e^{-\partial_{u_0}\partial_{\bar u_0}} A(u_0,\bar u_0), \qquad A(u_0,\bar u_0) = \langle\Omega_0(u_0)|A|\Omega_0(u_0)\rangle. \quad (50)
$$

In terms of the parameters (q_0, p_0), the same results are expressed as, with $|\Omega_0(q_0,p_0)\rangle \equiv |\Omega_0(u_0)\rangle$,

$$
A = \int_{\mathbb{R}^2} \frac{dq_0 dp_0}{2\pi\hbar}|\Omega_0(q_0,p_0)\rangle\, a(q_0,p_0)\,\langle\Omega_0(q_0,p_0)|, \quad (51)
$$

where

$$
a(q_0,p_0) = \exp\left(-\frac12\ell_0^2\partial_{q_0}^2 - \frac12\frac{\hbar^2}{\ell_0^2}\partial_{p_0}^2\right)\langle\Omega_0(q_0,p_0)|A|\Omega_0(q_0,p_0)\rangle. \quad (52)
$$

4 Fock Algebras for the Saturating Quantum States

4.1 Reversible Parametrisation Packages

Let us now address the quantum state content, $|\psi_0\rangle$, of the defining relation (13) for the saturated SR-UR of the Heisenberg algebra of quantum observables (Q, P), namely,

$$[(Q - q_0) - \lambda_0 (P - p_0)] |\psi_0\rangle = 0, \tag{53}$$

with

$$\lambda_0 = \frac{1}{(\Delta P)^2} \left(\frac{1}{2} \langle \{\bar{Q}, \bar{P}\}\rangle - \frac{1}{2} i\hbar \right) = (\Delta Q)^2 \frac{1}{\frac{1}{2} \langle \{\bar{Q}, \bar{P}\}\rangle + \frac{1}{2} i\hbar}. \tag{54}$$

Besides the complex variable u_0 already representing the real expectation values (q_0, p_0) given the physical scale ℓ_0, let us now introduce the angular parameter φ related to the possible (Q, P) correlation, such that $-\pi/2 \leq \varphi \leq +\pi/2$ and defined by

$$\cos\varphi = \frac{1}{\sqrt{1 + \frac{1}{\hbar^2} \langle \{\bar{Q}, \bar{P}\}\rangle^2}}, \quad \sin\varphi = \frac{\frac{1}{\hbar} \langle \{\bar{Q}, \bar{P}\}\rangle}{\sqrt{1 + \frac{1}{\hbar^2} \langle \{\bar{Q}, \bar{P}\}\rangle^2}}, \quad \tan\varphi = \frac{1}{\hbar} \langle \{\bar{Q}, \bar{P}\}\rangle. \tag{55}$$

Note that the saturated SR-UR is then expressed simply as

$$\Delta Q \, \Delta P = \frac{\hbar}{2} \frac{1}{\cos\varphi} = \frac{\hbar}{2} \sqrt{1 + \tan^2\varphi} \geq \frac{\hbar}{2}, \tag{56}$$

$$\left(\frac{\Delta Q}{\ell_0}\right) \left(\frac{\ell_0}{\hbar} \Delta P\right) = \frac{1}{2\cos\varphi} = \frac{1}{2} \sqrt{1 + \tan^2\varphi} \geq \frac{1}{2},$$

while the parameter λ_0 simplifies to

$$-\lambda_0 = \frac{i\hbar}{2(\Delta P)^2} \frac{1}{\cos\varphi} e^{i\varphi} = i \frac{\Delta Q}{\Delta P} e^{i\varphi}, \tag{57}$$

the latter expression thus also displaying explicitly the remaining fourth and last real (and positive) independent free parameter labelling the saturating states, namely the ratio $\Delta Q/\Delta P$. In particular we then have for the operator which annihilates the saturating quantum states

$$\frac{1}{\Delta Q} [(Q - q_0) - \lambda_0 (P - p_0)] = \left(\frac{Q - q_0}{\Delta Q} + i e^{i\varphi} \frac{P - p_0}{\Delta P} \right)$$

$$= \frac{1}{\sqrt{2}} \left(\frac{\ell_0}{\Delta Q} + \frac{\hbar e^{i\varphi}}{\ell_0 \Delta P} \right) b(u_0) + \frac{1}{\sqrt{2}} \left(\frac{\ell_0}{\Delta Q} - \frac{\hbar e^{i\varphi}}{\ell_0 \Delta P} \right) b^\dagger(u_0). \tag{58}$$

Consequently, when having in mind the reference Fock algebra $(a, a^\dagger)$ and in order to account for this last variable, $\Delta Q/\Delta P$ or $\Delta P/\Delta Q$, it proves useful to consider the following further definitions of properly normalised quantities, with $\rho_\pm \geq 0$ and $-\pi \leq \theta_\pm \leq +\pi$,

$$
\rho_\pm \, e^{i\theta_\pm} \equiv \frac{1}{2\sqrt{2}\cos\varphi} \left(\frac{\ell_0}{\Delta Q} \pm \frac{\hbar \, e^{i\varphi}}{\ell_0 \Delta P} \right) = \frac{1}{\sqrt{2}} \left(\left(\frac{\ell_0}{\hbar} \Delta P \right) \pm \left(\frac{\Delta Q}{\ell_0} \right) e^{i\varphi} \right),
\tag{59}
$$

so that

$$
\rho_\pm = \frac{1}{\sqrt{2}} \sqrt{ \left(\frac{\ell_0}{\hbar} \Delta P \right)^2 + \left(\frac{\Delta Q}{\ell_0} \right)^2 \pm 1 },
\tag{60}
$$

$$
\cos\theta_\pm = \frac{1}{\rho_\pm \sqrt{2}} \left[\left(\frac{\ell_0}{\hbar} \Delta P \right) \pm \cos\varphi \left(\frac{\Delta Q}{\ell_0} \right) \right],
\tag{61}
$$

$$
\sin\theta_\pm = \pm \frac{1}{\rho_\pm \sqrt{2}} \sin\varphi \left(\frac{\Delta Q}{\ell_0} \right),
$$

$$
\tan\theta_\pm = \frac{\pm \sin\varphi \left(\frac{\Delta Q}{\ell_0} \right)}{\left(\frac{\ell_0}{\hbar} \Delta P \right) \pm \cos\varphi \left(\frac{\Delta Q}{\ell_0} \right)} = \frac{\sin\varphi}{\cos\varphi \pm \frac{\ell_0^2}{\hbar} \frac{\Delta P}{\Delta Q}}.
\tag{62}
$$

Since $\rho_+^2 - \rho_-^2 = 1$, let us introduce finally a real parameter r such that $0 \leq r < +\infty$, defined by,

$$
\cosh r = \rho_+ = \frac{1}{\sqrt{2}} \sqrt{ \left(\frac{\ell_0}{\hbar} \Delta P \right)^2 + \left(\frac{\Delta Q}{\ell_0} \right)^2 + 1 } \geq 1,
$$

$$
\sinh r = \rho_- = \frac{1}{\sqrt{2}} \sqrt{ \left(\frac{\ell_0}{\hbar} \Delta P \right)^2 + \left(\frac{\Delta Q}{\ell_0} \right)^2 - 1 } \geq 0.
\tag{63}
$$

In terms of these quantities, the following notations prove to be useful later on as well,

$$
\zeta = e^{i\theta} \tanh r, \qquad z = r \, e^{i\theta}, \qquad e^{i\theta} = -e^{i(\theta_- - \theta_+)} = e^{i(\theta_- - \theta_+ \pm \pi)}.
\tag{64}
$$

Note that the complex variable z takes a priori all its values in the entire complex plane, while the complex variable ζ takes all its values inside the unit disk in the complex plane.

Hence given the physical scale ℓ_0 and any quantum state saturating the SR-UR, the associated real quantities q_0, p_0, $\langle \{ \bar{Q}, \bar{P} \} \rangle$, $\Delta Q > 0$ and $\Delta P > 0$, of which four are independent because of the property $(\Delta Q)^2 (\Delta P)^2 = \hbar^2 (1 + \langle \{ \bar{Q}, \bar{P} \} \rangle^2 / \hbar^2)/4$, determine in a unique manner through the above definitions the two independent complex quantities u_0 and z in the complex plane. These two complex variables, u_0 and z, thus label all SR-UR saturating quantum states.

Conversely, given the two complex variables u_0 and z taking any values in the complex plane, in terms of the physical scale ℓ_0 there corresponds to these parameters a SR-UR saturating quantum state, $|\psi_0\rangle$, whose relevant expectation values are constructed as follows. On the one hand for the Heisenberg observables Q and P, their expectation values are

$$q_0 = \frac{1}{\sqrt{2}} \ell_0 (u_0 + \bar{u}_0), \qquad p_0 = -\frac{i\hbar}{\ell_0 \sqrt{2}} (u_0 - \bar{u}_0), \qquad (65)$$

while on the other hand their uncertainties are such that

$$\left(\frac{\Delta Q}{\ell_0}\right)^2 + \left(\frac{\ell_0}{\hbar} \Delta P\right)^2 = \cosh 2r \geq 1, \qquad \left(\frac{\Delta Q}{\ell_0}\right)^2 - \left(\frac{\ell_0}{\hbar} \Delta P\right)^2 = \cos\theta \, \sinh 2r, \qquad (66)$$

namely,[9]

$$\left(\frac{\Delta Q}{\ell_0}\right)^2 = \frac{1}{2} (\cosh 2r + \cos\theta \, \sinh 2r), \qquad (67)$$

$$\left(\frac{\ell_0}{\hbar} \Delta P\right)^2 = \frac{1}{2} (\cosh 2r - \cos\theta \, \sinh 2r), $$

with thus the saturated SR-UR expressed as

$$(\Delta Q)^2 (\Delta P)^2 = \frac{1}{4}\hbar^2 \left(1 + \sin^2\theta \, \sinh^2 2r\right). \qquad (68)$$

Furthermore the (Q, P) correlation of these states $|\psi_0\rangle$ is then determined as

$$\frac{1}{\hbar}\langle\{\bar{Q}, \bar{P}\}\rangle = \tan\varphi = \sin\theta \, \sinh(2r), \qquad (69)$$

with

$$\cos\varphi = \frac{1}{\sqrt{1 + \sin^2\theta \, \sinh^2(2r)}}, \qquad \sin\varphi = \frac{\sin\theta \, \sinh(2r)}{\sqrt{1 + \sin^2\theta \, \sinh^2(2r)}}. \qquad (70)$$

The particular case of (Q, P) uncorrelated saturating states is worth a separate discussion. This situation, characterised by the vanishing correlation $\langle\{\bar{Q}, \bar{P}\}\rangle = 0$, corresponds to the phase value $\varphi = 0$. One then finds

$$\cos\theta_\pm = \text{sgn}\left(\frac{\ell_0}{\hbar} \Delta P \pm \frac{\Delta Q}{\ell_0}\right), \qquad \sin\theta_\pm = 0. \qquad (71)$$

---

[9] Note the identity $\cosh^2 2r - \cos^2\theta \, \sinh^2 2r = 1 + \sin^2\theta \, \sinh^2 2r$.

Consequently in such a case $\theta_+ = 0$, while the value for θ_- is determined as follows

$$\text{if} \quad \frac{\ell_0}{\hbar}\Delta P - \frac{\Delta Q}{\ell_0} > 0: \quad \theta_- = 0; \qquad \text{if} \quad \frac{\ell_0}{\hbar}\Delta P - \frac{\Delta Q}{\ell_0} < 0: \quad \theta_- = \pm\pi,$$

(72)

leading to the value for $\theta \equiv \theta_- - \theta_+ \pm \pi \pmod{2\pi}$ given as

$$\text{if} \quad \frac{\ell_0}{\hbar}\Delta P - \frac{\Delta Q}{\ell_0} > 0: \quad \theta = \pm\pi \pmod{2\pi}; \tag{73}$$

$$\text{if} \quad \frac{\ell_0}{\hbar}\Delta P - \frac{\Delta Q}{\ell_0} < 0: \quad \theta = 0 \pmod{2\pi}.$$

Nonetheless the value for r remains arbitrary,

$$\cosh r = \frac{1}{\sqrt{2}}\left(\frac{\ell_0}{\hbar}\Delta P + \frac{\Delta Q}{\ell_0}\right), \quad \sinh r = \frac{1}{\sqrt{2}}\left|\frac{\ell_0}{\hbar}\Delta P - \frac{\Delta Q}{\ell_0}\right|. \tag{74}$$

On the other hand, in terms of the (u_0, z) parametrisation (Q, P) uncorrelated saturating states correspond to either one of the two values $\theta = 0, \pm\pi \pmod{2\pi}$, thus leading to the quantities,

$$\text{if} \quad \theta = 0: \frac{\Delta Q}{\ell_0} = \frac{1}{\sqrt{2}}e^r, \quad \frac{\ell_0}{\hbar}\Delta P = \frac{1}{\sqrt{2}}e^{-r}, \quad \frac{\ell_0}{\hbar}\Delta P - \frac{\Delta Q}{\ell_0} = -\sqrt{2}\sinh r < 0;$$

$$\text{if} \quad \theta = \pm\pi: \frac{\Delta Q}{\ell_0} = \frac{1}{\sqrt{2}}e^{-r}, \quad \frac{\ell_0}{\hbar}\Delta P = \frac{1}{\sqrt{2}}e^r,$$

$$\frac{\ell_0}{\hbar}\Delta P - \frac{\Delta Q}{\ell_0} = \sqrt{2}\sinh r > 0. \tag{75}$$

Of course, these results are consistent with those derived above.

Given the latter expressions for $\Delta Q/\ell_0$ and $\ell_0\Delta P/\hbar$, it is clear why the parameter $r \geq 0$ is known as the squeezing parameter, while all such (Q, P) uncorrelated states then all saturate the H-UR rather than the SR-UR whatever the value for r. In addition, as the value for the correlation parameter θ or φ varies away from $\theta = 0, \pm\pi \pmod{2\pi}$ or $\varphi = 0 \, (-\pi/2 \leq \varphi \leq \pi/2)$, respectively, both these quantities remain limited within a finite interval whose width is set by the squeezing parameter r,

$$\frac{1}{\sqrt{2}}e^{-r} \leq \frac{\Delta Q}{\ell_0}, \frac{\ell_0}{\hbar}\Delta P \leq \frac{1}{\sqrt{2}}e^r. \tag{76}$$

In particular when $r = 0$, corresponding to $z = 0$ and thus to an irrelevant value for θ, one has the specific situation that $\Delta Q/\ell_0 = 1/\sqrt{2} = \ell_0\Delta P/\hbar$ in addition to the fact that $\langle\{\bar{Q}, \bar{P}\}\rangle = 0$, namely the fact that $\varphi = 0$, thereby leaving as only remaining free parameter the complex variable u_0 for these states which saturate the

H-UR rather than the SR-UR. This situation corresponds exactly to the displaced Fock vacua and coherent states $|\Omega_0(u_0)\rangle = D(u_0)|\Omega_0\rangle$ constructed in Sect. 3 out of the reference Fock vacuum $|\Omega_0\rangle$.

Consequently in this paper those quantum states that saturate the SR-UR are referred to generally as "squeezed states" (or squeezed coherent states, since they turn out to correspond to coherent states as well, as discussed hereafter). Note that if the parameter z is purely real, whether positive or negative (thus corresponding to $\theta = 0$ or $\theta = \pm\pi$ (mod 2π), respectively), such squeezed states have no (Q, P) correlation. While if z is strictly complex with $\theta \neq 0, \pm\pi$ (mod 2π) those squeezed states have a non-vanishing (Q, P) correlation. If the distinction needs to be emphasised, in this paper these situations will be referred to as "uncorrelated" and "correlated" squeezed states, respectively. Note that uncorrelated squeezed states saturate the H-UR, while correlated squeezed states saturate the SR-UR but not the H-UR. In the literature some authors reserve the term "squeezed states" specifically to uncorrelated squeezed states thus with z strictly real and which saturate the H-UR, while to emphasise the distinction correlated squeezed states are then referred to as "intelligent states" which saturate the SR-UR[11–13]. However given the considerations of this paper based on the SR-UR leading to this general class of squeezed states, whether (Q, P) correlated or not it seems preferable to refer to all of these as squeezed states. Furthermore when time evolution of such states is considered[10] the value for θ certainly evolves in time, thereby generating correlated squeezed states out of what could have been initially uncorrelated ones.

Another reason why it is legitimate to consider on a same footing correlated and uncorrelated squeezed states is the following fact. The general Schrödinger–Robertson uncertainty relation may also be expressed [3] in terms of the determinant of the covariance matrix of bi-correlations of observables (see Appendix 1),

$$\langle \bar{A}^2 \rangle \langle \bar{B}^2 \rangle - \langle \bar{A}\bar{B} \rangle \langle \bar{B}\bar{A} \rangle \geq 0, \tag{77}$$

$$(\Delta A)^2 (\Delta B)^2 - \left(\frac{1}{2} \langle \{\bar{A}, \bar{B}\} \rangle \right)^2 \geq \left(\frac{1}{2} \langle (-i)[A, B] \rangle \right)^2,$$

which in the case of the Heisenberg observables reads,

$$\langle \bar{Q}^2 \rangle \langle \bar{P}^2 \rangle - \langle \bar{Q}\bar{P} \rangle \langle \bar{P}\bar{Q} \rangle \geq 0, \qquad (\Delta Q)^2 (\Delta P)^2 - \hbar^2 \left(\frac{1}{2\hbar} \langle \{\bar{Q}, \bar{P}\} \rangle \right)^2 \geq \frac{1}{4}\hbar^2. \tag{78}$$

General squeezed states thus minimise the l.h.s. of these inequalities, whether correlated or uncorrelated, namely whether the parameter z is strictly complex or strictly real, respectively.

---

[10]In the simple situation of the harmonic oscillator of mass m and angular frequency ω, and by choosing then $\ell_0 = \sqrt{\hbar/(m\omega)}$, all these general squeezed states evolve coherently into one another with parameters u_0 and z whose time dependence is given by $u_0(t) = u_0 e^{i\omega t}$ and $z(t) = z e^{2i\omega t}$.

4.2   Correlated Squeezed Fock Algebras and Their Vacua

Coming back now to the operator (58) which annihilates the saturating states, note
that it may be expressed in the form

$$2\cos\varphi\, e^{i\theta_+}\left(\cosh r\, b(u_0) - e^{i\theta}\, \sinh r\, b^\dagger(u_0)\right) \tag{79}$$

$$= 2\cos\varphi\, e^{i\theta_+}\cosh r\left(b(u_0) - \zeta\, b^\dagger(u_0)\right).$$

Consequently let us now introduce the correlated displaced squeezed Fock algebra
generators defined as[11]

$$b(z, u_0) = \cosh r\, b(u_0) - e^{i\theta}\, \sinh r\, b^\dagger(u_0) \tag{80}$$

$$= \cosh r\, (a - u_0) - e^{i\theta}\, \sinh r\left(a^\dagger - \bar{u}_0\right),$$

$$b^\dagger(z, u_0) = -e^{-i\theta}\, \sinh r\, b(u_0) + \cosh r\, b^\dagger(u_0) \tag{81}$$

$$= -e^{-i\theta}\, \sinh r\, (a - u_0) + \cosh r\left(a^\dagger - \bar{u}_0\right),$$

which are such that

$$\left[b(z, u_0), b^\dagger(z, u_0)\right] = \mathbb{I}, \tag{82}$$

while a specific choice of overall phase factor has been effected for $b(z, u_0)$ and
$b^\dagger(z, u_0)$, consistent with the fact that $b(0, u_0) = b(u_0)$ and $b^\dagger(0, u_0) = b^\dagger(u_0)$.

Obviously the SR-UR saturating or squeezed quantum states are the normalised
Fock vacua of these displaced squeezed Fock algebras $(b(z, u_0), b^\dagger(z, u_0))$. Let us
denote these Fock states as $|\Omega_z(u_0)\rangle$ such that $\langle\Omega_z(u_0)|\Omega_z(u_0)\rangle = 1$ as well as
$b(z, u_0)|\Omega_z(u_0)\rangle = 0$. However one also observes that (hence the name of displaced
squeezed Fock algebra for $(b(z, u_0), b^\dagger(z, u_0))$),

$$b(z, u_0) = D(u_0)\, a(z)\, D^\dagger(u_0), \qquad b^\dagger(z, u_0) = D(u_0)\, a^\dagger(z)\, D^\dagger(u_0), \tag{83}$$

where the operators

$$a(z) = \cosh r\, a - e^{i\theta}\, \sinh r\, a^\dagger, \qquad a^\dagger(z) = -e^{-i\theta}\, \sinh r\, a + \cosh r\, a^\dagger, \tag{84}$$

define correlated squeezed Fock algebras such that

$$\left[a(z), a^\dagger(z)\right] = \mathbb{I}, \tag{85}$$

---

[11]Note the slight abuse of notation which is without consequence, which consists in denoting as
a dependence on z a dependence of $(b(z, u_0), b^\dagger(z, u_0))$ which is in fact separate in r and in $e^{i\theta}$
while $z = r e^{i\theta}$.

which are general Bogoliubov transformations of the reference Fock algebra $(a, a^\dagger)$ such that $(a(0), a^\dagger(0)) = (a, a^\dagger)$. Consequently if $|\Omega_z\rangle$ denote the normalised Fock vacua of the Fock algebras $(a(z), a^\dagger(z))$ for all $z \in \mathbb{C}$, such that $\langle \Omega_z | \Omega_z \rangle = 1$ and $a(z)|\Omega_z\rangle = 0$, the saturating squeezed states and thus also Fock vacua $|\Omega_z(u_0)\rangle$ are given as the displaced states of $|\Omega_z\rangle$,

$$|\Omega_z(u_0)\rangle = D(u_0)\,|\Omega_z\rangle, \qquad b(z, u_0)\,|\Omega_z(u_0)\rangle = D(u_0)\,a(z)\,|\Omega_z\rangle = 0. \quad (86)$$

Note that from the last of these two identities it follows that general squeezed states $|\Omega_z(u_0)\rangle$ with $u_0 \neq 0$ are also coherent states of the squeezed $(a(z), a^\dagger(z))$ Fock algebras. Indeed by introducing the quantities

$$u_0(z) \equiv \cosh r\, u_0 - e^{i\theta}\,\sinh r\, \bar{u}_0 = \cosh r\,(u_0 - \zeta\,\bar{u}_0)\,, \quad u_0(0) = u_0,$$
$$\bar{u}_0(z) \equiv -e^{-i\theta}\,\sinh r u_0 + \cosh r\,\bar{u}_0 = \cosh r\,(\bar{u}_0 - \bar{\zeta} u_0)\,, \quad \bar{u}_0(0) = \bar{u}_0, \quad (87)$$

one has

$$a(z)\,|\Omega_z(u_0)\rangle = u_0(z)\,|\Omega_z(u_0)\rangle, \quad (88)$$

as follows also from the identity,

$$b(z, u_0) = a(z) - u_0(z), \quad (89)$$

which shows that the $(b(z, u_0), b^\dagger(z, u_0))$ Fock algebras are shifted versions of the $(a(z), a^\dagger(z))$ Fock algebras.[12] As a matter of fact it may readily be checked that one has, independently from the value for z,

$$u_0(z)a^\dagger(z) - \bar{u}_0(z)a(z) = u_0 a^\dagger - \bar{u}_0 a, \quad (90)$$

so that

$$D(u_0) = e^{-\frac{i}{\hbar}q_0 P + \frac{i}{\hbar}p_0 Q} = e^{u_0 a^\dagger - \bar{u}_0 a} = e^{u_0(z)a^\dagger(z) - \bar{u}_0(z)a(z)}, \quad (91)$$

a property which thus explains the above results.

Inverting the Bogoliubov transformations (84), one finds

$$a = \cosh r\, a(z) + e^{i\theta}\,\sinh r\, a^\dagger(z), \qquad a^\dagger = e^{-i\theta}\,\sinh r\, a(z) + \cosh r\, a^\dagger(z), \quad (92)$$

hence likewise for the variables u_0 and $u_0(z)$,

$$u_0 = \cosh r\, u_0(z) + e^{i\theta}\,\sinh r\, \bar{u}_0(z), \qquad \bar{u}_0 = e^{-i\theta}\,\sinh r\, u_0(z) + \cosh r\, \bar{u}_0(z). \quad (93)$$

---

[12]In the same way that $b(u_0)|\Omega_0(u_0)\rangle = 0$, $a|\Omega_0(u_0)\rangle = u_0|\Omega_0(u_0)\rangle$ and $b(u_0) = a - u_0$, corresponding to the case with $z = 0$.

In terms of the Heisenberg observables, these definitions translate into

$$a(z) = \frac{1}{\sqrt{2}} \left(\cosh r - e^{i\theta} \sinh r \right) \frac{Q}{\ell_0} + \frac{i}{\sqrt{2}} \left(\cosh r + e^{i\theta} \sinh r \right) \frac{\ell_0}{\hbar} P,$$

$$a^\dagger(z) = \frac{1}{\sqrt{2}} \left(\cosh r - e^{-i\theta} \sinh r \right) \frac{Q}{\ell_0} - \frac{i}{\sqrt{2}} \left(\cosh r + e^{-i\theta} \sinh r \right) \frac{\ell_0}{\hbar} P, \quad (94)$$

with the inverse relations,

$$\frac{Q}{\ell_0} = \frac{1}{\sqrt{2}} \left(\cosh r + e^{-i\theta} \sinh r \right) a(z) + \frac{1}{\sqrt{2}} \left(\cosh r + e^{i\theta} \sinh r \right) a^\dagger(z),$$

$$\frac{\ell_0}{\hbar} P = -\frac{i}{\sqrt{2}} \left(\cosh r - e^{-i\theta} \sinh r \right) a(z) + \frac{i}{\sqrt{2}} \left(\cosh r - e^{i\theta} \sinh r \right) a^\dagger(z), \quad (95)$$

so that for the corresponding parameters $u_0(z)$, $\bar{u}_0(z)$, q_0 and p_0,

$$u_0(z) = \frac{1}{\sqrt{2}} \left(\cosh r - e^{i\theta} \sinh r \right) \frac{q_0}{\ell_0} + \frac{i}{\sqrt{2}} \left(\cosh r + e^{i\theta} \sinh r \right) \frac{\ell_0}{\hbar} p_0,$$

$$\bar{u}_0(z) = \frac{1}{\sqrt{2}} \left(\cosh r - e^{-i\theta} \sinh r \right) \frac{q_0}{\ell_0} - \frac{i}{\sqrt{2}} \left(\cosh r + e^{-i\theta} \sinh r \right) \frac{\ell_0}{\hbar} p_0, \quad (96)$$

while

$$\frac{q_0}{\ell_0} = \frac{1}{\sqrt{2}} \left(\cosh r + e^{-i\theta} \sinh r \right) u_0(z) + \frac{1}{\sqrt{2}} \left(\cosh r + e^{i\theta} \sinh r \right) \bar{u}_0(z),$$

$$\frac{\ell_0}{\hbar} p_0 = -\frac{i}{\sqrt{2}} \left(\cosh r - e^{-i\theta} \sinh r \right) u_0(z) + \frac{i}{\sqrt{2}} \left(\cosh r - e^{i\theta} \sinh r \right) \bar{u}_0(z). \quad (97)$$

4.3 Squeezed Fock Vacua and SR-UR Saturating Quantum States

Having understood that the SR-UR saturating states are the displaced coherent states of the squeezed Fock vacua $|\Omega_z\rangle$, namely $|\Omega_z(u_0)\rangle = D(u_0)|\Omega_z\rangle$, let us finally turn to the construction of the latter which are characterised by the condition that $a(z)|\Omega_z\rangle = 0$ with

$$a(z) = \cosh r \, a - e^{i\theta} \sinh r \, a^\dagger. \quad (98)$$

Given that the corresponding Bogoliubov transformation, linear in both generators of the reference Fock algebra $(a, a^\dagger)$, is unitary, necessarily it corresponds to a unitary operator acting on Hilbert space of the following form, defined up to an arbitrary global phase factor set here to a trivial value,

$$S(\alpha) = \exp\left(\frac{1}{2}\alpha a^{\dagger 2} - \frac{1}{2}\bar{\alpha}a^2\right), \qquad \alpha \in \mathbb{C}, \tag{99}$$

α being some complex parameter. The operator $S(\alpha)$ is thus such that

$$S^{\dagger}(\alpha) = S^{-1}(\alpha) = S(-\alpha), \qquad S(0) = \mathbb{I}. \tag{100}$$

A straightforward application of the BCH formula (172) in Appendix 2 then leads to the identities,

$$S(\alpha) a S^{\dagger}(\alpha) = \cosh \rho \, a - e^{i\phi} \sinh \rho \, a^{\dagger}, \tag{101}$$

$$S(\alpha) a^{\dagger} S^{\dagger}(\alpha) = - e^{-i\phi} \sinh \rho \, a + \cosh \rho \, a^{\dagger},$$

the parameter α being represented as $\alpha = \rho \, e^{i\phi}$ with $\rho \geq 0$.

Consequently by choosing $\alpha = z$, one finds for the squeezed Fock algebras $(a(z), a^{\dagger}(z))$,

$$a(z) = S(z) a S^{\dagger}(z), \quad a^{\dagger}(z) = S(z) a^{\dagger} S^{\dagger}(z), \quad \left[a(z), a^{\dagger}(z)\right] = \mathbb{I}, \tag{102}$$

while their normalised squeezed Fock vacua $|\Omega_z\rangle$ are constructed as follows out of the reference Fock vacuum $|\Omega_0\rangle$, since $a(z)S(z)|\Omega_0\rangle = S(z)a|\Omega_0\rangle = 0$,

$$|\Omega_z\rangle = S(z) |\Omega_0\rangle, \qquad \langle\Omega_z|\Omega_z\rangle = \langle\Omega_0|\Omega_0\rangle = 1. \tag{103}$$

Given the displacement operator $D(u_0)$, let us also introduce the operators

$$S(z, u_0) \equiv \exp\left(\frac{1}{2}z(a^{\dagger} - \bar{u}_0)^2 - \frac{1}{2}\bar{z}(a - u_0)^2\right), \tag{104}$$

which obey the following properties,[13]

$$D(u_0) S(z) = S(z, u_0) D(u_0), \qquad S(z, u_0) = D(u_0) S(z) D^{\dagger}(u_0). \tag{105}$$

Hence finally all normalised quantum states that saturate the Schrödinger–Robertson uncertainty relation for the Heisenberg observables (Q, P) are given by the algebraic representation

$$|\psi_0(z, u_0)\rangle \equiv |\Omega_z(u_0)\rangle = e^{u_0 a^{\dagger} - \bar{u}_0 a} \, e^{\frac{1}{2}z a^{\dagger 2} - \frac{1}{2}\bar{z}a^2} |\Omega_0\rangle \tag{106}$$

$$= D(u_0) S(z) |\Omega_0\rangle = S(z, u_0) D(u_0) |\Omega_0\rangle = S(z)D(u_0(z)) |\Omega_0\rangle,$$

---

[13]Note that because of (90), one also has the identity $D(u_0)S(z) = S(z)D(u_0(z))$ with $u_0(z) = \cosh r (u_0 - \zeta \bar{u}_0)$. The author thanks Victor Massart for a remark on this point.

$|\Omega_0\rangle$ being the normalised Fock vacuum of the reference Fock algebra $(a, a^\dagger)$. Note that this construction also fixes the absolute phase factor for all these saturating states, relative to the choice of phase made for the state $|\Omega_0\rangle$. The overall phase factor for the wave function of the saturating states, $\psi_0(q; z, u_0) \equiv \langle q|\psi_0(z, u_0)\rangle$ in Eq. (16), will be determined accordingly in Sect. 6.

5 Overcompleteness and Kernel Representation

In Sect. 3 two remarkable properties of the canonical coherent states, $|\Omega_0(u_0)\rangle$, were emphasised. Let us now consider how these properties extend to the general squeezed coherent states $|\Omega_z(u_0)\rangle$, beginning with the overcompleteness property.

As established in Eq. (44), given any normalised reference state $|\chi_0\rangle$, one has the following representation of the unit operator on the considered Hilbert space,

$$\int_{\mathbb{C}} \frac{du_0 \, d\bar{u}_0}{\pi} \, D(u_0)|\chi_0\rangle \, \langle\chi_0|D^\dagger(u_0) \tag{107}$$

$$= \int_{\mathbb{R}^2} \frac{dq_0 dp_0}{2\pi\hbar} \, D(q_0, p_0)|\chi_0\rangle \, \langle\chi_0|D^\dagger(q_0, p_0) = \mathbb{I}.$$

Hence by choosing $|\chi_0\rangle = |\Omega_z\rangle$ so that $D(u_0)|\chi_0\rangle = |\Omega_z(u_0)\rangle$, given any fixed value for $z \in \mathbb{C}$ one has the overcompleteness property for the SR-UR saturating states,

$$\int_{\mathbb{C}} \frac{du_0 \, d\bar{u}_0}{\pi} |\Omega_z(u_0)\rangle \, \langle\Omega_z(u_0)| = \int_{\mathbb{R}^2} \frac{dq_0 dp_0}{2\pi\hbar} |\Omega_z(u_0)\rangle \, \langle\Omega_z(u_0)| = \mathbb{I}, \tag{108}$$

which thus generalises the overcompleteness relation in Eq. (45) (which corresponds to the case $z = 0$). Note well, however, that this identity involves an integral over the entire complex plane only for the complex variable u_0, independently of the value for z which is fixed but arbitrary. Since the states $|\Omega_z\rangle = S(z)|\Omega_0\rangle$ involve those Fock states built from the reference Fock algebra $(a, a^\dagger)$ which include only an even number of the corresponding $a^\dagger$ Fock quanta and thereby span only half the Hilbert space under consideration, a similar identity involving rather such an integral only over the complex plane of z values but with a fixed value now for u_0 cannot apply. However, given any arbitrary normalisable and normalised integration measure $\mu(z, \bar{z})$ over the complex plane for all values of z provides still for a generalised form of overcompleteness relation involving then all the saturating states,

$$\int_{\mathbb{C}^2} \frac{du_0 \, d\bar{u}_0}{\pi} \frac{dz \, d\bar{z}}{\pi} \, \mu(z, \bar{z}) \, |\Omega_z(u_0)\rangle \, \langle\Omega_z(u_0)| = \mathbb{I}, \qquad \int_{\mathbb{C}} \frac{dz \, d\bar{z}}{\pi} \, \mu(z, \bar{z}) = 1. \tag{109}$$

Let us now consider the possibility of a diagonal kernel integral representation of operators. Given a fixed but arbitrary value for z, any finite order polynomial in the observables Q and P may be written as a linear combination of monomials which are expressed in normal ordered form with respect to the Fock algebra $(a(z), a^\dagger(z))$. Let us thus consider again the generating function of such normal ordered monomials, namely the operator $\exp(\alpha a^\dagger(z))\exp(-\bar\alpha a(z))$ with generating parameters $\alpha \in \mathbb{C}$ and $(-\bar\alpha = -\alpha^*)$. Following the same line of analysis as in Sect. 3, one has

$$e^{\alpha a^\dagger(z)} e^{-\bar\alpha a(z)} = e^{\alpha\bar\alpha} e^{-\bar\alpha a(z)} e^{\alpha a^\dagger(z)}$$

$$= \int_{\mathbb{C}} \frac{du_0 \, d\bar u_0}{\pi} e^{\alpha\bar\alpha} e^{-\bar\alpha a(z)} |\Omega_z(u_0)\rangle \langle\Omega_z(u_0)| e^{\alpha a^\dagger(z)}$$

$$= \int_{\mathbb{C}} \frac{du_0 \, d\bar u_0}{\pi} |\Omega_z(u_0)\rangle e^{\alpha\bar\alpha} e^{-\bar\alpha \bar u_0(z)} e^{\alpha \bar u_0(z)} \langle\Omega_z(u_0)| \quad (110)$$

$$= \int_{\mathbb{C}} \frac{du_0 \, d\bar u_0}{\pi} |\Omega_z(u_0)\rangle$$

$$\times \left[e^{-\partial_{u_0(z)}\partial_{\bar u_0(z)}} \langle\Omega_z(u_0)| e^{\alpha a^\dagger(z)} e^{-\bar\alpha a(z)} |\Omega_z(u_0)\rangle \right] \langle\Omega_z(u_0)|.$$

Consequently, any finite order polynomial in the Heisenberg observables Q and P may be given the following diagonal kernel integral representation, whatever the fixed but arbitrary value for the complex squeezing parameter z,

$$A = \int_{\mathbb{C}} \frac{du_0 \, d\bar u_0}{\pi} |\Omega_z(u_0)\rangle \, a(z, \bar z; u_0, \bar u_0) \, \langle\Omega_z(u_0)|, \quad (111)$$

where the diagonal kernel is defined as

$$a(z, \bar z; u_0, \bar u_0) = e^{-\partial_{u_0(z)}\partial_{\bar u_0(z)}} \langle\Omega_z(u_0)|A|\Omega_z(u_0)\rangle. \quad (112)$$

More generally given the normalised integration measure $\mu(z, \bar z)$, one may extend this representation to

$$A = \int_{\mathbb{C}^2} \frac{du_0 \, d\bar u_0}{\pi} \frac{dz \, d\bar z}{\pi} \mu(z, \bar z) |\Omega_z(u_0)\rangle \, a(z, \bar z; u_0, \bar u_0) \, \langle\Omega_z(u_0)|. \quad (113)$$

In the above representations the second order differential operator $\partial_{u_0(z)}\partial_{\bar u_0(z)}$ may also be expressed as

$$\partial_{u_0(z)}\partial_{\bar u_0(z)} = \frac{1}{2} e^{i\theta} \sinh 2r \, \partial_{u_0}^2 + \frac{1}{2} e^{-i\theta} \sinh 2r \, \partial_{\bar u_0}^2 + \cosh 2r \, \partial_{u_0}\partial_{\bar u_0} \quad (114)$$

$$= (\cosh 2r + \cos\theta \sinh 2r) \frac{1}{2} \ell_0^2 \, \partial_{q_0}^2 + (\cosh 2r - \cos\theta \sinh 2r) \frac{1}{2} \frac{\hbar^2}{\ell_0^2} \, \partial_{p_0}^2$$

$$+ \hbar \sin\theta \, \partial_{q_0} \partial_{p_0},$$

while it is worth noting that

$$du_0 \, d\bar{u}_0 = du_0(z) \, d\bar{u}_0(z). \tag{115}$$

Indeed, since (see (90))

$$|\Omega_z(u_0)\rangle = D(u_0)|\Omega_z\rangle = e^{u_0(z)a^\dagger(z) - \bar{u}_0(z)a(z)} |\Omega_z\rangle = e^{-\frac{1}{2}|u_0(z)|^2} e^{u_0(z)a^\dagger(z)} |\Omega_z\rangle, \tag{116}$$

the matrix element $\langle \Omega_z(u_0)|A|\Omega_z(u_0)\rangle$ is first a function of $u_0(z)$ and $\bar{u}_0(z)$ rather than directly a function of u_0 and $\bar{u}_0$ independently of the value for z.

6 Correlated Squeezed State Wavefunctions

6.1 Squeezed State Configuration Space Wave Functions

Having fully identified, in the form recalled hereafter, the quantum states that saturate the Schrödinger–Robertson uncertainty relation for the Heisenberg algebra of the observables Q and P, inclusive of their phase since that of the reference Fock vacuum has been specified,

$$|\psi_0(z, u_0)\rangle = |\Omega_z(u_0)\rangle = D(u_0) \, S(z) \, |\Omega_0\rangle, \tag{117}$$

we may reconsider the construction of the wave function representation of these states, say in configuration space.

In terms now of the notations and parametrisations introduced throughout the discussion, the expression for the wave functions of these states as determined in (16) and (17) reads[14]

$$\psi_0(q; z, u_0) \equiv \langle q|\Omega_z(u_0)\rangle = \left(\pi \ell_0^2\right)^{-1/4} (\cosh 2r + \cos\theta \sinh 2r)^{-1/4} \times$$

$$\times e^{i\varphi(z, u_0)} e^{\frac{i}{\hbar}qp_0} \exp\left(-\frac{1}{2} \frac{1 - i \sin\theta \sinh 2r}{\cosh 2r + \cos\theta \sinh 2r} \left(\frac{q - q_0}{\ell_0}\right)^2\right), \tag{118}$$

where $\varphi(z, u_0)$ is the phase factor still to be determined. Thus in particular, when $u_0 = 0$,

$$\langle q|\Omega_z(0)\rangle = \langle q|\Omega_z\rangle = \left(\pi \ell_0^2\right)^{-1/4} (\cosh 2r + \cos\theta \sinh 2r)^{-1/4} \times$$

$$\times e^{i\varphi(z, 0)} \exp\left(-\frac{1}{2} \frac{1 - i \sin\theta \sinh 2r}{\cosh 2r + \cos\theta \sinh 2r} \left(\frac{q}{\ell_0}\right)^2\right). \tag{119}$$

---

[14]Since one has the relations $\frac{i}{2\lambda_0\hbar} = -\frac{1}{\hbar}\frac{\Delta P}{\Delta Q}e^{-i\varphi} = -\frac{1}{2\ell_0^2}\frac{1 - i \sin\theta \sinh 2r}{\cosh 2r + \cos\theta \sinh 2r}$.

However since the displacement operator's action on Q eigenstates is such that

$$D(u_0)|q\rangle = e^{\frac{i}{2\hbar}q_0 p_0} e^{\frac{i}{\hbar}q p_0} |q + q_0\rangle, \tag{120}$$

$$\langle q|D(u_0) = \langle q|D^\dagger(-u_0) = \langle q - q_0| e^{-\frac{i}{2\hbar}q_0 p_0} e^{\frac{i}{\hbar}q p_0}, $$

one has

$$\langle q|\Omega_z(u_0)\rangle = \langle q|D(u_0)S(z)|\Omega_0\rangle = e^{-\frac{i}{2\hbar}q_0 p_0} e^{\frac{i}{\hbar}q p_0} \langle q - q_0|\Omega_z(0)\rangle, \tag{121}$$

which, given the above two expressions for $\langle q|\Omega_z(u_0)\rangle$ and $\langle q|\Omega_z(0)\rangle$, thus implies that

$$e^{i\varphi(z,u_0)} = e^{-\frac{i}{2\hbar}q_0 p_0} e^{i\varphi(z,0)}. \tag{122}$$

The final determination of the phase factor $\varphi(z, 0)$ is based now on the following relation between specific Fock state overlaps,

$$\langle \Omega_0|\Omega_z\rangle = \int_{-\infty}^{+\infty} dq \, \langle \Omega_0|q\rangle \langle q|\Omega_z\rangle. \tag{123}$$

The function $\langle q|\Omega_z\rangle$ is specified in (119) in terms of $e^{i\varphi(z,0)}$, while given the choice of phase for the reference Fock vacuum $|\Omega_0\rangle$ its own wave function was determined earlier on to be simply,

$$\langle q|\Omega_0\rangle = \left(\pi \ell_0^2\right)^{-1/4} e^{-\frac{1}{2}\frac{q^2}{\ell_0^2}}. \tag{124}$$

On the other hand, since the l.h.s. of the overlap (123) corresponds to $\langle \Omega_0|S(z)|\Omega_0\rangle$, clearly this latter quantity does not involve any phase factor left unspecified. Consequently the Gaussian integration in (123) determines the overall phase factor $\varphi(z, u_0)$ of the wave functions (118).

The evaluation of $\langle \Omega_0|\Omega_z\rangle$ is readily achieved by using the BCH formula (203) of Appendix 2 for the squeezing operator $S(z)$,

$$S(z) = e^{\frac{1}{2}\zeta a^{\dagger 2}} e^{\ln(1-|\zeta|^2)\frac{1}{2}(a^\dagger a + \frac{1}{2})} e^{-\frac{1}{2}\bar{\zeta}a^2}, \qquad \zeta = e^{i\theta} \tanh r, \quad z = r e^{i\theta}. \tag{125}$$

Hence,

$$\langle \Omega_0|\Omega_z\rangle = \langle \Omega_0|S(z)|\Omega_0\rangle = \left(1 - \tanh^2 r\right)^{1/4} = (\cosh r)^{-1/2}. \tag{126}$$

When combined with the normalisation factor in (119), the Gaussian integration in (123) leads to a factor which may be brought into the form of this last factor $(\cosh r)^{-1/2}$ being multiplied by a specific phase factor. In order to express the thereby determined phase factor $\varphi(z, 0)$, let us introduce two last angular parameters $\bar{\theta}_\pm(z)$ defined by

$$\cos \bar{\theta}_\pm(z) = \frac{\cosh r \pm \cos \theta \sinh r}{\sqrt{\cosh 2r \pm \cos \theta \sinh 2r}},$$

$$\sin \bar{\theta}_\pm(z) = \frac{\pm \sin \theta \sinh r}{\sqrt{\cosh 2r \pm \cos \theta \sinh 2r}}, \tag{127}$$

$$\tan \bar{\theta}_\pm(z) = \frac{\pm \sin \theta \sinh r}{\cosh r \pm \cos \theta \sinh r}, \tag{128}$$

and such that

$$\cos(\bar{\theta}_+(z) - \bar{\theta}_-(z)) = \frac{1}{\sqrt{\cosh^2 2r - \cos^2 \theta \sinh^2 2r}},$$

$$\sin(\bar{\theta}_+(z) - \bar{\theta}_-(z)) = \frac{\sin \theta \sinh 2r}{\sqrt{\cosh^2 2r - \cos^2 \theta \sinh^2 2r}}. \tag{129}$$

On completing the Gaussian integration in (123) (which requires some little work for simplifying some intermediate expressions), one then finally determines that

$$e^{i\varphi(z,0)} = e^{-\frac{i}{2}\bar{\theta}_+(z)}. \tag{130}$$

In conclusion the complete expression for the configuration space wave function representations of all the states that saturate the Schrödinger–Robertson uncertainty relation for the Heisenberg observables Q and P is given as

$$\psi_0(q; z, u_0)$$
$$\equiv \langle q|\Omega_z(u_0)\rangle = \left(\pi \ell_0^2\right)^{-1/4} (\cosh 2r + \cos \theta \sinh 2r)^{-1/4} \, e^{-\frac{i}{2}\bar{\theta}_+(z)} \times$$
$$\times e^{-\frac{i}{2\hbar} q_0 p_0} \, e^{\frac{i}{\hbar} q p_0} \exp\left(-\frac{1}{2} \frac{1 - i \sin \theta \sinh 2r}{\cosh 2r + \cos \theta \sinh 2r} \left(\frac{q - q_0}{\ell_0}\right)^2\right). \tag{131}$$

Note that because of the following identities,

$$\cosh r \pm e^{i\theta} \sinh r = \sqrt{\cosh 2r \pm \cos \theta \sinh 2r} \, e^{i\bar{\theta}_\pm},$$

$$\cosh r \pm e^{-i\theta} \sinh r = \sqrt{\cosh 2r \pm \cos \theta \sinh 2r} \, e^{-i\bar{\theta}_\pm}, \tag{132}$$

$$1 \pm i \sin \theta \sinh 2r = \left(\cosh r \pm e^{i\theta} \sinh r\right) \left(\cosh r \mp e^{-i\theta} \sinh r\right),$$

$$\cosh 2r \pm \cos \theta \sinh 2r = \left(\cosh r \pm e^{i\theta} \sinh r\right) \left(\cosh r \pm e^{-i\theta} \sinh r\right), \tag{133}$$

the same wave functions have also the equivalent representations,

$$\psi_0(q; z, u_0)$$

$$\equiv \langle q|\Omega_z(u_0)\rangle = \left(\pi \ell_0^2\right)^{-1/4} (\cosh 2r + \cos \theta \sinh 2r)^{-1/4} e^{-\frac{i}{2}\bar{\theta}_+(z)} \times$$

$$\times e^{-\frac{i}{2\hbar} q_0 p_0} e^{\frac{i}{\hbar} q p_0} \exp\left(-\frac{1}{2} \frac{\cosh r - e^{i\theta} \sinh r}{\cosh r + e^{i\theta} \sinh r} \left(\frac{q - q_0}{\ell_0}\right)^2\right), \tag{134}$$

and

$$\psi_0(q; z, u_0) \equiv \langle q|\Omega_z(u_0)\rangle = \left(\pi \ell_0^2\right)^{-1/4} (\cosh 2r + \cos \theta \sinh 2r)^{-1/4} e^{-\frac{i}{2}\bar{\theta}_+(z)} \times$$

$$\times e^{-\frac{i}{2\hbar} q_0 p_0} e^{\frac{i}{\hbar} q p_0} \exp\left(-\frac{1}{2} \sqrt{\frac{\cosh 2r - \cos \theta \sinh 2r}{\cosh 2r + \cos \theta \sinh 2r}}\right.$$

$$\left. \times e^{i(\bar{\theta}_-(z) - \bar{\theta}_+(z))} \left(\frac{q - q_0}{\ell_0}\right)^2\right). \tag{135}$$

Furthermore note that

$$(\cosh 2r \pm \cos \theta \sinh 2r)^{-1/4} e^{-\frac{i}{2}\bar{\theta}_\pm(z)} = \left(\cosh r \pm e^{i\theta} \sinh r\right)^{-1/2}, \tag{136}$$

a relation which invites us to consider finally the result in the following form:

$$\psi_0(q; z, u_0) \equiv \langle q|\Omega_z(u_0)\rangle = \left(\pi \ell_0^2\right)^{-1/4} \left(\cosh r + e^{i\theta} \sinh r\right)^{-1/2} \times \tag{137}$$

$$\times e^{-\frac{i}{2\hbar} q_0 p_0} e^{\frac{i}{\hbar} q p_0} \exp\left(-\frac{1}{2} \frac{\cosh r - e^{i\theta} \sinh r}{\cosh r + e^{i\theta} \sinh r} \left(\frac{q - q_0}{\ell_0}\right)^2\right).$$

6.2 The Fundamental Overlap $\langle \Omega_{z_2}(u_2)|\Omega_{z_1}(u_1)\rangle$ of Squeezed States

As a last quantity to be determined in this paper, let us consider the overlap of two arbitrary general squeezed coherent states, associated with the pairs of variables (z_2, u_2) and (z_1, u_1),

$$\langle \Omega_{z_2}(u_2)|\Omega_{z_1}(u_1)\rangle = \int_{-\infty}^{+\infty} dq \, \langle \Omega_{z_2}(u_2)|q\rangle \, \langle q|\Omega_{z_1}(u_1)\rangle. \tag{138}$$

Given the parameters u_2 and u_1, correspondingly one has the pairs of quantities (q_2, p_2) and (q_1, p_1), while related to z_2 and z_1 one has the remaining variables (r_2, θ_2) and (r_1, θ_1) such that

$$z_2 = r_2\, e^{i\theta_2}, \quad \zeta_2 = e^{i\theta_2}\tanh r_2; \quad z_1 = r_1\, e^{i\theta_1}, \quad \zeta_1 = e^{i\theta_1}\tanh r_1. \tag{139}$$

Even though a little tedious, the evaluation of the Gaussian integral in (138) is straightforward enough. It leads to the following equivalent expressions, by relying on a number of the identities pointed out above. In one form one finds

$$\langle \Omega_{z_2}(u_2)|\Omega_{z_1}(u_1)\rangle$$
$$= (\cosh r_2 \cdot \cosh r_1)^{-1/2}\left(1 - \bar{\zeta}_2\zeta_1\right)^{-1/2} e^{\frac{i}{2\hbar}(q_2 p_1 - q_1 p_2)}\, e^{-\frac{1}{4}G_{(2)}(2,1)}, \tag{140}$$

where the Gaussian quadratic form $G_{(2)}(2, 1)$ is given as

$$
\begin{aligned}
G_{(2)}(2, 1) = \frac{1}{1 - \bar{\zeta}_2\zeta_1}\Bigg[&\left(1 - \bar{\zeta}_2\right)\left(1 - \zeta_1\right)\left(\frac{q_2 - q_1}{\ell_0}\right)^2 - \\
&- 2i\left(\bar{\zeta}_2 - \zeta_1\right)\left(\frac{q_2 - q_1}{\ell_0}\right)\left(\frac{\ell_0}{\hbar}(p_2 - p_1)\right) \\
&+ \left(1 + \bar{\zeta}_2\right)\left(1 + \zeta_1\right)\left(\frac{\ell_0}{\hbar}(p_2 - p_1)\right)^2 \Bigg].
\end{aligned}
\tag{141}
$$

In terms of the variables (u_2, u_1) and (ζ_2, ζ_1), the same expression writes as

$$\langle \Omega_{z_2}(u_2)|\Omega_{z_1}(u_1)\rangle \tag{142}$$
$$= (\cosh r_2 \cdot \cosh r_1)^{-1/2}\left(1 - \bar{\zeta}_2\zeta_1\right)^{-1/2} e^{-\frac{1}{2}(u_2\bar{u}_1 - \bar{u}_2 u_1)}\, e^{-\frac{1}{4}G_{(2)}(2,1)},$$

with this time, in a further streamlined form for the Gaussian quadratic factor,

$$
\begin{aligned}
\frac{1}{2}G_{(2)}(2, 1) &\\
= \frac{1}{1 - \bar{\zeta}_2\zeta_1}&\left((1 + \bar{\zeta}_2\zeta_1)|u_2 - u_1|^2 - \bar{\zeta}_2(u_2 - u_1)^2 - \zeta_1(\bar{u}_2 - \bar{u}_1)^2\right) \\
= \frac{1}{1 - \bar{\zeta}_2\zeta_1}&\left((u_2 - u_1) - \zeta_2(\bar{u}_2 - \bar{u}_1)\right)^*\left((u_2 - u_1) - \zeta_1(\bar{u}_2 - \bar{u}_1)\right). \tag{143}
\end{aligned}
$$

That the dependence of this result on the different variables parametrising the SR-UR states $|\Omega_{z_2}(u_2)\rangle$ and $|\Omega_{z_1}(u_1)\rangle$ comes out as established above may be understood from the following two identities (for the first, see (125)),

$$S(z)|\Omega_0\rangle = (\cosh r)^{-1/2} e^{\frac{1}{2}\zeta a^{\dagger 2}}|\Omega_0\rangle, \tag{144}$$

$$D^{\dagger}(u_2)D(u_1) = D(-u_2)D(u_1) = e^{\frac{i}{2\hbar}(q_2 p_1 - q_1 p_2)} D(u_1 - u_2) \tag{145}$$

$$= e^{-\frac{1}{2}(u_2 \bar{u}_1 - \bar{u}_2 u_1)} D(u_1 - u_2),$$

which imply the relation,

$$\langle \Omega_{z_2}(u_2)|\Omega_{z_1}(u_1)\rangle \tag{146}$$

$$= (\cosh r_2 \cdot \cosh r_1)^{-1/2} e^{\frac{i}{2\hbar}(q_2 p_1 - q_1 p_2)} \langle \Omega_0|e^{\frac{1}{2}\bar{\zeta}_2 a^2} D(u_1 - u_2) e^{\frac{1}{2}\zeta_1 a^{\dagger 2}}|\Omega_0\rangle.$$

Hence the above evaluations have also established the corresponding general matrix element,

$$\langle \Omega_0|e^{\frac{1}{2}\bar{\zeta}_2 a^2} D(u) e^{\frac{1}{2}\zeta_1 a^{\dagger 2}}|\Omega_0\rangle \tag{147}$$

$$= \left(1 - \bar{\zeta}_2 \zeta_1\right)^{-1/2} \exp\left\{-\frac{1}{2}\frac{1}{1 - \bar{\zeta}_2 \zeta_1}\left(u - \bar{\zeta}_2 \bar{u}\right)^*\left(u - \zeta_1 \bar{u}\right)\right\}.$$

7 Conclusions

In this contribution to the present Workshop Proceedings we have explored the first step in the general programme outlined in the Introduction, in the case of the quantum observables Q and P defining the Heisenberg algebra. Namely, generally given a quantum system characterised by a set of quantum observables, one considers the set of quantum states that saturate the Schrödinger–Robertson uncertainty relation corresponding to this set of observables. Such states are the closest possible to displaying a classical behaviour of the quantum system while, being determined by a condition characteristic of coherent-like quantum states, they are parametrised by a collection of continuous variables and therefore define a specific submanifold within the Hilbert space of the quantum system. Correspondingly there arise specific geometric structures associated with this manifold, compatible with one another, namely both a quantum symplectic structure and a quantum Riemannian metric. It may even be possible to reconstruct the quantum dynamics of the system from that geometric data as well as a choice of Hamiltonian operator represented through its diagonal matrix elements for the saturating states, thereby offering a geometric formulation of quantum systems and their dynamics through a path integral representation.

This general programme is initiated herein, in the case of observables of the Heisenberg algebra for a single degree of freedom quantum system as an illustration. Correspondingly the saturating quantum states are the so-called and well-known general squeezed states, for which many properties and results were reviewed and presented together with quite many details and some original results, with the hope that some readers could become interested in taking part in such an exploration in the case of other possible choices of quantum observables and the ensuing saturating quantum states. For instance, the affine quantum algebra of scale transformations, $[Q, D] = i\hbar Q$ (say with $D = (QP + PQ)/2$), does also play an important role in quite many quantum systems [4, 6, 7, 14], with its own coherent states. To this author's best knowledge, the analogue states for the affine algebra of the squeezed states for the Heisenberg algebra remain to be fully understood. Other situations may be thought of as well, such as the operators and uncertainty relation related to the factorisation of a quantum Hamiltonian along the lines and methods of supersymmetric quantum mechanics, $H = A^\dagger A + E_0$.

Essential to such a programme is the evaluation of the overlap of the saturating quantum states given a set of quantum observables. In particular this quantity encodes the data necessary in identifying the inherent geometric structures, as well as in the construction of the quantum path integral of the system over the manifold in Hilbert space associated with the saturating quantum states. Usually overcompleteness relations ensue, implying that the overlap of saturating states determines a reproducing kernel representation of the Hilbert space.

In this contribution the discussion concludes with the evaluation of this reproducing kernel for the general squeezed states of the Heisenberg algebra which saturate the Schrödinger–Robertson uncertainty relation. We defer to a separate publication an analysis of the corresponding symplectic and Riemannian geometric structures, as well as of the path integral representation of the quantum system over the associated manifold of squeezed states, $|\Omega_z(u)\rangle = D(u)S(z)|\Omega_0\rangle$. All of these considerations are to follow from the quantities $\langle\Omega_{z_2}(u_2)|\Omega_{z_1}(u_1)\rangle$.

Thus in particular the overlap $\langle\Omega_{z_2}(u_2)|\Omega_{z_1}(u_1)\rangle$ determines a reproducing kernel representation of the Hilbert space of the Heisenberg algebra of observables Q and P. Indeed, given the generalised overcompleteness relation (109), obviously one has the property,

$$\langle\Omega_{z_2}(u_2)|\Omega_{z_1}(u_1)\rangle \tag{148}$$
$$= \int_{\mathbb{C}^2} \frac{du_3\, d\bar{u}_3}{\pi} \frac{dz_3\, d\bar{z}_3}{\pi} \mu(z_3, \bar{z}_3)\, \langle\Omega_{z_2}(u_2)|\Omega_{z_3}(u_3)\rangle\, \langle\Omega_{z_3}(u_3)|\Omega_{z_1}(u_1)\rangle.$$

We plan to report elsewhere on such applications and further developments of the results of the present contribution, as well as on the general programme outlined in the Introduction. This programme is offered here as a token of genuine and sincere appreciation for Professor Norbert Hounkonnou's constant interest and many scientific contributions of note, and certainly for his unswerving efforts as well towards the development of mathematical physics in Benin, in Western Africa, and on the African continent, to the benefit of the younger and future generations, and this on the occasion of this special COPROMAPH Workshop celebrating his sixtieth birthday.

Acknowledgements The author thanks Professor Frederik Scholtz and the National Institute for Theoretical Physics (NITheP, Stellenbosch node, South Africa) for their warm hospitality. This work is supported in part by the Institut Interuniversitaire des Sciences Nucléaires (I.I.S.N., Belgium), and by the Belgian Federal Office for Scientific, Technical and Cultural Affairs through the Interuniversity Attraction Poles (IAP) P6/11.

Appendix 1: Cauchy–Schwarz Inequality and Quantum Uncertainty Relations

In the first part of this Appendix, for the purpose of the present paper it proves useful to reconsider specific arguments leading to the Cauchy–Schwarz inequality. In the second part this inequality is applied to establish the Schrödinger–Robertson uncertainty relation (SR-UR) given any two quantum observables.

Let $|\psi_1\rangle$ and $|\psi_2\rangle$ be any two (normalisable and non-vanishing) quantum states, and consider their arbitrary complex linear combination, say in the form,

$$|\psi\rangle = |\psi_1\rangle + i\lambda e^{i\varphi}|\psi_2\rangle, \qquad \varphi, \lambda \in \mathbb{R}, \tag{149}$$

with φ a phase factor and λ a real parameter. Since the sesquilinear and hermitian inner product of Hilbert space is positive definite, the norm of the state $|\psi\rangle$ is positive definite whatever the values for these two parameters,

$$P(\lambda) \equiv \langle\psi|\psi\rangle = \lambda^2\langle\psi_2|\psi_2\rangle + i\lambda\left(e^{i\varphi}\langle\psi_1|\psi_2\rangle - e^{-i\varphi}\langle\psi_2|\psi_1\rangle\right) + \langle\psi_1|\psi_1\rangle \geq 0. \tag{150}$$

Note that the l.h.s. of this inequality is a real quadratic polynomial in $\lambda \in \mathbb{R}$ with real coefficients, $P(\lambda)$, of which the coefficient in λ^2 is strictly positive. Hence the parabolic graph of this function of λ lies entirely in the upper half plane and this polynomial has no real roots in λ, unless the state $|\psi\rangle$ itself vanishes identically in which case the two roots are degenerate and real for just a unique and specific set of values for the parameters φ and λ such that the parabola $P(\lambda)$ has its minimum just touching the horizontal coordinate axis in λ. Consequently the discriminant of this real quadratic form in λ is negative, namely

$$\langle\psi_1|\psi_1\rangle\langle\psi_2|\psi_2\rangle \geq -\frac{1}{4}\left(e^{i\varphi}\langle\psi_1|\psi_2\rangle - e^{-i\varphi}\langle\psi_2|\psi_1\rangle\right)^2 \geq 0. \tag{151}$$

This inequality is the tightest when the quantity on the r.h.s. of this relation, which is still a function of the parameter φ, reaches its maximal value. As readily established this maximum is obtained for a phase factor $\varphi = \varphi_0$ such that

$$e^{2i\varphi_0} = -\frac{\langle\psi_2|\psi_1\rangle}{\langle\psi_1|\psi_2\rangle}, \tag{152}$$

namely,

$$e^{i\varphi_0}\langle\psi_1|\psi_2\rangle = -e^{-i\varphi_0}\langle\psi_2|\psi_1\rangle, \tag{153}$$

$$ie^{i\varphi_0}\langle\psi_1|\psi_2\rangle = -ie^{-i\varphi_0}\langle\psi_2|\psi_1\rangle = \left(ie^{i\varphi_0}\langle\psi_1|\psi_2\rangle\right)^*.$$

Given this choice, the tightest discriminant inequality in (151) reduces to the well-known Cauchy–Schwarz inequality

$$\langle\psi_1|\psi_1\rangle\langle\psi_2|\psi_2\rangle \geq |\langle\psi_1|\psi_2\rangle|^2. \tag{154}$$

Having set the phase factor as $\varphi = \varphi_0$ the polynomial $P(\lambda)$ may be organised in the following form:

$$P(\lambda) = \langle\psi_2|\psi_2\rangle\left[\left(\lambda + ie^{i\varphi_0}\frac{\langle\psi_1|\psi_2\rangle}{\langle\psi_2|\psi_2\rangle}\right)^2 + \frac{\langle\psi_1|\psi_1\rangle\langle\psi_2|\psi_2\rangle - |\langle\psi_1|\psi_2\rangle|^2}{\langle\psi_2|\psi_2\rangle^2}\right] \geq 0. \tag{155}$$

Hence making now the additional choice $\lambda = \lambda_{CS}$ such that

$$\lambda_{CS} = -ie^{i\varphi_0}\frac{\langle\psi_1|\psi_2\rangle}{\langle\psi_2|\psi_2\rangle} = ie^{-i\varphi_0}\frac{\langle\psi_2|\psi_1\rangle}{\langle\psi_2|\psi_2\rangle}, \qquad ie^{i\varphi_0}\lambda_{CS} = -\frac{\langle\psi_2|\psi_1\rangle}{\langle\psi_2|\psi_2\rangle}, \tag{156}$$

λ_{CS} being indeed a real quantity on account of the properties in (153), one has

$$\langle\psi|\psi\rangle = P(\lambda_{CS}) = \langle\psi_2|\psi_2\rangle\frac{\langle\psi_1|\psi_1\rangle\langle\psi_2|\psi_2\rangle - |\langle\psi_1|\psi_2\rangle|^2}{\langle\psi_2|\psi_2\rangle^2} \geq 0. \tag{157}$$

Consequently, besides the Cauchy–Schwarz inequality (154), one also concludes that this inequality is saturated into a strict equality provided the states $|\psi_1\rangle$ and $|\psi_2\rangle$ are such that $|\psi\rangle = 0$ for these choices of parameters $\varphi = \varphi_0$ and $\lambda = \lambda_{CS}$, namely

$$|\psi_1\rangle - \frac{\langle\psi_2|\psi_1\rangle}{\langle\psi_2|\psi_2\rangle}|\psi_2\rangle = 0 \iff \langle\psi_1|\psi_1\rangle\langle\psi_2|\psi_2\rangle = |\langle\psi_1|\psi_2\rangle|^2. \tag{158}$$

Let now A and B be two arbitrary quantum observables, namely hermitian (and ideally, self-adjoint) operators acting on Hilbert space, $A^\dagger = A$ and $B^\dagger = B$, and consider an arbitrary (normalisable and non-vanishing) quantum state $|\psi_0\rangle$. Whatever choice of quantum operator $\mathcal{O}$, its expectation value for that state $|\psi_0\rangle$ is denoted as

$$\langle\mathcal{O}\rangle = \frac{\langle\psi_0|\mathcal{O}|\psi_0\rangle}{\langle\psi_0|\psi_0\rangle}. \tag{159}$$

In particular for the observables A and B we have their real valued expectation values

$$a_0 = \langle A \rangle, \qquad b_0 = \langle B \rangle, \qquad a_0, b_0 \in \mathbb{R}, \tag{160}$$

which are used to shift these observables as follows:

$$\bar{A} = A - a_0 \mathbb{I}, \qquad \bar{B} = B - b_0 \mathbb{I}, \qquad \bar{A}^\dagger = \bar{A}, \qquad \bar{B}^\dagger = \bar{B}, \tag{161}$$

such that $[\bar{A}, \bar{B}] = [A, B]$. Consequently we have

$$(\Delta A)^2 = \langle \bar{A}^2 \rangle, \qquad (\Delta B)^2 = \langle \bar{B}^2 \rangle. \tag{162}$$

In order to establish the SR-UR from the Cauchy–Schwarz inequality, let us consider the following two quantum states:

$$|\psi_1\rangle = \frac{1}{\sqrt{\langle \psi_0 | \psi_0 \rangle}} \bar{A} |\psi_0\rangle, \qquad |\psi_2\rangle = \frac{1}{\sqrt{\langle \psi_0 | \psi_0 \rangle}} \bar{B} |\psi_0\rangle, \tag{163}$$

which are such that

$$\langle \psi_1 | \psi_1 \rangle = (\Delta A)^2, \qquad \langle \psi_2 | \psi_2 \rangle = (\Delta B)^2, \tag{164}$$

$$\langle \psi_1 | \psi_2 \rangle = \langle \bar{A} \bar{B} \rangle, \qquad \langle \psi_2 | \psi_1 \rangle = \langle \bar{B} \bar{A} \rangle = \langle \bar{A} \bar{B} \rangle^*.$$

Consequently the Cauchy–Schwarz inequality (154) reads

$$(\Delta A)^2 (\Delta B)^2 \geq |\langle \bar{A} \bar{B} \rangle|^2, \qquad (\Delta A)^2 (\Delta B)^2 \geq \langle \bar{A} \bar{B} \rangle \langle \bar{B} \bar{A} \rangle. \tag{165}$$

Alternatively by expressing the quantity $\langle \bar{A} \bar{B} \rangle$ in terms of the commutator and anti-commutator of the operators $\bar{A}$ and $\bar{B}$ which then separate its real and imaginary parts,[15] namely

$$\langle \bar{A} \bar{B} \rangle = \langle \frac{1}{2} [\bar{A}, \bar{B}] + \frac{1}{2} \{\bar{A}, \bar{B}\} \rangle = \frac{1}{2} i \langle (-i) [\bar{A}, \bar{B}] \rangle + \frac{1}{2} \langle \{\bar{A}, \bar{B}\} \rangle, \tag{166}$$

$$\langle \bar{B} \bar{A} \rangle = \langle \bar{A} \bar{B} \rangle^* = -\frac{1}{2} i \langle (-i) [\bar{A}, \bar{B}] \rangle + \frac{1}{2} \langle \{\bar{A}, \bar{B}\} \rangle, \tag{167}$$

one obtains the inequality in the Schrödinger–Robertson form,

$$(\Delta A)^2 (\Delta B)^2 \geq \frac{1}{4} \langle (-i) [A, B] \rangle^2 + \frac{1}{4} \langle \{\bar{A}, \bar{B}\} \rangle^2. \tag{168}$$

---

[15] Note that $(-i)[\bar{A}, \bar{B}]$ and $\{\bar{A}, \bar{B}\}$ are hermitian (or even self-adjoint if A and B are self-adjoint) operators whose expectations values are thus real.

As a by-product one also derives the looser generalised Heisenberg uncertainty relation,

$$(\Delta A)^2 \, (\Delta B)^2 \geq \frac{1}{4} \langle (-i) \, [A, B] \rangle^2, \qquad (\Delta A) \, (\Delta B) \geq \frac{1}{2} | \langle (-i)[A, B] \rangle |. \tag{169}$$

Furthermore given (158) the SR-UR (168) is saturated, namely $(\Delta A)^2 \, (\Delta B)^2 = \langle \bar{A} \bar{B} \rangle \langle \bar{B} \bar{A} \rangle$, provided the state $|\psi_0\rangle$ is such that

$$\left(\bar{A} - \frac{\langle \bar{B} \bar{A} \rangle}{(\Delta B)^2} \, \bar{B} \right) |\psi_0\rangle = 0, \quad \left(\bar{A} - \lambda_0 \, \bar{B} \right) |\psi_0\rangle = 0, \tag{170}$$

$$(A - \lambda_0 B) |\psi_0\rangle = (\langle A \rangle - \lambda_0 \langle B \rangle) |\psi_0\rangle,$$

where the complex parameter λ_0 is given by $\lambda_0 = \langle \bar{B} \bar{A} \rangle / (\Delta B)^2 = (\Delta A)^2 / \langle \bar{A} \bar{B} \rangle$, namely

$$\lambda_0 = \frac{1}{(\Delta B)^2} \left(\frac{1}{2} \langle \{ \bar{A}, \bar{B} \} \rangle - \frac{1}{2} i \, \langle (-i) \, [A, B] \rangle \right) \tag{171}$$

$$= (\Delta A)^2 \, \frac{1}{\left(\frac{1}{2} \langle \{ \bar{A}, \bar{B} \} \rangle + \frac{1}{2} i \, \langle (-i) \, [A, B] \rangle \right)}.$$

Appendix 2: Baker–Campbell–Hausdorff Formulae

This Appendix is structured in three parts. The first recalls a basic Baker–Campbell–Hausdorff (BCH) formula. The second part discusses recent results established in [15] based on a construction of the most general BCH formula which is also outlined. Finally the third part applies these results to a SU(1,1) algebra directly related to the general squeezed coherent states arising as the states saturating the Schrödinger–Robertson uncertainty relation.

Given any two operators, A and B, the following basic BCH is well known,[16]

$$e^A \, B \, e^{-A} = B + [A, B] + \frac{1}{2!}[A, [A, B]] + \frac{1}{3!}[A, [A, [A, B]]] + \cdots = e^{ad \, A} \, B, \tag{172}$$

where the (Lie algebra) adjoint action of the operator A on an operator X is defined by

$$ad \, A \cdot X \equiv [A, X]. \tag{173}$$

---

[16]It suffices to consider the generating operator in λ, $e^{\lambda A} B e^{-\lambda A}$, expanded in series in λ.

This identify is to be used throughout hereafter. Note that it also implies

$$e^A e^B e^{-A} = e^{e^{ad\, A}\, B}, \qquad e^A e^B = e^{e^{ad\, A}\, B} e^A. \tag{174}$$

Hence in particular when $[A, B]$ commutes with both A and B, we have simply

$$e^A e^B = e^{[A,B]} e^B e^A. \tag{175}$$

In order to establish the general BCH formula, first let us consider some operator $A(\lambda)$ function of a parameter λ. Then the following identities apply,[17]

$$e^{-A(\lambda)} \frac{d}{d\lambda} e^{A(\lambda)} = \int_0^1 dt\, e^{-t A(\lambda)} \frac{d A(\lambda)}{d\lambda} e^{t A(\lambda)}$$

$$= \int_0^1 dt\, e^{-t\, ad\, A(\lambda)} \frac{d A(\lambda)}{d\lambda}$$

$$= \Phi\left(-ad\, A(\lambda)\right) \frac{d A(\lambda)}{d\lambda}, \tag{176}$$

where the function $\Phi(x)$ is given as

$$\Phi(x) = \int_0^1 dt\, e^{tx} = \sum_{n=0}^{\infty} \frac{1}{(n+1)!} x^n = \frac{e^x - 1}{x}. \tag{177}$$

This function being such that

$$\Phi(-\ln x) = \frac{x-1}{x \ln x} = \frac{1}{\Psi(x)}, \tag{178}$$

it proves useful to also introduce the function $\Psi(x)$ defined by[18]

$$\Psi(x) = \frac{x \ln x}{x - 1} = 1 + \sum_{n=1}^{\infty} \frac{(-1)^{n+1}}{n(n+1)} (x-1)^n, \qquad \Phi(-\ln x)\,\Psi(x) = 1. \tag{179}$$

Given two operators A and B, the general BCH formula provides an expression for the operator C defined by

$$C = \ln(e^A e^B), \qquad e^C = e^A e^B. \tag{180}$$

---

[17] As a reminder we have $\int_0^1 dt\, (1-t)^n\, t^m = n!\, m!/(n+m+1)!$ as well as $\int_0^1 dt\, t^n = 1/(n+1)$, hence in particular $\frac{d}{d\lambda} e^{A(\lambda)} = \int_0^1 dt\, e^{(1-t)A(\lambda)} \frac{d A(\lambda)}{d\lambda} e^{t A(\lambda)}$.

[18] Note that $\Psi(e^{-x}) = 1/\Phi(x) = x/(e^x - 1) = \sum_{n=0}^{\infty} B_n x^n/n!$ is a generating function for Bernoulli numbers, B_n. The author thanks Christian Hagendorf for pointing this out to him.

In order to establish this expression, let us introduce the generating operator $C(\lambda)$ such that

$$e^{C(\lambda)} = e^A \, e^{\lambda B}, \qquad C(\lambda) = \ln(e^A \, e^{\lambda B}), \qquad C(0) = A. \tag{181}$$

The adjoint action of the operator $C(\lambda)$ on any operator D is given as

$$e^{C(\lambda)} \, D \, e^{-C(\lambda)} = e^A \, e^{\lambda B} \, D \, e^{-\lambda B} \, e^{-A}, \quad \text{namely,} \quad e^{ad\,C(\lambda)} \, D = e^{ad\,A} \, e^{\lambda\,ad\,B} \, D, \tag{182}$$

which implies

$$e^{ad\,C(\lambda)} = e^{ad\,A} \, e^{\lambda\,ad\,B}, \qquad ad\,C(\lambda) = \ln(e^{ad\,A} \, e^{\lambda\,ad\,B}). \tag{183}$$

On the other hand, since

$$e^{-C(\lambda)} \, \frac{d}{d\lambda} e^{C(\lambda)} = e^{-\lambda B} \, e^{-A} \, \frac{d}{d\lambda} e^A \, e^{\lambda B}, \quad \text{namely,} \quad \Phi(-ad\,C(\lambda)) \frac{dC(\lambda)}{d\lambda} = B, \tag{184}$$

necessarily

$$\frac{dC(\lambda)}{d\lambda} = \Psi\left(e^{ad\,A} \, e^{\lambda\,ad\,B}\right) B. \tag{185}$$

Given the integration condition $C(0) = A$, finally the following general BCH formulae applies for the operator C,

$$C = \ln(e^A \, e^B) = A + \int_0^1 d\lambda \, \Psi\left(e^{ad\,A} \, e^{\lambda\,ad\,B}\right) B \tag{186}$$

$$= A + B - \int_0^1 d\lambda \sum_{n=1}^{\infty} \frac{1}{n(n+1)} \left(\mathbb{I} - e^{ad\,A} \, e^{\lambda\,ad\,B}\right)^n B$$

$$= A + B + \int_0^1 d\lambda \sum_{n=1}^{\infty} \frac{1}{n(n+1)} \left(\mathbb{I} - e^{ad\,A} \, e^{\lambda\,ad\,B}\right)^{n-1} \left(\frac{e^{ad\,A} - \mathbb{I}}{ad\,A}\right) [A, B].$$

In particular when $[A, B]$ commutes with both A and B, one has the well-known BCH formula,

$$C = \ln(e^A \, e^B) = A + B + \frac{1}{2}[A, B], \qquad e^A \, e^B = e^{A+B+\frac{1}{2}[A,B]} = e^{\frac{1}{2}[A,B]} \, e^{A+B}, \tag{187}$$

which implies again the result in (175).

It is the last form for the BCH formula in (187) which is the starting point of the recent analysis of [15] which manages to sum up the BCH formula in closed form in the case of operators A and B whose commutator is of the form,

$$[A, B] = uA + vB + c\mathbb{I}, \tag{188}$$

where u, v, and c are constant parameters.[19] Indeed in such a situation one has

$$ad\, A \cdot [A, B] = v\, [A, B], \quad ad\, B \cdot [A, B] = -u\, [A, B],$$
$$e^{ad\, A}\, [A, B] = e^{v}\, [A, B], \quad e^{\lambda\, ad\, B}\, [A, B] = e^{-\lambda u}\, [A, B], \tag{189}$$

which implies that

$$\ln(e^{A}\, e^{B}) = A + B + f(u, v)\, [A, B], \tag{190}$$

where $f(u, v)$ is a simple function determined from (187) in the form

$$f(u, v) = \frac{e^{v} - 1}{v} \int_{0}^{1} d\lambda \sum_{n=1}^{\infty} \frac{1}{n(n+1)} \left(1 - e^{v}\, e^{-\lambda v}\right)^{n-1}. \tag{191}$$

A direct evaluation finds for this function, which proves to be symmetric,

$$f(u, v) = \frac{ue^{u}(e^{v} - 1) - ve^{v}(e^{u} - 1)}{uv(e^{u} - e^{v})} = \frac{u(1 - e^{-v}) - v(1 - e^{-u})}{uv(e^{-v} - e^{-u})} = f(v, u), \tag{192}$$

with the distinguished value $f(0, 0) = 1/2$ (in agreement with (187)).

Finally given the reference Fock algebra of operators a and $a^{\dagger}$ introduced in Sect. 3, consider the operators

$$K_{0} = \frac{1}{2}\left(a^{\dagger}a + \frac{1}{2}\right), \quad K_{+} = \frac{1}{2}a^{\dagger 2}, \quad K_{-} = \frac{1}{2}a^{2}, \tag{193}$$

which generate a SU(1,1) algebra of transformations acting on the Hilbert space representing the Heisenberg algebra of observables Q and P,

$$[K_{0}, K_{\pm}] = \pm K_{\pm}, \quad [K_{-}, K_{+}] = 2K_{0}. \tag{194}$$

Independently of the representation which realises this SU(1,1) algebra, let us apply the result (190) to specific combinations of operators K_{0} and $K_{\pm}$ obeying this algebraic structure.

To begin with consider the following operator

$$e^{\alpha K_{+}}\, e^{\gamma K_{0}}\, e^{-\bar{\alpha} K_{-}}, \tag{195}$$

where α is an arbitrary complex parameter such that $|\alpha| < 1$ and $\gamma = \ln(1 - |\alpha|^{2})$. In order to apply (190), let us rewrite this operator in the form [16],

$$e^{\alpha K_{+}}\, e^{\gamma K_{0}}\, e^{-\bar{\alpha} K_{-}} = e^{\alpha K_{+}}\, e^{\gamma_{s} K_{0}}\, e^{\gamma_{-s} K_{0}}\, e^{-\bar{\alpha} K_{-}}, \tag{196}$$

---

[19]Note that for all practical purposes the results of [15] remain valid as stated if the term $c\mathbb{I}$ stands for an operator which commutes with both A and B.

where

$$\gamma_s = \ln(1 + s|\alpha|), \quad \gamma_s + \gamma_{-s} = \gamma_+ + \gamma_- = \gamma = \ln(1 - |\alpha|^2), \quad s = \pm 1. \tag{197}$$

Since we have

$$[\alpha K_+, \gamma_s K_0] = -\gamma_s (\alpha K_+), \qquad [\gamma_{-s} K_0, -\bar{\alpha} K_-] = -\gamma_{-s} (-\bar{\alpha} K_-), \tag{198}$$

the BCH formula (190) then applies separately to the first two, and the last two factors in (196). With the evaluation of the corresponding values for the function $f(u, v)$, one then finds

$$\ln(e^{\alpha K_+} e^{\gamma_s K_0}) = \frac{s}{|\alpha|} \gamma_s \alpha K_+ + \gamma_s K_0 \equiv \tilde{A},$$

$$\ln(e^{\gamma_{-s} K_0} e^{-\bar{\alpha} K_-}) = \frac{s}{|\alpha|} \gamma_{-s} \bar{\alpha} K_- + \gamma_{-s} K_0 \equiv \tilde{B}, \tag{199}$$

namely so far,

$$e^{\alpha K_+} e^{\gamma K_0} e^{-\bar{\alpha} K_-} = e^{\tilde{A}} e^{\tilde{B}}. \tag{200}$$

However the values for γ_s are chosen not only such that $\gamma_s + \gamma_{-s} = \gamma$ but also such that the BCH formula (190) may be applied once again to the latter product, which requires that the commutator of $\tilde{A}$ and $\tilde{B}$ be again a linear combination of these same two operators,

$$[\tilde{A}, \tilde{B}] = -\gamma_{-s} \tilde{A} - \gamma_s \tilde{B}. \tag{201}$$

The final evaluation of the BCH formula for (195) then reduces to the determination of the value $f(-\gamma_{-s}, -\gamma_s)$. In order to present this BCH formula, since $|\alpha| < 1$ let us introduce the following parameters, with $0 \le r < \infty$ and $-\pi \le \theta < +\pi$,

$$|\alpha| = \tanh r, \qquad \alpha = e^{i\theta} \tanh r, \qquad z = e^{i\theta} r. \tag{202}$$

One then has finally

$$e^{\alpha K_+} e^{\ln(1 - \tanh^2 r) K_0} e^{-\bar{\alpha} K_-} = e^{z K_+ - \bar{z} K_-}. \tag{203}$$

A similar procedure may be applied to the operator

$$e^{-\bar{\alpha} K_-} e^{-\gamma K_0} e^{\alpha K_+}. \tag{204}$$

However given the following inner automorphism of the SU(1,1) algebra,

$$K_0 \longleftrightarrow -K_0, \qquad K_+ \longleftrightarrow -K_-, \qquad K_- \longleftrightarrow -K_+, \tag{205}$$

from (203) one readily has (with then also $\alpha \leftrightarrow \bar{\alpha}$ and $z \leftrightarrow \bar{z}$)

$$e^{-\bar{\alpha} K_-} e^{-(1-\tanh^2 r) K_0} e^{\alpha K_+} = e^{z K_+ - \bar{z} K_-} = S(z) = e^{\alpha K_+} e^{\ln(1-\tanh^2 r) K_0} e^{-\bar{\alpha} K_-}.$$
(206)

The results (203) and (206), which thus apply to the squeezing operator $S(z)$ introduced in Sect. 4.3 are stated in [17] by establishing them in the defining representation of the SU(1,1) algebra. Here they are derived solely from the structure of the SU(1,1) algebra and independently of the representation of that algebra, by using the conclusions of [15].

References

1. W. Heisenberg, Zeit. für Phys. **43**, 172–198 (1927)
2. E. Schrödinger, Sitz. Preus. Acad. Wiss. (Phys.-Math. Klasse) **19**, 296–303 (1930) [translated to English in Bulg. J. Phys. **26**, 193 (1999) (e-print arXiv:quant-ph/9903100)]
3. H.P. Robertson, Phys. Rev. **34**, 163–164 (1929); H.P. Robertson, Phys. Rev. **35**, Abstract 40, p. 667 (1930); H.P. Robertson, Phys. Rev. **46**, 794 (1934)
4. J.R. Klauder, E.C.G. Sudarshan, *Fundamentals of Quantum Optics* (Dover, New York, 1968, 1996); J.R. Klauder, B.-S. Skagerstam, *Coherent States – Applications in Physics and Mathematical Physics* (World Scientific, Singapore, 1985)
5. M.M. Nieto, The discovery of squeezed states – in 1927. Preprint LA-UR-97-2760. e-print arXiv:quant-ph/9708012
6. J.R. Klauder, *Beyond Conventional Quantization* (Cambridge University Press, Cambridge, 2000); J.R. Klauder, *A Modern Approach to Functional Integration* (Birkhäuser, Springer, New York, 2010); J.R. Klauder, *Enhanced Quantization, Particles, Fields & Gravity* (World Scientific, Hackensack, 2015)
7. J.R. Klauder, Ann. Phys. **188**, 120 (1988)
8. J.P. Provost, G. Vallee, Commun. Math. Phys. **76**, 289 (1980)
9. A. Ashtekar, T.A. Schilling, Geometrical formulation of quantum mechanics. e-print arXiv:gr-qc/9706069
10. J.N. Kriel, H.J.R. van Zyl, F.G. Scholtz, J. High Energy Phys. **1511**, 140 (2015) (e-print arXiv:1509.02040 [hep-th])
11. D.A. Trifonov, J. Math. Phys. **34**, 100 (1993); D.A. Trifonov, J. Math. Phys. **35**, 2297 (1994); D.A. Trifonov, J. Phys. **A30**, 5941 (1997) (e-print arXiv:quant-ph/9701018); D.A. Trifonov, Phys. Scr. **58**, 246 (1998) (e-print arXiv:quant-ph/9705001); D.A. Trifonov, J. Opt. Soc. Am. **A17**, 2486 (2000) (e-print arXiv:quant-ph/0012072); D.A. Trifonov, Generalized intelligent states and SU(1,1) and SU(2) squeezing (2000) (e-print arXiv:quant-ph/0001028); D.A. Trifonov, Physics World **24**, 107 (2001) (eprint arXiv:physics/0105035)
12. A. Angelow, Evolution of Schrödinger uncertainty relation in quantum mechanics. e-print arXiv:0710.0670 [quant-ph]
13. C. Brif, Int. J. Theor. Phys. **36**, 1651 (1997) (e-print arXiv:quant-ph/9701003)
14. M. Fanuel, S. Zonetti, Europhys. Lett. **101**, 10001 (2013) (e-print arXiv:1203.4936 [hep-th])
15. A. Van-Brunt, M. Visser, J. Phys. A **48**, 225207 (2015) (e-print arXiv:1501.02506 [math-ph]); A. Van-Brunt, M. Visser, Explicit Baker-Campbell-Hausdorff formulae for some specific Lie algebras (2015). e-print arXiv:1505.04505 [math-ph]
16. M. Matone, J. High Energy Phys. **1505**, 113 (2015) (e-print arXiv:1502.06589 [math-ph])
17. A.M. Perelomov, Commun. Math. Phys. **26**, 222 (1972); A.M. Perelomov, *Generalized Coherent States and Their Applications* (Springer, Berlin, 1986)

The Role of the Jacobi Last Multiplier in Nonholonomic Systems and Locally Conformal Symplectic Structure

Partha Guha

To Mahouton Norbert Hounkonnou, collaborator, colleague and friend, dedicated to his 60th birthday with admiration and gratitude

Abstract In this pedagogic article we study the geometrical structure of non-holonomic system and elucidate the relationship between Jacobi's last multiplier (JLM) and nonholonomic systems endowed with the almost symplectic structure. In particular, we present an algorithmic way to describe how the two form and almost Poisson structure associated to nonholonomic system, studied by L. Bates and his coworkers (Rep Math Phys 42(1–2):231–247, 1998; Rep Math Phys 49(2–3):143–149, 2002; What is a completely integrable nonholonomic dynamical system, in Proceedings of the XXX symposium on mathematical physics, Toruń, 1998; Rep Math Phys 32:99–115, 1993), can be mapped to symplectic form and canonical Poisson structure using JLM. We demonstrate how JLM can be used to map an integrable nonholonomic system to a Liouville integrable system. We map the toral fibration defined by the common level sets of the integrals of a Liouville integrable Hamiltonian system with a toral fibration coming from a completely integrable nonholonomic system.

Keywords Jacobi last multiplier · Nonholonomic system · Conformal Hamiltonian · Liouville integrability · Torus fibration

P. Guha (✉)
IFSC, Universidade de São Paulo, São Carlos, SP, Brazil
S.N. Bose National Centre for Basic Sciences, Salt Lake City, Kolkata, India
e-mail: partha@bose.res.in

© Springer Nature Switzerland AG 2018
T. Diagana, B. Toni (eds.), *Mathematical Structures and Applications*,
STEAM-H: Science, Technology, Engineering, Agriculture,
Mathematics & Health, https://doi.org/10.1007/978-3-319-97175-9_12

1   Introduction

The Jacobi last multiplier (JLM) is a useful tool for deriving an additional first integral for a system of n first-order ODEs when $n-2$ first integrals of the system are known. Besides, the JLM allows us to determine the Lagrangian of a second-order ODE in many cases [15, 25, 31]. In his sixteenth lecture on dynamics Jacobi uses his method of the last multiplier [19, 20] to derive the components of the Laplace–Runge–Lenz vector for the two-dimensional Kepler problem. In recent years a number of articles have dealt with this particular aspect [10, 16, 24–26]. However, when a planar system of ODEs cannot be reduced to a second-order differential equation the question of interest arises whether the JLM can provide a mechanism for finding the Lagrangian of the system.

Let M be an even dimensional differentiable manifold endowed with a non-degenerate 2-form Ω, (M, Ω) is an almost symplectic manifold. An almost symplectic manifold (M, Ω) is called locally conformally symplectic (l.c.s.) manifold by Vaisman [29] if there is a global 1-form η, called the Lee form on M such that

$$d\Omega = \eta \wedge \Omega,$$

where $d\eta = 0$. (M, Ω) is globally conformally symplectic if the Lee form η is exact and when $\eta = 0$, then (M, Ω) is a symplectic manifold. The notion of locally conformally symplectic forms is due to Lee and, in more modern form, to Vaisman. Chinea et al. [8, 9] showed an extension of an observation made by I. Vaisman [29] that locally conformal symplectic manifolds can be seen as a natural geometrical setting for the description of time-independent Hamiltonian systems. In a seminal paper Wojkowski and Liverani [32] studied the Lyapunov spectrum in locally conformal Hamiltonian systems. It was demonstrated that Gaussian isokinetic dynamics, Nośe–Hoovers dynamics and other systems can be studied through locally conformal Hamiltonian systems. It must be noted that the conformal Hamiltonian structure appears in various dissipative dynamics as well as in the activator-inhibitor model connected to Turing pattern formation. It has been shown by Haller and Rybicki [18] that the Poisson algebra of a locally conformally symplectic manifold is integrable by making use of a convenient setting in global analysis. In this paper we explore the role of the Jacobi last multiplier in nonholonomic free particle motion and nonholonomic oscillator. These systems were studied extensively by L. Bates and his coworkers [2–5]. The two forms associated with these nonholonomic systems are not closed, in fact they satisfy l.c.s. condition. We apply JLM to such systems which guarantees that at least locally the symplectic form can be multiplied by a nonzero function to get a symplectic structure. In an interesting paper Bates and Cushman [4] compared the geometry of a toral fibration defined by the common level sets of the integrals of a Liouville integrable Hamiltonian system with a toral fibration coming from a completely integrable nonholonomic system. We apply JLM to study and compare these two toral fibrations. All the examples considered in this paper are taken from Bates et al. papers [2–5]. Relatively very little has been done when the flow is not complete. A quarter of a century ago, Flaschka [14] raised a number of questions concerning a simple class of integrable Hamiltonian systems in $\mathbf{R}^4$ for which the orbits lie on surfaces.

This paper is organized as follows. The first section recalls the definitions of the locally conformal symplectic structure and the Jacobi last multiplier. In Sect. 4 we study nonholonomic dynamics through an example—nonholonomic free particle motion, using constrained Lagrangian dynamics [7] and Bizyaev, Borisov, and Mamaev [6] method. We apply Jacobi last multiplier (JLM) method to transform nonholonomic dynamics into symplectic dynamics, a notion which, to our knowledge, does not appear explicitly in the literature. We study integrability property of the nonholonomic system in Sect. 5. The paper ends with a list of remarks regarding the further applications of JLM in nonholonomic systems. Finally, it is worthwhile to note that the first draft of this paper was circulated as an IHES preprint in 2013.

2   Preliminaries

We start with a brief review [17, 18, 29, 30] of the locally conformal symplectic structure. A differentiable manifold M of dimension $2n$ endowed with a non-degenerate 2-form ω and a closed 1-form η is called a locally conformally symplectic (l.c.s.) manifold if

$$d\omega + \omega \wedge \eta = 0. \tag{2.1}$$

The 1-form η is called the Lee form of ω [21]. This allows us to introduce the Lichnerowicz deformed differential operators

$$d_\eta : \Omega^*(M) \longrightarrow \Omega^{*+1}(M),$$

such that $d_\eta \theta = d\theta + \eta \wedge \theta$. Clearly $d_\eta^2 = 0$ and $d_\eta \omega = 0$. It must be worthwhile to note that l.c.s manifold is locally conformally equivalent to a symplectic manifold provided $\eta = df$ and $\omega = e^f \omega_0$, such that $d\omega_0 = 0$.

If (ω, η) is an l.c.s. structure on M and $f \in C^\infty(M, \mathbb{R})$, then $(e^f \omega, \eta - df) = (\omega', \eta')$ is again an l.c.s. structure on M then these two are conformally equivalent, and these two operators and Lee forms are cohomologous: $\eta' = \eta - df$. Hence d_η and $d_{\eta'}$ are gauge equivalent

$$d_{\eta'}(\beta) = (d_\eta - df \wedge)\beta = e^f \, d(e^{-f} \beta).$$

The r.h.s. is connected to Witten's differential. If $f \in C^\infty(M)$ and $t \geqslant 0$, Witten deformation of the usual differential $d_{tf} : \Omega^*(M) \longrightarrow \Omega^{*+1}(M)$ is defined by $d_{tf} = e^{tf} d e^{-tf}$, which means $d_{tf}\beta = d\beta + t\beta \wedge df$. Since d_η and $d_{\eta'}$ are gauge equivalent, the Lichnerowicz cohomology groups $H^*(\Omega^*(M), d_\eta)$ and

$H^*(\Omega^*(M), d_\eta)$ are isomorphic and the isomorphism is given by the conformal transformation $[\beta] \longmapsto [e^f \beta]$.

It is clear from the definition that d_η does not satisfy the Leibniz property:

$$d_\eta(\theta \wedge \psi) = (d + \eta\wedge)(\theta \wedge \psi) = d_\eta\theta \wedge \psi + (-1)^{deg\,\theta}\theta \wedge d\psi$$
$$= d\theta \wedge \psi + (-1)^{deg\,\theta}\theta \wedge d_\eta\psi.$$

For an l.c.s. manifold, we denote by

$$\mathrm{Diff}_c^\infty(M, \omega, \eta) := \{f \in \mathrm{Diff}_c^\infty(M) | (f^*\omega, f^*\eta) \simeq (\omega, \eta)\}$$

the group of compactly supported diffeomorphisms preserving the conformal equivalence class of (ω, η). The corresponding Lie algebra of vector fields is

$$\chi_c(M, \omega, \eta) := \{X \in \chi_c(M) \mid \exists c \in \mathbb{R} : L_X^\eta\omega = c\omega\},$$

where $L_X^\eta\beta = L_X\beta + \eta(X)\beta$. The Cartan magic formula for L_X^η is given by

$$L_X^\eta = d_\eta \circ i_X + i_X \circ d_\eta.$$

Here we list some of the important properties of the Lie derivative.

1. $L_X^\eta L_Y^\eta - L_X^\eta L_X^\eta = L_{[X,Y]}^\eta$.
2. $L_X^\eta d_\eta - d_\eta L_X^\eta = 0$
3. $L_X^\eta i_Y - i_Y L_X^\eta = 0$.
4. Let η_1 and η_2 be two Lee forms then $L_X^{\eta_1+\eta_2}(\theta \wedge \psi) = (L_X^{\eta_1}\theta) \wedge \psi + \theta \wedge (L_X^{\eta_2}\psi)$.

Let X and Y be the two conformal vector fields then $[X, Y]$ becomes the symplectic vector field. The proof of this claim is very simple, can easily show that $L_{[X,Y]}^\eta\omega = 0$.

2.1 Inverse Problem and the Jacobi Last Multiplier

We start with a brief introduction [10, 15, 24, 25, 31] of the Jacobi last multiplier and inverse problem of calculus of variations [22]. Consider a system of second-order ordinary differential equations

$$y_i'' = f_i(y_j, y_j') \qquad \text{for } 1 \leq i, j \leq n.$$

Geometrically these are the analytical expression of a second-order equation field Γ living on the first jet bundle $J^1\pi$ of a bundle $\pi : E \to \mathbb{R}$, so

$$\Gamma = y_i'\frac{\partial}{\partial y_i} + f_i(y_j, y_j')\frac{\partial}{\partial y_i'}.$$

The local formulation of the general inverse problem is the question for the existence of a non-singular multiplier matrix $g_{ij}(y, y')$, such that

$$g_{ij}(y''_j - f_j) \equiv \frac{d}{dt}\left(\frac{\partial L}{\partial y_i}\right) - \frac{\partial L}{\partial y'_i},$$

for some Lagrangian L. The most frequently used set of necessary and sufficient conditions for the existence of the g_{ij} are the so-called Helmholtz conditions due to Douglas [13, 27, 28].

Theorem 2.1 (Douglas [13]) *There exists a Lagrangian $L : TQ \to \mathbb{R}$ such that the equations are its Euler–Lagrange equations if and only if there exists a non-singular symmetric matrix g with entries g_{ij} satisfying the following three Helmholtz conditions:*

$$g_{ij} = g_{ji}, \qquad \widehat{\Gamma}(g_{ij}) = g_{ik}\Gamma^k_j + g_{jk}\Gamma^k_i,$$

$$g_{ik}\Phi^k_j = g_{jk}\Phi^k_i, \qquad \frac{\partial g_{ij}}{\partial y'_k} = \frac{\partial g_{ik}}{\partial y'_j},$$

$$\Gamma^k_j := -\frac{1}{2}\frac{\partial f_i}{\partial y'_j}, \qquad \Phi^k_i := -\frac{\partial f^k}{\partial x^i} - \Gamma^l_i\Gamma^k_l - \widehat{\Gamma}(\Gamma^k_i),$$

where $\widehat{\Gamma} = \frac{\partial}{\partial t} + y^i\frac{\partial}{\partial x^i} + f^i\frac{\partial}{\partial y^i}.$

When the system is one-dimensional we have $i = j = k = 1$ and then the three set of conditions become trivial and the fourth one reduces to one single P.D.E.

$$\Gamma(g) + g\frac{\partial f}{\partial v} \equiv v\frac{\partial g}{\partial x} + f\frac{\partial g}{\partial v} + g\frac{\partial f}{\partial v} = 0.$$

This is the equation defining the Jacobi multipliers, because $\operatorname{div}\Gamma = \frac{\partial f}{\partial v}$. The main equation can also be expressed as

$$\frac{dg}{dt} + g \cdot \operatorname{div}\Gamma = 0.$$

Then, the inverse problem reduces to find the function g (often denoted by μ) which is a Jacobi multiplier and L is obtained by integrating the function μ two times with respect to velocities.

An autonomous second-order differential equation $y'' = F(y, y')$ has associated a system of first-order differential equations

$$y' = v, \qquad v' = F(y, v) \tag{2.2}$$

whose solutions are the integral curves of the vector field in $\mathbb{R}^2$

$$\Gamma = v\frac{\partial}{\partial y} + F(y, v)\frac{\partial}{\partial v}. \tag{2.3}$$

A Jacobi multiplier μ for such a system must satisfy divergence free condition

$$\frac{\partial}{\partial y}(\mu v) + \frac{\partial}{\partial v}(\mu F) = 0,$$

which implies μ must be such that

$$v\frac{\partial \mu}{\partial y} + \frac{\partial \mu}{\partial v}F + \mu\frac{\partial F}{\partial v} = 0.$$

which taking into account $\frac{dM}{dx} = v\frac{\partial M}{\partial y} + F\frac{\partial M}{\partial v}$ above equation can be written as

$$\frac{d\log\mu}{dx} + \frac{\partial F}{\partial v} = 0. \tag{2.4}$$

The normal form of the differential equation determining the solutions of the Euler–Lagrange equation defined by the Lagrangian function $L(y, v)$ admits as a Jacobi multiplier the function

$$\mu = \frac{\partial^2 L}{\partial v^2}. \tag{2.5}$$

Conversely, if $\mu(y, v)$ is a last multiplier function for a second-order differential equation in normal form, then there exists a Lagrangian L for the system related to μ by the above equation.

Let L be such that condition $M = \frac{\partial^2 L}{\partial v^2}$ be satisfied, then

$$\frac{\partial L}{\partial v} = \int^v M(y, \zeta)d\zeta + \phi_1(y)$$

which yields

$$L(y, v) = \int^v dv' \int^{v'} M(y, \zeta)d\zeta + \phi_1(y)v + \phi_2(y).$$

Geometrical Interpretation of JLM Let M be a smooth, real, n-dimensional orientable manifold with fixed volume form Ω. Let $\dot{x}_i(t) = \gamma_i(x_1(t), \cdots, x_n(t))$, $1 \leqslant i \leqslant n$ generated by the vector field Γ and we consider the $(n-1)$-form $\Omega_\gamma = i_\Gamma\Omega$. The function $\mu \in C^\infty(M)$ is called a JLM of the ODE system generated by Γ, if $\mu\omega$ is closed, i.e.,

$$d(\mu\Omega_\gamma) = d\mu \wedge \Omega_\gamma + \mu d\Omega_\gamma.$$

This is equivalent to $\Gamma(\mu) + \mu.\,\text{div}\,\Gamma = 0$. Characterizations of the JLM can be obtained in terms of the deformed Lichnerowicz operator $d_\mu(\theta) = d\mu \wedge \theta + d\theta$, where the Lee form in terms of the last multiplier, i.e. $\eta = d\mu$. Hence, μ is a multiplier if and only if [11]

$$d(\mu\Omega_\gamma) \equiv d_\mu\Omega_\gamma + (m-1)d\Omega_\gamma = 0. \tag{2.6}$$

3 Nonholonomic Free Particle, Conformal Structure, and Jacobi Last Multiplier

Let us start with the discussion of Hamiltonian formulation of nonholonomic systems [6, 7]. Consider a mechanical system in 3D space with coordinates x, y, z. Let the coordinate z be cyclic. The motion takes place in the presence of a nonholonomic constraint which is given by

$$f = \dot{z} - y\dot{x} = 0. \tag{3.1}$$

We express the equation of motion in the form of Euler–Lagrange equations with undetermined multiplier λ

$$\frac{d}{dt}\left(\frac{\partial L}{\partial \dot{x}_i}\right) - \frac{\partial L}{\partial x_i} = \lambda\frac{\partial f}{\partial \dot{x}_i}, \qquad i = 1, 2, 3. \tag{3.2}$$

It is clear from the cyclic condition and definition of f that λ satisfies $\lambda = \frac{d}{dt}\left(\frac{\partial L}{\partial \dot{z}}\right)$.

We consider the motion of a free particle with unit mass subjected to a constraint (3.1) and the Lagrangian is $L = \frac{1}{2}(\dot{x}^2 + \dot{y}^2 + \dot{z}^2)$, although the results presented in this paper are quite general. We use (3.2) to obtain the equations of motion[1]

$$\dot{x} = p_x, \quad \dot{y} = p_y, \quad \dot{z} = p_z, \quad \dot{p}_x = -\lambda y, \quad \dot{p}_y = 0 \quad \dot{p}_z = \lambda. \tag{3.3}$$

Using the constraint equation $\dot{z} = y\dot{x}$ we can find

$$\lambda = p_x p_y - \lambda y^2, \qquad \text{or} \qquad \lambda = \frac{p_x p_y}{1 + y^2}$$

and this is equivalent to $\lambda = \left(\frac{\partial L}{\partial \dot{z}}\right)$. Hence eliminating the multiplier λ we obtain

$$\dot{x} = p_x, \quad \dot{y} = p_y, \quad \dot{p}_x = -y\frac{p_x p_y}{(1+y^2)}, \quad \dot{p}_y = 0. \tag{3.4}$$

---

[1]The physicist way of looking the constrained dynamics is different from our presentation, it is described by $L = \frac{1}{2}(\dot{x}^2 + \dot{y}^2 + \dot{z}^2) + \lambda(\dot{z} - y\dot{x})$, where momenta are given by $p_x = \dot{x} - \lambda y$, $p_y = \dot{y}$, $p_z = \dot{z} + \lambda$, $p_\lambda = 0$, The usual Dirac analysis of constraints then identifies the following two constraints, $\phi_1 = p_\lambda = 0$ $\phi_2 = p_z - yp_x - \lambda(1 + y^2)$ primary and secondary, respectively, which are second class, $\{\phi_1, \phi_2\} = (1 + y^2)$. It would be interesting to bridge the gap between these two methods.

3.1 Reduction, Constrained Hamiltonian and Nonholonomic Systems

Let $L_c(\mathbf{x}, \dot{\mathbf{x}})$ be the Lagrangian of the system after substituting the expression of $\dot{z}$ or $\dot{x}_3$. Thus we obtain a close system of equations for the variables $(\mathbf{x}, \dot{\mathbf{x}})$ and constraint $f = \dot{z} - y\dot{x} = 0$, given by

$$\frac{d}{dt}\left(\frac{\partial L_c}{\partial \dot{x}}\right) - \frac{\partial L_c}{\partial x} = \left(\frac{\partial L}{\partial \dot{z}}\right)^* \dot{y}, \qquad \frac{d}{dt}\left(\frac{\partial L_c}{\partial \dot{y}}\right) - \frac{\partial L_c}{\partial y} = -\left(\frac{\partial L}{\partial \dot{z}}\right)^* \dot{x}, \qquad (3.5)$$

where $\left(\frac{\partial L}{\partial \dot{z}}\right)^*$ means that the substitution $\dot{z}$ is made after the differentiation. This reduces to study the system with two degrees of freedom and preserves the energy integral

$$E = \frac{\partial L_c}{\partial \dot{x}}\dot{x} + \frac{\partial L_c}{\partial \dot{y}}\dot{y} - L_c.$$

Remark One can obtain the equations of motion (3.4) using the constrained Lagrangian. We now define the *constrained* Lagrangian by substituting the constraint equation $\dot{z} = y\dot{x}$ into Lagrangian:

$$L_c = \frac{1}{2}\left((1 + y^2)\dot{x}^2 + \dot{y}^2\right). \qquad (3.6)$$

The equations of motion can be obtained from the constrained Lagrangian $L_c(y, \dot{x}, \dot{y}) = L(\dot{x}, \dot{y}, y\dot{x})$ using chain rule. This is a special case of nonholonomic treatment given in Tony Bloch's book [7]. The general equations of motion for a nonholonomic system with the constraint equation $\dot{w} = -A_\alpha^a \dot{r}^\alpha$ in terms of constrained Lagrangian $L_c(r^\alpha, w^a, \dot{r}^\alpha) = L[(r^\alpha, w^a, \dot{r}^\alpha, -A_\alpha^a(r, w)\dot{r}^\alpha]$ are given as

$$\frac{d}{dt}\frac{\partial L_c}{\partial \dot{r}^\alpha} - \frac{\partial L_c}{\partial r^\alpha} + A_\alpha^a(r, w)\frac{\partial L_c}{\partial w^a} = -\frac{\partial L_c}{\partial \dot{w}^a}B_{\alpha\beta}^b r^\beta, \qquad (3.7)$$

where

$$B_{\alpha\beta}^b = \left(\frac{\partial A_\alpha^b}{\partial r^\beta} - \frac{\partial A_\beta^b}{\partial r^\alpha} + A_\alpha^a\frac{\partial A_\beta^b}{\partial w^a} - A_\beta^a\frac{\partial A_\alpha^b}{\partial w^a}\right).$$

Note that the system is holonomic if and only if the coefficients $B_{\alpha\beta}^b$ vanish.

The Lagrangian of the reduced system is $L_c = 1/2\left((1 + y^2)\dot{x}^2 + \dot{y}^2\right)$. Let S be the configuration space and $Le_{g_c} : TS \to T^*S$ be the Legendre transformation of the reduced system. Using Legendre transformation

$$m_i = \frac{\partial \tilde{L}}{\partial \dot{x}}, \qquad H = \sum_{i=1}^{2} m_i\dot{x}_i - L_c, \qquad i = 1, 2 \qquad (3.8)$$

we obtain the following system of equations

$$\dot{x}_i = \frac{\partial H}{\partial m_i}, \qquad \dot{m}_1 = -\frac{\partial H}{\partial x_1} + \frac{\partial H}{\partial m_2}S, \qquad \dot{m}_2 = -\frac{\partial H}{\partial x_2} - \frac{\partial H}{\partial m_2}S, \quad (3.9)$$

where $S = (\frac{\partial L}{\partial \dot{z}})^*$ and $i = (1, 2)$. Then the momenta corresponding to the reduced equations are given by

$$m_x = \frac{\partial L_c}{\partial \dot{x}} = (1 + y^2)\dot{x}, \qquad m_y = \frac{\partial L_c}{\partial \dot{y}}$$

and the corresponding Hamiltonian of the reduced system is given by

$$H_c = \frac{1}{2}\left(\frac{m_x^2}{1 + y^2} + m_y^2\right). \tag{3.10}$$

It is easy to find Hamiltonian equations from (3.9) as $\dot{x} = \frac{\partial H}{\partial m_x}$, $\dot{y} = \frac{\partial H}{\partial m_y}$, $\dot{m}_x = -\frac{\partial H}{\partial x} + \frac{\partial H}{\partial m_y}S = -0 + ym_y m_x/(1 + y^2)$, $\dot{m}_y = -\frac{\partial H}{\partial y} - \frac{\partial H}{\partial m_x}S = ym_x^2/(1 + y^2)^2 - ym_x^2/(1 + y^2)^2 = 0$. Here we tacitly use $S = \dot{z} = ym_x/(1 + y^2)$.

The new set of equations is given by

$$\dot{x} = \frac{m_x}{1 + y^2}, \quad \dot{y} = m_y, \quad \dot{m}_x = \frac{ym_x m_y}{1 + y^2}, \quad \dot{m}_y = 0. \tag{3.11}$$

The vector field

$$\Gamma = \frac{m_x}{1 + y^2}\partial_x + m_y\partial_y + \frac{ym_x m_y}{1 + y^2}\partial_{m_x} \tag{3.12}$$

satisfies

$$i_\Gamma \omega_{nh} = -dH_c,$$

where the two form is given by

$$\omega_{nh} = dm_x \wedge dx + dm_y \wedge dy - \frac{m_x y}{1 + y^2}dy \wedge dx. \tag{3.13}$$

Here ω_{nh} is the nondegenerate two form on phase space P, however it is not closed, i.e.,

$$d\omega_{nh} = \frac{ydy}{1 + y^2} \wedge dm_x \wedge dx = d\left(\frac{1}{2}\ln(1 + y^2)\right) \wedge \omega_{nh}. \tag{3.14}$$

The corresponding Poisson structure is given by

$$\{x, m_x\} = 1, \qquad \{y, m_y\} = 1, \qquad \{m_x, m_y\} = \frac{m_x y}{1 + y^2}, \qquad (3.15)$$

which does not satisfy Jacobi identity, it is known as almost Poisson structure. The (nonholonomic) Poisson bracket between two functions $f_i = f_i(x, y, m_x, m_y)$, $(i = 1, 2)$ is

$$\{f_1, f_2\}_{nh} = \frac{\partial f_1}{\partial x}\frac{\partial f_2}{\partial m_x} - \frac{\partial f_1}{\partial m_x}\frac{\partial f_2}{\partial x} + \frac{\partial f_1}{\partial y}\frac{\partial f_2}{\partial m_y} - \frac{\partial f_1}{\partial m_y}\frac{\partial f_2}{\partial y}$$
$$+ \frac{m_x y}{1 + y^2}\left(\frac{\partial f_1}{\partial m_x}\frac{\partial f_2}{\partial m_y} - \frac{\partial f_1}{\partial m_y}\frac{\partial f_2}{\partial m_x}\right).$$

The equations of motion may be given in terms of nonholonomic Poisson bracket

$$\dot{f} = \{f, H\}_{nh}, \qquad \forall f : M \to \mathbb{R}. \qquad (3.16)$$

A function $f : M \to \mathbb{R}$ is an integral of motion of the nonholonomic system if and only if it satisfies $\{f, H\}_{nh} = 0$.

Using these almost Poisson structures we can still do Hamiltonian dynamics as long as we are willing to give up the existence of canonical coordinates and the Jacobi identities for the Poisson brackets. We will subsequently see that the Jacobi last multiplier plays a crucial role to obtain the canonical coordinates and Poisson structures.

3.2 Hamiltonization and Reduction Using Jacobi Multiplier

Let us compute the JLM of the set of Eq. (3.4) from

$$\frac{d}{dt}\log \mu + \left(-\frac{y\dot{y}}{1 + y^2}\right) = 0,$$

thus we obtain

$$\mu = (1 + y^2)^{1/2}. \qquad (3.17)$$

It is worthwhile to note that if we compute the "JLM" of the set of Eq. (3.11) from $\frac{d}{dt}\log \mu + \left(\frac{y\dot{y}}{1+y^2}\right) = 0$, we obtain the inverse multiplier $\mu^{-1} = (1 + y^2)^{-1/2}$. It is obvious because we compute it on the *dual space*.

Using the Jacobi last multiplier (JLM) one can show that system (3.10) has an invariant measure that can be represented in the form $\mu(y)d\mathbf{x}d\mathbf{m}$. JLM is a smooth and positive function on the entire phase space, so it acts like a density of the invariant measure and satisfies the Liouville equation

$$\text{div}\,(\mu\Gamma) = 0,$$

where Γ stands for the vector field determined by system (3.10).

Proposition 3.1 *The function* $K = m_x/\sqrt{1+y^2} = \mu^{-1}m_x$ *is the integral of motion of the nonholonomic system, thus nonholonomic Poisson bracket with* H *vanishes,* $\{H, K\}_{nh} = 0$.

It follows directly from the set of Eq. (3.11).

3.3 Conformally Hamiltonian Formulation of Nonholonomic Systems

Let M be a symplectic manifold with symplectic form ω, when it is exact we write $\omega = d\theta$. For a function $H \in C^\infty(M)$ we denote its Hamiltonian vector field by X_H.

Definition 3.2 The diffeomorphism ϕ^a is conformal if $(\phi^a)^*\omega = \omega$ and corresponding to this flow the vector field Γ^a is said to be conformal with parameter $a \in \mathbb{R}$ if $L_{\Gamma^a}\omega = a\omega$.

It is clear

$$\frac{d}{dt}\phi_t^*\omega = \phi_t^* L_{\Gamma^a}\omega = a\phi_t^*\omega$$

which has a unique solution $\phi_t^*\omega = e^{at}\omega$.

The next proposition was given by McLachlan and Perlmutter [23].

Proposition 3.3 *Let* M *be a symplectic manifold with symplectic form* ω. *It admits a conformal vector field* $a \neq 0$ *if and only if* $\omega = -d\theta$.

(a) *Given a Hamiltonian* $H \in C^\infty(M)$, *the conformal Hamiltonian vector field* X_H^a *satisfies*

$$i_{X_H^a}\omega = dH - a\theta. \tag{3.18}$$

(b) *If* $H^1(M) = 0$, *then the set of conformal vector fields on* M *is given by* $\{X_H + cZ\} : H \in C^\infty(M)\}$, *where* Z *is defined by* $i_Z\omega = -\theta$ *and it is known as the Liouville vector field.*

If $H_1(M) = 0$, we know that every conformal vector field can be written as $X_H + cZ$ for some Hamiltonian and a unique $c \in \mathbb{R}$.

Let $\omega = dm_x \wedge dx + dm_y \wedge dy$ be the symplectic form. Then by contraction with respect to the Hamiltonian vector field we obtain

$$i_{X_H}\omega = -dH + \lambda\Big(\frac{\partial H}{\partial m_2}dx_1 - \frac{\partial H}{\partial m_1}dx_2\Big) \equiv -dH + \lambda\theta.$$

The vector field Z is tangent to the fibers is given by

$$Z = \frac{\partial H}{\partial m_2}\frac{\partial}{\partial m_1} - \frac{\partial H}{\partial m_1}\frac{\partial}{\partial m_2}, \qquad i_Z\omega = \theta.$$

Given the Hamiltonian $H_c \in C^\infty(M)$, the Hamiltonian vector field X_{H_c} corresponding to Hamiltonian H_c satisfies

$$i_{X_{H_c}}\omega^{nh} = dH_c - \theta, \quad \theta = \frac{m_x y}{1 + y^2}dx.$$

This yields a conformal vector field. Let $\omega = dm_x \wedge dx + dm_y \wedge dy$ be the symplectic form when the manifold equipped with coordinates (x, y, m_x, m_y). The conformal vector field is given by $X_{H_c} + Z$, where Z is defined by

$$i_Z\omega = -\theta, \qquad \text{where} \qquad Z = \frac{m_x y}{1 + y^2}\frac{\partial}{\partial m_x}. \tag{3.19}$$

4 Integrability of Nonholonomic Dynamics and Locally Conformally Symplectic Structure

In this section we unveil the connection between the Jacobi last multiplier, l.c.s. structure and integrability properties of nonholonomic dynamics.

Proposition 4.1 *The nonholonomic two form ω_{nh} and $\tilde{\omega}_{nh}$ satisfy locally conformal symplectic structure and the Lee form is $\eta = d\big(\log(1 + y^2)\big)^{1/2}\big) = d(\log\mu)$, where μ is the Jacobi's last multiplier.*

Proof It is straightforward to check

$$d\omega_{nh} = -\Big(\frac{ydy}{1 + y^2}\Big) \wedge dm_x \wedge dx$$

$$= d\Big(\log\frac{1}{2}(1 + y^2)\Big) \wedge \Big(dm_x \wedge dx + dm_y \wedge dy - \frac{m_x y}{1 + y^2}dy \wedge dx\Big) = \eta \wedge \omega_{nh}$$

and similarly for the other case.   □

The inverse multiplier plays an important role for changing locally conformal symplectic form ω_{nh} to symplectic form. In this process we find new momemta which satisfy canonical Poisson structure.

Proposition 4.2 *Let μ^{-1} be the inverse multiplier, then $\omega = \mu^{-1}\omega_{nh}$ is a symplectic form, given by*

$$\tilde{\omega} = d\tilde{m}_x \wedge dx + d\tilde{m}_y \wedge dy, \tag{4.1}$$

where the new momenta are

$$\tilde{m}_x = \mu^{-1}m_x = \frac{m_x}{\sqrt{1+y^2}} \qquad \tilde{m}_y = \frac{m_x}{\sqrt{1+y^2}}. \tag{4.2}$$

Proof By direct computation one obtains

$$\mu^{-1}\omega_{nh} = \frac{1}{\sqrt{1+y^2}}\left(dm_x \wedge dx + dm_y \wedge dy - \frac{m_x y}{1+y^2}dy \wedge dx\right)$$

$$= \frac{dm_x}{\sqrt{1+y^2}} \wedge dx + \frac{dm_y}{\sqrt{1+y^2}} \wedge dy - \frac{m_x y}{(1+y^2)^3/2}dy \wedge dx \equiv d\tilde{m}_x \wedge dx + d\tilde{m}_y \wedge dy.$$

$\square$

It is clear $d\tilde{\omega} = 0$ and the new momenta satisfy the canonical Poisson structure

$$\{x, \tilde{m}_x\} = 1, \qquad \{y, \tilde{m}_y\} = 1. \tag{4.3}$$

4.1 Role of Jacobi's Multiplier and Integrability of Nonholonomic Dynamical Systems

We now address the question of integrability of the nonholonomic systems as posed by Bates and Cushman [4, 12]. In their papers, they explored to what extent nonholonomic systems behave like an integrable system. The fundamental Liouville theorem states that it suffices to have n $\{f_1 = H, f_2, \cdots, f_n\}$ independent Poisson commuting functions to explicitly (i.e., by quadratures) integrate the equations of motion for generic initial conditions. Let $M_c = \{f_1 = c_1, \cdots, f_n = c_n\}$ be a common invariant level set, which is regular (i.e., $df_1, \cdots df_n$ are independent), compact and connected, then it is diffeomorphic to n-dimensional tori $\mathbb{T}^n = \mathbb{R}^n/\Lambda$, where Λ is a lattice in $\mathbb{R}^n$. These tori are known as the Liouville tori [1, 12]. In the neighborhood of M_c there exist canonical variables I, ϕ mod 2π, called action-angle variables which satisfy $\{\phi_i, I_j\} = \delta_{ij}$, $\{\phi_i, \phi_j\} = \{I_i, I_j\} = 0$, $i, j = 1, \cdots n$, such that the level sets of the actions $I, \cdots, I_n$ are invariant tori and $H = H(I_1, \ldots, I_n)$.

The vector fields $X_{f_1}, \cdots, X_{f_n}$ corresponding to the n integrals of motion $f_1, \cdots f_n$ are independent (it follows from the independency of differentials) and span the tangent spaces of $T_q M_c$ for all $q \in M_c$, since M_c is compact, hence X_{f_i}s are complete. The Poisson commutativity implies the commutativity of vector fields. In other words, the so-called invariant manifolds, which are the (generic) submanifolds traced out by the n commuting vector fields X_{f_i} are Liouville tori, the flow of each of the vector fields X_{f_i} is linear, so that the solutions of Hamilton's equations are quasi-periodic. A proof in the case of a Liouville integrable system on a symplectic manifold was given by Arnold [1].

We will soon figure out that the (reduced) nonholonomic problem which we are considered in this paper has two constants of motion H (Hamiltonian) and K, these are Poisson commuting. However, because the nonholonomic system does not satisfy the Jacobi identity, the associated vector fields X_H and X_K do not commute, i.e. $[X_H, X_K] \neq 0$, on the torus. So Bates and Cushman [4] asked if such system is integrable in some sense or how can it be converted to integrable systems.

4.2 JLM and Commuting of Vector Fields

It has been observed the reduced Hamiltonian equation of motion lies on the invariant manifold given by

$$K = \frac{m_x}{\sqrt{1 + y^2}}, \tag{4.4}$$

where K satisfies $\frac{dK}{dt} = 0$. The Hamiltonian vector field

$$X_K = \frac{1}{\sqrt{1 + y^2}} \frac{\partial}{\partial x} \tag{4.5}$$

satisfies $X_k \lrcorner \omega_{nh} = -dK$.

The Hamiltonian vector field X_H satisfies

$$L_{X_H} K = X_H(K) = 0, \tag{4.6}$$

which implies

$$\omega_{nh}(X_H, X_K) = X_K \lrcorner X_H \lrcorner \omega_{nh} = X_K \lrcorner \left(\frac{m_x}{1 + y^2} dm_x + m_y dm_y - \frac{m x^2 y}{(1 + y^2)^2} dy \right) = 0.$$

Next observe that the Lie bracket between vector fields X_H and X_K

$$[X_H, X_k] = -\frac{y m_x}{1 + y^2} X_K. \tag{4.7}$$

This has been demonstrated by Bates and Cushman the vector fields X_H and X_K do not commute on the torus, because the two form ω_{nh} is not closed. They try to seek an integrating factor g such that $[gX_K, X_H] = 0$. The next proposition addresses the value of g.

Proposition 4.3 *Let μ be the Jacobi last multiplier, then the modified vector field $\mu^{-1}X_K$ commutes with the Hamiltonian vector field X_H, i.e.,*

$$[\mu^{-1}X_K, X_h] = 0. \tag{4.8}$$

Proof We know that the JLM $\mu = \sqrt{1 + y^2}$, so that $\mu^{-1}X_K = \partial_x$. Hence we obtain $[\mu^{-1}X_K, X_h] = 0$. $\qquad\square$

5 Final Comments and Outlook

Our formalism can be easily extended to nonholonomic oscillator. In this case, Lagrangian is given by $L = \frac{1}{2}(\dot{x}^2 + \dot{y}^2 + \dot{z}^2) + \frac{1}{2}y^2$, subject to the nonholonomic constraint $\dot{z} = y\dot{x}$. The reduced system of equations are given by

$$\dot{x} = p_x, \quad \dot{y} = p_y, \quad \dot{p}_x = -\frac{y}{1 + y^2}p_x p_y, \quad \dot{p}_y = -y.$$

One can easily check that the last multiplier is $\mu = (1 + y^2)^{1/2}$. The two form associated to the reduced nonholonomic oscillator equation

$$\omega_{as} = (1 + y^2)dp_x \wedge dx + dp_y \wedge dy + yp_x dy \wedge dx$$

satisfies locally conformal symplectic structure, $d\omega_{as} + \eta \wedge \omega_{as} = 0$, where the Lee form $\eta = d\left(\log(1 + y^2)^{1/2}\right)$. Hence the inverse Jacobi's last multiplier transforms ω_{as} into a symplectic form

$$\mu^{-1}\omega_{as} \equiv \tilde{\omega} = d\tilde{p}_x \wedge dx + d\tilde{p}_y \wedge dy,$$

where the modified momenta are given by $\tilde{p}_x = \sqrt{1 + y^2}p_x$ and $p_y = \frac{p_y}{\sqrt{1+y^2}}$. Thus everything can be repeated here.

The application of the Jacobi Last Multiplier (JLM) for finding Lagrangians of any second-order differential equation has been extensively studied. It is known that the ratio of any two multipliers is a first integral of the system, in fact, it plays a role similar to the integrating factor for system of first-order differential equations. But so far, it has not been applied to nonholonomic systems. In this paper we have studied nonholonomic system endowed with a two form, which is closely related to locally conformal symplectic structure. We have applied JLM to map it to symplectic frame work. Also, we have shown how a toral fibration defined by the common level sets of integrable nonholonomic system, studied by Bates and Cushman, can be mapped to toral fibration defined of the integrals of a Liouville integrable Hamiltonian system.

There are some open problems popped up from this article. Firstly, it would be nice to study the time-dependent nonholonomic systems using JLM. Secondly, we have considered examples from the integrable domain, hence it would be great to apply JLM in nonintegrable domain.

Acknowledgements Most of the results herewith presented have been obtained in long-standing collaboration with Anindya Ghose Choudhury and Pepin Cariñena, which is gratefully acknowledged. The author wishes also to thank Gerardo Torres del Castillo, Clara Nucci, Peter Leach, Debasish Chatterji, Ravi Banavar, and Manuel de Leon for many enlightening discussions. This work was done mostly while the author was visiting IHES. He would like to express his gratitude to the members of IHES for their warm hospitality. The final part was done at IFSC, USP at Sao Carlos and the support of FAPESP is gratefully acknowledged with grant number 2016/06560-6.

References

1. V.I. Arnold, *Mathematical Methods of Classical Mechanics*. Graduate Texts in Mathematics, vol. 60 (Springer, New York, 1978). Translated from the Russian by K. Vogtmann and A. Weinstein,
2. L. Bates, Examples of singular nonholonomic reduction. Rep. Math. Phys. **42**(1–2), 231–247 (1998)
3. L. Bates, Problems and progress in nonholonomic reduction, in *XXXIII Symposium on Mathematical Physics*, Torun (2001). Rep. Math. Phys. **49**(2–3), 143–149 (2002)
4. L. Bates, R. Cushman, What is a completely integrable nonholonomic dynamical system, in *Proceedings of the XXX Symposium on Mathematical Physics*, Toruń (1998). Rep. Math. Phys. **44**(1–2), 29–35 (1999)
5. L. Bates, J. Sniatycki, Nonholonomic reduction. Rep. Math. Phys. **32**, 99–115 (1993)
6. I.A. Bizyaev, A.V. Borisov, I.S. Mamaev, Hamiltonization of elementary nonholonomic systems. Russ. J. Math. Phys. **22**(4), 444–453
7. A.M. Bloch, *Nonholonomic Mechanics and Control* (Springer, New York, 2003)
8. D. Chinea, M. de León, J.C. Marrero, Locally conformal cosymplectic manifolds and time-dependent Hamiltonian systems. Comment. Math. Univ. Carolin. **32**(2), 383–387 (1991)
9. D. Chinea, M. Domingo, M. de León, J.C. Marrero, Symplectic and cosymplectic foliations on cosymplectic manifolds. Publ. Inst. Math. (Beograd) (N.S.) **50**(64), 163–169 (1991)
10. A.G. Choudhury, P. Guha, B. Khanra, On the Jacobi last multiplier, integrating factors and the Lagrangian formulation of differential equations of the Painlevé-Gambier classification. J. Math. Anal. Appl. **360**(2), 651–664 (2009)
11. M. Crasmareanu, Last multipliers for multivectors with applications to Poisson geometry. Taiwan. J. Math. **13**(5), 1623–1636 (2009)
12. R. Cushman, L. Bates, *Global Aspects of Classical Integrable Systems* (Birkhauser, Basel, 1997)
13. J. Douglas, Solution to the inverse problem of the calculus of variations. Trans. Am. Math. Soc. **50**, 71–128 (1941)
14. H. Flaschka, A remark on integrable Hamiltonian systems. Phys. Lett. A **131**(9), 505–508 (1988)
15. A. Goriely, *Integrability and Nonintegrability of Dynamical Systems*. Advanced Series in Nonlinear Dynamics, vol. 19 (World Scientific, River Edge, 2001), xviii+415 pp.
16. P. Guha, A. Ghose Choudhury, Hamiltonization of higher-order nonlinear ordinary differential equations and the Jacobi last multiplier. Acta Appl. Math. **116**, 179–197 (2011)
17. S. Haller, T. Rybicki, On the group of diffeomorphisms preserving a locally conformal symplectic structure. Ann. Global Anal. Geom. **17**, 475–502 (1999)

18. S. Haller, T. Rybicki, Integrability of the Poisson algebra on a locally conformal symplectic manifold, in *The Proceedings of the 19th Winter School "Geometry and Physics"*, Srn (1999). Rend. Circ. Mat. Palermo (2) 63, 89–96 (2000)
19. C.G.J. Jacobi, Sul principio dell'ultimo moltiplicatore, e suo uso come nuovo principio generale di meccanica. Giornale Arcadico di Scienze, Lettere ed Arti **99**, 129–146 (1844)
20. C.G.J. Jacobi, Theoria novi multiplicatoris systemati aequationum differentialium vulgarium applicandi. J. Reine Angew. Math. **27**, 199–268 (1844); Ibid **29**, 213–279 and 333–376 (1845); Astrophys. J. **342**, 635–638, (1989)
21. H.-C. Lee, A kind of even-dimensional differential geometry and its applications to exterior calculus. Am. J. Math. **65**, 433–438 (1943)
22. J. Lopuszanski, *The Inverse Variational Problem in Classical Mechanics* (World Scientific, Singapore, 1999)
23. R.I. McLachlan, M. Perlmutter, Conformal Hamiltonian systems. J. Geom. Phys. **39**, 276–300 (2001)
24. M.C. Nucci, Jacobi last multiplier and Lie symmetries: a novel application of an old relationship. J. Nonlinear Math. Phys. **12**, 284–304 (2005)
25. M.C. Nucci, P.G.L. Leach, Jacobi's last multiplier and Lagrangians for multidimensional systems. J. Math. Phys. **49**, 073517 (2008)
26. M.C. Nucci, K.M. Tamizhmani, Lagrangians for biological systems (2011). arXiv:1108.2301v1
27. G.E. Prince, The inverse problem in the calculus of variation and ramifications, in *Geometric Approaches to Differential Equations*, ed. by P.J. Vassiliou, I.G. Lisle. Australian Mathematical Society Lecture Series, vol. 15 (Cambridge University Press, Cambridge, 2000)
28. W. Sarlet, The Helmholtz conditions revisited. A new approach to the inverse problems of Lagrangian dynamics. J. Phys A Math. Gen. **15**, 1503–1517 (1982)
29. I. Vaisman, Locally conformal symplectic manifolds. Int. J. Math. Math. Sci. **8**(3), 521–536 (1985)
30. H. Van Le, J. Vanzura, Cohomology theories on locally conformally symplectic manifolds (2011). math.SG 1111.3841v3
31. E.T. Whittaker, *A Treatise on the Analytical Dynamics of Particles and Rigid Bodies* (Cambridge University Press, Cambridge, 1988)
32. M.P. Wojtkowski, C. Liverani, Conformally symplectic dynamics and symmetry of the Lyapunov spectrum. Commun. Math. Phys. **194**, 47–60 (1998)

Non-perturbative Renormalization Group of a $U(1)$ Tensor Model

Vincent Lahoche and Dine Ousmane Samary

*To **Professor Mahouton Norbert Hounkonnou***

Abstract This paper aims at giving some comment on our new development on the functional renormalization group applied to the $U(1)$ tensor model previously studied in [Phys. Rev. D 95, 045013 (2017)]. Using the Wetterich non-perturbative equation, the flow of the couplings and mass parameter are discussed and the physical implication such as the asymptotically safety of the model is provided.

Keywords Random tensor models · Functional renormalization group · Flow equation

1 Introduction

Random Tensor Models [1–3] extends Matrix Models as promising candidates to understand Quantum Gravity in higher dimension, $D \geqslant 3$. Especially colored tensor models allow one to define probability measures on simplicial pseudo-manifolds, and they are considered as a convenient formalism for studying random geometries [3–8]. On the other hand, group field theory aims at describing a rudimentary phase of the geometry of spacetime, namely when this geometry is hypothetically still in a discrete form, or at least not yet continuous (the geometrogenesis scenario)

V. Lahoche
LaBRI, Univ. Bordeaux 351 cours de la Libération, Talence, France
e-mail: vincent.lahoche@labri.fr

D. Ousmane Samary (✉)
Max Planck Institute for Gravitational Physics, Albert Einstein Institute, Potsdam, Germany

Faculté des Sciences et Techniques/ICMPA-UNESCO Chair, Université d'Abomey-Calavi, Abomey-Calavi, Benin
e-mail: dine.ousmanesamary@aei.mpg.de

© Springer Nature Switzerland AG 2018
T. Diagana, B. Toni (eds.), *Mathematical Structures and Applications*,
STEAM-H: Science, Technology, Engineering, Agriculture,
Mathematics & Health, https://doi.org/10.1007/978-3-319-97175-9_13

[9–12]. It is also named "pre-geometric" phase of our spacetime. Recently Tensor models and group field theory have been combined to provide a new class of field theories so-called tensorial group field theory (TGFT). TGFTs improve the group field theories in order to allow for renormalization [13–17]. Moreover, it has been shown that several TGFT models are asymptotically free in the UV, in other words, near the Gaussian fixed point [17–26].

Much interest was focused on the FRG equation of various Matrix and TGFT models [20–29]. The flow equations also called the Wetterich equations were derived [30]. The fixed points were given and further evidence of asymptotically safety and asymptotically freedom were derived around these fixed points in the UV. The TGFT of the form T_5^6 on the $U(1)$ group with closure constraint is proved to be renormalizable [15]. The proof of this claim is performed using multi-scale analysis. The closure constraint also called gauge invariance condition can help to define the emergence of the metric on spacetime after phase transition and therefore makes this type of model relevant for the understanding the quantum theory of gravitation. This kind of model with closure constraint, namely the six-dimensional TGFT with quartic interactions is studied recently in [20] and [22]. The perturbative computation of the β-functions of the T_5^6 model is given in [19], in which we have showed that this model is asymptotically free in the UV. This result seems to be nonconsistent from the point of view of the FRG analysis. This work aims at giving our new contributions on the $U(1)$ TGFT such as the renormalization theorems and the functional renormalization group analysis to show the asymptotically safety of these type of models.

Our paper is organized as follows. Section 2 is devoted to present the model which is analyzed in this work, namely the T_5^6 model with closure constraints. In Sect. 3 the flow equations of the coupling constants and mass parameter are derived by using the dimensional renormalization parameters. In Sect. 4 we give the nontrivial fixed points and provide the numerical solution of the flow equations. The behavior of the model we studied in the vicinity of these fixed points is also given. The conclusion is made in Sect. 5.

2 $U(1)^d$ Tensorial Group Field Theory

In this section we give some basic notations and definitions of the TGFT that have been used in this note. Particularly we will use a power counting theorem to discuss the notion of canonical dimensions for each coupling. The canonical dimension allows us to make sense of the exponentiation of the action in the partition function.

We start by defining an action $S[\bar{\varphi}, \varphi]$ of TGFT that depends on the field φ and its conjugate $\bar{\varphi}$ acting on the compact Lie group G, i.e.,

$$\varphi : G^d \longrightarrow \mathbb{C}; \quad (g_1, \cdots, g_d) \longmapsto \varphi(g_1, \cdots, g_d). \tag{2.1}$$

We will always consider $G = U(1)$, but the analysis performed here can be extended to over groups. We are using the Fourier transformation of the field and are defining the momentum variable associated with the group elements $\vec{g} = (g_1, g_2, \cdots, g_d) \in U(1)^d$ as $\vec{p} = (p_1, p_2, \cdots, p_d) \in \mathbb{Z}^d$. Using the parametrization $g_k = e^{i\theta_k}$ we write

$$\varphi(g_1, \cdots, g_d) = \sum_{p_i \in \mathbb{Z}} \varphi(p_1, \cdots, p_d) e^{i \sum_k \theta_k p_k}, \qquad \theta_i \in [0, 2\pi), \qquad (2.2)$$

where we denote the Fourier transform of the field φ by $\varphi_{12\cdots d} =: \varphi(p_1, \cdots, p_d) =: \varphi_{\vec{p}}$ for simplicity. The functional action $S[\bar{\varphi}, \varphi]$ is written in general case as

$$S[\bar{\varphi}, \varphi] = \sum_{p_i} \bar{\varphi}_{\vec{p}} \, C^{-1}(\vec{p}, \vec{p}\,') \, \varphi_{\vec{p}} \prod_{i=1}^{D} \delta_{p_i p_i'} + S_{\text{int}}[\bar{\varphi}, \varphi] \qquad (2.3)$$

where C stands for the propagator and S_{int} collects all vertex contributions of the interaction.

Let $d\mu_C$ be the field measure associated with the covariance C, we have the relation

$$C(\vec{p}, \vec{p}\,') = \int d\mu_C \, \varphi_{\vec{p}} \bar{\varphi}_{\vec{p}\,'}, \qquad d\mu_C = \prod_{\vec{p}} d\bar{\varphi}_{\vec{p}} \, d\varphi_{\vec{p}} \, e^{-\bar{\varphi}_{\vec{p}} \, C^{-1}(\vec{p}, \vec{p}\,') \varphi_{\vec{p}}}. \qquad (2.4)$$

We introduce a cut-off Λ on $C_{\vec{p}, \vec{p}\,'}$, impose that to satisfy $|\vec{p}| \leqslant \Lambda$. The generating function or the vacuum–vacuum transition amplitude is

$$Z_\Lambda[J, \bar{J}] = e^{W_\Lambda[J, \bar{J}]} = \int d\mu_{C_\Lambda}(\bar{\varphi}, \varphi) e^{S_{\text{int}}[\bar{\varphi}, \varphi] + \langle \bar{J}, \varphi \rangle + \langle \bar{\varphi}, J \rangle} \qquad (2.5)$$

where the notation $\langle ., . \rangle$ means: $\langle \bar{J}, \varphi \rangle = \sum_{\vec{p} \in \mathbb{Z}^d} \bar{J}_{\vec{p}} \varphi_{\vec{p}}$, $d\mu_{C_\Lambda}$ is the Gaussian measure with the covariance C_Λ such that:

$$\int d\mu_{C_\Lambda} \varphi_{\vec{p}} \, \bar{\varphi}_{\vec{p}\,'} = \frac{e^{-(\vec{p}^2 + m^2)/\Lambda^2}}{\vec{p}^2 + m^2} \delta \left(\sum_{i=1}^{d} p_i \right) \delta_{\vec{p} \vec{p}\,'} = C_\Lambda(\vec{p}, \vec{p}\,') \qquad (2.6)$$

and the delta $\delta(\sum_{i=1}^{d} p_i)$ implements the closure constraint, see [13]. We keep in mind that we must send the cut-off to infinity in any circumstances. We define a model by its action at a high (UV) energy scale. The classical action S_{int} is defined as a sum of tensorial invariants [3]:

$$S_{int}[\bar{\varphi}, \varphi] = \sum_{b \in \mathcal{B}} \lambda_b \text{Tr}_b[\bar{\varphi}, \varphi]. \qquad (2.7)$$

A tensor invariant is a polynomial in the tensor φ and its conjugate $\bar\varphi$ which is invariant under the action of the tensor product of d independent copies of the unitary group $U(N)$. The sum is taken over a finite set $\mathcal{B}$ of such invariants d-bubbles [3] associated with the couplings λ_b. The interaction (2.7) of a tensor field theory in dimension $d = 5$ [15] is

$$S_{int}[\bar\varphi,\varphi] = \frac{\lambda_1}{2}\sum_{\ell=1}^{5}\sum_{\vec{p}_i}\mathcal{W}^{(\ell)}_{\vec{p}_1,\vec{p}_2,\vec{p}_3,\vec{p}_4}\,\varphi_{\vec{p}_1}\bar\varphi_{\vec{p}_2}\varphi_{\vec{p}_3}\bar\varphi_{\vec{p}_4}$$

$$+\frac{\lambda_2}{3}\sum_{\ell=1}^{5}\sum_{\vec{p}_i}\mathcal{X}^{(\ell)}_{\vec{p}_1,\vec{p}_2,\vec{p}_3,\vec{p}_4,\vec{p}_5,\vec{p}_6}\,\varphi_{\vec{p}_1}\bar\varphi_{\vec{p}_2}\varphi_{\vec{p}_3}\bar\varphi_{\vec{p}_4}\varphi_{\vec{p}_5}\bar\varphi_{\vec{p}_6}$$

$$+\lambda_3\sum_{\ell_i=1,i=1,2,3}^{5}\sum_{\vec{p}_i}\mathcal{Y}^{(\ell_1,\ell_2,\ell_3)}_{\vec{p}_1,\vec{p}_2,\vec{p}_3,\vec{p}_4,\vec{p}_5,\vec{p}_6}\,\varphi_{\vec{p}_1}\bar\varphi_{\vec{p}_2}\varphi_{\vec{p}_3}\bar\varphi_{\vec{p}_4}\varphi_{\vec{p}_5}\bar\varphi_{\vec{p}_6},\quad (2.8)$$

where the symbols $\mathcal{W}^{(\ell)}$, $\mathcal{X}^{(\ell)}$, and $\mathcal{Y}^{(\ell)}$ are products of delta functions associated with tensor invariant interactions, and $\lambda_i(\Lambda)$ are coupling constants. For instance:

$$\mathcal{W}^{(\ell)}_{\vec{p}_1,\vec{p}_2,\vec{p}_3,\vec{p}_4} = \delta_{p_{1\ell}p_{4\ell}}\delta_{p_{2\ell}p_{3\ell}}\prod_{j\neq\ell}\delta_{p_{1j}p_{2j}}\delta_{p_{3j}p_{4j}}. \quad (2.9)$$

Such a kernel is called *bubble* [3], and can be pictured graphically as a 6-colored bipartite regular graph, with black and white vertices corresponding, respectively, to the fields φ and $\bar\varphi$, and each line corresponding to a Kronecker delta.

$$\mathcal{W}^{(\ell)}(\mathbf{g}_1,\mathbf{g}_2,\mathbf{g}_3,\mathbf{g}_4) \quad = \quad$$

$$\mathcal{X}^{(\ell)}(\mathbf{g}_1,\mathbf{g}_2,\mathbf{g}_3,\mathbf{g}_4,\mathbf{g}_5,\mathbf{g}_6) \quad = \quad$$

(2.10)

$$\mathcal{Y}^{(\ell_1,\ell_2,\ell_3)}(\mathbf{g}_1,\mathbf{g}_2,\mathbf{g}_3,\mathbf{g}_4,\mathbf{g}_5,\mathbf{g}_6) \quad = \quad$$

$$\mathcal{X}^{(\ell)}(\mathbf{g}_1, \mathbf{g}_2, \mathbf{g}_3, \mathbf{g}_4, \mathbf{g}_5, \mathbf{g}_6) = \quad \begin{array}{c} S \searrow \quad \nearrow \bar{S} \\ S_1 \\ \downarrow \\ S_0 \end{array} \tag{2.11}$$

$$\mathcal{Y}^{(\ell_1,\ell_2,\ell_3)}(\mathbf{g}_1, \mathbf{g}_2, \mathbf{g}_3, \mathbf{g}_4, \mathbf{g}_5, \mathbf{g}_6) = \quad \begin{array}{c} \mathcal{L}_0 \\ \downarrow \\ \mathcal{L}_1 \\ \swarrow \quad \searrow \\ \mathcal{L} \quad\quad \bar{\mathcal{L}} \end{array} \tag{2.12}$$

Let $\mathcal{G}$ be a 2- or 4 or 6-point Feynman graph of model (2.8). Let be denoted by $\omega(\mathcal{G})$ the degree of the tensor graph $\mathcal{G}$, i.e.:

$$\omega(\mathcal{G}) = \sum_{J \text{jacket of } \mathcal{G}} g_J \tag{2.13}$$

where g_J is the genus of the jacket J. A simple way of computing the degree ω of a graph is to count its number F of faces. Let us pick an arbitrary orientation for all of the edges e and for all of the faces f. Then R is the rank of the incidence matrix ϵ_{fe}:

$$\epsilon_{fe} = \begin{cases} 1 & \text{if } e \in f \text{ and their orientation match} \\ -1 & \text{if } e \in f \text{ and their orientation do not match} \\ 0 & \text{otherwise.} \end{cases} \tag{2.14}$$

One can show that the rank R does not depend on the chosen orientation. We get the following results (see [15] for more detail):

Proposition 2.1 (Divergence Degree) *The degree of divergence ω_d of a $\varphi_d^n U(1)^d$ model is given by*

$$\omega_d(\mathcal{G}) = (-2L + F - R)(\mathcal{G}) \tag{2.15}$$

$$= -\frac{2}{(d-1)!}\big(\omega(\mathcal{G}) - \omega(\partial\mathcal{G})\big) - (C_{\partial\mathcal{G}} - 1) - \frac{d-3}{2}N(\mathcal{G}) + (d-1)$$

$$+ \frac{d-3}{2}nV(\mathcal{G}) - (d-1)V(\mathcal{G}) - R(\mathcal{G}) \tag{2.16}$$

where $V(\mathcal{G})$ is the number of vertices of $\mathcal{G}$, $N(\mathcal{G})$ its number of external legs, and $C_{\partial\mathcal{G}}$ is the number of connected components of its boundary graph $\partial\mathcal{G}$ and $\tilde{\omega}(\mathcal{G}) = \sum_{\tilde{J} \subset \mathcal{G}} g_{\tilde{J}}$ with $\tilde{J}$ the pinched jacket associated with a jacket J of $\mathcal{G}$, $C_{\partial\mathcal{G}}$ is the number of vertex-connected components of $\partial\mathcal{G}$.

Lemma 2.2 *Let $\mathcal{G}$ be a connected Feynman graph and $\mathcal{T}$ one of its spanning trees. If the rosette $\mathcal{G}/\mathcal{T}$ is fully melonic and*

Table 1 Classification of divergent graphs

N	$\widetilde{\omega}(\mathcal{G})$	$\omega(\partial\mathcal{G})$	$C_{\partial\mathcal{G}} - 1$	$\omega_d(\mathcal{G})$
2	0	0	0	2
4	0	0	0	1
4	0	0	1	0
6	0	0	0	0

$$F(\mathcal{G}) = (d-1)(L(\mathcal{G}) - V(\mathcal{G}) + 1), \tag{2.17a}$$

$$R(\mathcal{G}) = R_m(\mathcal{G}) = L(\mathcal{G}) - V(\mathcal{G}) + 1, \tag{2.17b}$$

$$2\omega_d(\mathcal{G}) = -(d-4)N(\mathcal{G}) + (d-4)nV(\mathcal{G}) - 2(d-2)V(\mathcal{G}) + 2(d-2). \tag{2.17c}$$

Proposition 2.3 *The divergent graphs of the models are classified in Table 1.*

Another important definition for our purpose concerns the notion of *canonical dimension*. We will only give the essential here, and the reader interested in the details may consult [22]. In our model, the divergence degree for an arbitrary Feynman graph $\mathcal{G}$ is given in proposition (2.1) Denoting by $n_i(\mathcal{G})$ the number of bubbles in $\mathcal{G}$ with $2i$ black and white nodes, the divergent sub- graphs are said to be melonic [7, 13], if and only if they satisfy the following relation:

$$F(\mathcal{G}) - R(\mathcal{G}) = 3\left(L(\mathcal{G}) - \sum_i n_i(\mathcal{G}) + 1\right) \tag{2.18}$$

which together with the topological relation $L(\mathcal{G}) = \sum_i in_i(\mathcal{G}) - N(\mathcal{G})/2$ leads to:

$$\omega_d(\mathcal{G}) = 3 - \frac{N(\mathcal{G})}{2} - 2n_1(\mathcal{G}) - n_2(\mathcal{G}), \tag{2.19}$$

where $N(\mathcal{G})$ denote the number of external lines of $\mathcal{G}$. For the rest $n_1(\mathcal{G}) = 0$. For $N = 4$, $\omega \leqslant 1$, the value 1 corresponding to melonic graphs with only 6-point interactions bubble. This conclusion indicates that perturbatively around the Gaussian Fixed Point (GFP), the coupling constant λ_1 scales as Λ for some cut-off Λ, and we associate a canonical dimension $[\lambda_1] = 1$ to this constant. In the same way, we deduce that for a generic coupling λ_b, associated to a melonic bubble with N_b external lines:

$$[\lambda_b] = 3 - \frac{N_b}{2} \tag{2.20}$$

giving explicitly:

$$[m] = 1 \qquad [\lambda_1] = 1 \qquad [\lambda_2] = [\lambda_3] = 0. \tag{2.21}$$

3 Functional Renormalization Group with Closure Constraint

The FRG method is based on the following deformation to our original partition function given in Eq. (2.5), i.e.,

$$Z_s[J, \bar{J}] = \int d\mu_{C_\Lambda}(\bar{T}, T) e^{S_{int}[\bar{T}, T] - \Delta S_s[\bar{T}, T] + \langle \bar{J}, T \rangle + \langle \bar{T}, J \rangle} \qquad (3.1)$$

where $T_{\vec{p}}$ denotes the mean field $T_{\vec{p}} := \frac{\partial \log Z_s}{\partial \bar{J}_{\vec{p}}}$, and is a gauge invariant field in the sense that: $T_{\vec{p}} = T_{\vec{p}} \delta\left(\sum_{i=1}^{5} p_i\right)$, and we have added to the original action an IR cut-off $\Delta S_s[\bar{T}, T]$, defined as:

$$\Delta S_s[\bar{T}, T] = \sum_{\vec{p} \in \mathbb{Z}^5} R_s(|\vec{p}|) \bar{T}_{\vec{p}} T_{\vec{p}}. \qquad (3.2)$$

The cut-off function R_s depends on the real parameter s playing the role of a running cut-off, and is chosen such that:

- $R_s(\vec{p}) \geqslant 0$ for all $\vec{p} \in \mathbb{Z}^d$ and $s \in (-\infty, +\infty)$.
- $\lim_{s \to -\infty} R_s(\vec{p}) = 0$, implying: $Z_{s=-\infty}[\bar{J}, J] = Z[\bar{J}, J]$. This condition ensures that the original model is in the family (3.1). Physically, it means that the original model is recovered when all the fluctuations are integrated out.
- $\lim_{s \to \ln \Lambda} R_s(\vec{p}) = +\infty$, ensuring that all the fluctuations are frozen when $e^s = \Lambda$. As a consequence, the bare action will be represented by the initial condition for the flow at $s = \ln \Lambda$.
- For $-\infty < s < \ln \Lambda$, the cut-off R_s is chosen so that $R_s(|p| > e^s) \ll 1$, a condition ensuring that the UV modes $|p| > e^s$ are almost unaffected by the additional cut-off term, while $R_s(|p| < e^s) \sim 1$, or $R_s(|p| < e^s) \gg 1$, will guarantee that the IR modes $|p| < e^s$ are decoupled.
- $\frac{d}{d|\vec{p}|} R_s(\vec{p}) \leqslant 0$, for all $\vec{p} \in \mathbb{Z}^d$ and $s \in (-\infty, +\infty)$, which means that high modes should not be suppressed more than low modes.

The equation describing the flow of the couplings, the so-called Wetterich equation has been established in [30] in the case of a theory with closure constraint: For a given cut-off R_s, the effective average action satisfies the following first order partial differential equation:

$$\partial_s \Gamma_s = \sum_{\vec{p} \in \mathbb{Z}^5} \partial_s R_s(|\vec{p}|) \cdot \left[\Gamma_s^{(2)} + R_k\right]^{-1}(\vec{p}, \vec{p}) \delta\left(\sum_{i=1}^{5} p_i\right). \qquad (3.3)$$

where Γ_s, the effective average action and is defined as the Legendre transform of the free energy $W_s := \ln[Z_s]$ as :

$$\Gamma_s[\bar{T}, T] + \sum_{\vec{p} \in \mathbb{Z}^5} R_s(|\vec{p}|)\bar{T}_{\vec{p}}T_{\vec{p}} := \langle \bar{J}, T \rangle + \langle \bar{T}, J \rangle - W_s[J, \bar{J}] \tag{3.4}$$

and

$$\Gamma_s^{(2)}(\vec{p}, \vec{p}') := \frac{\partial^2 \Gamma_s}{\partial T_{\vec{p}} \partial \bar{T}_{\vec{p}'}}. \tag{3.5}$$

The Wetterich flow equation is an exact differential equation which must be truncated, i.e. it must be projected to functions of few variables or even onto some finite-dimensional sub-theory space. In this section, we adopt the simplest truncation, consisting in a restriction to the essential and marginal coupling with respect to the perturbative power counting (i.e., whose canonical dimension is upper or equal to zero). As mentioned before, such a truncation makes sense as long as the anomalous dimension remains small, and a qualitative argument is the following. Let us define the anomalous dimension $\eta := \partial_s \ln(Z)$ (see Eq. (3.9) below). In the vicinity of a fixed point, η can reach to a non-zero value η_*. As a result, the effective propagator becomes:

$$\frac{Z^{-1}}{\vec{p}^2 + (m_s^2/Z)} \approx \frac{e^{-\eta_* s}}{\vec{p}^2 + m_*^2} \quad , \tag{3.6}$$

and then modifies the power counting (2.19), which becomes in the melonic sector (all the star-quantities refers to the non-Gaussian fixed point that we consider):

$$\omega_*(\mathcal{G}) = -(2 + \eta_*)L(\mathcal{G}) + (F(\mathcal{G}) - R(\mathcal{G}))$$

$$= 3 - \frac{N}{2}(1 - \eta_*) - 3\eta_* n_3 - (1 + 2\eta_*)n_2 - (2 + \eta_*)n_1 . \tag{3.7}$$

As a result, the canonical dimension (2.21) turns to be

$$[t_b]_* = 3 - \frac{N_b}{2}(1 - \eta_*) = [t_b] + \frac{N_b}{2}\eta_* \quad , \tag{3.8}$$

from which one can argue that, as long as $\eta_* \ll 1$, the classification in terms of relevant, irrelevant and marginal couplings remains unchanged, and the truncation around marginal couplings with respect to the perturbative power counting makes sense. Note that for more specific explanations the study of the critical exponent will help to prove whether or not the truncation given below equation should improve or not. Unlike the case of standard local field theory, each line here has several strands (the theory is non-local). The contractions in the loop of the tadpole concern only 4 strands out of 5. The last strand circulates freely, and corresponds to an external momentum. It is by developing on this external variable that we generate the contribution to the anomalous dimension η. Thus, the quantity $\mathcal{W}^{(\ell)}_{\vec{p}_1, \vec{p}_2, \vec{p}_3, \vec{p}_4}$ does not explicitly depend on the momenta. The dependence on the momenta is due to the non-locality of the interactions. Up to these considerations, our choice of truncation is the following:

$$\Gamma_k[\bar{T}, T] = \sum_{\vec{p} \in \mathbb{Z}^5} \left(Z(k)\vec{p}\,^2 + m^2(k) \right) T_{\vec{p}} \bar{T}_{\vec{p}} + \frac{\lambda_1(k)}{2} \sum_{\ell=1}^{5} \sum_{\vec{p}_i} W^{(\ell)}_{\vec{p}_1, \vec{p}_2, \vec{p}_3, \vec{p}_4} T_{\vec{p}_1} \bar{T}_{\vec{p}_2} T_{\vec{p}_3} \bar{T}_{\vec{p}_4}$$

$$+\frac{\lambda_2(k)}{3} \sum_{\ell=1}^{5} \sum_{\vec{p}_i} \mathcal{X}^{(\ell)}_{\vec{p}_1, \vec{p}_2, \vec{p}_3, \vec{p}_4, \vec{p}_5, \vec{p}_6} T_{\vec{p}_1} \bar{T}_{\vec{p}_2} T_{\vec{p}_3} \bar{T}_{\vec{p}_4} T_{\vec{p}_5} \bar{T}_{\vec{p}_6}$$

$$+\lambda_3(k) \sum_{\ell_i=1, i=1,2,3} \sum_{\vec{p}_i} \mathcal{Y}^{(\ell_1, \ell_2, \ell_3)}_{\vec{p}_1, \vec{p}_2, \vec{p}_3, \vec{p}_4, \vec{p}_5, \vec{p}_6} T_{\vec{p}_1} \bar{T}_{\vec{p}_2} T_{\vec{p}_3} \bar{T}_{\vec{p}_4} T_{\vec{p}_5} \bar{T}_{\vec{p}_6}. \qquad (3.9)$$

We derive the truncated flow equations for m^2, Z, and λ_i from the full Wetterich equation (3.3). We write the second derivative of Γ_k as:

$$\Gamma_k^{(2)}[\bar{T}, T](\vec{p}, \vec{p}\,') = \left(-Z(k)\vec{p}\,^2 + m^2(k) \right) \delta\left(\sum_{i=1}^{5} p_i \right) \delta_{\vec{p}\vec{p}'}$$

$$+ F_{k,(1)}[\bar{T}, T]_{\vec{p}, \vec{p}'} + F_{k,(2)}[\bar{T}, T]_{\vec{p}, \vec{p}'}$$

in such a way that all the field-dependent terms of order $2n$ are in $F_{k,(n)}$. In particular, $F_{k,(1)}$ depends on $\lambda_1(k)$, while $F_{k,(2)}$ depends on $\lambda_2(k)$ and $\lambda_3(k)$. For the regulator R_k, we adopt the Litim cut-off [31], in which we set $s = e^k$:

$$R_k(|\vec{p}|) = Z(k)(k^2 - \vec{p}\,^2)\Theta(k^2 - \vec{p}\,^2), \qquad (3.10)$$

and computing the first derivative with respect to k, we find:

$$k\partial_k R_k(|\vec{p}|) = \left\{ k\partial_k Z(k)(k^2 - \vec{p}\,^2) + 2Z(k)k^2 \right\}\Theta(k^2 - \vec{p}\,^2). \qquad (3.11)$$

Hence, we are now in a position to extract the flow equations for each couplings, which is the subject of the next section.

3.1 Flow Equations in the UV Regime

We will deduce the flow equation in the UV regime. In this regime, all the sums can be replaced by integration, following the arguments of [22], essentially because the divergences of the integral approximations are the same as the exact sums. The method consists of a formal expansion of the r.h.s of the Wetterich equation

Fig. 1 Contributions coming
from the 6-point interactions
to the 4-point interaction

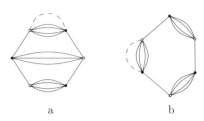

a   b

(3.3) in power of couplings, and an identification of the corresponding terms in the l.h.s. The r.h.s involves in general some contractions between the $F_{k(n)}$ and the effective propagator $\partial_k R_k$. And in this UV regime, only the melonic graphs contribute (Fig. 1). We get the flow equation of $m(k)$ as

$$k\partial_k m^2(k) = -\frac{4\pi}{3}\lambda_1(k)\frac{\eta(k)+5}{[Z(k)k^2+m^2(k)]^2}k^5 \qquad (3.12)$$

with the anomalous dimension $\eta(k)$ defined as:

$$\eta(k) := k\partial_k \ln(Z(k)) = \frac{5\pi}{2}\lambda_1(k)\frac{k^3}{[Z(k)k^2+m^2(k)]^2-\lambda_1(k)\frac{5}{6}\pi k^3}. \qquad (3.13)$$

Note that, in this case, the extraction of the local approximation in the UV limit brings up a very nice property of the melonic sector, called *traciality*. Traciality is a concept firstly introduced in a perturbative renormalization framework, ensuring that local approximation of high subgraphs makes sense in the TFGT context [13]. The contributions to $\lambda_1(k)$, $\lambda_2(k)$, and $\lambda_3(k)$ are given in Figs. 2, 3 and 4 and are explicitly written as

$$k\partial_k\lambda_1(k) = -(\lambda_2(k)+4\lambda_3(k))\frac{4\pi}{15}\frac{\eta(k)+5}{[Z(k)k^2+m^2(k)]^2}k^5$$

$$+ \lambda_1^2(k)\frac{4\pi}{15}\frac{\eta(k)+5}{[Z(k)k^2+m^2(k)]^3}k^5 \qquad (3.14)$$

$$k\partial_k\lambda_3(k) = \frac{16\pi}{15}\lambda_1(k)\lambda_3(k)\frac{\eta(k)+5}{[Z(k)k^2+m^2(k)]^3}k^5. \qquad (3.15)$$

$$k\partial\lambda_2(k)=\frac{24\pi}{15}\lambda_1(k)\lambda_2(k)\frac{\eta(k)+5}{[Z(k)k^2+m^2(k)]^3}k^5-\frac{12\pi}{15}\lambda_1^3(k)\frac{\eta(k)+5}{[Z(k)k^2+m^2(k)]^4}k^5.$$

$$(3.16)$$

Taking into account the canonical dimension the renormalized dimensionless couplings are defined as:

$$m(k) = \sqrt{Z(k)}k\bar{m} \qquad \lambda_1 = Z^2k\bar{\lambda}_1(k)$$

Fig. 2 Contribution to the
4-point interaction involved
two vertices

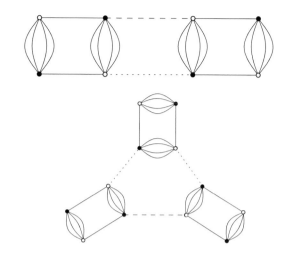

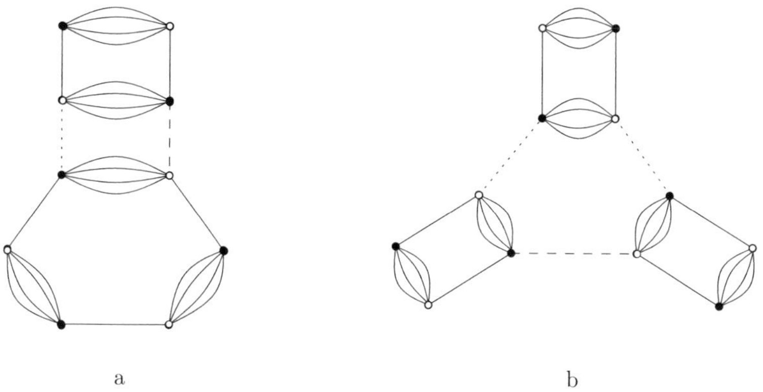

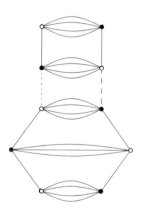

a   b

Fig. 3 Contributions to the flow of λ_2

Fig. 4 Contribution to the
flow of λ_3

$$\lambda_2(k) = Z^3(k)\bar{\lambda}_2(k) \qquad \lambda_3(k) = Z^3(k)\bar{\lambda}_3(k) \qquad (3.17)$$

Using the flow equations (3.13), (3.12), (3.14), (3.16), and (3.15), we find for the
dimensionless renormalized couplings the following autonomous system:

$$\eta(k) = \frac{5\pi}{2}\bar{\lambda}_1(k)\frac{1}{[1 + \bar{m}^2(k)]^2 - \bar{\lambda}_1(k)\frac{5}{6}\pi} \qquad (3.18)$$

$$\beta_{m^2} = -(2 + \eta)\bar{m}^2(k) - \frac{4\pi}{3}\bar{\lambda}_1(k)\frac{\eta(k) + 5}{[1 + \bar{m}^2(k)]^2} \qquad (3.19)$$

$$\beta_{\lambda_1} = -(1 + 2\eta)\bar{\lambda}_1(k) - (\bar{\lambda}_2 + 4\bar{\lambda}_3)\frac{4\pi}{15}\frac{\eta(k) + 5}{[1 + \bar{m}^2(k)]^2} + \bar{\lambda}_1^2(k)\frac{4\pi}{15}\frac{\eta(k) + 5}{[1 + \bar{m}^2(k)]^3}$$

$$\text{(3.20)}$$

$$\beta_{\lambda_3} = -3\eta\bar{\lambda}_3(k) + \frac{16\pi}{15}\bar{\lambda}_1\bar{\lambda}_3\frac{\eta(k) + 5}{[1 + \bar{m}^2(k)]^3}. \tag{3.21}$$

$$\beta_{\lambda_2} = -3\eta\bar{\lambda}_2(k) + \frac{24\pi}{15}\bar{\lambda}_1\bar{\lambda}_2\frac{\eta(k) + 5}{[1 + \bar{m}^2(k)]^3} - \bar{\lambda}_1^3\frac{12\pi}{15}\frac{\eta(k) + 5}{[1 + \bar{m}^2(k)]^4}, \tag{3.22}$$

with the definition: $\beta_\sigma := k\partial_k\bar{\sigma}, \sigma \in \{m^2, \lambda_1, \lambda_2, \lambda_3\}$.

4 Fixed Points in the UV Regime

At vanishing β-functions we obtain the fixed points. In the neighborhood of these fixed points, the stability is determined by the linearized system of β-functions. All these points are studied in detail in this section.

4.1 Vicinity of the Gaussian Fixed Point

The autonomous system describing the flow of the dimensionless couplings admits a trivial fixed point for the values $\bar{\lambda}_1 = \bar{\lambda}_2 = \bar{\lambda}_3 = \bar{m} = 0$ called Gaussian fixed point (GFP). Expanding our equations around this point, we find the reduced autonomous system:

$$\begin{cases} \beta_{m^2} \approx -2\bar{m}^2 - \frac{20\pi\bar{\lambda}_1}{3}, \\ \beta_{\lambda_1} \approx -\bar{\lambda}_1 - \frac{4\pi}{3}(\bar{\lambda}_2 + 4\bar{\lambda}_3)\left(1 + \frac{\pi}{2}\bar{\lambda}_1 + 2\bar{m}^2\right) - \frac{11\pi}{3}\bar{\lambda}_1^2, \\ \beta_{\lambda_2} \approx \frac{\pi}{2}\bar{\lambda}_1\bar{\lambda}_2, \\ \beta_{\lambda_3} \approx -\frac{13\pi}{6}\bar{\lambda}_1\bar{\lambda}_3 \end{cases} \tag{4.1}$$

and the anomalous dimension:

$$\eta(k) \approx \frac{5\pi\bar{\lambda}_1}{2}. \tag{4.2}$$

These equations give the qualitative behavior of the RG trajectories around the GFP. In order to study its stability, we compute the *stability matrix* $\beta_{ij} := \partial_i\beta_j \ i \in \{m^2, \lambda_1, \lambda_2, \lambda_3\}$, and evaluate each coefficient at the GFP. We find:

$$\beta_{ij}^{GFP} := \begin{pmatrix} -2 & 0 & 0 & 0 \\ -\frac{20\pi}{3} & -1 & 0 & 0 \\ 0 & -\frac{4\pi}{3} & 0 & 0 \\ 0 & -\frac{16\pi}{3} & 0 & 0 \end{pmatrix},$$

(4.3)

with eigenvalues $(-2, -1, 0, 0)$ and eigenvectors $e_1^{GFP} = (\frac{9}{160\pi^2}, \frac{3}{8\pi}, \frac{1}{4}, 1)$; $e_2^{GFP} = (0, \frac{3}{16\pi}, \frac{1}{4}, 1)$; $e_3^{GFP} = (0, 0, 0, 1)$; $e_4^{GFP} = (0, 0, 1, 0)$. One recall that the *critical exponents* are the opposite values of the eigenvalues of the β_{ij}, and that the fixed point can be classified following the sign of their critical exponents. Hence, we have two relevant directions in the UV, with critical exponents 2 and 1, and two marginal couplings with zero critical exponents. Moreover, note that the critical exponents are equal to the canonical dimension around the GFP. Finally, note that the previous system of equations admits other fixed points, or a line of fixed points $\bar{\lambda}_1 = \bar{m} = 0$; $\bar{\lambda}_3 = -\bar{\lambda}_2/4$, in addition to the Gaussian one. The same phenomenon happens away from the Gaussian fixed point. We will discuss this property in the next section.

For the moment, we are in a position to discuss the qualitative flow diagram around the Gaussian fixed point. First of all, note that all the coefficients of the β-function of the system (4.1) (i.e., the coefficients in the right-hand side of the system) are not negative. This fact seems to be a special feature of this model, meaning that the weight of the anomalous dimension does not dominate the vertex contribution. This fact is a first difference with respect to the similar non-Abelian ϕ^6 model studied in [14]. However, the analysis provided in this reference remains true, and the model is not asymptotically free. We will not repeat the complete analysis given in [14], but a qualitative argument is the following. Exploiting the fact that the hyperplans $\bar{\lambda}_2 = 0$ and $\bar{\lambda}_3 = 0$ are invariant under the flow, we can look at only a two-dimensional reduction of the complete system (4.1). We choose $\bar{\lambda}_2 = 0$, and plot the numerical integration of the reduced flow equation in Fig. 5 (on the left) below. In the domain, $\bar{\lambda}_3 > 0$, even if a given trajectory approaches of the Gaussian fixed point, $\bar{\lambda}_1$ reaches a negative value, and it is ultimately repelled for k sufficiently large. The same phenomenon occurs for $\bar{\lambda}_2$ in the plan $\bar{\lambda}_2 < 0$ (see Fig. 5 on the right). We will give additional precision about the issue of the asymptotic freedom and/or safety in the next section.

4.2   Non-Gaussian Fixed Points

Solving numerically the systems (3.12)–(3.15), we find some non-Gaussian fixed points, whose relevant characteristics are summarized in Table 2. In addition to the Gaussian one, the system admits a line of fixed points, LFP:

$$LFP = \{\bar{m}^2 = 0, \bar{\lambda}_1 = 0, \bar{\lambda}_2 = -4\bar{\lambda}_3\},$$

(4.4)

Table 2 Summary of the properties of the non-Gaussian fixed points

FP	$\bar{m}^2$	$\bar{\lambda}_1$	$\bar{\lambda}_2$	$\bar{\lambda}_3$	η	$\theta^{(1)}$	$\theta^{(2)}$	$\theta^{(3)}$	$\theta^{(4)}$
FP_1	-0.3	0.005	0.0009	-0.0002	-6.3	-299	56.1	-11.7	5.8
FP_2	-0.7	0.008	0.0006	-0.0002	0.76	$-7.4 - 1.9i$	$-7.4 + 1.9i$	3.34	-0.12
FP_3	-0.9	0.0007	$3.32.10^{-6}$	0.	1.3	-66.7	-42.63	-27.7	1.80
FP_4	-0.8	0.04	-0.02	0.	-5.9	-144.8	-14.4	-7.5	-5.4
FP_5	0.06	-0.006	0.002	0.	-0.04	1.9	1.09	-0.04	-0.01
FP_6	1.32	-0.5	-0.06	0.	-0.6	3.0	-1.23	-1.13	-0.39

Again, the critical exponents θ^i are the opposite values of the eigenvalues of the stability matrix:
$\beta_* =: diag(-\theta_*^1, -\theta_*^2, -\theta_*^3, -\theta_*^4)$

with critical exponents:

$$
\begin{cases}
\theta^{(1)} &= -2, \\
\theta^{(2)} &= 0, \\
\theta^{(3)} &= -\frac{1}{2}\left(1 + \sqrt{1 - \frac{128}{9}\pi^2\bar{\lambda}_2}\right), \\
\beta_{\lambda_3} &= -\frac{1}{2}\left(1 - \sqrt{1 - \frac{128}{9}\pi^2\bar{\lambda}_2}\right).
\end{cases}
\tag{4.5}
$$

The denominator of η, $D := [1 + \bar{m}^2(k)]^2 - \bar{\lambda}_1(k)\frac{5}{6}\pi$ introduces a singularity in the flow. At the Gaussian fixed point, and in a sufficiently small domain around, we have $D > 0$. But further away from the GFP, D may cancel, creating in the $(\bar{\lambda}_1, \bar{m}^2)$-plan a singularity line. The area below this line where $D < 0$ is thus disconnected from the region $D > 0$ connected to GFP. Then, we ignore for our purpose the fixed points in the disconnected region, for which $D < 0$. A direct computation shows that only the fixed points FP_2, FP_3, FP_5, and FP_6 are relevant for an analysis in the domain connected to the Gaussian fixed point.

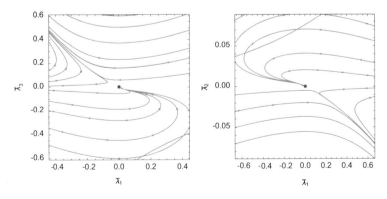

Fig. 5 Phase portrait in the plans $(\bar{\lambda}_1, \bar{\lambda}_3)$ for $\bar{\lambda}_2 = 0$ (on the left) and in the plan $(\bar{\lambda}_1, \bar{\lambda}_2)$, for $\bar{\lambda}_3 = 0$

Fig. 6 Qualitative behavior of the RG trajectories around an IR fixed point. The critical surface is spanned by the relevant directions in the IR, and the arrows are oriented toward the IR direction. This illustrates the scenario of asymptotically safety

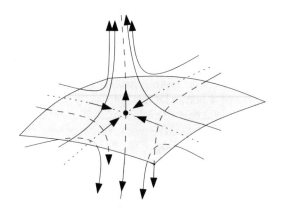

- The fixed points FP_2 and FP_3 are very similar. They have three irrelevant directions and one relevant direction in the UV. For each of these fixed points, the three irrelevant directions span a three-dimensional manifold on which trajectories run toward the fixed points in the IR, while the trajectories outside are repelled of this critical surface, as pictured in Fig. 6. This picture, the existence of a separatrix between two regions of the phase space is reminiscent of a critical behavior, with phase transition between a broken and a symmetric phase, and these separatrix are *IR-critical surfaces*. This interpretation is highlighted for the two fixed points in the zero momentum limit. Indeed, in both cases, the contributions in the effective action of the terms proportional to $\bar{\lambda}_2$ and $\bar{\lambda}_3$ can be neglected in comparison with the contributions to the first approximation of Ginsburg-Landau equation for ϕ^4 scalar complex theory. Note that for FP_2 two critical exponents are complex, providing some oscillations of the trajectories, and implying that the fixed point is an IR-attractor in the two-dimensional manifold spanned by the eigenvectors corresponding to these two critical exponents. Moreover, the fixed point FP_6 appears to be an IR fixed point, with coordinates of opposite sign.
- The fixed point FP_5 has two relevant and two irrelevant directions in the UV. The relevant directions in the UV span a two-dimensional manifold corresponding to a *UV-multicritical surface* [32]. Such a surface is interesting for the UV-completion of the theory. Indeed, all the trajectories in the surface are oriented toward the fixed point in the UV, while the dimension of the surface gives an interesting number of physical parameter, providing an evidence in favor of the *asymptotic safety*.
- Finally, we have the line of fixed point, for which we will distinguish four cases:

  1. In the domain $d_1 = \{\bar{\lambda}_2 < 0\}$ we have two relevant, one marginal and one irrelevant directions.

  2. At the point $d_2 = \{\bar{\lambda}_2 = 0\}$, we recover the GFP, with two relevant and two marginal directions.

3. In the domain $d_3 = \{\bar{\lambda}_2 \in]0, \left(\frac{3}{8\pi}\right)^2]\}$ we have three relevant and one marginal directions. One more time, this section of the critical line is interesting in view of the UV-completion of the theory and provides a supplementary evidence in favor of asymptotic safety. Indeed, in each point, the relevant directions in the UV span a three-dimensional UV-critical surface, in favor of the existence of a nontrivial asymptotically safe theory with three independent physical parameters. This line of fixed point has been recently discussed in [26] for a similar model improved by unconnected interaction bubbles.

4. In the domain $d_4 = \{\bar{\lambda}_2 > \left(\frac{3}{8\pi}\right)^2\}$, the situation is very reminiscent of the previous one. We have three eigenvalues with negative real part and one equal to zero. Hence, we have three relevant and one marginal directions. The only difference in comparison with the domain d_3 is that the eigenvalue has non-zero imaginary parts, giving some oscillations and attractor phenomena in the trajectories.

Finally, we briefly discuss the values of the anomalous dimensions. With our conventions, the couplings of the relevant operator are suppressed as a power of k in the UV limit $k \to \infty$. The couplings decrease when the trajectory goes away from the UV regime. However, the power law behavior is limited to the attractive region of the fixed point, far from its scaling regime it can deviate from the power law one. And we can evaluate this deviation. For instance, in the vicinity of FP_5, one deduce from (3.8) that the canonical dimension becomes:

$$[t_b]_{FP_5} \approx 3 - 1.6\frac{N_b}{2}, \tag{4.6}$$

from which we deduce that all the interactions of valence up or equal to four become inessentials. The same phenomenon occurs in the vicinity of FP_4, where all the interactions up to these of valence four become irrelevant/relevant (according to their convention or inessential/essential). On the contrary, at the fixed points FP_2 and FP_4 the anomalous dimension is positive, meaning that the power counting is improved with respect to the Gaussian one, and irrelevant operators are enhanced in the UV.

5 Concluding Remarks

In this work we have studied the functional renormalization group applied to a U(1) tensor model. The flow equations of the coupling constants and mass parameter are deduced. The nontrivial behavior around the fixed point is also given. We have compelling evidence that the model studied in this paper is asymptotically safe in the UV regime. Let us remark that it is possible to extend the truncation to higher orders. Also, over regulators maybe tested. In this case the convergence of the fixed points can be studied order by order. For more detail, see [24] and [25].

Acknowledgements D.O.S research at the Max-Planck Institute is supported by the Alexander von Humboldt foundation.

References

1. V. Rivasseau, The tensor track, III. Fortsch. Phys. **62**, 81 (2014). https://doi.org/10.1002/prop. 201300032 [arXiv:1311.1461 [hep-th]]
2. V. Rivasseau, The tensor track: an update. arXiv:1209.5284 [hep-th]
3. R. Gurau, J.P. Ryan, Colored tensor models - a review. SIGMA **8**, 020 (2012). https://doi.org/ 10.3842/SIGMA.2012.020 [arXiv:1109.4812 [hep-th]]
4. R. Gurau, Colored group field theory. Commun. Math. Phys. **304**, 69 (2011). https://doi.org/ 10.1007/s00220-011-1226-9 [arXiv:0907.2582 [hep-th]]
5. V. Rivasseau, Random Tensors and Quantum Gravity. arXiv:1603.07278 [math-ph]
6. D. Benedetti, R. Gurau, Symmetry breaking in tensor models. Phys. Rev. D **92**(10), 104041 (2015). https://doi.org/10.1103/PhysRevD.92.104041 [arXiv:1506.08542 [hep-th]]
7. R. Gurau, The complete 1/N expansion of colored tensor models in arbitrary dimension. Ann. Henri Poincare **13**, 399 (2012). https://doi.org/10.1007/s00023-011-0118-z [arXiv:1102.5759 [gr-qc]]
8. R. Gurau, The 1/N expansion of colored tensor models. Ann. Henri Poincare **12**, 829 (2011). https://doi.org/10.1007/s00023-011-0101-8 [arXiv:1011.2726 [gr-qc]]
9. D. Oriti, J.P. Ryan, J. Thurigen, Group field theories for all loop quantum gravity. New J. Phys. **17**(2), 023042 (2015). https://doi.org/10.1088/1367-2630/17/2/023042 [arXiv:1409.3150 [gr-qc]]
10. D. Oriti, A quantum field theory of simplicial geometry and the emergence of spacetime. J. Phys. Conf. Ser. **67**, 012052 (2007). https://doi.org/10.1088/1742-6596/67/1/012052 [hep-th/0612301]
11. C. Rovelli, Zakopane lectures on loop gravity. PoS QGQGS **2011**, 003 (2011). [arXiv:1102.3660 [gr-qc]]
12. C. Rovelli, Loop quantum gravity: the first twenty five years. Class. Quant. Grav. **28**, 153002 (2011). https://doi.org/10.1088/0264-9381/28/15/153002 [arXiv:1012.4707 [gr-qc]]
13. S. Carrozza, D. Oriti, V. Rivasseau, Renormalization of tensorial group field theories: Abelian U(1) models in four dimensions. Commun. Math. Phys. **327**, 603 (2014). https://doi.org/10. 1007/s00220-014-1954-8 [arXiv:1207.6734 [hep-th]]
14. S. Carrozza, Ann. Inst. Henri Poincaré Comb. Phys. Interact. **2**, 49–112 (2015). https://doi.org/ 10.4171/AIHPD/15 [arXiv:1407.4615 [hep-th]]
15. D. Ousmane Samary, F. Vignes-Tourneret, Just renormalizable TGFT's on $U(1)^d$ with gauge invariance. Commun. Math. Phys. **329**, 545 (2014). https://doi.org/10.1007/s00220-014-1930-3 [arXiv:1211.2618 [hep-th]]
16. J. Ben Geloun, V. Rivasseau, A renormalizable 4-dimensional tensor field theory. Commun. Math. Phys. **318**, 69 (2013). https://doi.org/10.1007/s00220-012-1549-1 [arXiv:1111.4997 [hep-th]]
17. J. Ben Geloun, D. Ousmane Samary, 3D tensor field theory: renormalization and one-loop β-functions. Ann. Henri Poincare **14**, 1599 (2013). https://doi.org/10.1007/s00023-012-0225-5 [arXiv:1201.0176 [hep-th]]
18. J. Ben Geloun, Two and four-loop β-functions of rank 4 renormalizable tensor field theories. Class. Quant. Grav. **29**, 235011 (2012). https://doi.org/10.1088/0264-9381/29/23/235011 [arXiv:1205.5513 [hep-th]]
19. D. Ousmane Samary, Beta functions of $U(1)^d$ gauge invariant just renormalizable tensor models. Phys. Rev. D **88**(10), 105003 (2013). https://doi.org/10.1103/PhysRevD.88.105003 [arXiv:1303.7256 [hep-th]]

20. J.B. Geloun, R. Martini, D. Oriti, Functional renormalisation group analysis of tensorial group field theories on $\mathbb{R}^d$. arXiv:1601.08211 [hep-th]
21. J.B. Geloun, R. Martini, D. Oriti, Functional renormalization group analysis of a tensorial group field theory on $\mathbb{R}^3$. Europhys. Lett. **112**(3), 31001 (2015). https://doi.org/10.1209/0295-5075/112/31001 [arXiv:1508.01855 [hep-th]]
22. D. Benedetti, V. Lahoche, Functional renormalization group approach for tensorial group field theory: a Rank-6 model with closure constraint. arXiv:1508.06384 [hep-th]
23. D. Benedetti, J. Ben Geloun, D. Oriti, Functional renormalisation group approach for tensorial group field theory: a rank-3 model. JHEP **1503**, 084 (2015). https://doi.org/10.1007/JHEP03(2015)084 [arXiv:1411.3180 [hep-th]]
24. V. Lahoche, D. Ousmane Samary, Functional renormalization group for the U(1)-T_5^6 tensorial group field theory with closure constraint. Phys. Rev. D **95**(4), 045013 (2017). https://doi.org/10.1103/PhysRevD.95.045013 [arXiv:1608.00379 [hep-th]]
25. S. Carrozza, V. Lahoche, Asymptotic safety in three-dimensional SU(2) group field theory: evidence in the local potential approximation. Class. Quant. Grav. **34**(11), 115004 (2017). https://doi.org/10.1088/1361-6382/aa6d90 [arXiv:1612.02452 [hep-th]]
26. J.B. Geloun, T.A. Koslowski, Nontrivial UV behavior of rank-4 tensor field models for quantum gravity. arXiv:1606.04044 [gr-qc]
27. P. Donà, A. Eichhorn, P. Labus, R. Percacci, Asymptotic safety in an interacting system of gravity and scalar matter. Phys. Rev. D **93**(4), 044049 (2016). https://doi.org/10.1103/PhysRevD.93.044049 [arXiv:1512.01589 [gr-qc]]
28. P. Donà, A. Eichhorn, R. Percacci, Consistency of matter models with asymptotically safe quantum gravity. Can. J. Phys. **93**(9), 988 (2015). https://doi.org/10.1139/cjp-2014-0574 [arXiv:1410.4411 [gr-qc]]
29. A. Eichhorn, T. Koslowski, Continuum limit in matrix models for quantum gravity from the functional renormalization group. Phys. Rev. D **88**, 084016 (2013). https://doi.org/10.1103/PhysRevD.88.084016 [arXiv:1309.1690 [gr-qc]]
30. C. Wetterich, Average action and the renormalization group equations. Nucl. Phys. B **352**, 529 (1991). https://doi.org/10.1016/0550-3213(91)90099-J
31. D.F. Litim, Optimization of the exact renormalization group. Phys. Lett. B **486**, 92 (2000). https://doi.org/10.1016/S0370-2693(00)00748-6 [hep-th/0005245]
32. R. Percacci, Asymptotic Safety (2008). arXiv: 0709.3851, [hep-th]

Ternary Z_2 and Z_3 Graded Algebras and Generalized Color Dynamics

Richard Kerner

Dedicated to Norbert Hounkonnou for his 60-th birthday

Abstract We discuss cubic and ternary algebras which are a direct generalization of Grassmann and Clifford algebras, but with Z_3-grading replacing the usual Z_2-grading. Combining Z_2 and Z_3 gradings results in algebras with Z_6 grading, which are also investigated. We introduce a universal constitutive equation combining binary and ternary cases.

Elementary properties and structures of such algebras are discussed, with special interest in low-dimensional ones, with two or three generators.

Invariant anti-symmetric quadratic and cubic forms on such algebras are introduced, and it is shown how the $SL(2, C)$ group arises naturally in the case of lowest dimension, with two generators only, as the symmetry group preserving these forms.

In the case of lowest dimension, with two generators only, it is shown how the cubic combinations of Z_3-graded elements behave like Lorentz spinors, and the binary product of elements of this algebra with an element of the conjugate algebra behave like Lorentz vectors.

Then Pauli's principle is generalized for the case of the Z_3 graded ternary algebras leading to cubic commutation relations. A generalized Dirac equation is introduced.

The model displays the color $SU(3)$ symmetry of strong interactions, as well as the $SU(2)$ and $U(1)$ symmetries giving rise to the Standard Model gauge fields.

Keywords Ternary algebras · Grassman and Cliffords algebras · Lorentz spinors · Pauli principle · Dirac equation · Standard model gauge fields

R. Kerner (✉)
Laboratoire de Physique Théorique de la Matière Condensée, Sorbonne-Universités, CNRS UMR 7600, Paris, France
e-mail: richard.kerner@upmc.fr

© Springer Nature Switzerland AG 2018
T. Diagana, B. Toni (eds.), *Mathematical Structures and Applications*,
STEAM-H: Science, Technology, Engineering, Agriculture,
Mathematics & Health, https://doi.org/10.1007/978-3-319-97175-9_14

1 Introduction

1.1 Z_2 and Z_3 Symmetries and Gradings

Of all symmetry groups characterizing physical phenomena and their mathematical models, the discrete groups seem to be most fundamental. Among those, the simplest discrete group Z_2 is omnipresent and plays a crucial role in fundamental interactions between elementary particles and fields: all theoretical models are checked by their response to the three representations of the Z_2 group, called "C" (charge conjugation, reflecting the symmetry between particles and anti-particles), "P" (parity, consisting in space reflection), and "T" (time reversion). Although in some situations parity or time reflection may be broken, all known phenomena are invariant under the simultaneous application of all these idempotents. This is often referred to as the "CPT"-theorem in elementary particle physics.

Another important manifestation of Z_2 symmetry in physics is the distinction between bosons and fermions, which in the language of quantum field theory corresponds to commutators (for bosons) or anti-commutators (for fermions) in the constitutive relations between the creation and annihilation operators:

$$a_i^\dagger a_k - a_k a_i^\dagger = \delta_{ik}, \quad a_i^\dagger a_k + a_k a_i^\dagger = \delta_{ik}. \tag{1}$$

for the Bose-Einstein or Fermi-Dirac statistics, respectively.

What we have here is the example of two distinct representations of the Z_2 symmetry group, the first trivial, another faithful. Let us analyze the structure of all possible representations of Z_2 in the complex plane.

All bilinear maps of vector spaces into complex numbers can be divided into irreducible symmetry classes according to the representations of the Z_2 group, e.g. symmetric, anti-symmetric, hermitian, or anti-hermitian:

1. The trivial representation defines the symmetric tensors:

$$S_{\pi(AB)} = S_{BA} = S_{AB},$$

2. The sign inversion defines the anti-symmetric tensors:

$$A_{\pi(CD)} = A_{DC} = -A_{CD},$$

3. The complex conjugation defines the hermitian tensors:

$$H_{\pi(AB)} = H_{BA} = \bar{H}_{AB},$$

4. $(-1)\times$ complex conjugation defines the anti-hermitian tensors.

$$T_{\pi(AB)} = T_{BA} = -\bar{T}_{AB},$$

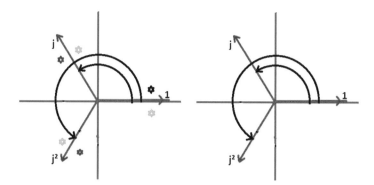

Fig. 1 Representation of the symmetry group S_3 (three rotations, including identity, and three reflections), and the complex representation of its Z_3 cyclic subgroup

The tri-linear mappings can be distinguished by their symmetry properties with respect to the permutations belonging to the Z_3 symmetry group.

There are several different representations of the action of the Z_3 permutation group on tensors with three indices. Consequently, such tensors can be divided into irreducible subspaces which are conserved under the action of Z_3 (Fig. 1).

There are three possibilities of an action of Z_3 being represented by multiplication by a complex number: the trivial one (multiplication by 1), and the two other representations, the multiplication by $j = e^{2\pi i/3}$ or by its complex conjugate, $j^2 = \bar{j} = e^{4\pi i/3}$.

$$T \in \mathscr{T}: \quad T_{ABC} = T_{BCA} = T_{CAB}, \tag{2}$$

$$\Lambda \in \mathscr{L}: \quad \Lambda_{ABC} = j\,\Lambda_{BCA} = j^2\,\Lambda_{CAB}, \tag{3}$$

$$\bar{\Lambda} \in \bar{\mathscr{L}}: \quad \bar{\Lambda}_{ABC} = j^2\,\bar{\Lambda}_{BCA} = j\,\bar{\Lambda}_{CAB}, \tag{4}$$

which can be called, respectively, totally symmetric, j-skew-symmetric, and j^2-skew-symmetric. The space of all tri-linear forms is the sum of three irreducible subspaces,

$$\Theta_3 = \mathscr{T} \oplus \mathscr{L} \oplus \bar{\mathscr{L}}$$

the corresponding dimensions being, respectively, $(N^3 + 2N)/3$ for $\mathscr{T}$ and $(N^3 - N)/3$ for $\mathscr{L}$ and for $\bar{\mathscr{L}}$.

Any three-form W^α_{ABC} mapping $\mathscr{A} \otimes \mathscr{A} \otimes \mathscr{A}$ into a vector space $\mathscr{X}$ of dimension k, $\alpha, \beta = 1, 2, \ldots k$, so that $X^\alpha = W^\alpha_{ABC}\,\theta^A\theta^B\theta^C$ can be represented as a linear combination of forms with specific symmetry properties, $W^\alpha_{ABC} = T^\alpha_{ABC} + \Lambda^\alpha_{ABC} + \bar{\Lambda}^\alpha_{ABC}$, with

$$T^\alpha_{ABC} := \frac{1}{3}\,(W^\alpha_{ABC} + W^\alpha_{BCA} + W^\alpha_{CAB}), \tag{5}$$

$$\Lambda^\alpha_{ABC} := \frac{1}{3}\,(W^\alpha_{ABC} + j\,W^\alpha_{BCA} + j^2\,W^\alpha_{CAB}), \tag{6}$$

$$\bar{\Lambda}^\alpha_{ABC} := \frac{1}{3}\,(W^\alpha_{ABC} + j^2\,W^\alpha_{BCA} + j\,W^\alpha_{CAB}), \tag{7}$$

As in the Z_2 case, the three symmetries above define irreducible and mutually orthogonal 3-forms.

Consequently, two different cubic commutation relations can be imposed on an associative algebra, say Λ-type and $\bar{\Lambda}$-type: for any three elements a, b, c belonging to algebra $\mathscr{A}_\Lambda$ we shall have $abc = j\ bca = j^2\ cab$, and for any three elements $\bar{a}, \bar{b}, \bar{c}$ belonging to algebra $\mathscr{A}_{\bar{\Lambda}}$ we shall have $\bar{a}\bar{b}\bar{c} = j^2\ \bar{b}\bar{c}\bar{a} = j\ \bar{c}\bar{a}\bar{b}$. The $\mathbb{Z}_2$-grading of ordinary (binary) algebras is well known and widely studied and applied (e.g., in the super-symmetric field theories in Physics [1, 2]). The Grassmann and Clifford algebras are perhaps the oldest and the best known examples of a $\mathbb{Z}_2$-graded structure. Other gradings are much less popular. The $\mathbb{Z}_3$-grading was introduced and studied in the paper [3]; the $\mathbb{Z}_N$ grading was discussed in [4]. An approach to ternary Clifford algebra based on ternary triples and a successive process of ternary Galois extensions is proposed in [5]. More general case of N-algebras, in which only the product of N elements is defined, was studied in [6].

2   Examples of Z_3-Graded Ternary Algebras

In the case of ternary algebras of type Λ or $\bar{\Lambda}$, the grade 1 is attributed to the generators θ^A and the grade 2 to the conjugate generators $\bar{\theta}^{\dot{B}}$. Consequently, their products acquire the grade which is the sum of grades of the factors modulo 3. When we consider an algebra including a ternary $\mathbb{Z}_3$-graded subalgebra and a binary $\mathbb{Z}_2$-graded one, we can quite naturally introduce a combination of the two gradings considered as a pair of two numbers, say (a, λ), with $a = 0, 1, 2$ representing the $\mathbb{Z}_3$-grade, and $\lambda = 0, 1$ representing the $\mathbb{Z}_2$ grade, $\lambda = 0, 1$. The first grades add up modulo 3, the second grades add up modulo 2. The six possible combined grades are then

$$(0,0), \quad (1,0), \quad (2,0), \quad (0,1), \quad (1,1) \quad \text{and} \quad (2,1). \tag{8}$$

To add up two of the combined grades amounts to adding up their first entries modulo 3, and their second entries modulo 2. Thus, we have

$$(2,1) + (1,1) = (3,2) \simeq (0,0), \quad \text{or} \quad ((2,1) + (1,0) = (3,1) \simeq (0,1), \quad \text{and so forth.}$$

Fig. 2 Representation of the cyclic group Z_6 in the complex plane

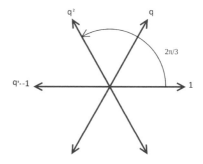

It is well known that the cartesian product of two cyclic groups $\mathbb{Z}_N \times \mathbb{Z}_n$, N and n being two prime numbers, is the cyclic group $\mathbb{Z}_{Nn}$ corresponding to the product of those prime numbers. This means that there is an isomorphism between the cyclic group $\mathbb{Z}_6$, generated by the *sixth* primitive root of unity, $q^6 = 1$, satisfying the equation

$$q + q^2 + q^3 + q^4 + q^5 + q^6 = 0.$$

This group can be represented on the complex plane, with $q = e^{\frac{2\pi i}{6}}$, as shown in Fig. 2:

The elements of the group $\mathbb{Z}_6$ represented by complex numbers multiply modulo 6, e. g. $q^4 \cdot q^5 = q^9 \simeq q^3$, etc. The six elements of $\mathbb{Z}_6$ can be put in the one-to-one correspondence with the pairs defining six elements of $\mathbb{Z}_3 \times \mathbb{Z}_2$ according to the following scheme:

$$(0, 0) \simeq q^0 = 1, \quad (2, 1) \simeq q, \quad (1, 0) \simeq q^2, \quad (0, 1) \simeq q^3, \quad (2, 0) \simeq q^4, \quad (1, 1) \simeq q^5. \tag{9}$$

The same result can be obtained directly using the representations of $\mathbb{Z}_3$ and $\mathbb{Z}_2$ in the complex plane. Taken separately, each of these cyclic groups is generated by one non-trivial element, the third root of unity $j = e^{\frac{2\pi i}{3}}$ for $\mathbb{Z}_3$ and $-1 = e^{\pi i}$ for $\mathbb{Z}_2$. It is enough to multiply these complex numbers and take their different powers in order to get all the six elements of the cyclic group $\mathbb{Z}_6$. One easily identifies then

$$-j^2 = q, \quad j = q^2, \quad -1 = q^3, \quad j^2 = q^4, \quad -j = q^5, \quad 1 = q^6.$$

This reminds the color symmetry in Quantum Chromodynamics, where exclusively the "white" combinations of three quarks and three anti-quarks, as well as the "white" quark-anti-quark pairs are declared observable. Replacing the word "white" by 0, we see that there are *two* vanishing linear combinations of *three* powers of q, and *three* pairs of powers of q that are also equal to zero. Indeed, we have:

$$q^2 + q^4 + q^6 = j + j^2 + 1 = 0, \quad \text{and} \quad q + q^3 + q^5 = -j^2 - 1 - j = 0, \tag{10}$$

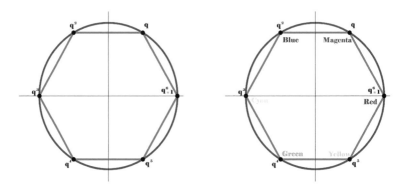

Fig. 3 Representation of the cyclic group Z_6 in the complex plane with three colors and three "anti-colors" attributed to even and odd powers of q, accordingly with colors attributed in Quantum Chromodynamics to quarks and to anti-quarks

as well as

$$q + q^4 = 0, \quad q^2 + q^5 = 0, \quad q^3 + q^6 = 0. \tag{11}$$

The $\mathbb{Z}_6$-grading should unite both $\mathbb{Z}_2$ and $\mathbb{Z}_3$ gradings, reproducing their essential properties. Obviously, the $\mathbb{Z}_3$ subgroup is formed by the elements 1, q^2 and q^4, while the $\mathbb{Z}_2$ subgroup is formed by the elements 1 and $q^3 = -1$. In what follows, we shall see that the associativity imposes many restrictions which can be postponed in the case of non-associative ternary structures (Fig. 3).

3   Ternary and Cubic Algebras

The usual definition of an algebra involves a linear space $\mathscr{A}$ (over real or complex numbers) endowed with a *binary* constitutive relations

$$\mathscr{A} \times \mathscr{A} \to \mathscr{A}. \tag{12}$$

In a finite dimensional case, dim $\mathscr{A}$ = N, in a chosen basis $\mathbf{e}_1, \mathbf{e}_2, \ldots, \mathbf{e}_N$, the constitutive relations (12) can be encoded in *structure constants* f_{ij}^k as follows:

$$\mathbf{e}_i \mathbf{e}_j = f_{ij}^k \, \mathbf{e}_k. \tag{13}$$

With the help of these structure constants all essential properties of a given algebra can be expressed, e.g. they will define a *Lie algebra* if they are anti-symmetric and satisfy the Jacobi identity:

$$f_{ij}^k = -f_{ji}^k, \quad f_{im}^k f_{jl}^m + f_{jm}^k f_{li}^m + f_{lm}^k f_{ij}^m = 0, \tag{14}$$

whereas an abelian algebra will have its structure constants symmetric, $f_{ij}^k = f_{ji}^k$.

Usually, when we speak of algebras, we mean *binary algebras*, understanding that they are defined via *quadratic* constitutive relations (13). On such algebras the notion of Z_2-grading can be naturally introduced. An algebra $\mathscr{A}$ is called a Z_2-graded algebra if it is a direct sum of two parts, with symmetric (abelian) and anti-symmetric product, respectively,

$$\mathscr{A} = \mathscr{A}_0 \oplus \mathscr{A}_1, \tag{15}$$

with *grade* of an element being 0 if it belongs to $\mathscr{A}_0$, and 1 if it belongs to $\mathscr{A}_1$. Under the multiplication in a Z_2-graded algebra the grades add up reproducing the composition law of the Z_2 permutation group: if the grade of an element A is a, and that of the element B is b, then the grade of their product will be $a + b$ *modulo* 2:

$$\text{grade}(AB) = \text{grade}(A) + \text{grade}(B). \tag{16}$$

A Z_2-graded algebra is called a Z_2-graded commutative if for any two homogeneous elements A, B we have

$$AB = (-1)^{ab} BA. \tag{17}$$

It is worthwhile to notice at this point that the above relationship can be written in an alternative form, with all the expressions on the left side as follows:

$$AB - (-1)^{ab} BA = 0, \quad \text{or} \quad AB + (-1)^{(ab+1)} BA = 0 \tag{18}$$

The equivalence between these two alternative definitions of commutation (anti-commutation) relations inside a Z_2-graded algebra is no more possible if by analogy we want to impose *cubic* relations on algebras with Z_3-symmetry properties, in which the non-trivial cubic root of unity, $j = e^{\frac{2\pi i}{3}}$ plays the role similar to that of -1 in the binary relations displaying a Z_2-symmetry [3].

The Z_3 cyclic group is an abelian subgroup of the S_3 symmetry group of permutations of three objects. The S_3 group contains *six* elements, including the group unit e (the identity permutation, leaving all objects in place: $(abc) \rightarrow (abc)$), the two cyclic permutations

$$(abc) \rightarrow (bca) \quad \text{and} \quad (abc) \rightarrow (cab),$$

and three odd permutations,

$$(abc) \rightarrow (cba), \quad (abc) \rightarrow (bac) \quad \text{and} \quad (abc) \rightarrow (acb).$$

There was a unique definition of *commutative* binary algebras given in two equivalent forms,

$$xy + (-1)yx = 0 \quad \text{or} \quad xy = yx. \tag{19}$$

In the case of cubic algebras [6] we have the following four generalizations of the notion of *commutative* algebras:

(a) Generalizing the first form of the commutativity relation (19), which amounts to replacing the -1 generator of $\mathbb{Z}_2$ by j-generator of $\mathbb{Z}_3$ and binary products by products of three elements, we get

$$S: \quad x^\mu x^\nu x^\lambda + j \, x^\nu x^\lambda x^\mu + j^2 \, x^\lambda x^\mu x^\nu = 0, \tag{20}$$

where $j = e^{\frac{2\pi i}{3}}$ is the primitive third root of unity.

(b) Another primitive third root, $j^2 = e^{\frac{4\pi i}{3}}$ can be used in place of the former one; this will define the conjugate algebra $\bar{S}$, satisfying the following cubic constitutive relations:

$$\bar{S}: \quad x^\mu x^\nu x^\lambda + j^2 \, x^\nu x^\lambda x^\mu + j \, x^\lambda x^\mu x^\nu = 0. \tag{21}$$

Clearly enough, both algebras are infinitely dimensional and have the same structure. Each of them is a possible generalization of infinitely dimensional algebra of usual commuting variables with a finite number of generators. In the usual $\mathbb{Z}_2$-graded case such algebras are just polynomials in variables $x^1, x^2, \ldots, x^N$; the algebras S and $\bar{S}$ defined above are also spanned by polynomials, but with different symmetry properties, and as a consequence, with different dimensions corresponding to a given power.

(c) Then we can impose the following "weak" commutation, valid only for cyclic permutations of factors:

$$S_1: \quad x^\mu x^\nu x^\lambda = x^\nu x^\lambda x^\mu \neq x^\nu x^\mu x^\lambda, \tag{22}$$

(d) Finally, we can impose the following "strong" commutation, valid for arbitrary (even or odd) permutations of three factors:

$$S_0: \quad x^\mu x^\nu x^\lambda = x^\nu x^\lambda x^\mu = x^\nu x^\mu x^\lambda \tag{23}$$

The four different associative algebras with cubic commutation relations can be represented in the following diagram, in which all arrows correspond to *surjective homomorphisms*. The commuting generators can be given the common grade 0.

Let us turn now to the $\mathbb{Z}_3$ generalization of anti-commuting generators, which in the usual homogeneous case with $\mathbb{Z}_2$-grading define Grassmann algebras. Here, too, we have four different choices:

(a) The "strong" cubic anti-commutation,

$$\mathscr{L}_0: \quad \Sigma_{\pi \in S_3}\, \theta^{\pi(A)}\theta^{\pi(B)}\theta^{\pi(C)} = 0, \tag{24}$$

i.e., the sum of *all* permutations of three factors, even and odd ones, must vanish.
(b) The somewhat weaker "cyclic" anti-commutation relation,

$$\mathscr{L}_1: \quad \theta^A\theta^B\theta^C + \theta^B\theta^C\theta^A + \theta^C\theta^A\theta^B = 0, \tag{25}$$

i.e., the sum of *cyclic* permutations of three elements must vanish. The same independent relation for the odd combination $\theta^C\theta^B\theta^A$ holds separately.
(c) The j-skew-symmetric algebra:

$$\mathscr{L}: \theta^A\theta^B\theta^C = j\,\theta^B\theta^C\theta^A. \tag{26}$$

and its conjugate algebra $\bar{\mathscr{L}}$, isomorphic with $\mathscr{L}$, which we distinguish by putting a bar on the generators and using dotted indices:
(d) The j^2-skew-symmetric algebra:

$$\bar{\mathscr{L}}: \quad \bar{\theta}^{\dot{A}}\bar{\theta}^{\dot{B}}\bar{\theta}^{\dot{C}} = j^2\bar{\theta}^{\dot{B}}\bar{\theta}^{\dot{C}}\bar{\theta}^{\dot{A}} \tag{27}$$

Both these algebras are finite dimensional. For j or j^2-skew-symmetric algebras with N generators the dimensions of their subspaces of given polynomial order are given by the following generating function:

$$H(t) = 1 + Nt + N^2 t^2 + \frac{N(N-1)(N+1)}{3} t^3, \tag{28}$$

where we include pure numbers (dimension 1), the N generators θ^A (or $\bar{\theta}^{\dot{B}}$), the N^2 independent quadratic combinations $\theta^A\theta^B$, and $N(N-1)(N+1)/3$ products of three generators $\theta^A\theta^B\theta^C$.

It is easy to see that all higher-order monomials starting from 4-th power must identically vanish if associativity holds:

$$\theta^A\theta^B\theta^C\theta^D = j\,\theta^B\theta^C\theta^A\theta^D = j^2\theta^B\theta^A\theta^D\theta^C = j^3\theta^A\theta^D\theta^B\theta^C = j^4\theta^A\theta^B\theta^C\theta^D. \tag{29}$$

As $j^4 = j \neq 1$, the expression $\theta^a\theta^B\theta^C\theta^D$ must identically vanish. The above four cubic generalization of Grassmann algebra are represented in the following diagram, in which all the arrows are surjective homomorphisms.

4 Examples of $\mathbb{Z}_3$-Graded Ternary Algebras

4.1 The $\mathbb{Z}_3$-Graded Analogue of Grassmann Algebra

Let us introduce N generators spanning a linear space over complex numbers, satisfying the following cubic relations [3, 7]:

$$\theta^A \theta^B \theta^C = j\, \theta^B \theta^C \theta^A = j^2\, \theta^C \theta^A \theta^B, \tag{30}$$

with $j = e^{2i\pi/3}$, the primitive root of 1. We have $1 + j + j^2 = 0$ and $\bar{j} = j^2$. It is worth to mention that there are no relations between binary products $\theta^A \theta^B$, i.e. all these products are linearly independent. Let us denote the algebra spanned by the θ^A generators by $\mathscr{A}$.

We shall also introduce a similar set of *conjugate* generators, $\bar{\theta}^{\dot{A}}$, $\dot{A}, \dot{B}, \ldots =$ $1, 2, \ldots, N$, satisfying similar condition with j^2 replacing j:

$$\bar{\theta}^{\dot{A}} \bar{\theta}^{\dot{B}} \bar{\theta}^{\dot{C}} = j^2\, \bar{\theta}^{\dot{B}} \bar{\theta}^{\dot{C}} \bar{\theta}^{\dot{A}} = j\, \bar{\theta}^{\dot{C}} \bar{\theta}^{\dot{A}} \bar{\theta}^{\dot{B}}, \tag{31}$$

Let us denote this algebra by $\bar{\mathscr{A}}$.

We shall endow the algebra $\mathscr{A} \oplus \bar{\mathscr{A}}$ with a natural $\mathbb{Z}_3$ grading, considering the generators θ^A as grade 1 elements, their conjugates $\bar{\theta}^{\dot{A}}$ being of grade 2. The grades add up modulo 3, so that the products $\theta^A \theta^B$ span a linear subspace of grade 2, and the cubic products $\theta^A \theta^B \theta^C$ being of grade 0. Similarly, all quadratic expressions in conjugate generators, $\bar{\theta}^{\dot{A}} \bar{\theta}^{\dot{B}}$ are of grade $2 + 2 = 4 \ (\mathrm{mod}\,3) = 1$, whereas their cubic products are again of grade 0, like the cubic products of θ^A's [8].

Combined with the associativity, these cubic relations impose finite dimension on the algebra generated by the $\mathbb{Z}_3$-graded generators. As a matter of fact, cubic expressions are the highest order that does not vanish identically. The proof is immediate:

$$\theta^A \theta^B \theta^C \theta^D = j\, \theta^B \theta^C \theta^A \theta^D = j^2\, \theta^B \theta^A \theta^D \theta^C = j^3\, \theta^A \theta^D \theta^B \theta^C = j^4\, \theta^A \theta^B \theta^C \theta^D, \tag{32}$$

and because $j^4 = j \neq 1$, the only solution is $\theta^A \theta^B \theta^C \theta^D = 0$.

4.2 The $\mathbb{Z}_3$ Graded Differential Forms

Instead of the usual exterior differential operator satisfying $d^2 = 0$, we can postulate its $\mathbb{Z}_3$-graded generalization satisfying

$$d^2 \neq 0, \quad d^3 f = 0$$

The first differential of a smooth function $f(x^i)$ is as usual

$$df = \partial_i f \, dx^i,$$

whereas the second differential is formally

$$d^2 f = (\partial_k \partial_i f) \, dx^k dx^i + (\partial_i f) \, d^2 x^i$$

We shall attribute the grade 1 to the 1-forms dx^i, $(i, j, k = 1, 2, \dots N)$, and grade 2 to the forms $d^2 x^i$, $(i, j, k = 1, 2, \dots N)$; under associative multiplication of these forms, the grades add up *modulo* 3

$$\text{grade}(\omega \, \theta) = \text{grade}(\omega) + \text{grade}(\theta) \ (modulo 3).$$

The Z_3-graded differential operator d has the following property, compatible with grading we have chosen:

$$d(\omega \, \theta) = (d\omega) \theta + j^{\text{grade}_\omega} \omega \, d\theta.$$

$$d^2 f = (\partial_i \partial_k f) dx^i dx^k + (\partial_i f) \, d^2 x^i,$$

$$d^3 f = (\partial_m \partial_i \partial_k f) dx^m dx^i dx^k + (\partial_i \partial_k f) d^2 x^i dx^k$$

$$+ j \, (\partial_i \partial_k f) dx^i d^2 x^k + (\partial_k \partial_i f) dx^k d^2 x^i + (\partial_i f) \, d^3 x^i.$$

equivalent with

$$d^3 f = (\partial_m \partial_i \partial_k f) dx^m dx^i dx^k + (\partial_i \partial_k f)[d^2 x^k dx^i - j^2 \, dx^i d^2 x^k] + (\partial_i f) \, d^3 x^i.$$

Consequently, assuming that $d^3 x^k = 0$ and $d^3 f = 0$, to make the remaining terms vanish we must impose the following commutation relations on the products of forms:

$$dx^i dx^k dx^m = j \, dx^k dx^m dx^i, \qquad dx^i d^2 x^k = j \, d^2 x^k dx^i,$$

therefore

$$d^2 x^k dx^i = j^2 \, dx^i d^2 x^k$$

As in the case of the abstract Z_3-graded Grassmann algebra, the fourth order expressions must vanish due to the associativity of the product:

$$dx^i dx^k dx^l dx^m = 0.$$

Consequently, we shall assume that also

$$d^2 x^i d^2 x^k = 0.$$

This completes the construction of algebra of Z_3-graded exterior forms (see, e.g., [4, 9, 10]).

4.3   Ternary Clifford Algebra

Let us introduce the following three 3×3 matrices:

$$Q_1 = \begin{pmatrix} 0 & 1 & 0 \\ 0 & 0 & j \\ j^2 & 0 & 0 \end{pmatrix}, \quad Q_2 = \begin{pmatrix} 0 & j & 0 \\ 0 & 0 & 1 \\ j^2 & 0 & 0 \end{pmatrix}, \quad Q_3 = \begin{pmatrix} 0 & 1 & 0 \\ 0 & 0 & 1 \\ 1 & 0 & 0 \end{pmatrix}, \tag{33}$$

and their hermitian conjugates

$$Q_1^{\dagger} = \begin{pmatrix} 0 & 0 & j \\ 1 & 0 & 0 \\ 0 & j^2 & 0 \end{pmatrix}, \quad Q_2^{\dagger} = \begin{pmatrix} 0 & 0 & j \\ j^2 & 0 & 0 \\ 0 & 1 & 0 \end{pmatrix}, \quad Q_3^{\dagger} = \begin{pmatrix} 0 & 0 & 1 \\ 1 & 0 & 0 \\ 0 & 1 & 0 \end{pmatrix}. \tag{34}$$

These matrices can be endowed with natural $\mathbb{Z}_3$-grading,

$$\text{grade}(Q_k) = 1, \quad \text{grade}(Q_k^{\dagger}) = 2, \tag{35}$$

The above matrices span a very interesting ternary algebra. Out of three independent $\mathbb{Z}_3$-graded ternary combinations, only one leads to a non-vanishing result. One can check without much effort that both j and j^2 skew ternary commutators do vanish:

$$\{Q_1, Q_2, Q_3\}_j = Q_1 Q_2 Q_3 + j Q_2 Q_3 Q_1 + j^2 Q_3 Q_1 Q_2 = 0,$$

$$\{Q_1, Q_2, Q_3\}_{j^2} = Q_1 Q_2 Q_3 + j^2 Q_2 Q_3 Q_1 + j Q_3 Q_1 Q_2 = 0,$$

and similarly for the odd permutation, $Q_2 Q_1 Q_3$. On the contrary, the totally symmetric combination does not vanish; it is proportional to the 3×3 identity matrix **1**:

$$Q_a Q_b Q_c + Q_b Q_c Q_a + Q_c Q_a Q_b = 3\, \eta_{abc}\, \mathbf{1}, \quad a, b, \ldots = 1, 2, 3. \tag{36}$$

with η_{abc} given by the following non-zero components:

$$\eta_{111} = \eta_{222} = \eta_{333} = 1, \quad \eta_{123} = \eta_{231} = \eta_{312} = 1, \quad \eta_{213} = \eta_{321} = \eta_{132} = j^2. \tag{37}$$

all other components vanishing. The relation (36) may serve as the definition of *ternary Clifford algebra* [5, 11].

Another set of three matrices is formed by the hermitian conjugates of Q_a, which we shall endow with dotted indices $\dot{a}, \dot{b}, \ldots = 1, 2, 3$: $Q_{\dot{a}} = Q_a^{\dagger}$ satisfying conjugate identities

$$Q_{\dot{a}} Q_{\dot{b}} Q_{\dot{c}} + Q_{\dot{b}} Q_{\dot{c}} Q_{\dot{a}} + Q_{\dot{c}} Q_{\dot{a}} Q_{\dot{b}} = 3 \eta_{\dot{a}\dot{b}\dot{c}} \mathbf{1}, \quad \dot{a}, \dot{b}, \ldots = 1, 2, 3. \qquad (38)$$

with $\eta_{\dot{a}\dot{b}\dot{c}} = \bar{\eta}_{cba}$.

It is obvious that any similarity transformation of the generators Q_a will keep the ternary anti-commutator (36) invariant. As a matter of fact, if we define $\tilde{Q}_b = P^{-1} Q_b P$, with P a non-singular 3×3 matrix, the new set of generators will satisfy the same ternary relations, because

$$\tilde{Q}_a \tilde{Q}_b \tilde{Q}_c = P^{-1} Q_a P P^{-1} Q_b P P^{-1} Q_c P = P^{-1} (Q_a Q_b Q_c) P,$$

and on the right-hand side we have the unit matrix which commutes with all other matrices, so that $P^{-1} \mathbf{1} P = \mathbf{1}$.

It is also worthwhile to note that the six matrices displayed in (33), (34) together with two traceless diagonal matrices

$$B = \begin{pmatrix} 1 & 0 & 0 \\ 0 & j & 0 \\ 0 & 0 & j^2 \end{pmatrix}, \quad B^{\dagger} = \begin{pmatrix} 1 & 0 & 0 \\ 0 & j^2 & 0 \\ 0 & 0 & j \end{pmatrix}$$

form the basis for certain representation of the SU(3), which was shown in the nineties by Kac [12].

5 Generalized $\mathbb{Z}_{\not{j}} \times \mathbb{Z}_{\not{j}}$-Graded Ternary Algebra

Let us suppose that we have binary skew-symmetric and ternary j-skew-symmetric products defined by corresponding structure constants:

$$\xi^{\alpha} \xi^{\beta} = -\xi^{\beta} \xi^{\alpha} \qquad (39)$$

$$\theta^A \theta^B \theta^C = j \, \theta^B \theta^C \theta^A \qquad (40)$$

The unifying ternary relation is of the type Λ_0, i.e.,

$$X^i X^j X^k + X^j X^k X^i + X^k X^i X^j + X^k X^j X^i + X^j X^i X^k + X^i X^k X^j = 0. \qquad (41)$$

It is obviously satisfied by both types of variables; the θ^A's by definition of the product, for which at this stage the associativity property can be not decided yet; the product of grassmannian ξ^{α} variables (39) on the contrary should be associative in order to make the formula (41) applicable.

It can be added that the cubic constitutive relation (40) satisfies a simpler condition with cyclic permutations only,

$$\theta^A \theta^B \theta^C + \theta^B \theta^C \theta^A + \theta^C \theta^A \theta^B = 0,$$

but the cubic products of grassmannian variables are invariant under even (cyclic) permutations, so that only the combination of all six permutations of $\xi^\alpha \xi^\beta \xi^\gamma$, like in (41) will vanish.

Now, if we want to merge the two algebras into a common one, we must impose the general condition (41) to the mixed cubic products (see [13]). These are of two types: $\theta^A \xi^\alpha \theta^B$ and $\xi^\alpha \theta^B \xi^\beta$, with two θ's and one ξ, or with two ξ's and one θ. These identities, all like (41) should follow from *binary* constitutive relations imposed on the *associative* products between one θ and one ξ variable.

Let us suppose that one has

$$\xi^\alpha \theta^B = \omega\, \theta^B \xi^\alpha \quad \text{and consequently} \quad \theta^A \xi^\beta = \omega^{-1} \, \xi^\beta \theta^A. \tag{42}$$

A simple exercise leads to the conclusion that in order to satisfy the general condition (41), the unknown factor ω must verify the equation $\omega + \omega^{-1} + 1 = 0$, or equivalently, $\omega + \omega^2 + \omega^3 = 0$. Indeed, we have, assuming the associativity:

$$\theta^A \xi^\alpha \theta^B = \omega^{-1} \, \xi^\alpha \theta^A \theta^B = \omega\, \theta^A \theta^B \xi^\alpha,$$

$$\theta^B \xi^\alpha \theta^A = \omega^{-1} \, \xi^\alpha \theta^B \theta^A = \omega\, \theta^B \theta^A \xi^\alpha.$$

From this we get, by transforming all the six products so that ξ^α should appear always on the first position in the monomials:

$$\theta^A \xi^\alpha \theta^B = \omega^{-1} \, \xi^\alpha \theta^A \theta^B, \quad \theta^A \theta^B \xi^\alpha = \omega^{-2} \, \xi^\alpha \theta^A \theta^B,$$

$$\theta^B \xi^\alpha \theta^A = \omega^{-1} \, \xi^\alpha \theta^B \theta^A, \quad \theta^B \theta^A \xi^\alpha = \omega^{-2} \, \xi^\alpha \theta^B \theta^A.$$

Adding up all permutations, even (cyclic) and odd alike, we get the following result:

$$\theta^A \xi^\alpha \theta^B + \xi^\alpha \theta^B \theta^A + \theta^B \theta^A \xi^\alpha + \theta^B \xi^\alpha \theta^A + \xi^\alpha \theta^A \theta^B + \theta^A \theta^B \xi^\alpha =$$

$$(1 + \omega + \omega^{-1}) \, \xi^\alpha \theta^A \theta^B + (1 + \omega + \omega^{-1}) \, \xi^\alpha \theta^B \theta^A. \tag{43}$$

The expression in (43) will identically vanish if $\omega = j = e^{\frac{2\pi i}{3}}$ (or $\omega = j^2$, which satisfies the same relation $j + j^2 + 1 = 0$).

The second type of cubic monomials, $\xi^\alpha \theta^B \xi^\beta$, satisfies the identity

$$\xi^\alpha \theta^B \xi^\delta + \theta^B \xi^\delta \xi^\alpha + \xi^\delta \xi^\alpha \theta^B + \xi^\delta \theta^B \xi^\alpha + \theta^B \xi^\alpha \xi^\delta + \xi^\alpha \xi^\delta \theta^B = 0 \tag{44}$$

no matter what the value of ω is chosen in the constitutive relation (42), the anti-symmetry of the product of two ξ's suffices. As a matter of fact, because we have $\xi^\alpha \xi^\delta = -\xi^\delta \xi^\alpha$, in the formula (44) the second term cancels the fifth term, and the third term is cancelled by the sixth one. What remains is the sum of the first and the fourth terms:

$$\xi^\alpha \theta^B \xi^\delta + \xi^\delta \theta^B \xi^\alpha.$$

Now we can transform both terms so as to put the factor θ to the first position; this will give

$$\xi^\alpha \theta^B \xi^\delta + \xi^\delta \theta^B \xi^\alpha = \omega \theta^B \xi^\alpha \xi^\delta + \omega \theta^B \xi^\delta \xi^\alpha = 0 \qquad (45)$$

because of the anti-symmetry between the two ξ's. This completes the construction of the $\mathbb{Z}_2 \times \mathbb{Z}_3$-graded extension of Grassmann algebra.

The existence of *two* cubic roots of unity, j and j^2, suggests that one can extend the above algebraic construction by introducing a set of *conjugate* generators, denoted for convenience with a bar and with dotted indices, satisfying conjugate ternary constitutive relation (31). The unifying condition of vanishing of the sum of all permutations (algebra of Λ_0-type) will be automatically satisfied.

But now we have to extend this condition to the triple products of the type $\theta^A \bar\theta^{\dot B} \theta^C$ and $\bar\theta^{\dot A} \theta^B \bar\theta^{\dot C}$. This will be achieved if we impose the obvious condition, similar to the one proposed already for binary combinations $\xi\theta$:

$$\theta^A \bar\theta^{\dot B} = -j \bar\theta^{\dot B} \theta^A, \quad \bar\theta^{\dot B} \theta^A = -j^2 \theta^A \bar\theta^{\dot B} \qquad (46)$$

The proof of the validity of the condition (41) for the above combinations is exactly the same as for the triple products $\xi^\alpha \theta^B \xi^\gamma$ and $\theta^A \xi^\delta \theta^B$.

We have also to impose commutation relations on the mixed products of the type

$$\xi^\alpha \bar\theta^{\dot B} \xi^\beta \quad \text{and} \quad \bar\theta^{\dot B} \xi^\beta \bar\theta^{\dot C}.$$

It is easy to see that like in the former case, it is enough to impose the commutation rule similar to the former one with θ's, namely

$$\xi^\alpha \bar\theta^{\dot B} = j^2 \bar\theta^{\dot B} \xi^\alpha \qquad (47)$$

Although we could stop at this point the extension of our algebra, for the sake of symmetry it seems useful to introduce the new set of conjugate variables $\bar\xi^{\dot\alpha}$ of the $\mathbb{Z}_2$-graded type. We shall suppose that they anti-commute, like the ξ^β's, and not only between themselves, but also with their conjugates, which means that we assume

$$\bar\xi^{\dot\alpha} \bar\xi^{\dot\beta} = -\bar\xi^{\dot\beta} \bar\xi^{\dot\alpha}, \quad \xi^\alpha \bar\xi^{\dot\beta} = -\bar\xi^{\dot\beta} \xi^\alpha. \qquad (48)$$

This ensures that the condition (41) will be satisfied by any ternary combination of the $\mathbb{Z}_2$-graded generators, including the mixed ones like

$$\bar{\xi}^{\dot{\alpha}} \xi^{\beta} \bar{\xi}^{\dot{\delta}} \quad \text{or} \quad \xi^{\beta} \bar{\xi}^{\dot{\alpha}} \xi^{\gamma} .$$

The dimensions of classical Grassmann algebras with n generators are well known: they are equal to 2^n, with subspaces spanned by the products of k generators having the dimension $C_k^n = n!/(n-k)!k!$. With $2n$ anti-commuting generators, ξ^{α} and $\bar{\xi}^{\dot{\beta}}$ we shall have the dimension of the corresponding Grassmann algebra equal to 2^{2n}.

It is also quite easy to determine the dimension of the $\mathbb{Z}_3$-graded generalizations of Grassmann algebras constructed above (see, e.g., in [7, 14, 15]). The $\mathbb{Z}_3$-graded algebra with N generators θ^A has the total dimension $N + N^2 + (N^3 - N)/3 = (N^3 + N^2 + 2N)/3$. The conjugate algebra, with the same number of generators, has the identical dimension. However, the dimension of the extended algebra unifying both these algebras is not equal to the square of the dimension of one of them because of the extra conditions on the mixed products between the generators and their conjugates, $\theta^A \bar{\theta}^{\dot{B}} = j\, \bar{\theta}^{\dot{B}} \theta^A$.

6 Two Distinct Gradings: $\mathbb{Z}_3 \times \mathbb{Z}_2$ Versus $\mathbb{Z}_6$

The natural choice for the $\mathbb{Z}_3$-graded algebra with cubic relations was to attribute grade 1 to the generators θ^A, and grade 2 to their conjugates $\bar{\theta}^{\dot{B}}$. All other expressions formed by products and powers of those got the well-defined grade, the sum of the grades of factors modulo 3. In a simple Cartesian product of two algebras, a $\mathbb{Z}_3$-graded with a $\mathbb{Z}_2$-graded one, the generators of the latter will be given grade 1, and their products will get automatically the grade which is the sum of the grades of factors modulo 2, which means that all the products and powers of generators ξ^{α} will acquire grade 1 or 0 according to the number and character of factors involved.

The mixed products of the type $\theta^A \xi^{\beta}$, $\xi^{\beta} \theta^B \theta^C$, etc. can be given the double $\mathbb{Z}_3 \times \mathbb{Z}_2$ grade according to (8). According to the isomorphism defined by (9), this is equivalent to a $\mathbb{Z}_6$-grading of the product algebra. As long as the algebra is supposed to be *homogeneous* in the sense that all the constitutive relations contain exclusively terms of *one and the same type*, like in the extension of Grassmann algebra discussed above, the supposed associativity does not impose any particular restrictions.

However, this is not the case if we consider the possibility of *non-homogeneous* constitutive equations, including terms of different nature, but with the same $\mathbb{Z}_6$-grade. The grading defined by (9) suggests a possibility of extending the constitutive relations by comparing terms of the type $\theta^A \theta^B \theta^C$, whose $\mathbb{Z}_6$-grade is 3, to the generators ξ^{α} having the same $\mathbb{Z}_6$-grade. This will lead to the following constitutive relations:

$$\theta^A \theta^B \theta^C = \rho^{ABC}{}_{\alpha}\, \xi^{\alpha} \quad \text{and} \quad \bar{\theta}^{\dot{A}} \bar{\theta}^{\dot{B}} \bar{\theta}^{\dot{C}} = \bar{\rho}^{\dot{A}\dot{B}\dot{C}}{}_{\dot{\alpha}}\, \bar{\xi}^{\dot{\alpha}} \tag{49}$$

with the coefficients (structure constants) $\rho^{ABC}{}_\alpha$ and $\bar{\rho}^{\dot{A}\dot{B}\dot{C}}{}_{\dot{\alpha}}$ displaying obvious symmetry properties mimicking the properties of ternary products of θ-generators with respect to cyclic permutations:

$$\rho^{ABC}{}_\alpha = j\,\rho^{BCA}{}_\alpha = j^2\,\rho^{CAB}{}_\alpha \quad \text{and} \quad \bar{\rho}^{\dot{A}\dot{B}\dot{C}}{}_{\dot{\alpha}} = j^2\,\bar{\rho}^{\dot{B}\dot{C}\dot{A}}{}_{\dot{\alpha}} = j\,\bar{\rho}^{\dot{C}\dot{A}\dot{B}}{}_{\dot{\alpha}}. \tag{50}$$

If all products are supposed to be associative, then we see immediately that the products between θ and ξ generators, as well as those between $\bar{\theta}$ and $\bar{\xi}$ generators must vanish identically, because of the vanishing of quartic products $\theta\theta\theta\theta = 0$ and $\bar{\theta}\bar{\theta}\bar{\theta}\bar{\theta} = 0$. This means that we must set

$$\theta^A\,\xi^\beta = 0, \quad \xi^\beta\theta^A = 0, \quad \text{as well as} \quad \bar{\theta}^{\dot{B}}\bar{\xi}^{\dot{\alpha}} = 0, \quad \bar{\xi}^{\dot{\alpha}}\bar{\theta}^{\dot{B}} = 0. \tag{51}$$

But now we want to unite the two gradings into a unique common one. Let us start by defining a ternary product of generators, not necessarily derived from an ordinary associative algebra. We shall just suppose the existence of ternary product of generators, displaying the j-skew symmetry property:

$$\{\theta^A, \theta^B, \theta^C\} = j\{\theta^B, \theta^C, \theta^A\} = j^2\{\theta^C, \theta^A, \theta^B\}. \tag{52}$$

and similarly, for the conjugate generators,

$$\{\bar{\theta}^{\dot{A}}, \bar{\theta}^{\dot{B}}, \bar{\theta}^{\dot{C}}\} = j^2\,\{\bar{\theta}^{\dot{B}}, \bar{\theta}^{\dot{C}}, \bar{\theta}^{\dot{A}}\} = j\,\{\bar{\theta}^{\dot{C}}, \bar{\theta}^{\dot{A}}, \bar{\theta}^{\dot{B}}\}. \tag{53}$$

Let us attribute the $\mathbb{Z}_6$-grade 1 to the generators θ^A. Then it is logical to attribute the $\mathbb{Z}_6$ grade 5 to the conjugate generators $\bar{\theta}^{\dot{B}}$, so that mixed products $\theta^A\bar{\theta}^{\dot{B}}$ would be of $\mathbb{Z}_6$ grade 0. Ternary products (52) are of grade 3, and ternary products of conjugate generators (53) are also of grade 3, because $5 + 5 + 5 = 15$, and 15 *modulo* $6 = 3$. But we have also $q^3 = -1$, which is the generator of the $\mathbb{Z}_2$-subalgebra of $\mathbb{Z}_6$. Therefore we should attribute the $\mathbb{Z}_6$-grade 3 to both kinds of the anti-commuting variables, ξ^α and $\bar{\xi}^{\dot{\beta}}$, because we can write their constitutive relations using the root q as follows:

$$\xi^\alpha\xi^\beta = -\xi^\beta\xi^\alpha = q^3\,\xi^\beta\xi^\alpha, \quad \bar{\xi}^{\dot{\alpha}}\bar{\xi}^{\dot{\beta}} = -\bar{\xi}^{\dot{\beta}}\bar{\xi}^{\dot{\alpha}} = q^3\,\bar{\xi}^{\dot{\beta}}\bar{\xi}^{\dot{\alpha}},$$

$$\xi^\alpha\bar{\xi}^{\dot{\beta}} = -\bar{\xi}^{\dot{\beta}}\xi^\alpha = q^3\,\bar{\xi}^{\dot{\beta}}\xi^\alpha, \tag{54}$$

On the other hand, we can consider products of θ with $\bar{\xi}$ and $\bar{\theta}$ with ξ:

$$\theta^A\bar{\xi}^{\dot{\alpha}} \quad \text{and} \quad \bar{\theta}^{\dot{B}}\xi^\beta$$

The first expression has the $\mathbb{Z}_6$-grade $1 + 3 = 4$, and the second product has the $\mathbb{Z}_6$-grade $5 + 3 = 8$ *modulo* $6 = 2$. Other products endowed with the same grade in our associative $\mathbb{Z}_6$-grade algebra are $\bar{\theta}^{\dot{A}}\bar{\theta}^{\dot{B}}$ (grade 4, because $5 + 5 = 10$ *modulo* $6 = 4$), and $\theta^A\theta^B$ (grade 2, because $1 + 1 = 2$).

This suggests that the following non-homogeneous constitutive relations can be proposed:

$$\theta^A \bar{\xi}^{\dot{\alpha}} = f^{A\dot{\alpha}}{}_{\dot{C}\dot{D}} \, \bar{\theta}^{\dot{C}} \bar{\theta}^{\dot{D}}, \quad \text{and} \quad \bar{\theta}^{\dot{A}} \xi^{\alpha} = \bar{f}^{\dot{A}\alpha}{}_{CD} \, \theta^C \theta^D, \tag{55}$$

where the coefficients should display the symmetry properties contravariant to those of the generators themselves, which means that we should have

$$f^{A\dot{\alpha}}{}_{\dot{C}\dot{D}} = j^2 \, f^{\dot{\alpha}A}{}_{\dot{C}\dot{D}} \quad \text{and} \quad \bar{f}^{\dot{A}\alpha}{}_{CD} = j \, \bar{f}^{\alpha\dot{A}}{}_{CD} \tag{56}$$

7 Low-Dimensional Algebras

Let us consider the simplest case of a $\mathbb{Z}_2$-graded algebra spanned by two generators ξ^{α}, $\alpha, \beta = 1, 2$. The anti-commutation property can be encoded in the invariant 2-form $\varepsilon_{\alpha\beta}$. We can obviously write

$$\varepsilon_{\alpha\beta} \xi^{\alpha} \xi^{\beta} = \varepsilon_{\beta\alpha} \xi^{\beta} \xi^{\alpha} = -\varepsilon_{\beta\alpha} \xi^{\alpha} \xi^{\beta},$$

from which we conclude that $\varepsilon_{\alpha\beta} = -\varepsilon_{\beta\alpha}$. We can choose the basis in which

$$\varepsilon_{11} = 0, \quad \varepsilon_{22} = 0, \quad \varepsilon_{12} = -\varepsilon_{21} = 1.$$

After a change of basis, $\xi^{\beta} \rightarrow S^{\alpha'}_{\beta} \xi^{\beta} = \eta^{\alpha'}$ the 2-form $\varepsilon_{\alpha\beta}$, as any tensor, also undergoes the inverse transformation:

$$S^{\alpha}_{\alpha'} S^{\beta}_{\beta'} \, \varepsilon_{\alpha\beta},$$

with $S^{\alpha}_{\beta'}$ the inverse matrix of the matrix $S^{\beta'}_{\beta}$. Whatever non-singular linear transformation $S^{\beta}_{\beta'}$ is chosen, the new components $\varepsilon_{\alpha'\beta'}$ remain anti-symmetric, but they have not necessarily the same values as those of $\varepsilon_{\alpha\beta}$. However, if we require that also in new basis

$$\varepsilon_{1'1'} = 0, \quad \varepsilon_{2'2'} = 0, \quad \varepsilon_{1'2'} = -\varepsilon_{2'1'} = 1,$$

then it is easy to show that this imposes extra condition on the 2×2 matrix $S^{\alpha'}_{\beta}$, namely that $\det S = 1$. This defines the $SL(2, \mathbb{C})$ group as the group of invariance of the subalgebra spanned by two anti-commuting generators ξ^{α}, $\alpha, \beta, \ldots = 1, 2$. This may be considered as the most elementary example of the spin–statistics relationship established by Pauli ([16], see also [17]).

Now let us turn to the invariance properties of the ternary subalgebra spanned by two generators θ^1, θ^2, satisfying homogeneous cubic cyclic j-skew-symmetric relations $\theta^A\theta^B\theta^C = j\,\theta^B\theta^C\theta^A$, and their conjugate counterparts $\bar{\theta}^{\dot{1}}$, $\bar{\theta}^{\dot{2}}$ satisfying homogeneous cubic cyclic j^2-skew-symmetric relations $\bar{\theta}^{\dot{A}}\bar{\theta}^{\dot{B}}\bar{\theta}^{\dot{C}} = j^2\,\bar{\theta}^{\dot{B}}\bar{\theta}^{\dot{C}}\bar{\theta}^{\dot{A}}$

We shall also impose *binary* constitutive relations between the generators θ^A and their conjugate counterparts $\bar{\theta}^{\dot{B}}$, making the choice consistent with the introduced $\mathbb{Z}_6$-grading

$$\theta^A\bar{\theta}^{\dot{B}} = -j\,\bar{\theta}^{\dot{B}}\theta^A, \qquad \bar{\theta}^{\dot{B}}\theta^A = -j^2\,\theta^A\bar{\theta}^{\dot{B}}. \tag{57}$$

Consider a tri-linear form ρ^α_{ABC}. We shall call this form $\mathbb{Z}_3$-invariant if we can write:

$$\rho^\alpha_{ABC}\,\theta^A\theta^B\theta^C = \frac{1}{3}\left[\rho^\alpha_{ABC}\,\theta^A\theta^B\theta^C + \rho^\alpha_{BCA}\,\theta^B\theta^C\theta^A + \rho^\alpha_{CAB}\,\theta^C\theta^A\theta^B\right] =$$

$$= \frac{1}{3}\left[\rho^\alpha_{ABC}\,\theta^A\theta^B\theta^C + \rho^\alpha_{BCA}\,(j^2\,\theta^A\theta^B\theta^C) + \rho^\alpha_{CAB}\,j\,(\theta^A\theta^B\theta^C)\right], \tag{58}$$

by virtue of the commutation relations (30). From this it follows that we should have

$$\rho^\alpha_{ABC}\,\theta^A\theta^B\theta^C = \frac{1}{3}\left[\rho^\alpha_{ABC} + j^2\,\rho^\alpha_{BCA} + j\,\rho^\alpha_{CAB}\right]\theta^A\theta^B\theta^C, \tag{59}$$

from which we get the following properties of the ρ-cubic matrices:

$$\rho^\alpha_{ABC} = j^2\,\rho^\alpha_{BCA} = j\,\rho^\alpha_{CAB}. \tag{60}$$

Even in this minimal and discrete case, there are covariant and contravariant indices: the lower and the upper indices display the inverse transformation property. If a given cyclic permutation is represented by a multiplication by j for the upper indices, the same permutation performed on the lower indices is represented by multiplication by the inverse, i.e. j^2, so that they compensate each other.

Similar reasoning leads to the definition of the conjugate forms $\bar{\rho}^{\dot{\alpha}}_{\dot{C}\dot{B}\dot{A}}$ satisfying the relations similar to (60) with j replaced by its conjugate, j^2:

$$\bar{\rho}^{\dot{\alpha}}_{\dot{A}\dot{B}\dot{C}} = j\,\bar{\rho}^{\dot{\alpha}}_{\dot{B}\dot{C}\dot{A}} = j^2\,\bar{\rho}^{\dot{\alpha}}_{\dot{C}\dot{A}\dot{B}} \tag{61}$$

In the simplest case of two generators, the j-skew-invariant forms have only two independent components:

$$\rho^1_{121} = j^2\,\rho^1_{211} = j\,\rho^1_{112},$$

$$\rho^2_{212} = j^2\,\rho^2_{122} = j\,\rho^2_{221},$$

and we can set

$$\rho^1_{121} = 1, \quad \rho^1_{211} = j, \quad \rho^1_{112} = j^2,$$

$$\rho^2_{212} = 1, \quad \rho^2_{122} = j, \quad \rho^2_{221} = j^2.$$

The constitutive cubic relations between the generators of the $\mathbb{Z}_3$-graded algebra can be considered as intrinsic if they are conserved after linear transformations with commuting (pure number) coefficients, i.e. if they are independent of the choice of the basis.

Let $U^{A'}_A$ denote a non-singular $N \times N$ matrix, transforming the generators θ^A into another set of generators, $\theta^{B'} = U^{B'}_B \theta^B$. In principle, the generators of the $\mathbb{Z}_2$-graded subalgebra ξ^α may or may not undergo a change of basis. Uniting the two subalgebras in one $\mathbb{Z}_6$-graded algebra suggests that a change of basis should concern all generators at once, both ξ^α and θ^A. This means a simultaneous change of basis

$$\xi^\alpha \to \tilde{\xi}^{\beta'} = S^{\beta'}_\alpha \xi^\alpha \, \xi^\alpha, \quad \theta^A \to \tilde{\theta}^{B'} = U^{B'}_A \theta^A. \tag{62}$$

It seems natural to identify the upper indices α, β appearing in the ρ-tensors with the indices appearing in the generators ξ^α of the $\mathbb{Z}_2$-graded subalgebra. Therefore, we are looking for the solution of the simultaneous invariance condition for the $\epsilon_{\alpha\beta}$ and ρ^α_{ABC} tensors:

$$\epsilon_{\alpha'\beta'} = S^\alpha_{\alpha'} S^\beta_{\beta'} \, \epsilon_{\alpha\beta}, \quad S^{\alpha'}_\beta \, \rho^\beta_{ABC} = U^{A'}_A U^{B'}_B U^{C'}_C \, \rho^{\alpha'}_{A'B'C'}, \tag{63}$$

so that in new basis the numerical values of both tensors remain the same as before, just like the components of the Minkowskian space-time metric tensor $g_{\mu\nu}$ remain unchanged under the Lorentz transformations. Notice that in the last formula above, (63), the matrix $S^{\alpha'}_\alpha$ is the inverse matrix for $S^\alpha_{\alpha'}$ appearing in the transformation of the basis ξ^β.

Now, $\rho^1_{121} = 1$, and we have two equations corresponding to the choice of values of the index α' equal to 1 or 2. For $\alpha' = 1'$ the ρ-matrix on the right-hand side is $\rho^{1'}_{A'B'C'}$, which has only three components,

$$\rho^{1'}_{1'2'1'} = 1, \quad \rho^{1'}_{2'1'1'} = j, \quad \rho^{1'}_{1'1'2'} = j^2,$$

which leads to the following equation:

$$S^{1'}_1 = U^{1'}_1 U^{2'}_2 U^{1'}_1 + j \, U^{2'}_1 U^{1'}_2 U^{1'}_1 + j^2 \, U^{1'}_1 U^{1'}_2 U^{2'}_1 = U^{1'}_1 (U^{2'}_2 U^{1'}_1 - U^{2'}_1 U^{1'}_2), \tag{64}$$

because $j + j^2 = -1$. For the alternative choice $\alpha' = 2'$ the ρ-matrix on the right-hand side is $\rho^{2'}_{A'B'C'}$, whose three non-vanishing components are

$$\rho^{2'}_{2'1'2'} = 1, \quad \rho^{2'}_{1'2'2'} = j, \quad \rho^{2'}_{2'2'1'} = j^2.$$

The corresponding equation becomes now:

$$S_1^{2'} = U_1^{2'} U_2^{1'} U_1^{2'} + j\, U_1^{1'} U_2^{2'} U_1^{2'} + j^2\, U_1^{2'} U_2^{2'} U_1^{1'} = U_1^{2'} (U_2^{1'} U_1^{2'} - U_1^{1'} U_2^{2'}).$$
(65)

The two remaining equations are obtained in a similar manner. We choose now the three lower indices on the left-hand side equal to another independent combination, (212). Then the ρ-matrix on the left-hand side must be ρ^2 whose component ρ_{212}^2 is equal to 1. This leads to the following equation when $\alpha' = 1'$:

$$S_2^{1'} = U_2^{1'} U_1^{2'} U_2^{1'} + j\, U_2^{2'} U_1^{1'} U_2^{1'} + j^2\, U_2^{1'} U_1^{1'} U_2^{2'} = U_2^{1'} (U_2^{1'} U_1^{2'} - U_1^{1'} U_2^{2'}),$$
(66)

and the fourth equation corresponding to $\alpha' = 2'$ is:

$$S_2^{2'} = U_2^{2'} U_1^{1'} U_2^{2'} + j\, U_2^{1'} U_1^{2'} U_2^{2'} + j^2\, U_2^{2'} U_1^{2'} U_2^{1'} = U_2^{2'} (U_1^{1'} U_2^{2'} - U_1^{2'} U_2^{1'}).$$
(67)

The determinant of the 2×2 complex matrix $U_B^{A'}$ appears everywhere on the right-hand side. Indeed Eq. (64) can be written in the form

$$S_1^{1'} = U_1^{1'} [\det(U)].$$
(68)

The remaining two equations are obtained in a similar manner, resulting in the following:

$$S_1^{2'} = -U_1^{2'} [\det(U)], \quad S_2^{1'} = -U_2^{1'} [\det(U)], \quad S_2^{2'} = U_2^{2'} [\det(U)].$$
(69)

The determinant of the 2×2 complex matrix $U_B^{A'}$ appears everywhere on the right-hand side. Taking the determinant of the matrix $S_\beta^{\alpha'}$ one gets immediately

$$\det(S) = [\det(U)]^3.$$
(70)

However, the U-matrices on the right-hand side are defined only up to the phase, which is due to the cubic character of the covariance relations (64)–(69), and they can take on three different values: 1, j, or j^2, i.e. the matrices $j\, U_B^{A'}$ or $j^2\, U_B^{A'}$ satisfy the same relations as the matrices $U_B^{A'}$ defined above. The determinant of U can take on the values 1, j, or j^2 if $det(S) = 1$

Another reason to impose the unitarity condition is as follows. It can be derived if we require the same behavior for the duals, ρ_β^{DEF}. This extra condition amounts to the invariance of the anti-symmetric tensor ϵ^{AB}, and this is possible only if the determinant of U-matrices is 1 (or j or j^2), because only cubic combinations of these matrices appear in the transformation law for ρ-forms.

We have determined the invariance group for the simultaneous change of the basis in our $\mathbb{Z}_6$-graded algebra. However, these transformations based on the $SL(2, \mathbb{C})$ groups combined with complex representation of the $\mathbb{Z}_3$ cyclic group keep invariant the binary constitutive relations between the $\mathbb{Z}_2$-graded generators ξ^α and the ternary constitutive relations between the $\mathbb{Z}_3$-graded generators alone, without mentioning their conjugates $\bar\xi^{\dot\alpha}$ and $\bar\theta^{\dot B}$.

Let us put aside for the moment the conjugate $\mathbb{Z}_2$-graded variables, and concentrate our attention on the conjugate $\mathbb{Z}_3$-graded generators $\bar\theta^{\dot A}$ and their commutation relations with θ^B generators, which we shall modify as:

$$\theta^A\,\bar\theta^{\dot B} = -j\,\bar\theta^{\dot B}\,\theta^A, \qquad \bar\theta^{\dot B}\,\theta^A = -j^2\,\theta^A\,\bar\theta^{\dot B}. \tag{71}$$

A similar covariance requirement can be formulated with respect to the set of 2-forms mapping the quadratic $\theta^A\,\bar\theta^{\dot B}$ combinations into a four-dimensional linear real space.

It is easy to see, by counting the independent combinations of dotted and undotted indices, that the symmetry (71) imposed on these expressions reduces their number to four: $(1\dot1)$, $(1,\dot2)$, $(2\ \dot1)$, $(2,\dot2)$, the conjugate combinations of the type $(\dot A\ B)$ being dependent on the first four because of the imposed symmetry properties.

Let us define two quadratic forms, $\pi^\mu_{A\dot B}$ and its conjugate $\bar\pi^\mu_{\dot B A}$

$$\pi^\mu_{A\dot B}\,\theta^A\bar\theta^{\dot B} \quad\text{and}\quad \bar\pi^\mu_{\dot B A}\,\bar\theta^{\dot B}\theta^A. \tag{72}$$

The Greek indices $\mu, \nu \dots$ take on four values, and we shall label them 0, 1, 2, 3.

The four tensors $\pi^\mu_{A\dot B}$ and their hermitian conjugates $\bar\pi^\mu_{\dot B A}$ define a bilinear mapping from the product of quark and anti-quark cubic algebras into a linear four-dimensional vector space, whose structure is not yet defined.

Let us impose the following invariance condition:

$$\pi^\mu_{A\dot B}\,\theta^A\bar\theta^{\dot B} = \bar\pi^\mu_{\dot B A}\,\bar\theta^{\dot B}\theta^A. \tag{73}$$

It follows immediately from (71) that

$$\pi^\mu_{A\dot B} = -j^2\,\bar\pi^\mu_{\dot B A}. \tag{74}$$

Such matrices are non-hermitian, and they can be realized by the following substitution:

$$\pi^\mu_{A\dot B} = j\,i\,\sigma^\mu_{A\dot B}, \qquad \bar\pi^\mu_{\dot B A} = -j^2\,i\,\sigma^\mu_{\dot B A} \tag{75}$$

where $\sigma^\mu_{A\dot B}$ are the unit 2 matrix for $\mu = 0$, and the three hermitian Pauli matrices for $\mu = 1, 2, 3$.

Again, we want to get the same form of these four matrices in another basis. Knowing that the lower indices A and $\dot B$ undergo the transformation with matrices $U^{A'}_B$ and $\bar U^{\dot A'}_{\dot B}$, we demand that there exist some 4×4 matrices $\Lambda^{\mu'}_\nu$ representing the transformation of lower indices by the matrices U and $\bar{U}$:

$$\Lambda^{\mu'}_{\nu}\, \pi^{\nu}_{A\dot{B}} = U^{A'}_{A}\, \bar{U}^{\dot{B}'}_{\dot{B}}\, \pi^{\mu'}_{A'\dot{B}'}, \tag{76}$$

It is clear that we can replace the matrices $\pi^{\nu}_{A\dot{B}}$ by the corresponding matrices $\sigma^{\nu}_{A\dot{B}}$, and this defines the vector (4×4) representation of the Lorentz group.

The first four equations relating the 4×4 real matrices $\Lambda^{\mu'}_{\nu}$ with the 2×2 complex matrices $U^{A'}_{B}$ and $\bar{U}^{\dot{A}'}_{\dot{B}}$ are as follows:

$$\Lambda^{0'}_{0} - \Lambda^{0'}_{3} = U^{1'}_{2}\, \bar{U}^{\dot{1}'}_{\dot{2}} + U^{2'}_{2}\, \bar{U}^{\dot{2}'}_{\dot{2}},$$

$$\Lambda^{0'}_{0} + \Lambda^{0'}_{3} = U^{1'}_{1}\, \bar{U}^{\dot{1}'}_{\dot{1}} + U^{2'}_{1}\, \bar{U}^{\dot{2}'}_{\dot{1}},$$

$$\Lambda^{0'}_{0} - i\Lambda^{0'}_{2} = U^{1'}_{1}\, \bar{U}^{\dot{1}'}_{\dot{2}} + U^{2'}_{1}\, \bar{U}^{\dot{2}'}_{\dot{2}},$$

$$\Lambda^{0'}_{0} + i\Lambda^{0'}_{2} = U^{1'}_{2}\, \bar{U}^{\dot{1}'}_{\dot{1}} + U^{2'}_{2}\, \bar{U}^{\dot{2}'}_{\dot{1}}.$$

The next four equations relating the 4×4 real matrices $\Lambda^{\mu'}_{\nu}$ with the 2×2 complex matrices $U^{A'}_{B}$ and $\bar{U}^{\dot{A}'}_{\dot{B}}$ are as follows:

$$\Lambda^{1'}_{0} - \Lambda^{1'}_{3} = U^{1'}_{2}\, \bar{U}^{\dot{2}'}_{\dot{2}} + U^{2'}_{2}\, \bar{U}^{\dot{1}'}_{\dot{2}},$$

$$\Lambda^{1'}_{0} + \Lambda^{1'}_{3} = U^{1'}_{1}\, \bar{U}^{\dot{2}'}_{\dot{1}} + U^{2'}_{1}\, \bar{U}^{\dot{1}'}_{\dot{1}},$$

$$\Lambda^{1'}_{1} - i\Lambda^{1'}_{2} = U^{1'}_{1}\, \bar{U}^{\dot{2}'}_{\dot{2}} + U^{2'}_{1}\, \bar{U}^{\dot{1}'}_{\dot{2}},$$

$$\Lambda^{1'}_{1} + i\Lambda^{1'}_{2} = U^{1'}_{2}\, \bar{U}^{\dot{2}'}_{\dot{1}} + U^{2'}_{2}\, \bar{U}^{\dot{1}'}_{\dot{1}}.$$

We skip the next two groups of four equations corresponding to the "spatial" indices 2 and 3, reproducing the same scheme as the last four equations with the space index equal to 1.

It can be checked that now $\det\,(\Lambda) = [\det U]^2\, [\det \bar{U}]^2$. The group of transformations thus defined is $SL(2, \mathbb{C})$, which is the covering group of the Lorentz group. With the invariant "spinorial metric" in two complex dimensions, ε^{AB} and $\varepsilon^{\dot{A}\dot{B}}$ such that $\varepsilon^{12} = -\varepsilon^{21} = 1$ and $\varepsilon^{\dot{1}\dot{2}} = -\varepsilon^{\dot{2}\dot{1}}$, we can define the contravariant components $\pi^{\nu\,A\dot{B}}$. It is easy to show that the Minkowskian space-time metric, invariant under the Lorentz transformations, can be defined as

$$g^{\mu\nu} = \frac{1}{2}\left[\pi^{\mu}_{A\dot{B}}\, \pi^{\nu\,A\dot{B}}\right] = diag(+, -, -, -). \tag{77}$$

Together with the anti-commuting spinors ψ^α the four real coefficients defining a Lorentz vector, $x_\mu \pi^\mu_{A\dot{B}}$, can generate now the supersymmetry via standard definitions of super-derivations. Let us then choose the matrices $\Lambda^{\alpha'}_\beta$ to be the usual spinor representation of the $SL(2, \mathbb{C})$ group, while the matrices $U^{A'}_B$ will be defined as follows:

$$U^{1'}_1 = j \Lambda^{1'}_1, \quad U^{1'}_2 = -j \Lambda^{1'}_2, \quad U^{2'}_1 = -j \Lambda^{2'}_1, \quad U^{2'}_2 = j \Lambda^{2'}_2, \tag{78}$$

the determinant of U being equal to j^2. Obviously, the same reasoning leads to the conjugate cubic representation of $SL(2, \mathbb{C})$ if we require the covariance of the conjugate tensor

$$\bar{\rho}^{\dot{\beta}}_{\dot{D}\dot{E}\dot{F}} = j\, \bar{\rho}^{\dot{\beta}}_{\dot{E}\dot{F}\dot{D}} = j^2\, \bar{\rho}^{\dot{\beta}}_{\dot{F}\dot{D}\dot{E}},$$

by imposing the equation similar to (63)

$$\Lambda^{\dot{\alpha}'}_{\dot{\beta}}\, \bar{\rho}^{\dot{\beta}}_{\dot{A}\dot{B}\dot{C}} = \bar{\rho}^{\dot{\alpha}'}_{\dot{A}'\dot{B}'\dot{C}'}\, \bar{U}^{\dot{A}'}_{\dot{A}}\, \bar{U}^{\dot{B}'}_{\dot{B}}\, \bar{U}^{\dot{C}'}_{\dot{C}}. \tag{79}$$

The matrix $\bar{U}$ is the complex conjugate of the matrix U, with determinant equal to j.

8 Ternary Dirac Equation

After the discovery of spin of the electron (the Stern-Gerlach experiment [18]), Pauli understood that a Schroedinger equation involving only one complex-valued wave function is not enough to take into account this new degree of freedom, and proposed to describe the dichotomic spin variable by introducing a two-component function forming a column on which hermitian matrices can act as linear operators [19].

The basis of complex traceless 2×2 hermitian matrices contains just three elements since then known as *Pauli matrices*:

$$\sigma_1 = \begin{pmatrix} 0 & 1 \\ 1 & 0 \end{pmatrix}, \quad \sigma_2 = \begin{pmatrix} 0 & -i \\ i & 0 \end{pmatrix}, \quad \sigma_1 = \begin{pmatrix} 1 & 0 \\ 0 & -1 \end{pmatrix},$$

which can be arranged in a single 3-covector $\boldsymbol{\sigma} = [\sigma_1, \sigma_2, \sigma_3]$.

The simplest linear relation between the operators of energy, mass, and momentum acting on a column vector (called *a Pauli spinor*) would read then:

$$\begin{pmatrix} E & 0 \\ 0 & E \end{pmatrix} \begin{pmatrix} \psi^1 \\ \psi^2 \end{pmatrix} = \begin{pmatrix} mc^2 & 0 \\ 0 & mc^2 \end{pmatrix} \begin{pmatrix} \psi^1 \\ \psi^2 \end{pmatrix} + c\, \boldsymbol{\sigma} \cdot \mathbf{p} \begin{pmatrix} \psi^1 \\ \psi^2 \end{pmatrix}, \tag{80}$$

where

$$\boldsymbol{\sigma} \cdot \mathbf{p} = \sigma_1 \, p^1 + \sigma_2 \, p^2 + \sigma_3 \, p^3 = \begin{pmatrix} p^3 & p^1 - i \, p^2 \\ p^1 + i \, p^2 & -p^3 \end{pmatrix}.$$

We can write (80) in a simplified manner, denoting the Pauli spinor by one letter ψ and treating the unit matrix symbolically like a number:

$$E \, \psi = mc^2 \, \psi + \boldsymbol{\sigma} \cdot \mathbf{p} \, \psi. \tag{81}$$

Such an equation is not invariant under Lorentz transformations. Indeed, by iterating, which amounts just to take the square of the same operator, we arrive at the following relation: between the operators of energy and momentum and the mass of the particle:

$$E^2 = m^2 c^4 + 2 \, mc^3 \mid \mathbf{p} \mid^2 \boldsymbol{\sigma} \cdot \mathbf{p} + c^2 \mathbf{p}^2, \tag{82}$$

instead of the relativistic relation

$$E^2 - c^2 \, \mathbf{p}^2 = m^2 c^4. \tag{83}$$

The double product in the expression for the energy squared can be removed if one introduces a second Pauli spinor satisfying a similar equation, and in such way that the two equations intertwine the two spinors. So let the first Pauli spinor be denoted by ψ_+ and the second one by ψ_-, and let them satisfy the following coupled system of equations:

$$E \, \psi_+ = mc^2 \, \psi_+ + \boldsymbol{\sigma} \cdot \mathbf{p} \, \psi_-,$$

$$E \, \psi_- = -mc^2 \, \psi_- + \boldsymbol{\sigma} \cdot \mathbf{p} \, \psi_+, \tag{84}$$

which coincides with the relativistic equation for the electron found by Dirac a few years later [20]. The relativistic invariance is now manifest, because due to the negative mass term in the second equation, the iteration leads to the separation of variables, and all the components satisfy the desired relation

$$[E^2 - c^2 \mathbf{p}^2] \psi_+ = m^2 c^4 \, \psi_+, \quad [E^2 - c^2 \mathbf{p}^2] \psi_- = m^2 c^4 \, \psi_-.$$

The side effect of this modification was the presence of solutions with negative mass, which led Dirac to the conclusion that the "holes" in the sea of such solutions could be interpreted as electrons with the same mass as the usual ones, but with opposite charge.

But at the time Pauli was considering the inclusion of spin of the electron in a quantum mechanical Schroedinger-like equation, the positron has not yet been discovered, and the introduction of negative mass states seemed absurd. This is why Pauli introduced the following non-relativistic equation:

$$E\ \psi = \left[\frac{1}{2m}(\boldsymbol{\sigma}\cdot(\mathbf{p}-e\mathbf{A}))^2 + eV(x)\right]\ \psi. = \left[\frac{1}{2m}(\mathbf{p}-e\mathbf{A})^2 + e\boldsymbol{\sigma}\cdot\mathbf{B} + eV(x)\right]\ \psi. \tag{85}$$

Later on it turned out that the Pauli equation (85) is the non-relativistic limit of the Dirac equation.

The two Eqs. (84) can be re-written using a matrix notation:

$$\begin{pmatrix} E & 0 \\ 0 & E \end{pmatrix}\begin{pmatrix} \psi_+ \\ \psi_- \end{pmatrix} = \begin{pmatrix} mc^2 & 0 \\ 0 & -mc^2 \end{pmatrix}\begin{pmatrix} \psi_+ \\ \psi_- \end{pmatrix} + \begin{pmatrix} 0 & c\,\boldsymbol{\sigma}\mathbf{p} \\ \boldsymbol{\sigma}\mathbf{p} & 0 \end{pmatrix}\begin{pmatrix} \psi_+ \\ \psi_- \end{pmatrix}, \tag{86}$$

where the entries in the energy operator and the mass matrix are in fact 2×2 identity matrices, as well as the sigma-matrices appearing in the last matrix, so that in reality the above equation represents the 4×4 Dirac equation, only in a different basis [20].

Now we want to describe three different two-component fields (which can be incidentally given the names of three colors [21, 22], the "red" one φ_+, the "blue" one χ_+, and the "green" one ψ_+); more explicitly,

$$\varphi_+ = \begin{pmatrix} \varphi_+^1 \\ \varphi_+^2 \end{pmatrix}, \quad \chi_+ = \begin{pmatrix} \chi_+^1 \\ \chi_+^2 \end{pmatrix}, \quad \psi_+ = \begin{pmatrix} \psi_+^1 \\ \psi_+^2 \end{pmatrix}, \tag{87}$$

In order to satisfy the required existence of anti-particles, we should introduce three "anti-colors," denoted by the "minus" underscript, corresponding to the opposite colors: "cyan" for φ_-, "yellow" for χ_-, and "magenta" for ψ_-; here, two, we have to do with two-component columns:

$$\varphi_- = \begin{pmatrix} \varphi_-^1 \\ \varphi_-^2 \end{pmatrix}, \quad \chi_- = \begin{pmatrix} \chi_-^1 \\ \chi_-^2 \end{pmatrix}, \quad \psi_- = \begin{pmatrix} \psi_-^1 \\ \psi_-^2 \end{pmatrix}, \tag{88}$$

all in all *twelve* components. This reflects the overall $Z_2 \times Z_2 \times Z_3$ symmetry: one Z_2 group corresponding to the spin-like dichotomic degree of freedom, describing two accessible states; the second Z_2 required to account for the particle–anti-particle symmetry, and the Z_3 group corresponding to color symmetry.

The "colors" should satisfy first order equations conceived in such a way that neither can propagate by itself, just like in the case of $\mathbf{E}$ and $\mathbf{B}$ components of Maxwell's tensor in electrodynamics, or the couple of two-component Pauli spinors which cannot propagate alone, but constitute one single entity, the four-component Dirac spinor [23].

This leaves little space for the choice of the system of intertwined equations; here is the ternary generalization of Dirac's equation, intertwining not only particles with anti-particles, but also the three "colors," in such a way that the entire system becomes invariant under the action of the $Z_2 \times Z_2 \times Z_3$ group:

$$E \, \varphi_+ = mc^2 \, \varphi_+ + c \, \boldsymbol{\sigma} \cdot \mathbf{p} \, \chi_-$$

$$E \, \chi_- = -j \, mc^2 \, \chi_- + c \, \boldsymbol{\sigma} \cdot \mathbf{p} \, \psi_+$$

$$E \, \psi_+ = j^2 \, mc^2 \, \psi_+ + c \, \boldsymbol{\sigma} \cdot \mathbf{p} \, \varphi_-$$

$$E \, \varphi_- = -mc^2 \, \varphi_- + c \, \boldsymbol{\sigma} \cdot \mathbf{p} \, \chi_+$$

$$E \, \chi_+ = j \, mc^2 \, \chi_+ + c \, \boldsymbol{\sigma} \cdot \mathbf{p} \, \psi_-$$

$$E \, \psi_- = -j^2 \, mc^2 \, \varphi_+ + c \, \boldsymbol{\sigma} \cdot \mathbf{p} \, \varphi_+ \tag{89}$$

$$\text{where} \quad \varphi_+ = \begin{pmatrix} \varphi_+^1 \\ \varphi_+^2 \end{pmatrix}, \quad \varphi_- = \begin{pmatrix} \varphi_-^1 \\ \varphi_-^2 \end{pmatrix}, \quad \chi_+ = \begin{pmatrix} \chi_+^1 \\ \chi_+^2 \end{pmatrix},$$

$$\chi_- = \begin{pmatrix} \chi_-^1 \\ \chi_-^2 \end{pmatrix}, \quad \psi_+ = \begin{pmatrix} \psi_+^1 \\ \psi_+^2 \end{pmatrix}, \quad \psi_- = \begin{pmatrix} \psi_-^1 \\ \psi_-^2 \end{pmatrix}, \tag{90}$$

on which Pauli sigma-matrices act in a natural way.

On the right-hand side, the mass terms form a diagonal matrix whose entries follow an ordered row of powers of the sixth root of unity $q = e^{\frac{2\pi i}{6}}$. Indeed, we have

$$m = q^6 m, \quad -jm = q^5 m, \quad j^2 m = q^4 m, \quad -m = q^3 m, \quad jm = q^2 m, \quad -j^2 m = qm.$$

Let us start the diagonalization of our system by deriving two third-order equations relating between them the φ_+ and φ_- fields. By iterating the E operator three times, we get the following equation:

$$E^3 \, \varphi_+ = m^3 c^6 \, \varphi_+ - 2j \, m^2 c^5 \, \boldsymbol{\sigma} \cdot \mathbf{p} \, \chi_- - 2j \, mc^3 \, \boldsymbol{\sigma} \cdot \mathbf{p} \, \psi_+ + |\, \mathbf{p} \,|^2 \, \boldsymbol{\sigma} \cdot \mathbf{p} \, \varphi_- \tag{91}$$

As one can see, at the third iteration diagonalization is not yet achieved because of the presence, besides the fields φ_+ and φ_-, of two other fields, namely ψ_+ and χ_-. Similar third-order equations are produced when we start the iteration from any of the five remaining components; in all cases, they contain four terms mixing other components. Total diagonalization is achieved only after the *sixth* iteration; the explicit calculus is quite tedious, but it can be performed using representation of our system in terms of tensor products of matrices.

The final result is extremely simple: all the components satisfy the same sixth-order equation,

$$E^6 \, \varphi_+ = m^6 c^{12} \, \varphi_+ + c^6 \, |\, \mathbf{p} \,|^6 \, \varphi_+,$$

$$E^6 \, \varphi_- = m^6 c^{12} \, \varphi_- + c^6 \, |\, \mathbf{p} \,|^6 \, \varphi_-. \tag{92}$$

The energy operator is obviously diagonal, and its action on the spinor-valued column-vector can be represented as a 6×6 operator valued unit matrix. The mass operator is diagonal, too, but its elements represent all powers of the sixth root of unity q, which are $q = -j^2$, $q^2 = j$, $q^3 = -1$, $q^2 = j^2$, $q^5 = -j$, and $q^6 = 1$. Finally, the momentum operator is proportional to a *circulant matrix* which mixes up all the components of the column vector.

On the basis in which the original system (89) was proposed, the matrix operators can be expressed as follows:

$$
M = \begin{pmatrix}
m & 0 & 0 & 0 & 0 & 0 \\
0 & -m & 0 & 0 & 0 & 0 \\
0 & 0 & jm & 0 & 0 & 0 \\
0 & 0 & 0 & -jm & 0 & 0 \\
0 & 0 & 0 & 0 & j^2 m & 0 \\
0 & 0 & 0 & 0 & 0 & -j^2 m
\end{pmatrix}, \quad
P = \begin{pmatrix}
0 & 0 & 0 & \sigma \cdot \mathbf{p} & 0 & 0 \\
0 & 0 & \sigma \cdot \mathbf{p} & 0 & 0 & 0 \\
0 & 0 & 0 & 0 & 0 & \sigma \cdot \mathbf{p} \\
0 & 0 & 0 & 0 & \sigma \cdot \mathbf{p} & 0 \\
0 & \sigma \cdot \mathbf{p} & 0 & 0 & 0 & 0 \\
\sigma \cdot \mathbf{p} & 0 & 0 & 0 & 0 & 0
\end{pmatrix}
\tag{93}
$$

In fact, the dimension of the two matrices M and P displayed in (93) above is 12×12: all the entries in the first one are proportional to the 2×2 identity matrix, so that in the definition one should read $\begin{pmatrix} m & 0 \\ 0 & m \end{pmatrix}$ instead of m, $\begin{pmatrix} jm & 0 \\ 0 & jm \end{pmatrix}$ instead of $j\,m$, etc. The entries in the second matrix P contain 2×2 Pauli's sigma-matrices, so that P is also a 12×12 matrix. The energy operator E is proportional to the 12×12 identity matrix.

Using a more rigorous mathematical language the three operators can be expressed in terms of tensor products of matrices of lower dimensions. Let us introduce the two following 3×3 matrices:

$$
B = \begin{pmatrix} 1 & 0 & 0 \\ 0 & j & 0 \\ 0 & 0 & j^2 \end{pmatrix} \quad \text{and} \quad Q_3 = \begin{pmatrix} 0 & 1 & 0 \\ 0 & 0 & 1 \\ 1 & 0 & 0 \end{pmatrix}
\tag{94}
$$

Then the 12×12 matrices M and P can be represented as the following tensor products:

$$
M = m\, B \otimes \sigma_3 \otimes \mathbb{1}_2, \qquad P = Q_3 \otimes \sigma_1 \otimes (\sigma \cdot \mathbf{p})
\tag{95}
$$

with as usual, $\mathbb{1}_2$, σ_1 and σ_3 denote the well-known 2×2 matrices

$$
\mathbb{1}_2 = \begin{pmatrix} 1 & 0 \\ 0 & 1 \end{pmatrix}, \quad \sigma_1 = \begin{pmatrix} 0 & 1 \\ 1 & 0 \end{pmatrix}, \quad \sigma_3 = \begin{pmatrix} 1 & 0 \\ 0 & -1 \end{pmatrix}.
$$

The energy operator, proportional to the 12×12 unit matrix, can be written in a similar manner as a product of three unit matrices, $\mathbb{1}_3 \otimes \mathbb{1}_2 \otimes \mathbb{1}_2$.

On the basis in which the functions are aligned in a column like in (90), the matrix operators take on another form, namely

$$M = m\,\sigma_3 \otimes B \otimes 1_2, \quad P = \sigma_1 \otimes Q_3 \otimes \sigma \cdot \mathbf{p} \tag{96}$$

Our system of twelve equations can be now encoded in the following form, using the matrices M and P defined above:

$$E\,\Psi = \left(c^2\,M + c\,P\right)\Psi. \tag{97}$$

Then we need to evaluate the sixth power of this operator acting on the vector column Ψ to prove that we have indeed $E^6 = m^6 c^{12} + c^6 \mid \mathbf{p} \mid^6$.

The fact that Eq. (97) is a *necessary* condition for the system (89) to be satisfied can be inferred from the following simple exercise. Let us write Eq. (97) in a slightly different form, by moving the mass matrix to the left-hand side. We get the formal equality between the actions of two matrices, when applied to the space of solutions of our system:

$$\left(E - c^2\,M\right)\Psi = c\,P\,\Psi. \tag{98}$$

We suppose that as far as Ψ satisfies the above Eq. (98), the determinants of the two matrices should be equal. On the right-hand side we can treat the operator $\sigma \cdot \mathbf{p}$ as a number whose square is just $\mathbf{p}^2$ multiplied by a 2×2 unit matrix. This enables us to reduce the problem to a 6×6 matrix form, forgetting for a while that the fields φ_+, φ_-, etc. are in fact two-component quantities. Then the 6×6 matrix on the left-hand side will read:

$$\begin{pmatrix} E - mc^2 & 0 & 0 & 0 & 0 & 0 \\ 0 & E + mc^2 & 0 & 0 & 0 & 0 \\ 0 & 0 & E - jmc^2 & 0 & 0 & 0 \\ 0 & 0 & 0 & E + jmc^2 & 0 & 0 \\ 0 & 0 & 0 & 0 & E - j^2 mc^2 & 0 \\ 0 & 0 & 0 & 0 & 0 & E + j^2 mc^2 \end{pmatrix},$$

while the matrix P is as given in (93).

The determinant of the diagonal matrix $E - c^2\,M$ is equal to the product of its six diagonal elements:

$$(E - mc^2)(E + mc^2)(E - jmc^2)(E + jmc^2)(E - j^2 mc^2)(E + j^2 mc^2) =$$

$$(E^2 - m^2 c^4)(E^2 - j^2 m^2 c^4)(E^2 - jm^2 c^4) = E^6 - m^6 c^{12},$$

while the determinant of the matrix $c\,P$ is easily proved to be $c^6 \mid \mathbf{p} \mid^6$, so that from equality of determinants we get the condition $E^6 - c^6 \mid \mathbf{p} \mid^6 = m^6 c^{12}$.

This, however, is not the proof that all the twelve components of the column-vector Ψ satisfy the unique differential equation of sixth order, resulting from the quantum correspondence principle. The rigorous proof is done by applying six times the matrices appearing on both sides of the equation and showing that the result is proportional to the unit 12×12 matrices multiplied by the three expressions appearing in (98), i.e. the sixth powers of energy, momentum, and mass.

Before we proceed to the discussion of the properties of general solutions of our system, we would like to draw attention to the similarity between this system, and the systems of Maxwell and Dirac equations *in vacuo*, i.e. describing free fields without any interactions. The Maxwell equations, which after gauge fixing on the 4-potential A_μ reduce to a system of four linear differential equations for four functions of t and $\mathbf{r}$. In general, the characteristic equation of such a system must be of the fourth order; however, due to the particular symmetry of Maxwell's equations, the characteristic equation is of the second order (in fact, it is a fourth-order equation which is a square of the second-order equation). This ensures the existence of a *single light cone* and the absence of bi-refringence in vacuum.

The same phenomenon occurs for the characteristic equation od the Dirac equation, which is also a system of four linear differential equations for the four components of a Dirac spinor. The diagonalization of the Dirac system occurs already at the level of the second order Klein-Gordon equation, although in the most general case of four linear differential equations the characteristic equation could be of the fourth order.

In our case we have as much as twelve linear equations imposed on twelve independent functions of time and space; therefore, one could expect a twelfth-order characteristic equation. However, due to the particular properties of Pauli's matrices and the symmetry of our system, the characteristic equation is of the sixth order only.

9 Solutions

The system (89) of twelve linear equations supposed to describe the dynamics of three intertwined fields was shown to be represented by a single matrix operator (98) acting on a 12-component vector: symbolically $E\Psi = (c^2 M + cP)\Psi$. By consecutive application of this matrix operator we are able to separate the variables and find the common equation of sixth order that is satisfied by each of the components:

$$E^6 \Psi = m^6 c^{12} \Psi + c^6 \mathbf{p}^6 \Psi. \tag{99}$$

Applying the quantum correspondence principle, the above equation relating mass, energy, and momentum (99) is transformed into a linear differential equation of the sixth order. Indeed, according to

$$E \to -i\hbar \frac{\partial}{\partial t}, \quad \mathbf{p} \to -i\hbar \nabla, \tag{100}$$

we get the following sixth-order partial differential equation to be satisfied by each of the components of the wave function Ψ:

$$-\hbar^6 \frac{\partial^6}{\partial t^6}\psi - m^6 c^{12}\psi = -\hbar^6 \Delta^3 \psi. \tag{101}$$

Let us write the algebraic expression relating mass, energy, and momentum (99) simply as follows:

$$E^6 - m^6 c^{12} = \mid \mathbf{p} \mid^6 c^6. \tag{102}$$

This equation can be factorized showing how it was obtained by subsequent action of the operators of the system (89):

$$E^6 - m^6 c^{12} = (E^3 - m^3 c^6)(E^3 + m^3 c^6) =$$

$$(E - mc^2)(jE - mc^2)(j^2 E - mc^2)(E + mc^2)(jE + mc^2)(j^2 E + mc^2) = \mid \mathbf{p} \mid^6 c^6.$$

Equation (101) can be solved by separation of variables; the time-dependent and the space-dependent factors have the same structure:

$$A_1 e^{\omega t} + A_2 e^{j \omega t} + A_3 e^{j^2 \omega t}, \quad B_1 e^{\mathbf{k}.\mathbf{r}} + B_2 e^{j \mathbf{k}.\mathbf{r}} + B_3 e^{j^2 \mathbf{k}.\mathbf{r}}$$

with ω and $\mathbf{k}$ satisfying the following dispersion relation:

$$\frac{\omega^6}{c^6} + \frac{m^6 c^6}{\hbar^6} = \mid \mathbf{k} \mid^6, \tag{103}$$

where we have identified $E = \hbar\omega$ and $\mathbf{p} = \hbar\mathbf{k}$.

Up to this point we follow exactly the way in which the Klein-Gordon equation is deduced from the Dirac equation as the common condition to be satisfied by each of the components of the Dirac spinor:

$$E^2\psi = m^2 c^4 \psi + c^2 \mathbf{p}^2 \psi \quad \rightarrow \quad -\hbar^2 \frac{\partial^2 \psi}{\partial t^2} = m^2 c^4 \psi - \hbar^2 \Delta \psi. \tag{104}$$

The solutions of (104) are sought in the plane wave form $\psi \sim e^{i(\omega t + \mathbf{k}\cdot\mathbf{r})}$. Due to the purely imaginary exponential, after such a substitution the Klein-Gordon equation reduces to the well-known algebraic condition

$$\hbar^2 \omega^2 = m^2 c^4 + \hbar^2 \mathbf{k}^2, \tag{105}$$

which coincides with the previously established relation between the energy, momentum, and mass due to the correspondence $E = \hbar\omega$ and $\mathbf{p} = \hbar\mathbf{k}$ introduced by de Broglie. The relation (103) is invariant under the action of $Z_2 \times Z_3 = Z_6$ symmetry, because to any solution with given real ω and $\mathbf{k}$ one can add solutions with ω replaced by $j\omega$ or $j^2\omega$, $j\mathbf{k}$ or $j^2\mathbf{k}$, as well as $-\omega$; there is no need to introduce also $-\mathbf{k}$ instead of $\mathbf{k}$ because the vector $\mathbf{k}$ can take on all possible directions covering the unit sphere.

The nine complex solutions with positive frequency ω as well as with $j\ \omega$ and $j^2\ \omega$ obtained by the action of the Z_3-group can be displayed in a compact manner in form of a 3×3 matrix. The inclusion of the essential Z_2-symmetry ensuring the existence of *anti-particles* leads to the nine similar solutions with negative ω. The two matrices are displayed below:

$$\begin{pmatrix} e^{\omega t+\mathbf{k}\cdot\mathbf{r}} & e^{\omega t+j\mathbf{k}\cdot\mathbf{r}} & e^{\omega t+j^2\mathbf{k}\cdot\mathbf{r}} \\ e^{j\omega t+\mathbf{k}\cdot\mathbf{r}} & e^{j\omega t+j\mathbf{k}\cdot\mathbf{r}} & e^{j\omega t+j^2\mathbf{k}\cdot\mathbf{r}} \\ e^{j^2\omega t+\mathbf{k}\cdot\mathbf{r}} & e^{j^2\omega t+\mathbf{k}\cdot\mathbf{r}} & e^{j^2\omega t+j^2\mathbf{k}\cdot\mathbf{r}} \end{pmatrix}, \quad \begin{pmatrix} e^{-\omega t-\mathbf{k}\cdot\mathbf{r}} & e^{-\omega t-j\mathbf{k}\cdot\mathbf{r}} & e^{-\omega t-j^2\mathbf{k}\cdot\mathbf{r}} \\ e^{-j\omega t-\mathbf{k}\cdot\mathbf{r}} & e^{-j\omega t-j\mathbf{k}\cdot\mathbf{r}} & e^{-j\omega t-j^2\mathbf{k}\cdot\mathbf{r}} \\ e^{-j^2\omega t-\mathbf{k}\cdot\mathbf{r}} & e^{-j^2\omega t-\mathbf{k}\cdot\mathbf{r}} & e^{-j^2\omega t-j^2\mathbf{k}\cdot\mathbf{r}} \end{pmatrix}$$
$$(106)$$

and their nine *real* linear combinations can be represented in the following 3×3 matrix of functions as follows:

$$\begin{pmatrix} e^{\omega t+\mathbf{k}\cdot\mathbf{r}} & e^{\omega t-\frac{\mathbf{k}\cdot\mathbf{r}}{2}}\cos(\mathbf{K}\cdot\mathbf{r}) & e^{\omega t-\frac{\mathbf{k}\cdot\mathbf{r}}{2}}\sin(\mathbf{K}\cdot\mathbf{r}) \\ e^{-\frac{\omega t}{2}+\mathbf{k}\cdot\mathbf{r}}\cos\Omega t & e^{-\frac{\omega t}{2}-\frac{\mathbf{k}\cdot\mathbf{r}}{2}}\cos(\Omega t - \mathbf{K}\cdot\mathbf{r}) & e^{-\frac{\omega t}{2}-\frac{\mathbf{k}\cdot\mathbf{r}}{2}}\cos(\Omega t - \mathbf{K}\cdot\mathbf{r}) \\ e^{-\frac{\omega t}{2}+\mathbf{k}\cdot\mathbf{r}}\sin\Omega t & e^{-\frac{\omega t}{2}-\frac{\mathbf{k}\cdot\mathbf{r}}{2}}\sin(\Omega t - \mathbf{K}\cdot\mathbf{r}) & e^{-\frac{\omega t}{2}-\frac{\mathbf{k}\cdot\mathbf{r}}{2}}\sin(\Omega t - \mathbf{K}\cdot\mathbf{r}) \end{pmatrix},$$
$$(107)$$

where $\Omega = \frac{\sqrt{3}}{2}\omega$ and $\mathbf{K} = \frac{\sqrt{3}}{2}\mathbf{k}$; the same can be done with the conjugate solutions (with $-\omega$ instead of ω). A similar matrix, of course, can be produced for the alternative negative ω choice.

The functions displayed in the matrix do not represent a wave; however, one can produce a propagating solution by forming certain cubic combinations, e.g.

$$e^{\omega t+\mathbf{k}\cdot\mathbf{r}}\,e^{-\frac{\omega t}{2}-\frac{\mathbf{k}\cdot\mathbf{r}}{2}}\cos(\Omega t - \mathbf{K}\cdot\mathbf{r})\,e^{-\frac{\omega t}{2}-\frac{\mathbf{k}\cdot\mathbf{r}}{2}}\sin(\Omega t - \mathbf{K}\cdot\mathbf{r}) = \frac{1}{2}\sin(2\Omega t - 2\mathbf{K}\cdot\mathbf{r}).$$

What we need now is a multiplication scheme that would define triple products of non-propagating solutions yielding propagating ones, like in the example given above, but under the condition that the factors belong to three distinct subsets (which can be later on identified as "colors").

Before we proceed farther, let us remind that the set of six independent functions is expected to generate the most general solution of our sixth-order differential equation. Therefore, among the nine functions displayed in the above matrices (106), as well as in the real basis (107), three are superfluous.

Indeed, the determinants of the two complex matrices (106), as well as that of the real matrix (107), identically vanish. Their lower 2×2 minors are also zero, which confirms the idea that only *six* out of nine functions are independent. In principle, we could pick up any six functions, but for symmetry reasons we shall remove the diagonal ones. The remaining six functions are displayed in the truncated matrix:

$$\begin{pmatrix} 0 & e^{\omega t - \frac{\mathbf{k} \cdot \mathbf{r}}{2}} \cos(\mathbf{K} \cdot \mathbf{r}) & e^{\omega t - \frac{\mathbf{k} \cdot \mathbf{r}}{2}} \sin(\mathbf{K} \cdot \mathbf{r}) \\ e^{-\frac{\omega t}{2} + \mathbf{k} \cdot \mathbf{r}} \cos \Omega t & 0 & e^{-\frac{\omega t}{2} - \frac{\mathbf{k} \cdot \mathbf{r}}{2}} \cos(\Omega t - \mathbf{K} \cdot \mathbf{r}) \\ e^{-\frac{\omega t}{2} + \mathbf{k} \cdot \mathbf{r}} \sin \Omega t \ \ e^{-\frac{\omega t}{2} - \frac{\mathbf{k} \cdot \mathbf{r}}{2}} \sin(\Omega t - \mathbf{K} \cdot \mathbf{r}) & 0 \end{pmatrix},$$

$$(108)$$

In what follows, we shall choose the Cartesian system of space coordinates with its x-axis aligned with the vector $\mathbf{k}$, so that in all the six remaining functions displayed in the real matrix (107) we can replace the scalar product $\mathbf{k} \cdot \mathbf{r}$ by kx, and $\mathbf{K} \cdot \mathbf{r}$ by Kx, with $K = \frac{\sqrt{3}}{2} k$.

With this in mind, let us display the six independent solutions in the following two groups of three:

$$F_1 = e^{-\frac{\omega t}{2} + kx} \sin \Omega t, \qquad F_2 = e^{-\frac{\omega t}{2} + kx} \cos \Omega t,$$

$$G_1 = e^{\omega t - \frac{kx}{2}} \sin Kx, \qquad G_2 = e^{\omega t - \frac{kx}{2}} \cos Kx,$$

$$H_1 = e^{-\frac{\omega t}{2} - \frac{kx}{2}} \sin(\Omega t - Kx), \qquad H_2 = e^{-\frac{\omega t}{2} - \frac{kx}{2}} \cos(\Omega t - Kx). \qquad (109)$$

Neither of the six functions above can represent a freely propagating wave: even the last two functions H_1 and H_2 contain, besides the running sinusoidal waves, the real exponentials which have a damping effect. (The wave cannot penetrate distances greater than a few wavelengths, and can last only for times comparable with few oscillations.) However, we shall show that certain *cubic* expressions can represent a freely propagating wave, without any damping factors. Taking a closer look at the six solutions displayed in (109), we see that the only way to get rid of the real exponents present in all those functions, but different damping factors, is to form cubic expressions constructed with three functions labelled with three *different* letters. Here is the exhaustive list of *eight* admissible cubic combinations:

$$F_1 \, G_1 \, H_1, \qquad\qquad F_2 \, G_1 \, H_1;$$

$$F_1 \, G_1 \, H_2, \qquad\qquad F_2 \, G_1 \, H_2;$$

$$F_1 \, G_2 \, H_1, \qquad\qquad F_2 \, G_2 \, H_1;$$

$$F_1 \, G_2 \, H_2, \qquad\qquad F_2 \, G_2 \, H_2;$$

But these expressions still contain, besides running waves with double frequency 2Ω, undesirable functions like $\sin \Omega t$ or $\cos Kx$. To take an example, we have

$$F_1 \, G_2 \, H_2 = \sin \Omega t \, \cos Kx \, \cos(\Omega t - Kx) =$$

$$\frac{1}{2} \, [\sin(\Omega t + Kx) + \sin(\Omega t - Kx)] \cos(\Omega t - Kx) =$$

$$\frac{1}{4} \, \sin(2\Omega t - 2Kx) + \frac{1}{4} \, \sin(2\Omega t) + \frac{1}{4} \, \sin(2Kx).$$

The explicit expressions, in terms of the trigonometric functions, of the eight independent cubic expressions displayed above, are quite cumbersome. Here we give the final result, showing that there are only *two* combinations of cubic products of solutions of the generalized ternary Dirac equation that represent running waves, which are the following:

$$F_1 G_2 H_2 + F_1 G_1 H_1 - F_2 G_1 H_2 + F_2 G_2 H_1 = \sin(2\Omega t - 2Kx), \qquad (110)$$

$$F_2 G_2 H_2 + F_2 G_1 H_1 + F_1 G_1 H_2 - F_1 G_2 H_1 = \cos(2\Omega t - 2Kx). \qquad (111)$$

The symmetry of these expressions appears better when grouped as follows:

$$F_1(G_2 H_2 + G_1 H_1) + F_2(G_2 H_1 - G_1 H_2) = \sin(2\Omega t - 2Kx), \qquad (112)$$

$$F_2(G_2 H_2 + G_1 H_1) + F_1(G_1 H_2 - G_2 H_1) = \cos(2\Omega t - 2Kx). \qquad (113)$$

Similarly two running waves are produced by forming corresponding cubic combinations of negative frequency solutions obtained by substituting $-\omega$ instead of ω and $-k$ instead of k. The four running waves so obtained could represent freely propagating Dirac spinor if the dispersion relation relating ω and k was the usual quadratic one, but here it is not. So we are still unable to produce a Dirac particle from cubic combinations of solutions of our sixth-order system.

Functions describing running waves can also be obtained via binary combinations of solutions belonging to classes with opposite sign of ω. The *"singlets"* corresponding to product of a solution with its "anti-solution," i.e. the mirror image obtained by the mere change of signs in the exponentials will not give anything interesting, because such products are equal to a constant, e.g.,

$$e^{j\omega t + j^2 kx} \cdot e^{-j\omega t - j^2 kx} = e^0 = 1.$$

But with two functions which are not complex conjugates the result is different:

$$e^{j\omega t + j^2 kx} \cdot e^{-j^2 \omega t - jkx} = e^{(j - j^2)\,\omega t - (j^2 - j)\,kx} = e^{i(\Omega t - Kx)}, \qquad (114)$$

with $\Omega = \frac{\sqrt{3}}{2}\omega, \qquad K = \frac{\sqrt{3}}{2}k$. An alternative choice of two elementary solutions leads to a free wave running in the opposite direction:

$$e^{-j^2 \omega t - j^2 kx} \cdot e^{j\omega t + jkx} = e^{i(\Omega t + Kx)}.$$

10 Relativistic Invariance

Let us rewrite the matrix operator generating the system (89) when it acts on the column vector containing twelve components of three "color" fields,

$$E \, \mathbb{1}_2 \otimes \mathbb{1}_3 \otimes \mathbb{1}_2 = mc^2 \, \sigma_3 \otimes B \otimes \mathbb{1}_2 + \sigma_1 \otimes Q_3 \otimes c \sigma \cdot \mathbf{p}$$

in a slightly different way, with energy and momentum operators on the left-hand side, and the mass operator on the right-hand side:

$$E \, \mathbb{1}_2 \otimes \mathbb{1}_3 \otimes \mathbb{1}_2 - \sigma_1 \otimes Q_3 \otimes \sigma \cdot \mathbf{p} = mc^2 \, \sigma_3 \otimes B \otimes \mathbb{1}_2 \tag{115}$$

Following a similar procedure known from the treatment of the standard Dirac equation, let us transform this equation so that the mass operator becomes proportional to the unit matrix. To do so, let us multiply Eq. (115) from the left by the matrix $\sigma_3 \otimes B^\dagger \otimes \mathbb{1}_2$. Now we get the following equation which enables us to interpret the energy and the momentum as the components of a Minkowskian four-vector $c \, p^\mu = [E, \ c\mathbf{p}]$:

$$E \, \sigma_3 \otimes B^\dagger \otimes \mathbb{1}_2 - i\sigma_2 \otimes j^2 \, Q_2 \otimes c \sigma \cdot \mathbf{p} = mc^2 \, \mathbb{1}_2 \otimes \mathbb{1}_3 \otimes \mathbb{1}_2, \tag{116}$$

where we used the fact that under matrix multiplication, $\sigma_3 \sigma_3 = \mathbb{1}_2$, $B^\dagger B = \mathbb{1}_3$, and $B^\dagger Q_3 = j^2 \, Q_2$.

One can check by direct computation that the sixth power of this operator gives the same result as before,

$$\left[E \, \sigma_3 \otimes B^\dagger \otimes \mathbb{1}_2 - i\sigma_2 \otimes j^2 \, Q_2 \otimes c \sigma \cdot \mathbf{p} \right]^6 = \left[E^6 - c^6 \mathbf{p}^6 \right] \mathbb{1}_{12} = m^6 c^{12} \mathbb{1}_{12} \tag{117}$$

Equation (116) can be written in a concise manner using the Minkowskian indices and the usual pseudo-scalar product of two four-vectors as follows:

$$\Gamma^\mu \, p_\mu = mc \, \mathbb{1}_{12}, \quad \text{with} \quad p^0 = \frac{E}{c}, \quad p^k = mc \frac{dx^k}{ds}. \tag{118}$$

with 12×12 matrices Γ^μ, $\mu = 0, 1, 2, 3$ defined as follows:

$$\Gamma^0 = \sigma_3 \otimes B^\dagger \otimes \mathbb{1}_2, \quad \Gamma^k = -i\sigma_2 \otimes j^2 \, Q_2 \otimes \sigma^k \tag{119}$$

The four 12×12 matrices do not satisfy usual anti-commutation relations similar to those of the 4×4 Dirac matrices γ^μ, i.e. $\gamma^\mu \gamma^\nu + \gamma^\nu \gamma^\mu = 2 \, g^{\mu\nu} \, \mathbf{1}_4$.

Nevertheless, the system of equations satisfied by the 12-dimensional wave function Ψ

$$-i\hbar \, \Gamma^\mu \, \partial_\mu \, \Psi = mc\Psi \tag{120}$$

is a hyperbolic one, and has the same light cone as the Klein-Gordon equation. To corroborate this statement, let us first consider the massless case,

$$- i\hbar\, \Gamma^\mu\, \partial_\mu\, \Psi = 0. \tag{121}$$

Assuming the general solution as usual in the form of an exponential function $e^{k_\mu x^\mu}$, we can replace the derivations by the components of the wave 4-vector k^μ, and take the sixth power of the matrix $\Gamma^\mu k_\mu$. The resulting dispersion relation in the dual space is shown to be

$$k_0^6 - \mid \mathbf{k} \mid^6 = \left(k_0^2 - \mid \mathbf{k} \mid^2\right)\left(k_0^2 - j \mid \mathbf{k} \mid^2\right)\left(k_0^2 - j^2 \mid \mathbf{k} \mid^2\right)$$

$$= \left(k_0^2 - \mid \mathbf{k} \mid^2\right)\left(k_0^4 + k_0^2 \mid \mathbf{k} \mid^2 + \mid \mathbf{k} \mid^4\right) = 0. \tag{122}$$

Here first factor defines the usual relativistic light cone, while next factor of degree four is strictly positive (besides the origin 0), which means that the system has only one characteristic surface which is the same for all massless fields. Each of the three factors remains invariant under a different representation of the $SL(2, \mathbf{C})$ group.

Let us introduce the following three matrices representing the same four-vector k^μ:

$$K_3 = \begin{pmatrix} k_0 & k_x \\ k_x & k_0 \end{pmatrix}, \quad K_1 = \begin{pmatrix} k_0 & jk_x \\ jk_x & k_0 \end{pmatrix}, \quad K_2 = \begin{pmatrix} k_0 & j^2 k_x \\ j^2 k_x & k_0 \end{pmatrix}, \tag{123}$$

whose determinants are, respectively,

$$\det K_1 = k_0^2 - j^2 k_x^2, \quad \det K_2 = k_0^2 - jk_x^2, \quad \det K_3 = k_0^2 - k_x^2. \tag{124}$$

Let us notice that only the third matrix K_3 is hermitian, and corresponds to a *real* space-time vector k^μ, while neither of the remaining two matrices K_1 and K_2 is hermitian; however, one is the hermitian conjugate of another.

In what follows, we shall replace the absolute value of the wave vector $\mid \mathbf{k} \mid$ by a single spatial component, say k_x, because for any given four-vector $k^\mu = [k_0, \mathbf{k}]$ we can choose the coordinate system in such a way that its x-axis shall be aligned along the vector $\mathbf{k}$. Then it is easy to check that if we set

$$\begin{pmatrix} k'_0 \\ k'_x \end{pmatrix} = \begin{pmatrix} \cosh u & \sinh u \\ \sinh u & \cosh u \end{pmatrix}\begin{pmatrix} k_0 \\ k_x \end{pmatrix}, \tag{125}$$

then $\quad \begin{pmatrix} \cosh u & j^2 \sinh u \\ j \sinh u & \cosh u \end{pmatrix}\begin{pmatrix} k_0 \\ j\, k_x \end{pmatrix} = \begin{pmatrix} k'_0 \\ j\, k'_x \end{pmatrix} \quad$ and

$$\begin{pmatrix} \cosh u & j \sinh u \\ j^2 \sinh u & \cosh u \end{pmatrix} \begin{pmatrix} k_0 \\ j^2 k_x \end{pmatrix} = \begin{pmatrix} k'_0 \\ j^2 k'_x \end{pmatrix}, \tag{126}$$

where the transformed 4-vectors are given by the following expressions, for each case above:

$$i) \quad k'_0 = k_0 \cosh u + k_x \sinh u, \quad k'_x = k_0 \sinh u + k_x \cosh u$$

$$ii) \quad k'_0 = k_0 \cosh u + j^2 k_x \sinh u, \quad k'_x = j \, k_0 \sinh u + k_x \cosh u$$

$$iii) \quad k'_0 = k_0 \cosh u + j \, k_x \sinh u, \quad k'_x = j^2 k_0 \sinh u + k_x \cosh u$$

Let us now introduce a 6×6 matrix composed with the above three 2×2 matrices:

$$\begin{pmatrix} 0 & k_0 \, \mathbb{1}_2 + \mathbf{k} \cdot \boldsymbol{\sigma} & 0 \\ 0 & 0 & k_0 \, \mathbb{1}_2 + j \, \mathbf{k} \cdot \boldsymbol{\sigma} \\ k_0 \, \mathbb{1}_2 + j^2 \mathbf{k} \cdot \boldsymbol{\sigma} & 0 & 0 \end{pmatrix} \tag{127}$$

or, more explicitly,

$$K = \begin{pmatrix} 0 & 0 & k_0 \, k_x & 0 & 0 \\ 0 & 0 & k_x & k_0 & 0 & 0 \\ 0 & 0 & 0 & 0 & k_0 & jk_x \\ 0 & 0 & 0 & 0 & jk_x & k_0 \\ k_0 & j^2 k_x & 0 & 0 & 0 & 0 \\ j^2 k_x & k_0 & 0 & 0 & 0 & 0 \end{pmatrix} \tag{128}$$

It is easy to check that

$$\det \ K = (\det K_1) \cdot (\det K_2) \cdot (\det K_3)$$

$$= (k_0^2 - k_x^2)(k_0^2 - j^2 k_x^2)(k_0^2 - jk_x^2) = k_0^6 - k_x^6. \tag{129}$$

It is also remarkable that the determinant remains the same on the basis in which the ternary Dirac operator was proposed, namely when we consider the matrix

$$K = \begin{pmatrix} k_0 & 0 & 0 & k_x & 0 & 0 \\ 0 & k_0 & k_x & 0 & 0 & 0 \\ 0 & 0 & k_0 & 0 & 0 & jk_x \\ 0 & 0 & 0 & k_0 & jk_x & 0 \\ 0 & j^2 k_x & 0 & 0 & k_0 & 0 \\ j^2 k_x & 0 & 0 & 0 & 0 & k_0 \end{pmatrix} \tag{130}$$

Let us show now that the spinorial representation of Lorentz boosts can be applied to each of the three matrices K_1, K_2, and K_3 separately, keeping their determinants unchanged. As a matter of fact, besides the well-known formula:

$$\begin{pmatrix} \cosh\frac{u}{2} & \sinh\frac{u}{2} \\ \sinh\frac{u}{2} & \cosh\frac{u}{2} \end{pmatrix} \begin{pmatrix} k_0 & k_x \\ k_x & k_0 \end{pmatrix} \begin{pmatrix} \cosh\frac{u}{2} & \sinh\frac{u}{2} \\ \sinh\frac{u}{2} & \cosh\frac{u}{2} \end{pmatrix} = \begin{pmatrix} k'_0 & k'_x \\ k'_x & k'_0 \end{pmatrix}, \tag{131}$$

with

$$k'_0 = k_0\,\cosh u + k_x\,\sinh u, \quad k'_x = k_0\,\sinh u + k_x\,\cosh u. \tag{132}$$

which becomes apparent when we remind that $\cosh^2\frac{u}{2} + \sinh^2\frac{u}{2} = \cosh u$ and $2\sinh\frac{u}{2}\,\cosh\frac{u}{2} = \sinh u$, keeping unchanged the Minkowskian square invariant: $k'^2_0 - k'^2_x = k^2_0 - k^2_x$, we have also two transformations of the same kind which keep invariant the "complexified" Minkowskian squares appearing as factors in the sixth-order expression $k^6_0 - k^6_x$, namely

$$k^2_0 - j\,k^2_x \quad \text{and} \quad k^2_0 - j^2\,k^2_x.$$

Notice that the expressions above can be identified as the determinants of the following 2×2 matrices:

$$k^2_0 - j\,k^2_x = \det\begin{pmatrix} k_0 & j^2 k^2_x \\ j^2\,k_x & k_0 \end{pmatrix}, \quad k^2_0 - j^2\,k^2_x = \det\begin{pmatrix} k_0 & j k^2_x \\ j\,k_x & k_0 \end{pmatrix}. \tag{133}$$

It is easy to check that we have:

$$\begin{pmatrix} \cosh\frac{u}{2} & \sinh\frac{u}{2} \\ \sinh\frac{u}{2} & \cosh\frac{u}{2} \end{pmatrix} \begin{pmatrix} k_0 & j k_x \\ j k_x & k_0 \end{pmatrix} \begin{pmatrix} \cosh\frac{u}{2} & \sinh\frac{u}{2} \\ \sinh\frac{u}{2} & \cosh\frac{u}{2} \end{pmatrix} = \begin{pmatrix} k'_0 & j\,k'_x \\ j\,k'_x & k'_0 \end{pmatrix}, \tag{134}$$

with $k'_0 = k_0\,\cosh u + j\,k'_x\,\sinh u$, so that $k'^2_0 - jk'^2_x = k^2_0 - jk^2_x$, as well as

$$\begin{pmatrix} \cosh\frac{u}{2} & \sinh\frac{u}{2} \\ \sinh\frac{u}{2} & \cosh\frac{u}{2} \end{pmatrix} \begin{pmatrix} k_0 & j^2\,k_x \\ j^2\,k_x & k_0 \end{pmatrix} \begin{pmatrix} \cosh\frac{u}{2} & \sinh\frac{u}{2} \\ \sinh\frac{u}{2} & \cosh\frac{u}{2} \end{pmatrix} = \begin{pmatrix} k'_0 & j^2\,k'_x \\ j^2\,k'_x & k'_0 \end{pmatrix}, \tag{135}$$

This shows that spinorial representation of the $SL(2, \mathbf{C})$ group acts on the generalized Dirac matrix operator through three peculiar representations associated with the elements of the Z_3 group. Each of these representations acts separately on one ordinary 4-vector and two complex conjugate 4-vectors:

$$[k_0, \mathbf{k}], \quad [k_0, \mathbf{k}], \quad \text{and} \quad [k_0, \mathbf{k}].$$

11 Propagators

Let us introduce the Fourier transform of a real function of one variable, and the inverse Fourier transform as follows [24]:

$$\hat{f}(k) = \int_{-\infty}^{\infty} f(x)\, e^{ikx}\, dx, \quad f(x) = \frac{1}{2\pi} \int_{-\infty}^{\infty} \hat{f}(k)\, e^{-ikx}\, dk. \tag{136}$$

In this convention, the constant function $f(x) = 1$ is transformed into the Dirac delta function $\delta(k)$ multiplied by 2π.

In terms of their Fourier transforms, linear differential operators of any order are represented by corresponding algebraical expressions multiplying the Fourier transform of the unknown function. The Fourier transform of the Green function is then given by the inverse of this expression, for example, the Fourier transform of the Green function of the Klein-Gordon operator is defined as

$$\hat{G}(k^\mu) = \frac{1}{k_0^2 - \mathbf{k}^2 - mu^2},$$

(with $\mu = \frac{mc}{\hbar}$). The Fourier transform of Green's function for the Dirac equation is a 4×4 matrix:

$$\hat{D}(k^\mu) = \frac{\gamma^\mu k_\mu + m\, \mathbb{1}_4}{-k_0^2 + \mathbf{k}^2 - m^2},$$

because quite obviously one has

$$(\gamma^\mu k_\mu + m\, \mathbb{1}_4)(\gamma^\mu k_\mu - m\, \mathbb{1}_4) = -k_0^2 + \mathbf{k}^2 - m^2\, \mathbb{1}_4.$$

The ternary generalization of Dirac's equation being written in the most compact form as in (121), in terms of Fourier transforms it becomes

$$\left(\Gamma^\mu\, k_\mu - m\, \mathbb{1}_{12} \right) \hat{\Psi}(k) = 0. \tag{137}$$

The sixth power of the matrix $\Gamma^\mu k_\mu$ is diagonal and proportional to m^6, so that we have

$$\left(\Gamma^\mu k_\mu \right)^6 - m^6\, \mathbb{1}_{12} = \left(k_0^6 - \mid \mathbf{k} \mid^6 - m^6 \right) \mathbb{1}_{12} = 0. \tag{138}$$

Now we have to find the inverse of the matrix $\left(\Gamma^\mu\, k_\mu - m\, \mathbb{1}_{12} \right)$. To this effect, let us note that the sixth-order expression on the left-hand side in (138) can be factorized as follows:

$$\left(\Gamma^\mu k_\mu\right)^6 - m^6 = \left(\left(\Gamma^\mu k_\mu\right)^2 - m^2\right) \left(\left(\Gamma^\mu k_\mu\right)^2 - j\, m^2\right) \left(\left(\Gamma^\mu k_\mu\right)^2 - j^2\, m^2\right).$$

(139)

The first factor is in turn the product of two linear expressions, one of which is the ternary Dirac operator:

$$\left(\Gamma^\mu k_\mu\right)^6 - m^6$$
$$= \left(\Gamma^\mu k_\mu - m\right) \left(\Gamma^\mu k_\mu + m\right) \left(\left(\Gamma^\mu k_\mu\right)^2 - j\, m^2\right) \left(\left(\Gamma^\mu k_\mu\right)^2 - j^2\, m^2\right).$$

(140)

Therefore the inverse of the Fourier transform of the ternary Dirac operator is given by the following matrix:

$$\left[\left(\Gamma^\mu k_\mu\right)^6 - m^6\right]^{-1} = \frac{\left(\Gamma^\mu k_\mu + m\right) \left(\left(\Gamma^\mu k_\mu\right)^2 - j\, m^2\right) \left(\left(\Gamma^\mu k_\mu\right)^2 - j^2\, m^2\right)}{\left(k_0^6 - \mid \mathbf{k} \mid^6 - m^6\right)}.$$

(141)

It takes almost no effort to prove that the numerator can be given a more symmetric form. Taking into account that

$$\left(\left(\Gamma^\mu k_\mu\right)^2 - j\, m^2\right) \left(\left(\Gamma^\mu k_\mu\right)^2 - j^2\, m^2\right) = \left(\Gamma^\mu k_\mu\right)^4 + m^2 \left(\Gamma^\mu k_\mu\right)^2 + m^4,$$

we find that

$$\left(\Gamma^\mu k_\mu + m\right) \left(\left(\Gamma^\mu k_\mu\right)^2 - j\, m^2\right) \left(\left(\Gamma^\mu k_\mu\right)^2 - j^2\, m^2\right) =$$

$$\left(\Gamma^\mu k_\mu\right)^5 + m \left(\Gamma^\mu k_\mu\right)^4 + m^2 \left(\Gamma^\mu k_\mu\right)^3 + m^3 \left(\Gamma^\mu k_\mu\right)^2 + m^4 \left(\Gamma^\mu k_\mu\right) + m^5,$$

so that the final expression can be written in a concise form as

$$\left[\left(\Gamma^\mu k_\mu\right)^6 - m^6\right]^{-1} = \frac{\sum_{s=0}^{5} m^s \left(\Gamma^\mu k_\mu\right)^{(5-s)}}{\left(k_0^6 - \mid \mathbf{k} \mid^6 - m^6\right)}.$$

(142)

In the massless case, the operator equation whose Green's function we want to evaluate reduces to

$$\left[\frac{1}{c^6}\frac{\partial^6}{\partial t^6} - \left(\frac{\partial^2}{\partial x^2} + \frac{\partial^2}{\partial y^2} + \frac{\partial^2}{\partial z^2}\right)^3\right] G(t, \mathbf{r}) = \delta^4(x) = \delta(ct)\delta(x)\delta(y)\delta(z).$$

Using the Fourier transformation method, we can write:

$$\left[\frac{\omega^6}{c^6} - |\mathbf{k}|^6\right] \hat{G}(k_0, \mathbf{k}) = 1, \quad \text{where } k_0 = \frac{\omega}{c}, \tag{143}$$

from which we get

$$\hat{G}(k_0, \mathbf{k}) = \frac{1}{k_0^6 - |\mathbf{k}|^6} + \Phi(k_0, \mathbf{k}), \tag{144}$$

where $\Phi(k_0, \mathbf{k})$ is a solution of the homogeneous equation,

$$\left[k_0^6 - |\mathbf{k}|^6\right] \Phi(k_0, \mathbf{k}) = 0 \quad \rightarrow \quad \Phi(k_0, \mathbf{k}) = \delta(k_0^6 - |\mathbf{k}|^6). \tag{145}$$

The sixth-order polynomial $k_0^6 - |\mathbf{k}|^6$ can be split into the product of three second-order factors as follows:

$$k_0^6 - |\mathbf{k}|^6 = (k_0^2 - |\mathbf{k}|^2)(k_0^2 - j|\mathbf{k}|^2)(k_0^6 - j^2|\mathbf{k}|^2), \tag{146}$$

each of which being a product of two linear expressions with opposite signs of $|\mathbf{k}|$:

$$(k_0^2 - |\mathbf{k}|^2) = (k_0 + |\mathbf{k}|)(k_0 - |\mathbf{k}|),$$

$$(k_0^2 - j|\mathbf{k}|^2) = (k_0 + j^2|\mathbf{k}|)(k_0 - j^2|\mathbf{k}|),$$

$$(k_0^2 - j^2|\mathbf{k}|^2) = (k_0 + j|\mathbf{k}|)(k_0 - j|\mathbf{k}|),$$

so that the sixth-order expression appearing in (143) can be decomposed into a product of six linear terms. Let us represent the inverse of this expression appearing in (144) as a sum of three fractions with second-order expressions in their denominators:

$$\frac{1}{k_0^6 - |\mathbf{k}|^6} = \frac{1}{3|\mathbf{k}|^4}\left[\frac{1}{k_0^2 - |\mathbf{k}|^2} + \frac{j}{k_0^2 - j|\mathbf{k}|^2} + \frac{j^2}{k_0^2 - j^2|\mathbf{k}|^2}\right], \tag{147}$$

which is to be compared with the usual Fourier inverse of the d'Alembert operator:

$$\frac{1}{k_0^2 - |\mathbf{k}|^2} = \frac{1}{2|\mathbf{k}|^2}\left[\frac{1}{k_0 - |\mathbf{k}|} - \frac{1}{k_0 + |\mathbf{k}|}\right] \tag{148}$$

The difference in the order of the equation leads to the difference in the algebraic structure of the polynomial representing the equation for the Fourier transform. Its inverse displays not just two, but as much as *six* simple poles displayed in Fig. 4: In the case of the usual d'Alembertian several different Green's functions can be obtained by taking the inverse Fourier transform of the (148). The most widely used

Fig. 4 The six simple poles
of the integral representation
of zero-mass propagator of
the sixth-order equation

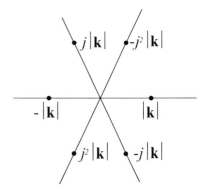

is the *retarded* Green's function, proportional to the well-knwon expression

$$G_{ret}(x^\mu) = \frac{\delta(ct - kr)}{4\pi r} \tag{149}$$

Now we have to perform the following integration (in spherical coordinates):

$$G(x^\mu) = \frac{1}{16\pi^4} \int_0^{2\pi} d\varphi \int_0^\pi \sin\theta d\theta \int_0^\infty \mathbf{k}^2 d \mid \mathbf{k} \mid \int_{-\infty}^\infty dk_0 e^{-i(k_0 ct - |\mathbf{k}|r \cos\theta)} \hat{G}(k^\mu), \tag{150}$$

with $\hat{G}(k^\mu)$ given by the expression (147). Each of its three terms contains an inverse of quadratic expression resembling the usual d'Alembertian, but multiplied by the extra factor $\mid \mathbf{k} \mid^{-4}$, and with $\mid \mathbf{k}^2 \mid$ appearing with factors 1, j and j^2. In what follows, we shall write k instead of $\mid \mathbf{k} \mid$ when there is no risk of ambiguity. Supposing that $G(k^\mu)$ is spherically symmetric, the integration over $d\varphi$ gives just the factor 2π. Next, we can perform integration over $d\theta$, factorizing the only term depending on θ, which is $e^{ikr \cos\theta}$. This integral gives

$$\int_0^\pi e^{ikr \cos\theta} \sin\theta d\theta = \int_{-1}^1 e^{ikru} du = \frac{2 \sin kr}{kr}. \tag{151}$$

Here we have the sum of three contributions:

$$\hat{G}_1 = \frac{1}{3 \mid \mathbf{k} \mid^4} \left[\frac{1}{k_0^2 - \mid \mathbf{k} \mid^2} \right] \quad \hat{G}_2 = \frac{1}{3 \mid \mathbf{k} \mid^4} \left[\frac{j}{k_0^2 - j \mid \mathbf{k} \mid^2} \right]$$

$$\hat{G}_3 = \frac{1}{3 \mid \mathbf{k} \mid^4} \left[\frac{j^2}{k_0^2 - j^2 \mid \mathbf{k} \mid^2} \right], \tag{152}$$

All three look like the Fourier transforms of the usual d'Alembert operator, but are marred by a severe infrared singularity due to the common factor $\mid \mathbf{k} \mid^{-4}$. They can be evaluated by performing the contour integrations in the complex plane with

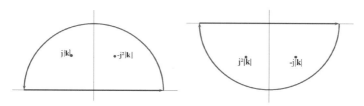

Fig. 5 Left: The upper contour, containing the poles at $j \mid \mathbf{k} \mid$ and $-j^2 \mid \mathbf{k} \mid$, for $t < 0$; Right: The lower contour, containing the poles at $-j \mid \mathbf{k} \mid$ and $j^2 \mid \mathbf{k} \mid$, for $t > 0$

respect to the variable k_0, as in the usual case: The resulting integrals yield the following expressions:

$$-2\pi i \, Y(t) \left[j \, e^{ijkct} - j^2 \, e^{ij^2kct} \right] \quad 2\pi i \, Y(-t) \left[j^2 \, e^{ij^2kct} - j \, e^{-ijkct} \right] \quad (153)$$

Substituting explicit expressions for all complex numbers appearing in these expressions, we get two real functions:

$$2\pi \, Y(t) \, e^{-\frac{\sqrt{3}}{2}kct} \left[\sqrt{3} \cos \frac{k}{2} ct + \sin \frac{k}{2} ct \right],$$

$$2\pi \, Y(-t) \, e^{\frac{\sqrt{3}}{2}kct} \left[\sqrt{3} \cos \frac{k}{2} ct - \sin \frac{k}{2} ct \right],$$

Both expressions contain the damping factor $e^{\frac{\sqrt{3}}{2}kct}$ which is absent in the first contribution proportional to the usual d'Alembertian. But as all three components mix together, all will acquire these damping factors and fade away very quickly (Fig. 5).

These expressions multiply the Fourier transforms of each of the three k-dependent parts of Fourier images of Green's function, G_2 and G_3, while the first one, G_1 similar to the usual d'Alembertian, has to be multiplied by $2\pi \, Y(t) \, \sin(kct)$. Before performing the last integration over dk, they should be multiplied by the factor $\frac{2\pi}{kr} \sin(kr)$.

If it were not for the extra $\mid \mathbf{k} \mid^{-4}$ factor, the subsequent calculus would follow the usual computation of the Green's function by performing the last integral with respect to $k^2 \, dk$. Here the integral is very strongly divergent.

However, the Green function we are looking for can be used to investigate propagation properties if we restrict the class of smooth functions describing the sources. As a matter of fact, it is enough to consider a rapidly decreasing function multiplied by the fourth power of k, e.g.

$$\hat{f}(k) = \mid \mathbf{k} \mid^4 e^{-a^2 k^2}, \quad \text{with} \quad \text{real} \quad a, \quad (154)$$

then the Fourier transform of the convolution of $G(x)$ with the original source function $f(x)$ will behave as the convolution of a much more "regular" Green function whose Fourier image is equal to $\mid \mathbf{k} \mid^4 \hat{G}(k)$ with a Gaussian source which is the inverse Fourier transform of $e^{-a^2 k^2}$ proportional to another Gaussian function $e^{-\frac{x^2}{2a^2}}$. On a source whose Fourier transform is given by (154) the propagator acts as a Green's function given by the following expression:

$$G(x^\mu) \simeq \frac{\delta(k(ct-r))}{r} + \frac{j\,\delta(k(ct-j\,r))}{r} + \frac{j^2\,\delta(k(ct-j^2\,r))}{r}.$$

12 Conclusion

We have presented a consistent *ternary* generalization of both Pauli's exclusion principle and the Dirac equation. Both novelties are based on the extension of fundamental symmetry including the cyclic group Z_3, besides the two fundamental groups Z_2 present in the usual Dirac equation of the electron. Adding of Z_3 to the set of fundamental symmetries enabled us to include *color* in the picture, and formulate a generalization of Dirac's equation acting on a 12-component column containing the equivalent of three Dirac spinors; however, their properties are radically different.The generalized Z_3 Pauli's principle makes possible coexistence of three particle states, two similar and one different, contrary to the usual Z_2 case when only two different states of opposite spin can coexist.

The "ternary" Dirac equation also displays very unusual properties. Once diagonalized, it leads to the common sixth-order equation which all the components must satisfy. The equation of sixth order gives rise to propagators with very strong infrared singularity; nevertheless, we can get some insight into the way fields propagate if the sources we choose display a sufficiently fast decaying behavior at space infinity,

Perhaps the more interesting feature of this construction appears when we consider the overall symmetry of this generalization of the Dirac equation. Let us rewrite again the first-order matrix operator:

$$E\,\sigma_3 \otimes B^\dagger \otimes \mathbb{1}_2 - i\sigma_2 \otimes j^2\,Q_2 \otimes c\boldsymbol{\sigma}\cdot\mathbf{p} = mc^2\,\mathbb{1}_2 \otimes \mathbb{1}_3 \otimes \mathbb{1}_2, \tag{155}$$

Let us look for possible similarity transformations performed on the matrices entering this definition, keeping invariant the result of the sixth iteration, i.e. the matrix equation $(E^6 - c^6 \mid \mathbf{p} \mid^6)\,\mathbb{1}_{12}$.

Let us remind that performing the same similarity transformation on the σ-matrices appearing in the first factors in (155) will not change their commutation properties: if we set

$$\sigma_k \to \sigma_k' = M\sigma_k M^{-1}, \quad \text{then} \quad [\sigma_k', \sigma_m'] = M\,[\sigma_k, \sigma_m]\,M^{-1}.$$

The matrices that keep the Lie algebra structure generated by the three Pauli matrices belong to the $SU(2)$ group.

Similarly, we can perform the same similarity transformation on the two 3×3 matrices appearing as the second factor in the tensor product defining the ternary Dirac operator:

$$B^\dagger \to \tilde{B}^\dagger = P B^\dagger P^{-1}, \quad Q_2 \to \tilde{Q}_2 = P Q_2 P^{-1}.$$

The Lie group that preserves all commutation relations necessary to keep the result invariant is the $SU(3)$ group.

Finally, the last factor in the tensor products (155) is either the 2×2 identity matrix, or has its σ-matrices multiplied scalarly by the momentum vector $\mathbf{p}$. Nevertheless, the last factor can be multiplied by one of the elements of the Z_6 group, represented by one of the entire powers of the sixth root of unity q without changing the result of the diagonalization. These three symmetries span the full symmetry of the Standard Model.

The minimal coupling between the Dirac particles (electrons and positrons) with the electromagnetic field is obtained by inserting the four-potential A_μ into the Dirac equation:

$$\gamma^\mu (p_\mu - e\, A_\mu)\, \psi = m\, \psi. \tag{156}$$

Ternary generalization of Dirac's equation, when expressed in the form (120) with explicit Minkowskian indices, offers a similar possibility to introduce gauge fields. The particular structure of 12×12 matrices Γ_μ makes possible the accommodation of three types of gauge fields, corresponding to three factors from which the tensor product results. The overall gauge field can be decomposed into a sum of three contributions: the $SU(3)$ gauge field $\lambda_a B_\mu^a$, with λ_a, $a = 1, 2, ..8$ denoting the eight 3×3 traceless Gell-Mann matrices, the $SU(2)$ gauge field $\sigma_k A_\mu^k$, $k = 1, 2, 3$ and the electric field potential A_μ^{em}. We propose to insert each of these gauge potentials into a common 12×12 matrix as follows:

The strong interaction gauge potential is aligned on the $SU(3)$ matrix basis:

$$B_\mu = \mathbb{1}_2 \otimes \lambda_a B_\mu^a \otimes \mathbb{1}_2, \quad a, b, .. = 1, 2, \ldots, 8.$$

appearing as the second factor in the tensor product;

The $SU(2)$ weak interaction potential A_μ^i aligned along the three σ-matrices of the first tensorial factor

$$\sigma_k A_\mu^k \otimes \mathbb{1}_3, \quad i, k, .. = 1, 2, 3.$$

appearing in the first factor of the tensor product; and finally the electromagnetic field potential

$$A_\mu^{em} \mathbb{1}_2$$

aligned along the unit 2×2 matrix, appearing in the last factor of the tensor product, so that the overall expression for the gauge potential becomes:

$$\mathscr{A}_\mu = \mathbb{1}_2 \otimes \lambda_a B_\mu^a \otimes \mathbb{1}_2 + \sigma_k A_\mu^k \otimes \mathbb{1}_3 \otimes \mathbb{1}_2 + \mathbb{1}_2 \otimes \mathbb{1}_3 \otimes A_\mu^{em} \mathbb{1}_2 \tag{157}$$

This scheme is in agreement with the no-go theorem stipulating that the only way to combine the Lorentz symmetry with internal symmetries is a trivial direct product of groups (as shown in [25, 26]). It would be interesting to try to solve ternary Dirac equation with some special form of gauge potentials fixed in advance, describing the field created by two quarks acting on the third one.

The approach presented here is still at its first stages [27, 28], and many of its aspects need further development. A lot of questions raised here remain without answer yet, and we hope that new developments will see the light, as we do hope, in a not very distant future.

Acknowledgements I am greatly indebted to Michel Dubois-Violette, Viktor Abramov, and Karol Penson for many discussions and constructive criticism. I would like to express my sincere thanks to Jan-Willem van Holten, Yuri Dokshitser, Paul Sorba, and Jürg Frölich for important suggestions and remarks.

References

1. J. Wess, B. Zumino, Nucl. Phys. B **70**, 39–50 (1974)
2. R. Kerner, Commun. Math. Phys. **91**(2), 213–234 (1983)
3. R. Kerner, C. R. Acad. Sci. Paris. **10**, 1237–1240 (1991)
4. M. Dubois-Violette, R. Kerner, Acta Math. Univ. Comenianae **LXV 14**(2), 175–188 (1996)
5. A. Trovon, O. Suzuki, Adv. Appl. Clifford Algebr. (2015). https://doi.org/10.1007/s00006-015-0565-6
6. L. Vainerman, R. Kerner, J. Math. Phys. **37**(5), 2553–2665 (1996)
7. R. Kerner, J. Math. Phys. **33**(1), 403–4011 (1992)
8. R. Kerner, in *Proceedings of the 23-rd ICGTMP Colloquium, Dubna 2000* (2001). arXiv:math-ph/0011023
9. M. Dubois-Violette, $d^N = 0$: Generalized homology. K-Theory **14**(4), 371–404 (1996)
10. V. Abramov, R. Kerner, J. Math. Phys. **41**(8), 5598–5614 (2000)
11. R. Kerner, O. Suzuki, Int. J. Geom. Methods Mod. Phys. **09**, 1261007 (2012)
12. V. Kac, *Infinite Dimensional Lie Algebras* (Cambridge University Press, Cambridge, 1994)
13. V. Abramov, R. Kerner, O. Liivaapuu, 2015 - arXiv preprint. arXiv:1512.02106, 2015 - arxiv.org
14. V. Abramov, R .Kerner, B. Le Roy, J. Math. Phys. **38**(3), 1650–1669 (1997)
15. R. Kerner, Classical Quantum Gravity **14**(1A), A203–A225 (1997)
16. W. Pauli, The connection between spin and statistics. Phys. Rev. **58**, 716–722 (1940)
17. F.J. Dyson, J. Math. Phys. **8**, 1538–1545 (1967)
18. W. von Gerlach, O. Stern, Z. Phys. **8**, 110 (1921); *ibid*: **9**, 349 (1922)
19. W. Pauli, Z. Phys. **43**, 601–623 (1927)
20. P.A.M. Dirac, Proc. R. Soc. (Lond.) A **117**, 610–624 (1928); *ibid* A **118**, 351–361 (1928)
21. M. Gell-Mann, Y. Ne'eman, *The Eightfold Way* (Benjamin, New York, 1964)
22. S. Okubo, J. Math. Phys. **34**, 3273 (1993); *ibid*, 3292 (1993)
23. H.J. Lipkin, *Frontiers of the Quark Model*. Weizmann Inst. pr. WIS-87–47-PH (1987)

24. H. Bremerman, *Distributions, Complex Variables and Fourier Transforms* (Addison-Wesley, Reading, 1965)
25. L. O'Raifeartaigh, Phys. Rev. **139**, B1052 (1965)
26. S. Coleman, J. Mandula, Phys. Rev. **159**, 1251 (1967)
27. R. Kerner, in *Symmetries and Groups in Contemporary Physics*, ed. by C. Bai, J.-P. Gazeau, M.-L. Ge (World Scientific, Singapore, 2013), pp. 283–288
28. R. Kerner, *Algebra, Geometry and Mathematical Physics*, ed. by A. Makhlouf, E. Paal. Springer Proceedings in Mathematics and Statistics Series, vol. 85 (Springer, Berlin, 2014), pp. 617–637

Pseudo-Solution of Weight Equations in Neural Networks: Application for Statistical Parameters Estimation

Vincent J. M. Kiki, Villévo Adanhounme, and Mahouton Norbert Hounkonnou

Abstract An algebraic approach for representing multidimensional nonlinear functions by feedforward neural networks is implemented for the approximation of smooth batch data containing input–output of the hidden neurons and the final neural output of the network. The training set is associated with the adjustable parameters of the network by weight equations that may be compatible or incompatible.Then we have obtained the exact input weight of the nonlinear equations and the approximated output weight of the linear equations using the conjugate gradient method with an adaptive learning rate. Using the multi-agent system as the different rates of traders of five regions in the Republic of Benin smuggling the fuel from the Federal Republic of Nigeria and the computational neural networks, one can predict the average rates of fuel smuggling traders thinking of this activity in terms of its dangerous character and those susceptible to give up this activity, respectively. This information enables the planner or the decision-maker to compare alternative actions, to select the best one for ensuring the retraining of these traders.

Keywords Function approximation · Conjugate gradient method · Adaptive training

1 Introduction

Artificial Neural Networks have been widely used in many application areas in recent years and have shown their strength in solving complex problems in Artificial Intelligence as well as in Statistics. Although many different models of neural

V. J. M. Kiki (✉)
Ecole Nationale d'Economie Appliquée et de Management, Université d'Abomey-Calavi, Cotonou, Benin

V. Adanhounme · M. N. Hounkonnou
University of Abomey-Calavi, International Chair in Mathematical Physics and Applications (ICMPA), Cotonou, Benin
e-mail: norbert.hounkonnou@cipma.uac.bj

© Springer Nature Switzerland AG 2018
T. Diagana, B. Toni (eds.), *Mathematical Structures and Applications*,
STEAM-H: Science, Technology, Engineering, Agriculture,
Mathematics & Health, https://doi.org/10.1007/978-3-319-97175-9_15

networks have been proposed, multilayered FNNs are the commonest [1]. In this paper, we propose to use the Artificial Feedforward Neural Network in statistical applications for estimating specific statistical parameters. Those are proportions of contraband fuel sellers in Benin that have a certain behavior. It should be noted that this is a population without sampling frame, and thus a population whose parameters are not identifiable. In this case, the conventional estimation methods are irrelevant [2] and [3]. As in [2], many other authors, believe that in this case, we obtain better estimators by using artificial neural networks, and especially the feedforward networks. More specifically, according to[2], the neural network, in its operation, produces solutions to very complex problems by adjusting weight coefficients in a fairly consistent learning scheme.

These authors showed that when the parameters of the studied population are not identifiable (as in the case of populations without sampling frame), the theoretical statistical methods (such as maximum likelihood, ordinary least squares, and moments methods) lose their relevance. They advocate in this case the use of neural networks for constructing the prediction intervals, and therefore the necessary estimators.

There are two main categories of neural networks, namely supervised learning neural networks and unsupervised learning neural networks including the Multilayer Feedforward Neural Networks. The latter is the one used in this research.

The statistical parameters that we estimate using neural networks in this research are exclusively descriptive proportions of an opinion poll. An application was made to the case of contraband fuel sellers. What seems very interesting in this application is the fact that an artificial neural network is, like biological neurons, an interconnected set of neurons acting towards solving complex problems.

For example, when a physical person records information, those (biological) neurons are activated depending on the importance (input weights) of such information in an intelligent thought process. This information undergoes changes before being issued with a certain degree of efficiency (output weights) resulting in decisions or actions from this person. Similarly, when it comes to an artificial neural network, the information is introduced with input weights w_i in an input layer.

With activation functions, this information is processed in hidden layers before it comes out with output weights v_i towards the output layer in the form of results. Some artificial neural networks, in terms of their functioning, are classified in the category of statistical applications. Thus, just as a neural network can be used to solve specific problems in artificial intelligence, so can it be used in statistical applications to test hypotheses or make estimates, and therefore produce results in statistical inference.

Many studies highlight the wide variety of application areas of neural networks. Those include, as regards the use of neural networks in artificial intelligence, very recent works such as those published by Kokkinos and Margaritis [4] that made use of neural networks for static hand gesture recognition on granular neural networks [5].

Besides studies on artificial intelligence methods, there are numerous relevant studies on statistical applications using neural networks.

The authors in [6] published the paper on Nonlinear Time Series Modeling by using neural networks. This is an innovative approach in which the proposed model is a combination of several other stationary and non-stationary linear models. It's a model that works thanks in part to the adjustment of the weighting coefficients of inputs and outputs. Another very interesting research in statistical application dedicated in February 2012 to confidence and prediction intervals for determining the number of nodes in artificial neural networks was carried out by Hwang et al. [2]. The authors show that here the classical statistical theory for constructing prediction intervals is inappropriate. The reason they mention is that the parameters are not identifiable. They establish that this inadequacy can be solved by using neural network. In their work, the authors in [2] proceeded to the construction of confidence intervals and asymptotically valid predictions. They propose a more or less operational approach that allows using prediction intervals for choosing the number of nodes in the network. They have successfully applied this method to a specific case of electrical load prediction.

Another model for statistical prediction was published by Poli and Jones [7]. This is a nonlinear statistical prediction using neural networks. Their study is based on a stochastic model featuring a multilayer feedforward architecture with random connections between units and noisy response functions.

The proportions estimated by using neural networks in this paper are those of the contraband fuel sellers who are aware of the dangerous nature of their trade on the one hand, and those who are in favor of abandoning this activity on the other hand.

In order to train an FNN, supervised training is probably the most frequently used technique. The training process is an incremental adaptation of connection weights that propagate information between simple processing units called neurons. The neurons are arranged in layers and connections between the neurons of one layer and those of the next exist. Also, one can use an algebraic training [1] which is the approach for approximating multidimensional nonlinear functions by FNNs based on available input–output. Typically, training involves the numerical optimization of the error between the data and the actual network's performance with respect to its adjustable parameters or weights. Considerable effort has put into developing techniques for accelerating the convergence of these optimization-based training algorithms [8–10]. Another line of research has focused on the mathematical investigation of network's approximation properties [11–15]. The latter results provide few practical guidelines for implementing the training algorithms, and they cannot be used to evaluate the properties of the solutions obtained by numerical optimization. The algebraic training approach provides a unifying framework that can be used both to train the network and to investigate their approximation properties. The data are associated with the adjustable parameters by means of neural network input–output. Hence the nonlinear training process and related approximation properties can be investigated via linear algebra. For the first time to our best knowledgde of the literature, the approach as an extension of typical neural network methodology we propose in this paper has been applied to train a feedforward neural network that predicts the average rate of sellers smuggling the fuel from the Federal Republic of Nigeria to the Republic of Benin, considering this trade as dangerous and susceptible to give up trading.

Taking account of the historical, geographical, and sociological realities of the different peoples of Benin, we aimed to study the behavior of the fuel sellers in five main regions of Benin smuggling the fuel from the Federal Republic of Nigeria. Furthermore Porto-Novo, Cotonou, Tchaourou, and Parakou are located at the neighborhood of the border Benin-Nigeria and Natitingou is the main region in the North of Benin where the economic activities are well developed. From this point of view, these regions remain of a great importance for studying the behavior of the fuel smuggling traders and are worthy of interest for estimating the statistical parameters derived from this sample.

Indeed, in order to execute the survey one selects 70 interviewers through five regions, namely Natitingou, Parakou, Tchaourou, Porto Novo, and Cotonou where the smuggling fuel is well spread. The goal of the interview is to provide the collection of answers to the two following questions: Are you conscious of the dangerous character of smuggling fuel trade on the shelves, in the streets end, in the houses? Do you get involved with giving up trading the smuggling fuel if the government of Benin decides your retraining? By virtue of the above two questions, one interviewed 660 fuel smuggling sellers in Benin in 2013 from 15 to 31 May, during 17 days following the steps: 40 interviewees per day during the first 16 days; 20 interviewees during the last day. In this paper, the Neural Networks have been applied efficiently for the prediction of the average rates of traders in five regions of the Republic of Benin smuggling the fuel from the Federal Republic of Nigeria who think of this trade in terms of its dangerous character and susceptible to give up trading. The remaining part of the paper is organized as follows: In Sect. 2 we propose an algebraic training algorithm with an adaptive learning rate based on the conjugate gradient method [16, 17]. In Sect. 3, a system containing five variables denoting the rates of traders of five regions in the Republic of Benin smuggling the fuel from the Federal Republic of Nigeria are considered and experiments, simulation results are presented. The last two sections contain the discussion and the conclusion.

2 Framework

2.1 Development of the Algebraic Approach

The objective is to approximate smooth scalar functions of q inputs [1]:

$$h, g_i : \mathbb{R}^q \to \mathbb{R} \quad i = 1 \cdots s \tag{2.1}$$

using a feedforward neural network. Typically, the functions to be approximated are unknown analytically, but a precise set of input-to-nodes samples $\{x^k, v_i^k\}_{k=1,\cdots,p}$ and a precise set of input–output samples $\{x^k, u^k\}_{k=1,\cdots,p}$ can be generated as follows:

$$v_i^k = g_i(x^k) \quad , \quad u^k = h(x^k) \quad \forall \quad k \tag{2.2}$$

These sets of samples are referred to as training sets. We assume that:

- the scalar output y_i of the i-th neuron belonging to the hidden layer is computed as:

$$y_i = f[x^T w_i + \theta_i] \quad , \quad w_i = [w_{i1} \cdots w_{iq}]^T, \tag{2.3}$$

where w_{ij}, $1 \leq j \leq q$ are the weights connecting q inputs to the i-th hidden neuron of the network, θ_i is the input bias of the i-th neuron, and f the logistical function of the hidden layer defined as follows:

$$f(\tau) = \frac{1}{1 + e^{-\tau}} \tag{2.4}$$

- the final scalar output of the network z is computed as a nonlinear transformation of the weighted sum of the input-to-node variables n_i with $i = 1, \cdots, s$:

$$z = v^T f[wx + \theta] + \lambda \tag{2.5}$$

where

$$f[n] = [f(n_1) \cdots f(n_s)]^T; \tag{2.6}$$

$$n = wx + \theta = [n_1 \cdots n_s]^T \tag{2.7}$$

$$w = (w_{ij})_{1 \leqslant i \leqslant s; 1 \leqslant j \leqslant q}; \quad \theta = [\theta_1 \cdots \theta_s]^T \tag{2.8}$$

and λ the output bias.

Remark In the general case, one can choose s hidden layer activation functions f_i, $1 \leqslant i \leqslant s$. The activation function for the output neuron is chosen as the identity function.

The computational neural network matches the training set $\{x^k, v_i{}^k, u^k\}_{k=1,\cdots,p}$ exactly if, given the input x^k, it produces $v_i{}^k$ and u^k as follows:

$$y_i(x^k) = v_i{}^k; \quad z(x^k) = u^k \tag{2.9}$$

which leads to

$$v_i{}^k = f[(x^k)^T w_i^k + \theta_i] \tag{2.10}$$

$$u^k = v^T f[w^k x^k + \theta] + \lambda \tag{2.11}$$

where v_i, $1 \leq i \leq s$ are the weights connecting s hidden neurons to the output neuron. Grouping the known elements $v_i{}^k$ and u^k from the training set in the vectors $v_i = [v_i{}^1 \cdots v_i{}^p]^T;$ $u = [u^1 \cdots u^p]^T$ Eqs. (2.10) and (2.11) can be written using matrix notation

$$v_i = f[n_i] \tag{2.12}$$

$$u = Sv + \Lambda \tag{2.13}$$

which are referred to as weight equations, where

$$n_i = [n_i{}^1 \cdots n_i{}^p]^T; \quad n_i{}^k = (x^k)^T w_i^k + \theta_i; \tag{2.14}$$

$$f[n_i] = [f[n_i{}^1] \cdots f[n_i{}^p]]^T; \quad \Lambda = [\lambda \cdots \lambda]^T; \tag{2.15}$$

$$S = \begin{pmatrix} f[n_1{}^1] & f[n_2{}^1] & \cdots & f[n_s{}^1] \\ f[n_1{}^2] & f[n_2{}^2] & \cdots & f[n_s{}^2] \\ \cdots & \cdots & & \cdots \\ f[n_1{}^p] & f[n_2{}^p] & \cdots & f[n_s{}^p] \end{pmatrix}$$

2.2 Proposed Algorithm

In this subsection we aim at solving the weight equations (2.12) and (2.13), where the unknowns are w_i and v. For this purpose, we define a pseudo-solution of the system of linear equations, for which we will prove the existence and the uniqueness.

Consider the system (2.12) of linear equations

$$A(k)w_i^k = b_i^k, \quad w_i^k \in \mathbf{R}^q, \quad b_i^k \in \mathbf{R}, \quad i = 1, \ldots, s, \quad k = 1, \ldots, p \tag{2.16}$$

where $A(k) = (x^k)^T$ are the $1 \times q$ matrixes, $b_i^k = f^{-1}(v_i^k) - \theta_i$, $f^{-1}(\xi) = \ln\left(\frac{\xi}{1-\xi}\right)$, $\mathbf{R}^q$ and $\mathbf{R}$ are two q, 1-dimensional vector spaces, respectively. For fixed k, $A(k)$ is constant and we set $A = A(k)$, $w_i = w_i^k$, $b_i = b_i^k$. In the sequel, i and k are considered as fixed.

Definition 2.1 *A pseudo-solution of the system (2.16) is called the vector with the least norm belonging to the set of vectors w_i such that the error function $\Omega(w_i) = [b_i - Aw_i]^2$ can be minimized.*

Consider the vector space $\mathbf{R}^q$ endowed with the inner product defined as follows: $\langle \xi, \eta \rangle = \xi^T \eta$. The following statement holds [18]

Theorem 21 *The system of linear equation (2.16) admits a unique pseudo-solution.*

Proof Let us consider the error function Ω defined as follows

$$\Omega(w_i) = (b_i - Aw_i)^T(b_i - Aw_i), \quad w_i \in \mathbf{R}^q \tag{2.17}$$

This function can reach its extremum only at the points where $d\Omega(w_i) = 0$, i.e.,

$$A^T A w_i = A^T b_i \tag{2.18}$$

First of all, we have to prove the compatibility of the system (2.18), independently of the fact that the system (2.16) is compatible or not. Taking into account that the matrix $A^T A$ is symmetric, we can write the homogeneous adjoint system associated with the system (2.18) in the form

$$A^T A \eta = 0, \quad \eta \in \mathbf{R}^q \tag{2.19}$$

Then, for every nonzero solution of (2.19) we can write the relation

$$\eta^T A^T A \eta = (A\eta)^T (A\eta) = 0 \tag{2.20}$$

which leads to

$$A\eta = 0 \Longrightarrow \eta^T (A^T b_i) = 0. \tag{2.21}$$

This last relation means that the assumption of Fredholm's theorem (see Appendix) is satisfied; therefore, the system (2.18) is compatible.

Now, let us prove that the infimum of the function Ω is reached on the set

$$\Gamma = \left\{ w_i \in \mathbf{R}^q : A^T A w_i = A^T b_i \right\}.$$

Indeed, for $w_{i0} \in \mathbf{R}^q$, $w_{i0} + \delta w_i \in \mathbf{R}^q$ we have

$$\Omega(w_{i0} + \delta w_i) = \Omega(w_{i0}) - 2(\delta w_i)^T A^T (b_i - A w_{i0}) + (A\delta w_i)^T (A\delta w_i) \tag{2.22}$$

yielding

$$\Omega(w_{i0} + \delta w_i) \geqslant \Omega(w_{i0}), \quad w_{i0} \in \Gamma, \quad \forall \delta w_i. \tag{2.23}$$

Conversely, the function $\Omega : w_i \mapsto \Omega(w_i)$ can reach its infimum only at the local extremum points where $d\Omega(w_i) = 0$, i.e. there follows the system (2.18). Finally we obtain that the infimum of the function Ω is reached on the set Γ.

In order to prove the existence and the uniqueness of the pseudo-solution we define the sets $K \subset \mathbf{R}^q$ and $L \subset \mathbf{R}^q$ as follows:

$$K = \{z \in \mathbf{R}^q : Az = 0\}, \quad L = \left\{ A^T b_i \in \mathbf{R}^q, \ b_i \in \mathbf{R} \right\}.$$

The condition $\xi \in L$ means that the system $A^T \chi = \xi$, $\forall \chi \in \mathbf{R}$ is compatible.

On the other hand, according to Fredholm's theorem the latter system is compatible if and only if $\forall z \in K, z^T \xi = 0$ which means that $K = L^\perp$. For the subsets L and K of the Euclidean space $\mathbf{R}^q$, we can write

$$w_i = w_{i0} + z$$

where $w_{i0} \in L$, $z \in K$. So, every vector $w_i \in \Gamma$ can be expressed in the form

$$w_i = w_{i0} + w_{i1}, \quad w_{i0} \in L, \quad w_{i1} \in K$$

since

$$A^T A w_i = A^T A(w_{i0} + w_{i1}) = A^T (A w_{i0}) + A^T (A w_{i1}) = A^T b_i.$$

If η is another vector of Γ, then $\eta = w_{i0} \in L$ because of the uniqueness of the orthogonal projection

$$\eta = (\eta - w_i + w_{i1}) + w_{i0}$$

with $(\eta - w_i) + w_{i1} \in K$. Therefore, for all vectors of Γ, the vector w_{i0} is the same.

For an arbitrary vector $w_i \in \Gamma$, we can write

$$||w_i||^2 = (w_{i0} + w_{i1})^T (w_{i0} + w_{i1}) = ||w_{i0}||^2 + ||w_{i1}||^2 \tag{2.24}$$

where $w_{i1}^T w_{i0} = 0$; it follows that $||w_i|| \geqslant ||w_{i0}||$. Therefore there exists a unique common vector w_i belonging to the sets Γ and L which has the minimal norm. In other words, the unique pseudo-solution of the system (2.16) is of the form

$$w_i = A^T z \tag{2.25}$$

where z is the solution of a system

$$AA^T z = b_i. \tag{2.26}$$

Replacing A^T by x^k, w_i by w_i^k, and b_i^k by $f^{-1}(v_i^k) - \theta_i$, we can write the pseudo-solution of the system as

$$w_i^k = x^k z \tag{2.27}$$

where z is the solution of the equation

$$(x^k)^T x^k z = f^{-1}(v_i^k) - \theta_i. \tag{2.28}$$

Finally we get

$$w_i^k = \frac{f^{-1}[v_i^k] - \theta_i}{\langle x^k, x^k \rangle} x^k \tag{2.29}$$

for fixed k, where $\langle ., . \rangle$ denotes the inner product; that ends the proof. $\square$

As the weights w_i^k are known from the relation (2.29), the $p \times s$ matrix S is known. Without loss of generality we can choose $\Lambda = 0$ [1] and the system (2.13) takes the form

$$Sv = u \tag{2.30}$$

As mentioned before, the system (2.30) admits a unique pseudo-solution $v^\star$.

In order to construct the sequence of approximated solutions $\{v^m\}$ of the system (2.30) convergent to the unique pseudo-solution $v^\star$, we use the conjugate gradient method for the minimization of the function Ω defined by $\Omega(v) = |u - Sv|^2$, giving the following iteration [16]:

$$S^T Sv^\star = S^T u \; ; \tag{2.31}$$

$$v^{m+1} = v^m + \alpha_m d^m; \quad m = 0, 1, 2, 3, \ldots \tag{2.32}$$

where v^m is the current point, d^m a searched direction, and α_m the step length.

Various choices of the direction d^m give rise to distinct algorithms. A broad class of methods uses $-d^m = \nabla\Omega(v^m)$ as a searched direction and the step length α_m is provided either by the relation

$$\min_{\alpha_m > 0} \Omega\left(v^m - \alpha_m \nabla\Omega(v^m)\right) \tag{2.33}$$

or by Wolfe's conditions [9]. The widely used gradient-based training algorithm, termed batch back-propagation (BP), minimizes the error function using the following steepest descent method with constant, heuristically chosen learning rate α:

$$v^{m+1} = v^m - \alpha \nabla\Omega(v^m). \tag{2.34}$$

Clearly, the behavior of any algorithm depends on the choice of the step length not less than the choice of the searched direction. It is well known that pure gradient descent methods with fixed learning rate tend to be inefficient [10]. The proposed algorithm is an adaptive learning rate algorithm based on the conjugate gradient method. The motivation for this choice is that it provides the fast convergence of the approximation method which yields the iteration [16] and [17]:

$$v^{m+1} = v^m - \frac{\langle Sd^m; Sv^m - u \rangle}{\langle Sd^m; Sd^m \rangle} d^m; \tag{2.35}$$

where

$$d^0 = 2[S^T Sv^0 - S^T u], \tag{2.36}$$

$$d^m = 2[S^T Sv^m - S^T u] + \frac{\langle S^T Sv^m - S^T u; S^T Sv^m - S^T u \rangle}{\langle S^T Sv^{m-1} - S^T u; S^T Sv^{m-1} - S^T u \rangle} d^{m-1}, \tag{2.37}$$

where $m = 1, 2, 3, \cdots$.

Furthermore the learning rate α_m can be written in the form:

$$\alpha_m = \frac{\langle Sd^m; Sv^m - u \rangle}{\langle Sd^m; Sd^m \rangle} \tag{2.38}$$

3 Experiments and Results

Modelling has become a very important tool in the modern science and research. Scientists use modelling to test hypotheses, to evaluate the performance of systems, to explore some fields that are difficult to assess by experimentation. Many researches on plant growth and physical system modellings [19–23], on estimation of model parameters [24], on motion estimation [25, 26], on statistical properties of transactional databases [27], and on biological population growth modelling were done during the few past years. Those works are based on various mathematical approaches. A Sequential Learning Neural Network (SLNN) is applied in [28] to agriculture. De Reffye and his team [29] have built a model based on the probability theory.

Since a statistical model is often constructed in order to make conclusions, predictions, or inferences from data, an estimate of statistical parameters can be obtained using the computational neural network matching the data set provided by statistical model to predict future data as accurately as possible. In this work we established the new formulas suitable for computing the weights as the pseudo-solutions of the system of linear equations compatible or not. These results can be used in the other branches of mathematics. In this paper, on the basis of daily sampling surveys we carry out the rates of traders obtained during 17 days of five regions in the Republic of Benin, namely Natitingou (region F_1), Parakou (region F_2), Tchaourou (region F_3), Porto-Novo (region F_4), and Cotonou (region F_5) where the smuggling fuel trade from the Federal Republic of Nigeria is widespread. The method we propose in this paper has been applied in this section to train a feedforward neural network that predicts the average rates of traders of these five regions in the Republic of Benin who think of the fuel smuggling trade in terms of its dangerous character and those susceptible to give up the fuel smuggling trade.

3.1 Rates of Traders Thinking of the Trade in Terms of Its Dangerous Character

Considering the rates of traders of five regions in the Republic of Benin who think of the fuel smuggling trade in terms of its dangerous character, we have a multi-agent system which contains five variables :

- the rate of traders in the region F_1 ;
- the rate of traders in the region F_2 ;

- the rate of traders in the region F_3 ;
- the rate of traders in the region F_4 ;
- the rate of traders in the region F_5.

In order to construct an approximately unbiased estimator of the rate of traders, we use the estimate of average rate defined as follows [30]:

$$\sigma = \frac{1}{5} \sum_{i=1}^{5} x_i \qquad (3.1)$$

where x_i, $1 \leqslant i \leqslant 5$ is the rate of traders in the region F_i.

The data set during 17 days gets the form:

$$\{x^k = (x_1^k, x_2^k, x_3^k, x_4^k, x_5^k)^T; \ u^k = \sigma^k; \ v_i^k; \ k = 1, \ldots, 17\} \qquad (3.2)$$

where x_i^k is the rate of traders in the region F_i at a day k, $v_i^{\ k}$ are chosen randomly in the interval $[0.5; 1[$.

Putting

$$w_i^k = (w_{i1}^k, w_{i2}^k, w_{i3}^k, w_{i4}^k, w_{i5}^k)^T \qquad (3.3)$$

Eq. (2.29) can be written

$$w_{i1}^k = \frac{f^{-1}(v_i^k) - \theta_i}{(x_1^k)^2 + (x_2^k)^2 + (x_3^k)^2 + (x_4^k)^2 + (x_5^k)^2} x_1^k \qquad (3.4)$$

$$w_{i2}^k = \frac{f^{-1}(v_i^k) - \theta_i}{(x_1^k)^2 + (x_2^k)^2 + (x_3^k)^2 + (x_4^k)^2 + (x_5^k)^2} x_2^k \qquad (3.5)$$

$$w_{i3}^k = \frac{f^{-1}(v_i^k) - \theta_i}{(x_1^k)^2 + (x_2^k)^2 + (x_3^k)^2 + (x_4^k)^2 + (x_5^k)^2} x_3^k \qquad (3.6)$$

$$w_{i4}^k = \frac{f^{-1}(v_i^k) - \theta_i}{(x_1^k)^2 + (x_2^k)^2 + (x_3^k)^2 + (x_4^k)^2 + (x_5^k)^2} x_4^k \qquad (3.7)$$

$$w_{i5}^k = \frac{f^{-1}(v_i^k) - \theta_i}{(x_1^k)^2 + (x_2^k)^2 + (x_3^k)^2 + (x_4^k)^2 + (x_5^k)^2} x_5^k \qquad (3.8)$$

Choosing $\theta_i = -1$ the 17×5 matrix S is known and choosing the initial approximated solution v^0 randomly we can apply Eq. (2.35).

The network consists of three layers whose input layer has five nodes and the number of hidden nodes is 5; in this case, the training is suitable since too few hidden nodes limit a network's generalization capabilities while too many hidden nodes can result in overtraining or memorization by the network. The output layer consists

Table 1 Simulation results

k	x_1^k	x_2^k	x_3^k	x_4^k	x_5^k	σ^k	MSE	n	α_m
1	0.941	0.906	0.865	0.830	0.823	0.67	0.26	1	0.012
2	0.98	0.911	0.865	0.804	0.916	0.75	0.26	1	0.012
3	0.947	0.910	0.812	0.804	0.915	0.72	0.26	1	0.012
4	0.963	0.906	0.833	0.833	0.920	0.97	0.26	1	0.012
5	0.963	0.860	0.812	0.828	0.920	0.78	0.26	1	0.012
6	0.947	0.941	0.920	0.820	0.820	0.72	0.26	1	0.012
7	0.980	0.917	0.865	0.820	0.915	0.95	0.26	1	0.012
8	0.989	0.906	0.920	0.804	0.680	0.94	0.26	1	0.012
9	0.947	0.990	0.833	0.833	0.678	0.78	0.26	1	0.012
10	0.989	0.991	0.842	0.828	0.681	0.97	0.26	1	0.012
11	0.941	0.980	0.920	0.804	0.823	0.78	0.26	1	0.012
12	0.963	0.989	0.812	0.868	0.890	0.77	0.26	1	0.012
13	0.941	0.860	0.812	0.828	0.823	0.97	0.26	1	0.012
14	0.947	0.906	0.865	0.804	0.823	0.75	0.26	1	0.012
15	0.963	0.861	0.834	0.832	0.689	0.95	0.26	1	0.012
16	0.941	0.820	0.865	0.804	0.916	0.73	0.26	1	0.012
17	0.963	0.906	0.865	0.833	0.864	0.81	0.26	1	0.012

of a single node representing the estimate of the average rate of fuel smuggling traders who think of this activity in terms of its dangerous character. SOFTWARE R is used for the implementation. Simulations show that the choice of the proper initial weights is important for algorithms convergence. Results are shown in Table 1 where σ^k, $1 \ldots, 17$ are the average rates, MSE is the mean square error, α_m is the variable learning rate, and n the total number of epochs to be trained. For the variable learning rate around 0.012 and for MSE around 0.26, the stability of the computation is obtained and the average rates σ^k, $1 \ldots, 17$ belong to the interval [0.67; 0.97].

3.2 Rates of Traders Susceptible to Give Up Trading

Our method can also be applied to the prediction of the average rate of traders susceptible to give up the fuel smuggling trade. We have a system containing five variables :

- the rate of traders in the region F_1 ;
- the rate of traders in the region F_2 ;
- the rate of traders in the region F_3 ;
- the rate of traders in the region F_4 ;
- the rate of traders in the region F_5.

In order to construct an approximately unbiased estimator of the rate of traders, we use the estimate of average rate defined as follows [30]:

$$\delta = \frac{1}{5} \sum_{i=1}^{5} \widetilde{x}_i \qquad (3.9)$$

where $\widetilde{x}_i$, $1 \leqslant i \leqslant 5$ is the rate of traders of the region F_i.

During 17 days the data set gets the form :

$$\{\widetilde{x}^k = (\widetilde{x}_1^k, \widetilde{x}_2^k, \widetilde{x}_3^k, \widetilde{x}_4^k, \widetilde{x}_5^k)^T; \quad u^k = \delta^k; \quad v_i^k; \quad k = 1, \dots, 17\} \qquad (3.10)$$

where $\widetilde{x}_i^k$ is the rate of traders in the region F_i at a day k, $v_i{}^k$ are chosen randomly in the interval $[0.5; 1[$. Putting

$$\widetilde{w}_i^k = (\widetilde{w}_{i1}^k, \widetilde{w}_{i2}^k, \widetilde{w}_{i3}^k, \widetilde{w}_{i4}^k, \widetilde{w}_{i5}^k)^T \qquad (3.11)$$

Eq. (2.29) can be written

$$\widetilde{w}_{i1}^k = \frac{f^{-1}(v_i^k) - \theta_i}{(\widetilde{x}_1^k)^2 + (\widetilde{x}_2^k)^2 + (\widetilde{x}_3^k)^2 + (\widetilde{x}_4^k)^2 + (\widetilde{x}_5^k)^2} \widetilde{x}_1^k \qquad (3.12)$$

$$\widetilde{w}_{i2}^k = \frac{f^{-1}(v_i^k) - \theta_i}{(\widetilde{x}_1^k)^2 + (\widetilde{x}_2^k)^2 + (\widetilde{x}_3^k)^2 + (\widetilde{x}_4^k)^2 + (\widetilde{x}_5^k)^2} \widetilde{x}_2^k \qquad (3.13)$$

$$\widetilde{w}_{i3}^k = \frac{f^{-1}(v_i^k) - \theta_i}{(\widetilde{x}_1^k)^2 + (\widetilde{x}_2^k)^2 + (\widetilde{x}_3^k)^2 + (\widetilde{x}_4^k)^2 + (\widetilde{x}_5^k)^2} \widetilde{x}_3^k \qquad (3.14)$$

$$\widetilde{w}_{i4}^k = \frac{f^{-1}(v_i^k) - \theta_i}{(\widetilde{x}_1^k)^2 + (\widetilde{x}_2^k)^2 + (\widetilde{x}_3^k)^2 + (\widetilde{x}_4^k)^2 + (\widetilde{x}_5^k)^2} \widetilde{x}_4^k \qquad (3.15)$$

$$\widetilde{w}_{i5}^k = \frac{f^{-1}(v_i^k) - \theta_i}{(\widetilde{x}_1^k)^2 + (\widetilde{x}_2^k)^2 + (\widetilde{x}_3^k)^2 + (\widetilde{x}_4^k)^2 + (\widetilde{x}_5^k)^2} \widetilde{x}_5^k \qquad (3.16)$$

One can apply the same process used in the previous implementation. Results are shown in Table 2. For the variable learning rate around 0.004 and for MSE around 0.21, the stability of the computation is obtained and the average rates δ^k, $k = 1, \dots, 17$ belong to the interval $[0.63; 0.99]$.

4 Discussion

Using the algebraic approach we obtain the weight equations which can be compatible or not. The pseudo-solutions for these equations are the generalization of the solutions obtained in [1]. The approach with these solutions can be used efficiently for the identification and control of dynamical systems, mapping the

Table 2 Simulation results

k	$\widetilde{x}_1^k$	$\widetilde{x}_2^k$	$\widetilde{x}_3^k$	$\widetilde{x}_4^k$	$\widetilde{x}_5^k$	δ^k	MSE	n	α_m
1	0.941	0.969	0.946	0.965	0.946	0.97	0.21	1	0.004
2	0.98	0.958	0.946	0.980	0.961	0.92	0.21	1	0.004
3	1.000	0.970	0.933	0.960	0.951	0.71	0.21	1	0.004
4	0.989	0.969	0.933	0.960	0.940	0.94	0.21	1	0.004
5	0.980	0.977	0.917	0.983	0.960	0.73	0.21	1	0.004
6	0.989	0.971	0.917	0.981	0.989	0.83	0.21	1	0.004
7	0.989	0.958	0.946	0.960	0.982	0.88	0.21	1	0.004
8	0.980	0.969	0.960	0.981	0.946	0.80	0.21	1	0.004
9	0.980	0.959	0.933	0.960	0.960	0.71	0.21	1	0.004
10	0.989	0.975	0.947	0.983	0.895	0.69	0.21	1	0.004
11	0.989	0.977	0.920	0.980	0.890	0.97	0.21	1	0.004
12	0.980	0.956	0.946	0.961	0.892	0.86	0.21	1	0.004
13	0.990	0.958	0.926	0.983	0.946	0.63	0.21	1	0.004
14	0.991	0.977	0.920	1.000	0.946	0.99	0.21	1	0.004
15	0.980	0.971	0.922	0.990	0.892	0.81	0.21	1	0.004
16	0.980	0.969	0.946	0.989	0.964	0.81	0.21	1	0.004
17	0.989	0.969	0.946	0.981	0.983	0.74	0.21	1	0.004

input–output representation of an unknown system and its control law [1]. Using the conjugate gradient method we construct the sequence convergent to the pseudo-solution, in other words we predict the average rates of fuel smuggling traders who think of this trade in terms of its dangerous character and the average rates of traders susceptible to give up the fuel smuggling trade. The simulations show that the method with the adaptive learning rate is more stable and converge very fast.

As the smuggling fuel is stored not suitably, its volatile contaminant components may evaporate to become components of the gaseous phase (air) present in the void space. Therefore the quality of this fuel has been continuously deteriorating, causing much concern to both suppliers and users. The prediction of the average rates of these fuel smuggling traders derived from the statistical models enables the planner or the decision-maker to compare alternatives action, to select the best one for ensuring the retraining of these traders.

5 Conclusion

The techniques developed in this paper match input–output information approximately or exactly by neural networks. The adjustable parameters or weights are determined by solving algebraic equations and by using the conjugate gradient method. The algorithms used are derived based on the exact and approximated solutions of input–output weight equations. Their effectiveness is demonstrated by training feedforward neural networks which produce average rates of the smuggling traders thinking of this trade in terms of its dangerous character and susceptible to give up this trade. The experimentations show that our combination of the algebraic approach and the fast convergent conjugate gradient method is a useful approach to solve many complex problems.

Acknowledgements This work is partially supported by the ICTP through the OEA-ICMPA-Prj-15. The ICMPA is in partnership with the Daniel Iagolnitzer Foundation (DIF), France.
The authors would like to thank the Konida National Foundation for Scientific Researches (KNFSR) for its financial support.

Appendix

Theorem 51 (Fredholm) *The system (2.16) is compatible if and only if every solution of homogeneous adjoint system*

$$A^T \eta = 0 \qquad \eta \in \mathbf{R}^q \tag{5.1}$$

satisfies the equation

$$b_i \eta = 0 \tag{5.2}$$

References

1. S. Ferrari, R.F. Stengel, Smooth function approximation using neural networks. IEEE Trans. Neural Netw. **16**(1), 24–38 (2005)
2. G.-B. Hwang, X. Ding, H. Zhou, R. Zhang, Extreme learning machine for regression and multiclass classification. IEEE Trans. Syst. Main Cybern. Path B Cybern. **42**(2), 513–529 (2012)
3. M. Wilhelm, Echantillonnage boule de neige: la méthode de sondage déterminé par les répondants. Office fédéral de la statistique, Neuchatel (2014)
4. Y. Kokkinos, G. Margaritis, Breaking ties of plurality voting in ensembles of distributing neural networks classifiers using soft max accumulation. Parallel and Distributed Processing Laboratory, Department of Applied Informatics, University of Macedonia (2014)
5. H. Hasan, S. Abdul-kareem, Human-computer interaction using vision based hand gesture recognition system: a survey, in *Neural Computing and Application (NCA)* (Springer, Berlin, 2014), pp. 251–261
6. M. Suárez-Fariñas et al., Local global neural networks: a new approach for nonlinear time series modeling. J. Am. Stat. Assoc. **99**(468), 1092–1107 (2004)
7. A. Poli, R.D. Jones, A neural networks model for prediction. J. Am. Stat. Assoc. **89**, 117–121 (2012)
8. D. Rumelhart et al., Learning representations by back-propagating errors. Nature **323**, 533–536 (1986)
9. P.H. Wolfe, Convergence conditions for ascend methods. SIAM Rev. **11**, 226–235 (1969)
10. E. Polak, *Optimization: Algorithms and Consistent Approximations* (Springer, Berlin, 1997)

11. R.A. Jacobs, Increased rates of convergence through learning rate adaptation. Neural Netw. **1**(4), 295–308 (1988)
12. A.K. Rigler et al., Rescaling of variables in back-propagation learning. Neural Netw. **3**(5), 561–573 (1990)
13. A.N. Kolmogorov, On the representation of continuous function of several variables by superposition of continuous functions of one variable and addition. Dokl. Akad. Nauk SSSR **114**, 953–956 (1957)
14. K. Hornik et al., Multi-layer feeforward networks are universal approximators. Neural Netw. **2**, 359–366 (1989)
15. A.R. Baron, Universal approximation bounds for superposition of a sigmoidal functions. IEEE Trans. Inf. Theory **39**(3), 930–945 (1993)
16. F.L. Vassiliev, Numerical Methods for the optimization problems. Nauk, Moscow (1988)(in Russian)
17. V. Adanhounme, T.K. Dagba, S.A. Adedjouma, Neural smooth function approximation and prediction with adaptive leraning rate, in *Transactions on CCI VII*. Lecture Notes in Computer Science, vol. 7270 (Springer, Berlin, 2012), pp. 103–118
18. D. Beklemichev, Cours de géométrie analytique et d'algèbre linéaire. Editions Mir, Moscou (1988)
19. T.K. Dagba, V. Adanhounmè, S.A. Adédjouma, Modélisation de la croissance des plantes par la méthode d'apprentissage supervisé du neurone, in Premier colloque de l'UAC des sciences, cultures et technologies, mathématiques, Abomey-Calavi, pp. 245–250 (2007)
20. J.-M. Dembelé, C. Cambier, Modélisation multi-agents de systèmes physiques: application à l'érosion cotière, in CARI'06, Cotonou, pp. 223–230 (2006)
21. T. Fourcaud, Analyse du comportement mécanique d'une plante en croissance par la méthode des éléments finis. PhD thesis, Université de Bordeaux 1, Talence (1995)
22. A. Rostand-Mathieu, Essai sur la modélisation des interactions entre la croissance et le développement d'une plante, cas du modèle greenlab. Ph.D thesis, Ecole Centrale de Paris (2006)
23. L. Wu, F.-X. Le Dimet, P. De Reffye, B.-G. Hu, A new mathematical formulation for plant structure dynamics, in *CARI'06*, Cotonou, Bénin, pp. 353–360 (2006)
24. S.W. Jang, M. Pomplun, H. Choi, Adaptive robust estimation of model parameters from motion vectors, in *International Conference on Rought Sets and Current Trends in Computing*, Banff, vol. 2005, pp. 438–441 (2001)
25. A. Hamosfakidis, Y. Paker, A novel hexagonal search algorithm for fast block matching motion estimation. EURASIP J. Appl. Signal Process. 2002 **6**, 595–600 (2002). Hindawi Publishing Corporation
26. F. Moschetti, A statistical approach of motion estimation. Ph.D. Thesis, Ecole Polytechnique Fédérale de Lausanne (2001)
27. S. Orlando, P. Palmerini, R. Perego, Statistical properties of transactional databases, in *ACM Symposium on Applied Computing*, Nicosia, Cyprus (2004)
28. C. Deng, F. Xiong, Y. Tan, Z. He, Sequential learning neural network and its application in agriculture, in *IEEE International Joint Conference on Neural Networks*, vol. 1, pp. 221–225 (1998)
29. P. De Reffye, C. Edelin, M. Jaeger, La modélisation de la croissance des plantes. La Recherche **20**(207), 158–168 (1989)
30. A.A. Borovkov, Statistique mathématique. Editions Mir, Moscou (1987)

A Note on Curvatures and Rank 2 Seiberg–Witten Invariants

Fortuné Massamba

To Professor Norbert M. Hounkonnou on the occasion of his 60th Birthday

Abstract In this paper, we investigate rank 2 Seiberg–Witten equations which were introduced and studied in Massamba and Thompson (J Geom Phys 56:643–665, 2006). We derive some lower bounds for certain curvature functionals on the space of Riemannian metrics of a smooth compact 4-manifold with non-trivial rank 2 Seiberg–Witten invariants. Existence of Einstein and anti-self-dual metrics on some compact oriented 4-manifolds is also discussed.

Keywords Seiberg–Witten equation · Weyl curvature · Kähler metric · Einstein metric · Anti-self-dual metric

2010 Mathematics Subject Classification: 81T13; 70S15; 58D27

1 Introduction

One of the goals of modern Riemannian geometry is to understand the relationship between topology and curvature. In this note, the subject matter is the existence of Einstein metrics on 4-manifolds. We follow LeBrun's analysis of the constraints on curvature of any metric which arise if the manifold in question has a non-trivial Seiberg–Witten invariant [6, 7].

F. Massamba (✉)
School of Mathematics, Statistics and Computer Science, University of KwaZulu-Natal, Scottsville, South Africa
e-mail: Massamba@ukzn.ac.za

© Springer Nature Switzerland AG 2018
T. Diagana, B. Toni (eds.), *Mathematical Structures and Applications*,
STEAM-H: Science, Technology, Engineering, Agriculture,
Mathematics & Health, https://doi.org/10.1007/978-3-319-97175-9_16

Let (X, g) be an oriented, compact, smooth Riemannian manifold. Recall that such a manifold is said to be Einstein if its Ricci curvature r is a constant multiple of the metric, that is

$$r = \lambda g.$$

As is known, not every smooth compact oriented 4-manifold admits such a metric. A well known obstruction is given by the following result due to N. Hitchin and J. Thorpe (see [12]):

If a compact, oriented 4-manifold X satisfies $2\chi(X) < 3|\tau(X)|$, then X does not admit an Einstein metric. Moreover, if $2\chi(X) = 3|\tau(X)|$, then X admits no Einstein metric unless it is either flat or a $K3$ surface or an Enriques surface or the quotient of an Enriques surface by a free anti-holomorphic involution.

Here $\chi(X)$ and $\tau(X)$ denote the Euler characteristic and the signature of X, respectively. The Gauss-Bonnet-like formula

$$(2\chi \pm 3\tau)(X) = \frac{1}{4\pi^2} \int_X \left(\frac{R^2}{24} + 2|W^\pm|^2 - \frac{|r_0|^2}{2} \right) d\mu, \tag{1.1}$$

implies Hitchin-Thorpe's inequality because Einstein metrics are characterized by the vanishing of r_0, and this is the only negative term in the above integrand. In the formula, R, r_0, W^+, W^- denote the scalar, trace-free Ricci, self-dual Weyl, and anti-self-dual Weyl curvature tensors of a Riemannian metric, respectively.

This result has been improved by Lebrun (see [6] or [7] for more details), using carefully estimates on the L^2-norm of the scalar-curvature tensor R and the L^2-norm of the self-dual part of the Weyl tensor W^+ arising from the rank 1 Seiberg–Witten equations. To obtain these estimates LeBrun in [6, 7] used the fact that such an X admits irreducible solutions to the rank 1 SW equations for every metric.

In this paper, we follow LeBrun's approach but in the rank 2 case. The estimates found in our case, fortunately, do lead to new obstructions to the existence of Einstein metrics for the special case of Spin$_\mathbb{C}$ structure chosen by LeBrun [6, 7].

This paper is organized as follows. In the next section we recall the rank 2 Seiberg–Witten equations. Estimates on the curvatures (scalar curvature, Weyl curvature, etc.,) are found and compared with those of LeBrun (see [6, 7]) in the third section.

2   The Seiberg–Witten Equations

We fix an oriented, compact, Riemannian 4-manifold (g, X). By Theorem 2.9 in [8, p. 16], X admits a Spin$_\mathbb{C}$-structure. This can be described as a choice of rank two complex vector bundles, which we write as $S^\pm \otimes L$. Note that the bundles $S^+ \otimes L$ and $S^- \otimes L$ exist as genuine vector bundles even though the factors S^+, S^-, and L do not unless X is a spin manifold. The sections of $S^+ \otimes L$ and $S^- \otimes L$ are called *spinor fields* of positive or negative chirality, respectively. In general case L is not well-defined, but $L^{\otimes 2}$ is (see [10] for more details).

Now, let $\mathbf{E}$ be rank r vector bundle on X. Denote the self-dual 2-forms by $\Omega^2_+(X)$. Let ψ be a section of $W^+ = \text{Spin}_{\mathbb{C}}(X) \otimes \mathbf{E}$. For ϕ and λ in W^+, let $q : W^+ \times W^+ \to \Omega^2_+(X) \otimes \text{End }\mathbf{E}$ be the trace free part of the endomorphism

$$\theta \mapsto <\theta, \phi> \lambda.$$

Let Φ be a section of $\text{End (End }\mathbf{E})$. The equations of interest are

$$F^+_A + \Phi . q(\psi, \psi) = 0,$$
$$D\!\!\!\!/_{\Gamma+A}\psi = 0,$$
$$d_A \Phi = 0, \qquad (2.1)$$

where A is a connection on $\mathbf{E}$, Γ is the Levi-Civita connection on $\text{Spin}_{\mathbb{C}}$, and d_A is the covariant derivative on $\text{End (End }\mathbf{E})$ with the connection induced from that on $\mathbf{E}$.

One possible solution to these equations would have Φ equal to a scalar times the identity endomorphism. In this case, on a Kähler manifold, the equations become (up to a perturbation) equivalent to a set of equations discussed in [3]. Those equations are shown to have a notion of stability.

If Φ is not proportional to the identity endomorphism then, to have a solution to the last equation in (2.1), the bundles must split. The equations that we consider in this paper correspond to such a situation. In this case, we have

$$\text{Spin}_{\mathbb{C}}(X) \otimes \mathbf{E} = \bigoplus_i \left(L_1^{E_{i1}} \otimes \cdots \otimes L_r^{E_{ir}} \otimes S^+ \right).$$

The r line bundles L_i, $i = 1, \ldots, r$, on X are considered so that $S^+ \otimes L_i$ are $\text{Spin}_{\mathbb{C}}$ structures on X for all i. However, the $\text{Spin}_{\mathbb{C}}$ structures of interest are

$$\mathbf{L}_i \otimes S^+ = L_1^{E_{i1}} \otimes \cdots \otimes L_r^{E_{ir}} \otimes S^+,$$

and neither the spin bundle S^+ nor the line bundles L_i need exist. Here the matrix E_{ij} may well be the identity matrix, though in general we only demand that $\det E \neq 0$, that the entries be integers and they are such that the $L_i^{\otimes 2}$ are honest line bundles. Summing over all tuples $(L_1, \ldots, L_r)$ for a general matrix E means that one does not sum over all possible tuples of $\text{Spin}_{\mathbb{C}}$ structures on X. However, for $E \in SL(r, \mathbb{Z})$ then one does sum over all such tuples of $\text{Spin}_{\mathbb{C}}$ structures.

Let $2A_i$ be connections on the line bundles $L_i^{\otimes 2}$, with an abuse of language we will say that the A_i are connections on L_i. The connection forms are $\sqrt{-1}A_i$ so that the A_i are real. Denote by M_i charged positive chirality spinors, that is sections of the bundles $S^+ \otimes \mathbf{L}_i$. The rank r Seiberg–Witten equations are

$$F_{A_i}^+ + \sum_j D^{ij} q(M_j, M_j) = 0, \tag{2.2}$$

$$\displaystyle{\not{D}}(\mathbf{A}_i) \, M_i = 0. \tag{2.3}$$

where, locally, $q_{\mu\nu}(M_i, M_i) = \frac{\sqrt{-1}}{2} \left(\overline{M}_i \sigma_{\mu\nu} M_i \right)$, and $\mathbf{A}_i = \sum_j E_{ij} A_j$ is a connection on L_i. Here $\sigma_{\mu\nu} = \frac{1}{2}[\gamma_\mu, \gamma_\nu]$, where the gamma matrices satisfy $\{\gamma_\mu, \gamma_\nu\} = 2g_{\mu\nu}$.

The choice of Φ in (2.1) fixes the matrix D. The equations under consideration were proposed in the context of studying the Rozansky–Witten invariants on a 3-manifold, Y, [1]. Higher rank equations of this type should correspond to higher rank Rozansky–Witten invariants, that is to higher order LMO or Casson invariants [5]. One would expect that considering these equations on $X = Y \times S^1$ one would get something like the Euler characteristic of a suitable Floer theory. This was part of our motivation for studying the higher rank case on a 4-manifold.

The two matrices E and D that appear in the equations are related. That relation is dictated by wishing to emulate the use of the Weitzenböck trick to get a vanishing theorem as in the case of the rank 1 Seiberg–Witten equation [14]. The condition on the matrices is that $D^{-1}.E$ be a symmetric positive definite matrix. In fact the matrix D need not have integer entries.

Though not strictly necessary we impose the further condition that D^{-1} have integral entries. With this assumption in hand we can write (2.2) as

$$F_{\mathbf{B}}^+ = -q(\mathbf{M}, \mathbf{M}), \tag{2.4}$$

with $\mathbf{B} = D^{-1}.A$, and so that B_i is a connection on $L_1^{\otimes D_{i1}^{-1}} \otimes \cdots \otimes L_r^{\otimes D_{ir}^{-1}}$.

Note that conformal classes of a metric on X yield related equations. Denote the Dirac operator and sections on (g, X) by $\not{D}$ and M_i (as above) and those on $(e^\rho g, X)$ by $\not{D}^\rho$ and M_i^ρ. The rank r SW equations on $(e^\rho g, X)$ are

$$F_{\mathbf{B}}^+ = -q(\mathbf{M}^\rho, \mathbf{M}^\rho,), \quad \not{D}(\mathbf{A}_i)^\rho M_i^\rho = 0,$$

with $M_i^\rho = e^{-3\rho/2} M_i$. Note that the Hödge star operator acting on 2-forms is conformal invariant and so the $+$ superscript is the same for (g, X) and $(e^{2\rho} g, X)$. Therefore, the equations for $(e^{2\rho} g, X)$ are $F_{\mathbf{B}}^+ = -e^{-\rho} q(\mathbf{M}, \mathbf{M}), \not{D}(\mathbf{A}_i) M_i = 0$. Let $\mathcal{G}_i$ denote the gauge group of bundle automorphisms of L_i. The space of gauge transformations, $\mathcal{G}$, is the product of these spaces of bundle automorphisms,

$$\mathcal{G} = \mathcal{G}_1 \times \cdots \times \mathcal{G}_r.$$

Each of the $\mathcal{G}_i$ is a copy of Map$(X, U(1))$ and their complexifications are copies of Map$(X, \mathbb{C}^*)$. The space of solutions $\mathcal{M}(\mathbf{L}, \mathbf{h})$ to the rank r SW equations is left invariant under $\mathcal{G}$. By moduli space we mean the space of solutions to the rank r

SW equations modulo gauge transformations. The virtual dimension of this moduli space is $d = -(2\chi + 3\tau)/2 + (E^T.E)^{ij}c_1(L_i)c_1(L_j)$. The basic classes are $x = (x_1, \ldots, x_r) = (-c_1(\mathbf{L}_1^{\otimes 2}), \ldots, -c_1(\mathbf{L}_r^{\otimes 2}))$. Note that in the special case that $E_{ij} = \delta_{ij}$ the equations decouple. In that case the basic invariants are essentially r-tuples of the usual SW basic classes.

Next, we give the Weitzenböck formula which is needed in the sequel of this paper. We begin with a squaring argument.

Set

$$s_{i\mu\nu} = F_{\mu\nu}^{i+} + \frac{\sqrt{-1}}{2} \sum_{j=1}^{2} D^{ij} \left(\overline{M}_j \sigma_{\mu\nu} M_j \right),$$

$$k_i = \slashed{D}(E_{ij}A_j) M_i,$$

and for a solution to the SW equations we must have

$$\int_X d^4x \sqrt{g} \sum_{i=1}^{2} \left(\frac{1}{2} G^{ij} s_j.s_i + |k_i|^2 \right) = 0, \tag{2.5}$$

with $G = E^T.D^{-1}$ a symmetric and positive definite matrix. The explicit form of $G^{ij} s_j . s_i$ and $|k_i|^2$ are

$$G^{ij} s_j.s_i = G^{ij} F^{j+} F^{i+} + \sqrt{-1} G^{ij} F^{i+} \sum_{k=1}^{2} D^{jk}\overline{M}_k \sigma M_k$$

$$- \frac{1}{4} G^{ij} (\sum_{k=1}^{2} D^{jk}\overline{M}_k \sigma M_k)(\sum_{l=1}^{2} D^{il}\overline{M}_l \sigma M_l),$$

$$\int_X d^4x \sqrt{g}|k_i|^2 = \int_X d^4x \sqrt{g}(|D M_i|^2 - \frac{\sqrt{-1}}{2} \sum_{j=1}^{2} E_{ij} F^{j+}\overline{M}_i \sigma M_i + \frac{1}{4} R |M_i|^2).$$

Then, Weitzenböck formula is given by Massamba and Thompson [9]:

$$\slashed{D}(E_{ij}A^j)^2 M_i = D^\mu D_\mu M_i + \frac{\sqrt{-1}}{2} \sum_{j=1}^{2} E_{ij} F_{\mu\nu}^j \cdot \sigma^{\mu\nu} M_i - \frac{1}{4}R M_i. \tag{2.6}$$

With the relation (2.6), we find that (2.5) becomes

$$\int_X d^4x \sqrt{g} \sum_{i=1}^{2} \left(\frac{1}{2} G^{ij} s_j.s_i + |k_i|^2 \right)$$

$$= \int_X d^4x \sqrt{g} \sum_{i,j=1}^{2} \left(\frac{1}{2} G^{ij} F^{i+} F^{j+} - \frac{1}{8} \sum_{\mu,\nu} \overline{M}_i \sigma_{\mu\nu} M_i \, B_{ij} \overline{M}_j \sigma^{\mu\nu} M_j \right.$$

$$\left. + \delta_{ij} |D M_i|^2 + \frac{1}{4} R \, \delta_{ij} |M_i|^2 \right) = 0, \tag{2.7}$$

with $B = D^T.E^T = D^T.G.D$ a positive definite symmetric matrix. At this point the choice of the matrix G becomes evident. It was chosen, so that the "mixed term" $F_+.\overline{M}\sigma M.E$ would drop out of the equations. One of the terms of the last expression above will be specified by using the Fierz identity (see [9]) for the gamma matrices given by :

$$\frac{1}{2} (\sigma_{\mu\nu})^\alpha_\epsilon (\sigma^{\mu\nu})^\gamma_\beta = -4\delta^\alpha_\beta \delta^\gamma_\epsilon + \delta^\alpha_\epsilon \delta^\gamma_\beta + (\gamma_\mu)^\alpha_\epsilon (\gamma^\mu)^\gamma_\beta + (\gamma_5)^\alpha_\epsilon (\gamma^5)^\gamma_\beta$$

$$- (\gamma_\mu \gamma_5)^\alpha_\epsilon (\gamma^\mu \gamma_5)^\gamma_\beta. \tag{2.8}$$

We have

$$-\frac{1}{8} \sum_{\mu,\nu} \overline{M}_j \sigma_{\mu\nu} M_j \cdot \overline{M}_k \sigma^{\mu\nu} M_k = \frac{1}{2} \left(2|\overline{M}_j M_k|^2 - |M_j|^2 |M_k|^2 \right). \tag{2.9}$$

It is now easy to check that

$$|B_{12}| |M_1|^2 |M_2|^2 \geqslant \left(2|\overline{M}_1 M_2|^2 - |M_1|^2 |M_2|^2 \right) B_{12} \geqslant -|B_{12}| |M_1|^2 |M_2|^2,$$

so that

$$-\frac{1}{8} \sum_{j,k=1}^{2} \sum_{\mu,\nu} \overline{M}_j \sigma_{\mu\nu} M_j \, B_{jk} \overline{M}_k \sigma^{\mu\nu} M_k$$

$$\geqslant \frac{1}{2} \left(\sqrt{B_{11}} |M_1|^2 - \sqrt{B_{22}} |M_2|^2 \right)^2 + \left(\sqrt{B_{11} B_{22}} - |B_{12}| \right) |M_1|^2 |M_2|^2,$$

where $\sqrt{B_{ij}}$ always represents the positive root. We can cast (2.5) in the form of an inequality,

$$\int_X d^4x \sqrt{g} \left(\frac{1}{2} \sum_{i,j=1}^{2} G_{ij} F^{i+} F^{j+} + \sum_{i=1}^{2} |D M_i|^2 + \frac{1}{2} (\sqrt{B_{11}} |M_1|^2 - \sqrt{B_{22}} |M_2|^2)^2 \right.$$

$$\left. + (\sqrt{B_{11} B_{22}} - |B_{12}|) |M_1|^2 |M_2|^2 + \frac{1}{4} R \sum_{i=1}^{2} |M_i|^2 \right) \leqslant 0. \tag{2.10}$$

One immediate consequence of (2.10) is that, just as for the usual SW invariants, there are no solutions apart from the trivial ones if R is non-negative. In [9], the authors used (2.10) to prove that the moduli space $\mathcal{M}(\mathbf{L}, \mathbf{h})$ of interest is compact. They also proved that it is orientable [9, Proposition 4.3].

If X is Kähler one has decompositions $S^+ \otimes \mathbf{L}_i = (K_X^{1/2} \otimes \mathbf{L}_i) \oplus (K_X^{-1/2} \otimes \mathbf{L}_i)$ where, as before, neither $K_X^{\pm 1/2}$ nor $\mathbf{L}_i$ necessarily exist. Denote the components of M_i in $K_X^{1/2} \otimes \mathbf{L}_i$ by α_i and those in $K_X^{-1/2} \otimes \mathbf{L}_i$ by $\sqrt{-1}\,\overline{\beta}_i$. In this the Dirac operator is $\slashed{D}(\mathbf{A}_i) = \sqrt{2}\left(\overline{\partial}_{\mathbf{A}_i} + \overline{\partial}_{\mathbf{A}_i}^*\right)$. The perturbed SW equations are

$$F_{\mathbf{B}_i}^{(2,0)} = \alpha_i \beta_i - \eta_i$$

$$\omega \wedge F_{\mathbf{B}_i} = \frac{1}{2}\omega^2 \left(|\alpha_i|^2 - |\beta_i|^2\right)$$

$$\overline{\partial}_{\mathbf{A}_i}\alpha_i = -\sqrt{-1}\,\overline{\partial}_{\mathbf{A}}^*\,\overline{\beta}_i, \tag{2.11}$$

where the $\mathbf{B}_i = \sum_j D_{ij}^{-1} A_j$ are connections on the bundles $\mathcal{L}_i = L_1^{D_{i1}^{-1}} \otimes L_2^{D_{i2}^{-1}}$.

Does the system of equations (2.11) admit solutions? The answer is affirmative and this was proven in [9]. What we have in [9] is that given a pair of holomorphic sections to $K_X^{1/2} \otimes \mathbf{L}_i$ on the Kähler manifold (ω, X) we are guaranteed a solution to the rank 2 Seiberg–Witten equations on $(e^{2\rho}\omega, X)$.

3 Curvature Estimates

In this section, we derive new L^2 estimates for combinations of the Weyl and scalar curvatures of certain Riemannian 4-manifolds using the Gauss-Bonnet like formula (4.3). Unfortunately, as we will see, these bounds are somewhat weaker than the estimates previously found in [6, 7]. Many of the important consequences of the rank 2 Seiberg–Witten theory stem from the fact that Eqs. (2.2) and (2.3) implied the Weitzenböck formula of Proposition 2.6. This proposition allows us to obtain the following useful inequalities.

Lemma 3.1 *If (A_1, A_2, M_1, M_2) is a solution of (2.2), (2.3) on a compact oriented Riemannian 4-manifold (X, g), then*

$$\int_X \left(-R\,|M_1|^4 - 4\,|M_1|^2|\nabla M_1|^2\right) \geq 4\,B_{11} \int_X |M_1|^6 - 4\,|B_{12}| \int_X |M_1|^4|M_2|^2, \tag{3.1}$$

$$\int_X \left(-R\,|M_2|^4 - 4\,|M_2|^2|\nabla M_2|^2\right) \geq 4\,B_{22} \int_X |M_2|^6 - 4\,|B_{12}| \int_X |M_1|^2|M_2|^4, \tag{3.2}$$

where $|\,.\,|$ is the point-wise norm determined by g.

Proof Note that we have $\int_X d^4x \sqrt{g} |k_i|^2 f = 0$, for any bounded function f. We can expand the formula to obtain

$$\int_X d^4x \sqrt{g} (|D\, M_i|^2 - \frac{\sqrt{-1}}{2} \sum_{j=1}^{2} E_{ij}\, F^{j+} \overline{M_i} \sigma\, M_i + \frac{1}{4} R\, |M_i|^2)\, f$$

$$= \int_X d^4x \sqrt{g} (|D\, M_i|^2 - \frac{1}{4} \sum_{j=1}^{2} B_{ij}\, \overline{M_i} \sigma\, M_i . \overline{M_j} \sigma\, M_j + \frac{1}{4} R\, |M_i|^2)\, f = 0.$$

Now set $f = |M_i|^2$ and make use of (2.9) and the inequality that follows it to obtain the lemma. $\qquad\square$

We will see in a moment that the inequalities (3.1) and (3.2) imply some estimates for the Weyl curvature of suitable 4-manifolds. But before doing this, we make a digression on the Gauss-Bonnet like formula. Our interest being on the $\text{Spin}_{\mathbb{C}}$-structures determined by the integral complex structure J_X on X, that is,

$$c_1(J_X) = -\frac{1}{2} c_1(K_X), \tag{3.3}$$

the line bundles $\mathbf{L}_i^{\otimes 2}$ become the anti-canonical line bundle $K_X^{-1} = \Lambda^{0,2}(X)$. From (3.3), we have

$$c_1(\mathbf{L}_i)^2 = \frac{1}{4}(2\chi + 3\tau)(X), \tag{3.4}$$

and for this reason we consider the formula (4.3) with the plus sign on the left-hand side. It is the self-dual Weyl tensor which then appears on the right-hand side of the equation. Let us recall that the self-dual Weyl tensor, W^+, at a point x of (X, g), may be viewed as the trace-free endomorphism $W^+(x) : \Lambda_x^+ \longrightarrow \Lambda_x^+$ of self-dual 2-forms at x. Let $\omega(x)$ denote the lowest eigenvalue of this endomorphism and notice that is automatically a Lipschitz continuous function $\omega : X \longrightarrow (-\infty, 0]$ (see [6] for more details).

Theorem 3.2 *Let (X, g) be a compact oriented Riemannian 4-manifold on which there is a solution of the Rank 2 Seiberg–Witten equations. Let $c_1(\mathbf{L}_i)$, with $i = 1, 2$, be the first Chern classes of the bundles $\mathbf{L}_i$ and $c_1^+(\mathbf{L}_i)$ denote their self-dual parts with respect to g. Then, under conditions $B_{11} \geqslant |B_{12}|$ and $B_{22} \geqslant |B_{12}|$ on the symmetric matrix $(B_{ij})_{1 \leq i,j \leq 2}$, the metric g satisfies*

$$V^{1/3} \left(\int_X \left| \frac{2}{3} R_g + 2\omega_g \right|^3 d\mu_g \right)^{2/3} \geq \frac{16\pi^2}{(\det B)^2} \left(\kappa_1 |c_1^+(\mathbf{L}_1)| - \kappa_2 |c_1^+(\mathbf{L}_2)| \right)^2$$

$$+ \frac{32\pi^2}{(\det B)^2} (\kappa_1 \cdot \kappa_2 - \kappa^2) |c_1^+(\mathbf{L}_1)| \cdot |c_1^+(\mathbf{L}_2)|,$$

where

$$\kappa_1^2 = B_{22}^2 (B_{11} - |B_{12}|)^2 + B_{21}^2 (B_{22} - |B_{12}|)^2,$$

$$\kappa_2^2 = B_{12}^2 (B_{11} - |B_{12}|)^2 + B_{11}^2 (B_{22} - |B_{12}|)^2,$$

$$\kappa^2 = |B_{22} B_{12}| (B_{11} - |B_{12}|)^2 + |B_{21} B_{11}| (B_{22} - |B_{12}|)^2,$$

$V = \mathrm{Vol}(X, g) = \int_X d\mu_g$ *denotes the total volume of* (X, g), $d\mu_g$ *is the volume form of* g, *and* $|c_1^+(\mathbf{L}_i)| := \sqrt{(c_1^+(\mathbf{L}_i))^2}$.

The proof of this proposition needs the following lemmas.

Lemma 3.3 ([2]) *Any self-dual 2-form* θ *on any oriented 4-manifold satisfies*

$$\left(d + d^*\right)^2 \theta = \nabla^* \nabla \theta - 2W^+(\theta, \cdot) + \frac{R}{3} \theta. \tag{3.5}$$

Lemma 3.4 *For the self-dual 2-forms* $\theta = q(M, M)$, *we have*

$$|\nabla \theta|^2 = 4|M|^2 \cdot |\nabla M|^2 + (\nabla \overline{M} \cdot M - \overline{M} \cdot \nabla M)^2$$

so that $|\nabla \theta|^2 \leq 4|M|^2 \cdot |\nabla M|^2$.

Proof By definition, the self-dual 2-form θ is given by $\theta = q(M, M) = \frac{\sqrt{-1}}{2} \overline{M} \sigma M$ and its gradient is given by

$$\nabla \theta = \frac{\sqrt{-1}}{2} \left\{ \nabla \overline{M} \sigma M + \overline{M} \sigma \nabla M \right\}.$$

Since θ is real, so is $\nabla \theta$ and we have

$$|\nabla \theta|^2 = -\frac{1}{4} \left(\nabla \overline{M} \sigma M\right)^2 - \frac{1}{4} \left(\overline{M} \sigma \nabla M\right)^2 - \frac{1}{2} \left(\nabla \overline{M} \sigma M\right) \left(\overline{M} \sigma \nabla M\right). \tag{3.6}$$

From Fierz identity given in (2.8), we can obtain

$$\frac{1}{2} \left(\nabla \overline{M} \sigma M\right)^2 = \frac{1}{2} \left(\nabla \overline{M} \sigma_{\mu\nu} M\right) \left(\nabla \overline{M} \sigma^{\mu\nu} M\right)$$

$$= \frac{1}{2} \left((\nabla \overline{M})_\alpha (\sigma_{\mu\nu})_\epsilon^\alpha M^\epsilon\right) \left((\nabla \overline{M})_\gamma (\sigma^{\mu\nu})_\beta^\gamma M^\beta\right)$$

$$= (\nabla \overline{M})_\alpha M^\epsilon (\nabla \overline{M})_\gamma M^\beta \left\{ \frac{1}{2} (\sigma_{\mu\nu})_\epsilon^\alpha (\sigma^{\mu\nu})_\beta^\gamma \right\}$$

$$= (\nabla \overline{M})_\alpha M^\epsilon (\nabla \overline{M})_\gamma M^\beta \left\{ -4\delta_\beta^\alpha \delta_\epsilon^\gamma + \delta_\epsilon^\alpha \delta_\beta^\gamma + (\gamma_\mu)_\epsilon^\alpha (\gamma^\mu)_\beta^\gamma \right.$$

$$+(\gamma_5)^\alpha_\epsilon (\gamma^5)^\gamma_\beta - (\gamma_\mu \gamma_5)^\alpha_\epsilon (\gamma^\mu \gamma_5)^\gamma_\beta \Big\}$$
$$= -2 \left(\nabla \overline{M} \cdot M \right)^2. \tag{3.7}$$

Likewise,

$$\frac{1}{2} \left(\overline{M} \sigma \nabla M \right)^2 = -2 \left(\overline{M} \cdot \nabla M \right)^2, \tag{3.8}$$

and $\frac{1}{2} \left(\nabla \overline{M} \sigma M \right) \left(\overline{M} \sigma \nabla M \right) = -4|\nabla M|^2 |M|^2 + 2|\nabla \overline{M} \cdot M|^2 \tag{3.9}$

Putting (3.7)–(3.9) into (3.6), one obtains

$$|\nabla \theta|^2 = \left(\nabla \overline{M} \cdot M \right)^2 + \left(\overline{M} \cdot \nabla M \right)^2 + 4|\nabla M|^2 |M|^2 - 2|\nabla \overline{M} \cdot M|^2$$
$$= 4|\nabla M|^2 |M|^2 + \left(\nabla \overline{M} \cdot M \right)^2 + \left(\overline{M} \cdot \nabla M \right)^2 - 2 \left(\nabla \overline{M} \cdot M \right) \left(\overline{M} \cdot \nabla M \right)$$
$$= 4|M|^2 \cdot |\nabla M|^2 + (\nabla \overline{M} \cdot M - \overline{M} \cdot \nabla M)^2.$$

Since the term $\nabla \overline{M} \cdot M - \overline{M} \cdot \nabla M$ is pure imaginary, $(\nabla \overline{M} \cdot M - \overline{M} \cdot \nabla M)^2 \leq 0$ and $|\nabla \theta|^2 \leq 4|M|^2 \cdot |\nabla M|^2$, which completes the proof.

Proof of Theorem 3.2 Any self-dual 2-forms θ_i on any oriented 4-manifold satisfy (3.5). It follows, by integration and the fact that the function ω is the smallest eigenvalue, that

$$- \int_X 2\omega_g \, |\theta_i|^2 \, d\mu_g \geq \int_X -\frac{R_g}{3} |\theta_i|^2 \, d\mu_g - \int_X |\nabla \theta_i|^2 \, d\mu_g$$

and hence

$$- \int_X \left(\frac{2}{3} R_g + 2\omega_g \right) |\theta_i|^2 \, d\mu_g \geq \int_X -R_g \, |\theta_i|^2 \, d\mu_g - \int_X |\nabla \theta_i|^2 \, d\mu_g.$$

On the other hand, for the particular self-dual 2-forms given by $\theta_i = q(M_i, M_i)$, with the inequality in Lemma 3.4, allows us to obtain

$$- \int_X \left(\frac{2}{3} R_g + 2\omega_g \right) |M_i|^4 \, d\mu_g \geq \int_X \left(-R|M_i|^4 \, d\mu_g - 4|M_i|^2 \cdot |\nabla M_i|^2 \right) d\mu_g.$$

The inequalities (3.1) and (3.2), the Holder inequality, and a tedious calculation yield

$$\left(\int_X \left| \frac{2}{3} \mathfrak{G}_g \right|^3 d\mu_g \right)^{1/3} \geq 2(B_{11} - |B_{12}|) \left(\int_X |M_1|^6 \, d\mu_g \right)^{1/3}$$

$$+ 2(B_{22} - |B_{12}|) \left(\int_X |M_2|^6 \, d\mu_g \right)^{1/3},$$

where $\mathfrak{G}_g = R_g + 3\,\omega_g$.

At this point, we want to take the square of the inequality. Under the conditions $B_{11} \geqslant |B_{12}|$ and $B_{22} \geqslant |B_{12}|$, the right-hand side of the inequality is positive semi-definite we can square both sides and the inequality remains valid. Do that and use once more the Holder inequality to obtain

$$V^{1/3} \left(\int_X \left| \frac{2}{3} \mathfrak{G}_g \right|^3 d\mu \right)^{2/3} \geq 4 \, (B_{11} - |B_{12}|)^2 \left(\int_X |M_1|^4 \, d\mu \right)$$

$$+ 4(B_{22} - |B_{12}|)^2 \left(\int_X |M_2|^4 \, d\mu \right). \qquad (3.10)$$

Writing the norms of $|M_i|^4$ as functions of the self-dual parts of the first Chern classes $c_1(\mathbf{L}_i)$, we easily deduce that the positive constants κ_1, κ_2, and κ are as given in the proposition. $\qquad \square$

There is an alternative inequality that one can obtain and that is useful, namely,

Theorem 3.5 *With the conditions of the previous proposition with $B_{22} > 0$, and on taking $c_1(\mathbf{L}_1) = c_1(\mathbf{L}_2) = c_1(\mathbf{L})$ we have that*

$$V^{1/3} \left(\int_X \left| \frac{2}{3} R_g + 2\omega_g \right|^3 d\mu_g \right)^{2/3} \geq 64\pi^2 \left(\frac{B_{22} - B_{12}}{B_{22} + |B_{12}|} \right)^2 (c_1^+(\mathbf{L}))^2,$$

with a similar formula on exchanging $B_{11} \leftrightarrow B_{22}$.

These are the keys to everything that follows, its direct utilities are limited by the fact they are L^3, rather than L^2, estimates. Fortunately, we will be able to extract L^2 estimates by means of a conformal rescaling trick, the general idea of which is drawn from LeBrun [6] or [7], which follows from the standard proof of the Yamabe problem in the form developed by Gursky [4]. We have

Lemma 3.6 ([6]) *Let (X, γ) be a compact oriented 4-manifold with a fixed smooth conformal class of Riemannian metrics. Suppose, moreover, that γ does not contain a metric of positive scalar curvature. Then, for any $\alpha \in \,]0, 1]$, there is a metric $g_\gamma \in \gamma$ of differentiability class $C^{2,\alpha}$ for which $R + 3\omega$ is a non-positive constant.*

Let γ be a smooth conformal class of some adapted metric g on a smooth oriented 4-manifold X such that there is a Spin$_\mathbb{C}$-structure on X for which the rank 2 SW equations (2.2), (2.3) have a solution for every metric γ. Then γ does not contain any metrics of positive scalar curvature. This can be easily seen using the Weitzenböck formula in Proposition 2.6. Lemma 3.6 therefore implies and tells us that the conformal class γ contains a metric g_γ for which that the function $\mathfrak{G}_{g_\gamma} = R_{g_\gamma} + 3\omega_{g_\gamma}$ is a non-positive constant.

Proposition 3.7 [6] *Let g be the conformally related metric $g = u^2 g_\gamma$, with u a positive C^2 function, then*

$$\int_X \left(\frac{2}{3} R_g + 2\omega_g \right)^2 d\mu_g \geq \int_X \left(\frac{2}{3} R_{g_\gamma} + 2\omega_{g_\gamma} \right)^2 d\mu_{g_\gamma}.$$

Lemma 3.8 *The smallest eigenvalue ω of W^+ is bounded as $-\sqrt{\frac{2}{3}}|W^+|_g \leq \omega_g$.*

Proof The proof uses only the fact that the self-dual Weyl tensor W^+ is trace-free. ☐

Corollary 3.9 (to Theorem 3.2) *Let (X, γ) be a smooth compact oriented Riemannian 4-manifold, as in Lemma 3.6, on which there is a solution of the Rank 2 Seiberg–Witten equations. Let $c_1(\mathbf{L}_i)$, with $i = 1, 2$, be the first Chern classes of the bundles $\mathbf{L}_i$ and $c_1^+(\mathbf{L}_i)$ denote their self-dual parts with respect to g_γ. Then, with B_{ij}, κ_1, κ_2, and κ as given in Theorem 3.2*

$$\frac{1}{4\pi^2} \int_X \left(\frac{R_g^2}{24} + 2|W^+|_g^2 \right) d\mu_g \geq \frac{1}{3(\det B)^2} \left(\kappa_1 \cdot |c_1^+(\mathbf{L}_1)| - \kappa_2 \cdot |c_1^+(\mathbf{L}_2)| \right)^2$$

$$+ \frac{2}{3(\det B)^2} \left(\kappa_1 \cdot \kappa_2 - \kappa^2 \right) |c_1^+(\mathbf{L}_1)| \cdot |c_1^+(\mathbf{L}_2)|.$$

(3.11)

Proof We have the following:

$$\left(\int_X |\mathfrak{G}_g|^2 \right)^{1/2} \geq \left(\int_X |\mathfrak{G}_{g_\gamma}|^2 \right)^{1/2} = \left(\int_X |\mathfrak{G}_{g_\gamma}|^3 \right)^{1/3}$$

the first inequality follows from Proposition 3.7 while the equality follows from Lemma 3.6. One now applies Theorem 3.2 for the metric g_γ to get the bound involving the Chern classes. Lemma 3.8 allows us to exchange the eigenvalue for the norm of W^+. ☐

The same line of argument gives us the following:

Corollary 3.10 (to Theorem 3.5) *Let (X, γ) be a smooth compact oriented Riemannian 4-manifold, as in Lemma 3.6, on which there is a solution of the Rank 2 Seiberg–Witten equations with $c_1(\mathbf{L}_1) = c_1(\mathbf{L}_2) = c_1(\mathbf{L})$ and let $c_1^+(\mathbf{L})$ denote the self-dual part with respect to g_γ. Then*

$$\frac{1}{4\pi^2} \int_X \left(\frac{R_g^2}{24} + 2|W^+|_g^2 \right) d\mu_g \geq \frac{4}{3} \min\{\lambda_1, \lambda_2\}|c_1^+(\mathbf{L})|^2,$$

(3.12)

where $\lambda_1 = \left(\dfrac{B_{11} - B_{12}}{B_{11} + |B_{12}|} \right)^2$ and $\lambda_2 = \left(\dfrac{B_{22} - B_{12}}{B_{22} + |B_{12}|} \right)^2$.

LeBrun, [6, Proposition 3.1] finds the following bound:

$$\frac{1}{4\pi^2} \int_X \left(\frac{R_g^2}{24} + 2|W^+|_g^2 \right) d\mu \geq \frac{2}{3} \left(2\chi(X) + 3\tau(X) \right).$$

Given that for zero-dimensional moduli space, we have that $c_1(\mathbf{L}) = \pm c_1(K_X)$ and that $|c_1^+(\mathbf{L})|^2 \geqslant |c_1(\mathbf{L})|^2$ we have that Corollary 3.10 gives the bound,

$$\frac{1}{4\pi^2} \int_X \left(\frac{R_g^2}{24} + 2|W^+|_g^2 \right) d\mu_g \geq \frac{1}{3} \min\{\lambda_1, \lambda_2\} \left(2\chi(X) + 3\tau(X)\right).$$

As $(B_{22} - B_{12})^2/(B_{22} + |B_{12}|)^2 \leqslant 1$ and $(B_{11} - B_{12})^2/(B_{11} + |B_{12}|)^2 \leqslant 1$, we see that our bound is somewhat weaker than that of LeBrun.

Many bounds may be obtained from the rank 2 SW equations. Thus, under suitable conditions on the symmetric matrix $(B_{ij})_{1 \leq i, j \leq 2}$, we shall not exclude the fact that it may exist other bounds which could be stronger than that those of LeBrun.

4   Einstein Metrics

In this section, we deal with non-existence results for Einstein metrics on suitable smooth compact 4-manifolds.

Let X be a complex surface $X = Y \sharp l \overline{\mathbb{C}P}^2$ obtained by blowing up $l > 0$ points from a complex surface Y. We have

$$H^2(Y \sharp l \overline{\mathbb{C}P}^2, \mathbb{Z}) = H^2(Y, \mathbb{Z}) \oplus \bigoplus_{p=1}^{l} H^2(\overline{\mathbb{C}P}^2, \mathbb{Z}) = H^2(Y, \mathbb{Z}) \oplus \mathbb{Z}^{\oplus l}.$$

Lemma 4.1 *Let K_Y be the canonical bundle on surface Y, and $X = Y \sharp \overline{\mathbb{C}P}^2$, with E a generator of $H^2(\overline{\mathbb{C}P}^2, \mathbb{Z})$ which is an exceptional divisor such that $E^2 = -1$. Then, the first Chern class of the canonical line bundle K_X on X is $c_1(K_X) = \pi^*(c_1(K_Y)) \pm E$, where π is the natural projection $X \longrightarrow Y$.*

Proof Note that $c_1(K_X) = \pi^*(c_1(K_Y)) + a E$, with a some constant. Using the adjunction formula [11], we obtain the desired result. □

More generally, we have:

Lemma 4.2 *The first Chern class of the canonical line bundle on $X = Y \sharp l \overline{\mathbb{C}P}^2$ is $c_1(K_X) = \pi^*(c_1(K_Y)) + \sum_{p=1}^{l} \pm E_p$, where $E_1, \ldots, E_l$ are the exceptional divisors in X, introduced on blowing up l times, with $E_p \cdot E_{p'} = -\delta_{pp'}$, $p, p' = 1, 2, \cdots, l$. Moreover, $2c_1(J_X) = 2 c_1(J_Y) + \sum_{p=1}^{l} \pm E_p$.*

Proof It follows from a straightforward calculation using (3.3). □

If $E_1, \ldots, E_l$ are the exceptional divisors in X, introduced on blowing up l times, with $E_p \cdot E_{p'} = -\delta_{pp'}$, $p, p' = 1, 2, \cdots, l$, the complex structure J_X has Chern class,

$$2c_1(J_X) = 2c_1(J_Y) + \sum_{p=1}^{l} \pm E_p.$$

By a result in [9], this is a monopole class of X for each choice of signs and the basis classes are of the form (x_1, x_2) with

$$x_i = -2 \sum_j E_{ij} c_1(L_j) = -2c_1(\mathbf{L}_i),$$

that is,

$$\sum_j E_{ij} c_1(L_j) = c_1(\mathbf{L}_i), \tag{4.1}$$

with the right-hand side being an integral class. We now fix our choice of signs so that

$$c_1^+(J_Y) \cdot (\pm E_p) \geq 0,$$

for each p, with respect to the decomposition induced by the given metric g. Using (3.3), one has

$$[c_1^+(J_X)]^2 = (c_1^+(J_Y) + \frac{1}{2} \sum_{p=1}^{l} \pm E_p)^2$$

$$= [c_1^+(J_Y)]^2 + 2 \sum_p^{l} c_1^+(J_Y) \cdot (\pm E_p) + (\sum_{p=1}^{l} \pm E_p)^2$$

$$\geq [c_1^+(J_Y)]^2$$

$$\geq \frac{1}{4}(2\chi + 3\tau)(Y). \tag{4.2}$$

This leads to the following result.

Theorem 4.3 *Let* (Y, J_Y) *be a compact complex surface with* $b_2^+(Y) > 1$, *and let* (X, J_X) *be the complex surface obtained from* Y *by blowing up* $l > 0$ *points on which there is a solution of the Rank 2 Seiberg–Witten equations with* $c_1(\mathbf{L}_1) = c_1(\mathbf{L}_2) = c_1(\mathbf{L})$. *Then any Riemannian metric* g *on the 4-manifold* $X = Y \sharp l \overline{\mathbb{C}P}^2$ *satisfy*

$$\frac{1}{4\pi^2} \int_X \left(\frac{R_g^2}{24} + 2|W^+|_g^2 \right) d\mu_g \geq \frac{1}{3} \min\{\lambda_1, \lambda_2\}(2\chi(Y) + 3\tau(Y)),$$

where $\lambda_1 = \left(\dfrac{B_{11} - B_{12}}{B_{11} + |B_{12}|} \right)^2$ *and* $\lambda_2 = \left(\dfrac{B_{22} - B_{12}}{B_{22} + |B_{12}|} \right)^2$.

Assume that $(2\tau + 3\tau)(Y) > 0$, since otherwise the result follows from the Hitchin-Thorpe inequality.

Now

$$(2\chi + 3\tau)(X) = \frac{1}{4\pi^2} \int_X \left(\frac{R_g^2}{24} + 2|W^+|_g^2 - \frac{|r_0|^2}{2} \right) d\mu_g, \qquad (4.3)$$

for any metric on g on X. If g is an Einstein metric, the trace-free part r_0 of the Ricci curvature vanishes, and we then have

$$(2\chi + 3\tau)(Y) - l = (2\chi + 3\tau)(X)$$

$$= \frac{1}{4\pi^2} \int_X \left(\frac{R_g^2}{24} + 2|W^+|_g^2 \right) d\mu_g$$

$$\geq \frac{1}{3} \min\{\lambda_1, \lambda_2\}(2\chi(Y) + 3\tau(Y)), \qquad (4.4)$$

by Theorem 4.3. If M carries an Einstein metric, it therefore follows that

$$l \leq \frac{3 - \min\{\lambda_1, \lambda_2\}}{3} (2\chi(Y) + 3\tau(Y)). \qquad (4.5)$$

Therefore,

Theorem 4.4 *Let (Y, J_Y) be a compact complex surface with $b_2^+(Y) > 1$, and let (X, J_X) be the complex surface obtained from Y by blowing up $l > 0$ points on which there is a solution of the Rank 2 Seiberg–Witten equations with $c_1(\mathbf{L}_1) = c_1(\mathbf{L}_2) = c_1(\mathbf{L})$. Then the smooth compact 4-manifold X does not admit any Einstein metrics if*

$$l > \frac{3 - \min\{\lambda_1, \lambda_2\}}{3} (2\chi(Y) + 3\tau(Y)), \qquad (4.6)$$

where λ_1 and λ_2 are defined as in Theorem 4.3.

Since $\lambda_1 \leq 1$ and $\lambda_2 \leq 1$, it is easy to check that the number $3 - \min\{\lambda_1, \lambda_2\}$ in (4.6) is bounded below by 2, that is,

$$3 - \min\{\lambda_1, \lambda_2\} \geq 2.$$

Theorem 4.5 *Let (Y, J_Y) be a compact complex surface with $b_2^+(Y) > 1$, and let (X, J_X) be the complex surface obtained from Y by blowing up $l > 0$ points on which there is a solution of the Rank 2 Seiberg–Witten equations with $c_1(\mathbf{L}_1) = c_1(\mathbf{L}_2) = c_1(\mathbf{L})$. If X is Kähler, every Riemannian metric g on the 4-manifold $X = Y \sharp l \overline{\mathbb{CP}}^2$ satisfies*

$$\frac{1}{4\pi^2} \int_X R_g^2 d\mu_g \geq \frac{8}{3} \min\{\lambda_1, \lambda_2\}(2\chi(Y) + 3\tau(Y)),$$

where λ_1 and λ_2 are defined as in Theorem 4.3.

In [13], Taubes has shown that for any smooth compact orientable X^4, there is an integer l_0 such that $X = Y \sharp l \overline{\mathbb{C}P}^2$ admits metrics with $W^+ = 0$ provided that $l \geq l_0$. In particular, there are many anti-self-dual 4-manifolds with rank 1 non-trivial Seiberg–Witten invariants. If there are non-trivial rank 2 Seiberg–Witten invariants, then, for such manifolds, we get an interesting scalar-curvature estimate.

Proposition 4.6 *Let* (X, J_X) *be a compact anti-self-dual 4-manifold with a non-zero rank 2 Seiberg–Witten invariant with* $c_1(\mathbf{L}_1) = c_1(\mathbf{L}_2) = c_1(\mathbf{L})$. *Then every Riemannian metric* g *on the 4-manifold* X *satisfies*

$$\frac{1}{4\pi^2} \int_X R_g^2 d\mu_g \geq 8 \min\{\lambda_1, \lambda_2\}(2\chi(Y) + 3\tau(Y)),$$

where λ_1 *and* λ_2 *are defined as in Theorem 4.3.*

One might instead ask whether $X = Y \sharp l \overline{\mathbb{C}P}^2$ admits anti-self-dual Einstein metrics, since Taubes' theorem [13] tells us that anti-self-dual (but non-Einstein) metrics exist when l is very large. Using Proposition 4.6, we have:

Theorem 4.7 *Let* (Y, J_Y) *be a compact complex surface with* $b_2^+(Y) > 1$, *and let* (X, J_X) *be the complex surface obtained from* Y *by blowing up* $l > 0$ *points on which there is non-trivial Rank 2 Seiberg–Witten invariant with* $c_1(\mathbf{L}_1) = c_1(\mathbf{L}_2) = c_1(\mathbf{L})$. *Then,* $X = Y \sharp l \overline{\mathbb{C}P}^2$ *cannot admit anti-self-dual metrics.*

Acknowledgements The author would like to express his sincere gratitude to Professor Norbert M. Houkonnou for his continuous support and invaluable friendship over years. He is also grateful to G. Thompson for invaluable discussions and support. Finally, the author thanks the referee for his/her valuable comments and suggestions. This work is based on the research supported wholly/in part by the National Research Foundation of South Africa (Grant no: 95931).

References

1. M. Blau, G. Thompson, On the relationship between the Rozansky–Witten and the 3-dimensional Seiberg–Witten invariants. Adv. Theor. Math. Phys. **5**, 483–498 (2001)
2. J.-P. Bourguignon, Les variétés de dimension 4 à signature non nulle dont la courbure est harmonique sont d'Einstein. Invent. Math. **63**, 263–286 (1981)
3. S. Bradlow, G. Daskalopoulos, O. García-Prada, R. Wentworth, Stable augmented bundles over Riemann surfaces, in *Vector Bundles in Algebraic Geometry*, ed. by N.J. Hitchin, P.E. Newstead, W.M. Oxbury (Cambridge University Press, Cambridge, 1995)
4. M.J. Gursky, Four-manifolds with $\delta W^+ = 0$ and Einstein constants of the sphere. Math. Ann. **318**, 417–431 (2000)
5. N. Habbeger, G. Thompson, The Universal perturbative quantum 3-manifold pnvariant, Rozansky–Witten invariants and the generalized Casson invariant. Acta Math. Vietnam. **33**(3), 363–419 (2008)
6. C. LeBrun, Ricci curvature, minimal volumes, and Seiberg-Witten theory. Invent. Math. **145**, 279–316 (2001)

7. C. LeBrun, Curvature and Smooth topology in dimension four. Séminaire et Congrès **4**, 179–200 (2002)
8. M. Marcolli, *Seiberg-Witten Gauge Theory*. Texts and Readings in Mathematics, vol. 17 (Hindustan Book Agency, New Delhi, 1999)
9. F. Massamba, G. Thompson, On a system of Seiberg-Witten equations. J. Geom. Phys. **56**, 643–665 (2006)
10. J.D. Moore, *Lectures on Seiberg-Witten Invariants*. Lecture Notes in Mathematics, vol. 1629 (Springer, Berlin, 1996)
11. L.I. Nicolaescu, *Notes on Seiberg-Witten Theory*. Graduate Studies in Mathematics, vol. 28 (American Mathematical Society, Providence, RI, 2000)
12. A. Sambusetti, An obstruction to the existence of the Einstein metrics on four-manifolds. Math. Ann. **311**, 533–548 (1998)
13. C.H. Taubes, The existence of anti-self-dual conformal structures. J. Differ. Geom. **36**, 163–253 (1992)
14. E. Witten, Monopoles and 4-manifolds. Math. Res. Lett. **1**, 769–796 (1994)

Shape Invariant Potential Formalism for Photon-Added Coherent State Construction

Komi Sodoga, Isiaka Aremua, and Mahouton Norbert Hounkonnou

Abstract An algebro-operator approach, called shape invariant potential method, of constructing generalized coherent states for photon-added particle system is presented. Illustration is given on Pöschl–Teller potentials.

Keywords Shape invariant potentials · Photon-added coherent states · Pöschl-Teller I Potential · Density operator

1 Introduction

Coherent states (CS) play an important role in many fields of quantum mechanics since their early days. These states were first introduced by Schrödinger [36] in 1926 for the harmonic oscillator. Then followed decades of intensive works in order to extend the CS concept to other types of exactly solvable systems [5, 6, 18, 19, 21, 30]. It was shown in the 1980s that a large class of solvable potentials are characterized by a single property, i.e., a discrete reparametrization invariance, called shape invariance [11, 14, 17, 25], introduced in the framework of the supersymmetric quantum mechanics (SUSY QM) [10, 24]. It was then shown that shape invariant potentials (SIP) [11, 25] have an underlying algebraic structure

K. Sodoga (✉) · I. Aremua
Université de Lomé, Faculté des Sciences, Département de Physique, Laboratoire de Physique des Matériaux et de Mécanique Appliquée, Lomé, Togo

University of Abomey-Calavi, International Chair in Mathematical Physics and Applications (ICMPA), Cotonou, Benin
e-mail: ksodoga@univ-lome.tg; iaremua@univ-lome.tg

M. N. Hounkonnou
University of Abomey-Calavi, International Chair in Mathematical Physics and Applications (ICMPA), Cotonou, Benin
e-mail: norbert.hounkonnou@cipma.uac.bj

© Springer Nature Switzerland AG 2018
T. Diagana, B. Toni (eds.), *Mathematical Structures and Applications*,
STEAM-H: Science, Technology, Engineering, Agriculture,
Mathematics & Health, https://doi.org/10.1007/978-3-319-97175-9_17

and the associated Lie algebras were identified [2, 7]. Using this algebraic structure, a general definition of CS for shape invariant potentials was introduced by different authors [2, 16].

In 1980s, a new class of nonclassical states, known as photon-added coherent states (PACS), was introduced by Agarwal and Tara [1]. These states which are intermediate states between CS and Fock states are constructed by repeated applications of the creation operator on an ordinary CS. Since the works of Agarwal and Tara, PACS were intensively studied, as shown in the different extensions [15, 22, 26, 31, 32, 37]. Besides, PACS have various applications in quantum optics, quantum information and quantum computation [8, 12] (and references therein).

In a recent work [38], we constructed photon-added CS for SIP and investigated different cases following the Infeld-Hull [24] classification.

In the present paper, we aim at providing a rigorous mathematical formulation of the CS and their photon-added counterparts for SIP. We apply this formalism to Pöschl–Teller potentials of great importance in atomic physics. The diagonal P-representation of the density operator ρ is elaborated with thermal expectation values. This computation gives value on the use of Meijer G-functions. Novel results are obtained and discussed.

The paper is organized as follows. In Sect. 2, we review the concepts of SUSY QM factorization, give the algebraic formulation of the shape invariance condition, and define the generalized shape invariant potential coherent states (SIPCS). In Sect. 3, we construct the photon-added shape invariant potentials coherent states (PA-SIPCS) by successive applications of the raising operator on the SIP-CS. We calculate the inner product of two different PA-SIPCS in order to show that the obtained states are not mutually orthogonal. In contrast, we prove that these states are normalized. The resolution of unity is checked. Finally, we study the thermal statistical properties of the PA-SIPCS in terms of the Mandel's Q-parameter. In Sect. 4, Pöschl–Teller potentials are investigated as illustration. We end, in Sect. 5, with some concluding remarks.

2   Mathematical Formulation of SUSYQM: Integrability Condition and Coherent State Construction

In this section, we introduce the SUSY QM factorization method [20] (and references therein), give the integrability condition, known as shape invariance condition, and define the associated generalized CS.

Let $\mathcal{H} = L^2(]a, b[, \, dx)$ be the Hilbert space with the inner product defined by:

$$\langle u, v \rangle := \int_a^b \bar{u}(x)v(x)dx, \quad \forall \, u, v \in \mathcal{H}, \tag{2.1}$$

where $\bar{u}$ is the complex conjugate of u. Consider on $\mathcal{H}$ the one-dimensional bound-state Hamiltonian ($\hbar = 2m = 1$)

$$H = -\frac{d}{dx}^2 + V(x), \quad x \in]a, b[\subset \mathbb{R} \tag{2.2}$$

with the domain

$$\mathcal{D}(H) = \left\{ u \in \mathcal{H}, \quad -u'' + Vu \in \mathcal{H} \right\}, \tag{2.3}$$

where V is a real continuous function on $]a, b[$. Let us denote E_n and Ψ_n the eigenvalues and eigenfunctions of H, respectively. Let the first-order differential operator A be defined by:

$$A = \frac{d}{dx} + W(x), \text{ with the domain } \mathcal{D}(A) = \left\{ u \in \mathcal{H}, \quad u' + Wu \in \mathcal{H} \right\}, \tag{2.4}$$

$W = -\frac{d}{dx}[\ln(\Psi_0)]$ is a real continuous functions on $]a, b[$. The adjoint operator $A^\dagger$ of A is defined on [39]:

$$\mathcal{D}(A^\dagger) = \{ u \in \mathcal{H} | \exists \tilde{v} \in \mathcal{H} : \langle Au, v \rangle = \langle u, \tilde{v} \rangle \ \forall u \in \mathcal{D}(A) \}, \quad A^\dagger v = \tilde{v}. \tag{2.5}$$

We infer $\mathcal{D}(A)$ dense in $\mathcal{H}$ since $H^{1,2}(]a, b[, \rho(x)dx)$ is dense in $\mathcal{H}$, and $H^{1,2}(]a, b[, \rho(x)dx) \subset \mathcal{D}(H)$, where $H^{m,n}(\Omega)$ is the Sobolev spaces of indices (m, n). We assume that the operator A is closed in $\mathcal{H}$. The explicit expression of $A^\dagger$ is given through the following theorem.

Theorem 2.1 *Suppose the following boundary condition:*

$$u(x) \, v(x) \Big|_a^b = 0, \quad \forall \, u \in \mathcal{D}(A) \text{ and } v \in \mathcal{D}(A^\dagger), \tag{2.6}$$

is verified. Then the operator $A^\dagger$ can be written as

$$A^\dagger = \left[-\frac{d}{dx} + W(x) \right]. \tag{2.7}$$

Proof The proof follows as:

$$\langle A\bar{u}, v \rangle \equiv \int_a^b \left[\bar{u}'(x) + W(x)\bar{u}(x) \right] v(x)dx$$

$$= \bar{u}(x)v(x) \Big|_a^b + \int_a^b \bar{u}(x) \left[-\frac{d}{dx} + W(x) \right] v(x)dx$$

$$= \langle \bar{u}, A^\dagger v \rangle \quad \text{for any} \quad u \in \mathcal{D}(A), \quad v \in \mathcal{D}(A^\dagger). \quad \blacksquare$$

Let H_1 and H_2 be the product operators $A^\dagger A$ and $A A^\dagger$, respectively, with the corresponding domains

$$\mathcal{D}(H_1) = \left\{ u \in \mathcal{D}(A), \, v = Au \in \mathcal{D}(A^\dagger) \text{ and } A^\dagger v \in \mathcal{H} \right\},$$

$$\mathcal{D}(H_2) = \left\{ u \in \mathcal{D}(A^\dagger), \ v = A^\dagger u \in \mathcal{D}(A) \text{ and } Av \in \mathcal{H} \right\}. \tag{2.8}$$

Remark that

$$H^{1,2}\left(\,]a, b[, \ dx\right) \subset \mathcal{D}(A) \subset \mathcal{D}(A^\dagger).$$

Then

$$\mathcal{D}(H_1), \mathcal{D}(H_2) \supset H^{2,2}\left(\,]a, b[, \ dx\right).$$

We infer then that $\mathcal{D}(H_1)$ and $\mathcal{D}(H_2)$ are dense in $\mathcal{H}$. The following theorem gives additional conditions on W so that the operator H factorizes in terms of A and $A^\dagger$.

Theorem 2.2 *Suppose that the function W verifies the Riccati type equation:*

$$V - E_0 = W^2 - W'. \tag{2.9}$$

Then the operators $H_{1,2}$ are self-adjoint, and:

$$H_1 = A^\dagger A = H - E_0 = -\frac{d^2}{dx^2} + W^2 - W',$$

$$H_2 = A A^\dagger = -\frac{d^2}{dx^2} + W^2 + W'. \tag{2.10}$$

Proof The operators $A^\dagger A$ and $A A^\dagger$ are self-adjoint since A and $A^\dagger$ are mutually adjoint and A is closed with $\mathcal{D}(A)$ dense in $\mathcal{H}$. From the definitions (2.4) and (2.7) of the differential operators A and $A^\dagger$, we have the following products

$$A^\dagger A = -\frac{d^2}{dx^2} + (W^2 - W'), \quad A A^\dagger = -\frac{d^2}{dx^2} + (W^2 + W').$$

Equation (2.9) are readily deduced from the above relations and (2.10). ∎

We can rewrite the operators $H_{1,2}$ as:

$$H_{1,2} = -\frac{d^2}{dx^2} + V_{1,2}, \quad \text{where} \quad V_{1,2} = W^2 \mp W'. \tag{2.11}$$

In SUSY QM terminology, $H_{1,2}$ are called SUSY partner Hamiltonians; $V_{1,2}$ are called SUSY partner potentials, and the function W is called the superpotential.

Now let us establish some results showing that the eigenvalues of partner Hamiltonians are positive definite ($E_n^{1,2} \geq 0$) and isospectral, i.e, they have almost the same energy eigenvalues, except for the ground state energy of H_1 [10].

Proposition 2.3 *The eigenvalues of H_1 and H_2 are non negative*

$$E_n^{(1)} \geq 0, \quad E_n^{(2)} \geq 0.$$

Proof Let $E_n^{(1)}$ be an eigenvalue of H_1 corresponding to the eigenfunction $\Psi_n^{(1)}$. In Dirac notation, this reads as $H_1|\Psi_n^{(1)}\rangle = E_n^{(1)}|\Psi_n^{(1)}\rangle$. Then $\langle\Psi_n^{(1)}|A^\dagger A|\Psi_n^{(1)}\rangle = E_n^{(1)}\langle\Psi_n^{(1)}|\Psi_n^{(1)}\rangle$, i.e, $||A|\Psi_n^{(1)}\rangle||^2 = E_n^{(1)}|||\Psi_n^{(1)}\rangle||^2$. Therefore $E_n^{(1)} \geq 0$, since $||A|\Psi_n^{(1)}\rangle||^2 \geq 0$ and $|||\Psi_n^{(1)}\rangle||^2 \geq 0$. Similarly, one can show that $E_n^{(2)} \geq 0$. ∎

Proposition 2.4 *Let $|\Psi_n^{(1)}\rangle$ and $|\Psi_n^{(2)}\rangle$ be the normalized eigenstates of H_1 and H_2 associated to the eigenvalues $E_n^{(1)}$ and $E_n^{(2)}$, respectively. Then*

$$A|\Psi_n^{(1)}\rangle = 0 \Longleftrightarrow E_n^{(1)} = 0, \quad A^\dagger|\Psi_n^{(2)}\rangle = 0 \Longleftrightarrow E_n^{(2)} = 0.$$

Proof

$$A|\Psi_n^{(1)}\rangle = 0 \Longleftrightarrow ||A|\Psi_n^{(1)}\rangle||^2 = 0$$
$$\Longleftrightarrow \langle\Psi_n^{(1)}|A^\dagger A|\Psi_n^{(1)}\rangle = 0$$
$$\Longleftrightarrow E_n^{(1)}\langle\Psi_n^{(1)}|\Psi_n^{(1)}\rangle = 0$$
$$\Longleftrightarrow E_n^{(1)} = 0.$$

By analogy, one can show that $A^\dagger|\Psi_n^{(2)}\rangle = 0 \Longleftrightarrow E_n^{(2)} = 0$. ∎

As a consequence of this proposition $E_0^{(1)} = 0$, since $A\Psi_0^{(1)} = A\Psi_0 = 0$.

Proposition 2.5 *If H_1 admits a normalized eigenstate $|\Psi_0^{(1)}\rangle$ so that $E_0^{(1)} = 0$, then H_2 does not admit a normalized eigenstate $|\Psi_0^{(2)}\rangle$ corresponding to the eigenvalue $E_0^{(2)} = 0$.*

Proof If $E_0^{(1)} = 0$, then from the Proposition 2.4, $A\Psi_0^{(1)} = 0$. We deduce from this that

$$AA^\dagger A\Psi_0^{(1)} = H_2(A\Psi_0^{(1)}) = 0. \tag{2.12}$$

Suppose that there exists a normalizable eigenstate $|\Psi_0^{(2)}\rangle$ of H_2 corresponding to $E_0^{(2)} = 0$. It follows from (2.12) that $|\Psi_0^{(2)}\rangle \propto A\Psi_0^{(1)} = 0$, what is inconsistent. ∎

This proposition shows that H_2 cannot possess a normalized state $\left|\Psi_0^{(2)}\right\rangle$ corresponding to the eigenvalues $E_0^{(2)} = 0$, since $E_0^{(1)} = 0$, what means $E_0^{(2)} \neq 0$.

Proposition 2.6 *Let $|\Psi_n^{(1)}\rangle$ and $|\Psi_n^{(2)}\rangle$ be normalized eigenstates of H_1 and H_2, respectively, such that $A|\Psi_n^{(1)}\rangle \neq 0$, $A^\dagger|\Psi_n^{(2)}\rangle \neq 0$ and the corresponding eigenvalues are, respectively, $E_n^{(1)} \neq 0$ and $E_n^{(2)} \neq 0$. Then $E_n^{(1)}$ is also an eigenvalue of H_2 associated to the eigenstate*

$$|\Psi_{n-1}^{(2)}\rangle = (E_n^{(1)})^{-1/2}A|\Psi_n^{(1)}\rangle;$$

$E_n^{(2)}$ is also an eigenvalue of H_1 associated to the eigenstate

$$|\Psi_{n+1}^{(1)}\rangle = (E_n^{(2)})^{-1/2}A^\dagger|\Psi_n^{(2)}\rangle.$$

Proof We have $H_1|\Psi_n^{(1)}\rangle = E_n^{(1)}|\Psi_n^{(1)}\rangle$. From this, we deduce that $AA^\dagger(A|\Psi_n^{(1)}\rangle) = E_n^{(1)}(A|\Psi_n^{(1)}\rangle)$, or

$$H_2(A|\Psi_n^{(1)}\rangle) = E_n^{(1)}(A|\Psi_n^{(1)}\rangle), \tag{2.13}$$

i.e., $A|\Psi_n^{(1)}\rangle$ is an eigenstate of H_2 associated to the eigenvalue $E_n^{(1)}$. Similarly, $H_2|\Psi_n^{(2)}\rangle = E_n^{(2)}|\Psi_n^{(2)}\rangle$ implies $A^\dagger A(A^\dagger|\Psi_n^{(2)}\rangle) = E_n^{(2)}(A^\dagger|\Psi_n^{(2)}\rangle)$, i.e,

$$H_1(A^\dagger|\Psi_n^{(2)}\rangle) = E_n^{(2)}(A^\dagger|\Psi_n^{(2)}\rangle). \tag{2.14}$$

This means that $A^\dagger|\Psi_n^{(2)}\rangle$ is an eigenstate of H_1 with the eigenvalue $E_n^{(2)}$. The eigenvalues of H_1 being non degenerate (since we consider only bound state of H_1), it follows that there exists a unique normalized eigenstate $|\Psi_k^{(1)}\rangle$ of H_1, up to a multiplicative constant, corresponding to an eigenvalue $E_k^{(1)}$ such that $|\Psi_k^{(1)}\rangle = cA^\dagger|\Psi_n^{(2)}\rangle$. The normalization constant c is given by $c = (E_n^{(2)})^{-1/2}$. We have

$$|\Psi_k^{(1)}\rangle = (E_n^{(2)})^{-1/2}A^\dagger|\Psi_n^{(2)}\rangle. \tag{2.15}$$

It follows from (2.15) that

$$\begin{aligned}
H_1|\Psi_k^{(1)}\rangle &= (E_n^{(2)})^{-1/2}H_1\left(A^\dagger\left|\Psi_n^{(2)}\right\rangle\right) \\
&= (E_n^{(2)})^{-1/2}E_n^{(2)}\left(A^\dagger\left|\Psi_n^{(2)}\right\rangle\right) \qquad \text{from} \quad (2.14) \\
&= E_n^{(2)}|\Psi_k^{(1)}\rangle \qquad\qquad\qquad\quad \text{from} \quad (2.15).
\end{aligned}$$

Then $H_1|\Psi_k^{(1)}\rangle = E_k^{(1)}|\Psi_k^{(1)}\rangle = E_n^{(2)}|\Psi_k^{(1)}\rangle$. It follows from this that $E_k^{(1)} = E_n^{(2)}$. Since $E_0^{(1)} \neq E_0^{(2)}$ ($E_0^{(1)} = 0$ and $E_0^{(2)} \neq 0$), a simplest solution of the index equation is $k = n + 1$. Hence

$$\begin{cases} E_{n+1}^{(1)} = E_n^{(2)} \\ |\Psi_{n+1}^{(1)}\rangle = (E_n^{(2)})^{-1/2}(A^\dagger|\Psi_n^{(2)}\rangle). \end{cases} \tag{2.16}$$

One can similarly show from (2.13) that

$$\begin{cases} E_n^{(2)} = E_{n+1}^{(1)} \\ |\Psi_n^{(2)}\rangle = (E_{n+1}^{(1)})^{-1/2}A|\Psi_{n+1}^{(1)}\rangle. \end{cases} \tag{2.17}$$

∎

It follows from these propositions that the eigenvalues of H_1 and H_2 are positive definite $(E_n^{1,2} \geq 0)$, and the partner Hamiltonians are isospectral, i.e., they have almost the same energy eigenvalues, except for the ground state energy of H_1 which is missing in the spectrum of H_2. The spectra are linked as [10]:

$$E_n^{(2)} = E_{n+1}^{(1)}, \quad E_0^{(1)} = 0, \quad n = 0, 1, 2, \dots,$$
$$\Psi_n^{(2)} = \left[E_{n+1}^{(1)} \right]^{(-1/2)} A \Psi_{n+1}^{(1)},$$
$$\Psi_{n+1}^{(1)} = \left[E_n^{(2)} \right]^{(-1/2)} A^\dagger \Psi_n^{(2)}. \tag{2.18}$$

Hence, if the eigenvalues and eigenfunctions of one of the partner, say H_1, are known, one can immediately derive the eigenvalues and eigenfunctions of H_2.

However, the above relations (2.18) only give the relationship between the eigenvalues and eigenfunctions of the two partner Hamiltonians, but do not allow to determine their spectra. A condition of an exact solvability is known as the shape invariance condition; that is, the pair of SUSY partner potentials $V_{1,2}$ are similar in shape and differ only in the parameters that appear in them. Gendenshtein states the shape invariance condition as [10, 17]

$$V_2(x; a_1) = V_1(x; a_2) + \mathcal{R}(a_1), \tag{2.19}$$

where a_1 is a set of parameters and a_2 is a function of a_1, $(a_2 = f(a_1))$, and $\mathcal{R}(a_1)$ is the non-vanishing remainder independent of x. In such a case, the eigenvalues and the eigenfunctions of H_1 can explicitly be deduced [17]. If this Hamiltonian H_1 has p $(p \geq 1)$ bound states with eigenvalues $E_n^{(1)}$, and eigenfunctions $\Psi_n^{(1)}$ with $0 \leq n \leq p - 1$, the starting point of constructing the spectra is to generate a hierarchy of $(p - 1)$ Hamiltonians $H_2, \dots H_p$ such that the mth member of the hierarchy (H_m) has the same spectrum as H_1 except that the first $m - 1$ eigenvalues of H_1 are missing in the spectrum of H_m [10]. In order m, $(m = 2, 3, \dots p)$, we have partner Hamiltonians

$$H_m(x; a_1) = A_m^\dagger(x; a_1) A_m(x; a_1) + E_0^{(m)} = -\frac{d^2}{dx^2} + V_m(x; a_1),$$

$$H_{m+1}(x; a_1) = A_m(x; a_1) A_m^\dagger(x; a_1) + E_0^{(m)} = -\frac{d^2}{dx^2} + V_{m+1}(x; a_1),$$

the spectra of which are related as

$$E_n^{(m+1)} = E_{n+1}^{(m)}, \quad \Psi_n^{(m+1)} = \left(E_{n+1}^{(m)} - E_0^{(m)} \right)^{-1/2} A_m \Psi_{n+1}^{(m)}.$$

In terms of the spectrum of H_1 we have

$$E_n^{(m)} = E_{n+1}^{(m-1)} = E_{n+2}^{(m-2)} = \cdots = E_{n+m-1}^{(1)} \tag{2.20}$$

$$\Psi_n^{(m)} = \left(E_{n+m-1}^{(1)} - E_{m-2}^{(1)}\right)^{-1/2} \cdots \left(E_{n+m-1}^{(1)} - E_0^{(1)}\right)^{-1/2} A_{m-1} \cdots A_1 \Psi_{n+m-1}^{(1)}(x; a_1).$$

Theorem 2.7 *The eigenvalues of H_1 are given by [10, 17]*

$$E_n^{(1)} = \sum_{k=1}^{n} \mathcal{R}(a_k). \tag{2.21}$$

Proof Consider the partner Hamiltonians H_m and H_{m+1} of the hierarchy of Hamiltonians constructed from H_1. If the partner potentials are shape invariant, we can write

$$\begin{aligned}
V_{m+1}(x; a_1) &= V_m(x; a_2) + \mathcal{R}(a_1) \\
&= V_{m-1}(x; a_3) + \mathcal{R}(a_2) + \mathcal{R}(a_1) \\
&= V_{m-2}(x; a_4) + \mathcal{R}(a_3) + \mathcal{R}(a_2) + \mathcal{R}(a_1) \\
&\vdots \\
&= V_2(x; a_m) + \mathcal{R}(a_{m-1}) + \mathcal{R}(a_{m-2}) + \cdots + \mathcal{R}(a_1) \\
&= V_1(x; a_{m+1}) + \sum_{k=1}^{m} \mathcal{R}(a_k).
\end{aligned}$$

It follows from the above that $H_m(x; a_1) = H_1(x; a_m) + \sum_{k=1}^{m-1} \mathcal{R}(a_k)$. Hence $E_0^{(m)} = \sum_{k=1}^{m-1} \mathcal{R}(a_k)$. From Eq. (2.18), $E_0^{(m)} = E_{m-1}^{(1)}$. Then $E_{m-1}^{(1)} = \sum_{k=1}^{m-1} \mathcal{R}(a_k)$, i.e, $E_n^{(1)} = \sum_{k=1}^{n} \mathcal{R}(a_k)$. ∎

Theorem 2.8 *The normalized eigenfunctions of H_1 are given by [11]*

$$\Psi_n(x; a_1) = \left\{ \prod_{k=1}^{n} \left(\sum_{p=1}^{k} \mathcal{R}(a_p) \right) \right\}^{-1/2} A^{\dagger}(x; a_1) \cdots A^{\dagger}(x; a_n) \Psi_0^{(1)}(x; a_{n+1}). \tag{2.22}$$

Proof From the shape invariance condition (2.19), we deduce the following relation between the eigenfunctions of the partner Hamiltonians H_1 and H_2

$$\Psi_n^{(2)}(x; a_1) = \Psi_n^{(1)}(x; a_2) \tag{2.23}$$

We know from (2.16) that

$$
\begin{aligned}
\Psi_{n+1}^{(1)}(x; a_1) &= (E_n^{(2)})^{-1/2} A^\dagger(x; a_1) \Psi_n^{(2)}(x; a_1) \\
&= (E_n^{(2)})^{-1/2} A^\dagger(x; a_1) \Psi_n^{(1)}(x; a_2) \quad \text{from} \quad (2.23) \\
&= (E_n^{(2)})^{-1/2}(E_{n-1}^{(2)})^{-1/2} A^\dagger(x; a_1) A^\dagger(x; a_2) \Psi_{n-1}^{(2)}(x; a_3) \\
&= \vdots \\
&= (E_n^2)^{-1/2} \cdots (E_0^2)^{-1/2} A^\dagger(x; a_1) \cdots A^\dagger(x; a_{n+1}) \Psi_0^{(1)}(x; a_{n+2}).
\end{aligned}
$$

It deduces from above equations that

$$
\Psi_n^{(1)}(x; a_1) = \left\{ \prod_{k=1}^{n} \left(\sum_{p=1}^{k} \mathcal{R}(a_p) \right) \right\}^{-1/2} A^\dagger(x; a_1) \cdots A^\dagger(x; a_n) \Psi_0^{(1)}(x; a_{n+1}). \tag{2.24}
$$

∎

The shape invariance condition (2.19) can be rewritten in terms of the factorization operators defined in Eqs. (2.4)–(2.7),

$$
A(a_1) A^\dagger(a_1) = A^\dagger(a_2) A(a_2) + \mathcal{R}(a_1), \tag{2.25}
$$

where a_2 is a function of a_1. Here, we only consider the translation class of shape invariance potentials, that is the case where the parameters a_1 and a_2 are related as $a_2 = a_1 + \eta$ [11] and the potentials are known in closed form. The scaling class [25] is not treated here since the potentials, in this case, can only be written as Taylor expansion.

Introducing a reparametrization operator T_η defined as

$$
T_\eta : \mathcal{H} \longrightarrow \mathcal{H} \qquad T_\eta \Phi(x; a_1) := \phi(x; a_1 + \eta) = \Phi(x; a_2) \tag{2.26}
$$

which replaces a_1 with a_2 in a given operator [7]

$$
T_\eta \mathcal{O}(a_1) T_\eta^{-1} = \mathcal{O}(a_1 + \eta) := \mathcal{O}(a_2), \tag{2.27}
$$

and the operators

$$
B_-, B_+ : \mathcal{H} \longrightarrow \mathcal{H} \quad B_+ = A^\dagger(a_1) T_\eta, \quad B_- = T_\eta^\dagger A(a_1), \tag{2.28}
$$

with the domains

$$
\mathcal{D}(B_-) = \left\{ u \in \mathcal{H}, \quad v = u' + W u \in \mathcal{H} \quad \text{and} \quad T_\eta^\dagger v \in \mathcal{H} \right\} \tag{2.29}
$$

$$
\mathcal{D}(B_+) = \left\{ u \in \mathcal{H}, \quad v = T_\eta u \in \mathcal{H} \quad \text{and} \quad -v' + W v \in \mathcal{H} \right\}. \tag{2.30}
$$

The Hamiltonian factorizes in terms of the new operators as follow:

$$H - E_0 = H_1 = A^\dagger(a_1)A(a_1) = B_+ B_-$$

(2.31)

where

$$[B_-, B_+] = \mathcal{R}(a_0) \quad , \quad B_- |\Psi_0\rangle = 0 .$$

(2.32)

The states $B_+^n |\Psi_0\rangle$ are eigenfunctions of H with eigenvalues E_n, ie,

$$H(B_+^n |\Psi_0\rangle) = \underbrace{\left[\sum_{k=1}^n \mathcal{R}(a_k)\right]}_{E_n} B_+^n |\Psi_0\rangle$$

(2.33)

$B_\pm$ act as raising and lowering operators:

$$B_+ |\Psi_n\rangle = \sqrt{E_{n+1}} |\Psi_{n+1}\rangle \quad , B_- |\Psi_n\rangle = \sqrt{\mathcal{R}(a_0) + E_{n-1}} |\Psi_{n-1}\rangle .$$

(2.34)

To define shape invariant potential CS, Balantekin et al. [2] introduced the right inverse of B_- as: $B_- B_-^{-1} = \mathbb{1}$ and the left inverse H^{-1} of H such that: $H^{-1} B_+ = B_-^{-1}$. The SIPCS defined by

$$|z\rangle = \sum_{n=0}^n (z B_-^{-1})^n |\Psi_0\rangle$$

(2.35)

are eigenstates of the lowering operator B_-:

$$B_- |z\rangle = z |z\rangle .$$

(2.36)

A generalization of the SIPCS (2.35) was done as [2]:

$$|z; a_j\rangle = \sum_{n=0}^\infty \left\{z Z_j B_-^{-1}\right\}^n |\Psi_0\rangle , \quad z, Z_j \in \mathbb{C}$$

(2.37)

where $Z_j \equiv Z_{(a_j)} \equiv Z(a_1, a_2, \ldots)$. Observing that $B_-^{-1} Z_j = Z_{j+1} B_-^{-1}$, and from

$$Z_{j-1} = T^\dagger(a_1) Z_j T(a_1)$$

(2.38)

one can readily show that

$$(z Z_j B_-^{-1})^n = z^n \prod_{k=0}^{n-1} Z_{j+k} B_-^{-n} .$$

(2.39)

Using (2.39), one can straightforwardly deduce that (2.37) are eigenstates of B_-:

$$B_- \left| z; a_j \right\rangle = z Z_{j-1} \left| z; a_j \right\rangle . \tag{2.40}$$

Observing that

$$B_-^{-n} \left| \Psi_0 \right\rangle = C_n \left| \Psi_n \right\rangle , \quad C_n = \left[\prod_{k=1}^{n} \left(\sum_{s=k}^{n} \mathcal{R}(a_s) \right) \right]^{-1/2} \tag{2.41}$$

and using (2.41), the normalized form of the CS (2.37) can be obtained as:

$$\left| z; a_r \right\rangle = \mathcal{N}(|z|^2; a_r) \sum_{n=0}^{\infty} \frac{z^n}{h_n(a_r)} \left| \Psi_n \right\rangle , \tag{2.42}$$

where we used the shorthand notation $a_r \equiv [\mathcal{R}(a_1), \mathcal{R}(a_2), \ldots, \mathcal{R}(a_n); a_j, a_{j+1}, \ldots, a_{j+n-1}]$. The expansion coefficient $h_n(a_r)$ and the normalization constant $\mathcal{N}(|z|^2; a_r)$ are:

$$h_n(a_r) = \frac{\sqrt{\prod_{k=1}^{n} \left(\sum_{s=k}^{n} \mathcal{R}(a_s) \right)}}{\prod_{k=0}^{n-1} Z_{j+k}} \quad \text{for } n \geq 1, \; h_0(a_r) = 1,$$

$$\mathcal{N}(x; a_r) = \left[\sum_{n=0}^{\infty} \frac{x^n}{|h_n(a_r)|^2} \right]^{-1/2} . \tag{2.43}$$

It is shown [2] that these states (2.37) fulfill the standard properties of label continuity, overcompleteness, temporal stability and action identity.

3 Construction of Photon-Added Coherent States for Shape Invariant Systems

In this section, a construction of PA-SIPCS [38], and their physical and mathematical properties are presented.

3.1 Definition of the PA-SIPCS

Let $\mathfrak{H}_m$ be the Hilbert subspace of $\mathfrak{H}$ defined as follows:

$$\mathfrak{H}_m := span\,\{|\Psi_{n+m}\rangle\}_{n\geq 0}\,. \tag{3.1}$$

By successive applications of the raising operator B_+ on the generalized SIPCS (2.36), we can obtain photon-added shape invariant potential CS (PA-SIPCS) denoted by $|z; a_r\rangle_m$:

$$|z; a_r\rangle_m := (B_+^m)\,|z; a_r\rangle \tag{3.2}$$

where m is a positive integer standing for the number of added quanta or photons.

It is worth mentioning that the first m eigenstates $|\Psi_n\rangle$, $n = 0, 1, \ldots, m-1$ are absent from the wavefunction $|z; a_r\rangle_m \in \mathfrak{H}_m$. Therefore, from the orthonormality relation satisfied by the states $|\Psi_n\rangle$, the overcompleteness relation fulfilled by the identity operator on $\mathfrak{H}_m$, denoted by $\mathbb{1}_{\mathfrak{H}_m}$, is written as [31, 37]

$$\mathbb{1}_{\mathfrak{H}_m} = \sum_{n=m}^{\infty} |\Psi_n\rangle\,\langle\Psi_n| = \sum_{n=0}^{\infty} |\Psi_{n+m}\rangle\,\langle\Psi_{n+m}|\,. \tag{3.3}$$

Here, $\mathbb{1}_{\mathfrak{H}_m}$ is only required to be a bounded positive operator with a densely defined inverse [3].

From (2.27) and using the relations $B_+\mathcal{R}(a_{n-1}) = \mathcal{R}(a_n)B_+$ and $B_+|\Psi_n\rangle = \sqrt{E_{n+1}}\,|\Psi_{n+1}\rangle$, we obtain the PA-SIPCS as:

$$|z; a_r\rangle_m = \mathcal{N}_m(|z|^2; a_r) \sum_{n=0}^{\infty} \frac{z^n}{K_n^m(a_r)}\,|\Psi_{n+m}\rangle \tag{3.4}$$

where the expansion coefficient takes the form:

$$K_n^m(a_r) = \frac{\left[\displaystyle\prod_{k=m+1}^{n+m}\left(\sum_{s=k}^{n+m}\mathcal{R}(a_s)\right)\right]^{1/2}}{\left[\displaystyle\prod_{k=m}^{n+m-1} Z_{j+k}\right]\cdot\left[\displaystyle\prod_{k=1}^{m}\left(\sum_{s=k}^{n+m}\mathcal{R}(a_s)\right)\right]^{1/2}}\,, \tag{3.5}$$

and the normalization constant $\mathcal{N}_m(|z|^2; a_r)$ is given by:

$$\mathcal{N}_m(|z|^2; a_r) = \left(\sum_{n=0}^{\infty} \frac{|z|^{2n}}{|K_n^m(a_r)|^2}\right)^{-1/2}\,. \tag{3.6}$$

The inner product of two different PA-SIPCS $|z; a_r\rangle_m$ and $\left|z'; a_r\right\rangle_{m'}$

$$_{m'}\left\langle z'; a_r \mid z; a_r\right\rangle_m = \mathcal{N}_{m'}(|z'|^2; a_r)\mathcal{N}_m(|z|^2; a_r)$$

$$\sum_{n,n'=0}^{\infty} \frac{z'^{\star n'} z^n}{K_{n'}^{m'\star}(a_r)K_n^m(a_r)}\left\langle \Psi_{n'+m'} \mid \Psi_{n+m}\right\rangle \tag{3.7}$$

does not vanish. Indeed, due to the orthonormality of the eigenstates $|\Psi_n\rangle$, the inner product (3.7) can be rewritten as

$$_{m'}\left\langle z'; a_r \mid z; a_r\right\rangle_m = \mathcal{N}_{m'}(|z'|^2; a_r)\mathcal{N}_m(|z|^2; a_r)z'^{\star(m-m')} \sum_{n=0}^{\infty} \frac{(z'^{\star}z)^n}{K_{n+m-m'}^{m'\star}(a_r)\, K_n^m(a_r)}, \tag{3.8}$$

showing that the PA-SIPCS are not mutually orthogonal.

3.2 Label Continuity

In the Hilbert space $\mathfrak{H}$, the PA-SIPCS $|z, a_r\rangle_m$ are labeled by m and z. The label continuity condition can then be stated as:

$$|z - z'| \to 0 \text{ and } |m - m'| \to 0 \Longrightarrow \||z, a_r\rangle_m - |z', a_r\rangle_{m'}\|^2$$
$$= 2\left[1 - \mathcal{R}e\left(_{m'}\left\langle z'; a_r \mid z; a_r\right\rangle_m\right)\right] \to 0. \tag{3.9}$$

This is satisfied by the states $|z, a_r\rangle_m$, since from Eqs. (3.6), (3.8), we see that

$$m \to m' \quad \text{and} \quad z \to z' \Longrightarrow {}_{m'}\left\langle z'; a_r \mid z; a_r\right\rangle_m \to 1. \tag{3.10}$$

Therefore the PA-SIPCS $|z, a_r\rangle_m$ are continuous in their labels.

3.3 Overcompleteness

We check the realization of the resolution of the identity in the Hilbert space (3.1) with the identity operator defined as (3.3):

$$\int_{\mathbb{C}} d^2z \, |z; a_r\rangle_m \, \omega_m(|z|^2; a_r) \, _m\langle z; a_r| = \mathbb{1}_{\mathfrak{H}_m}. \tag{3.11}$$

Inserting the definition (3.4) of the PA-SIPCS $|z; a_r\rangle_m$ into Eq. (3.11) yields, after taking the angular integration of the diagonal matrix elements:

$$\int_0^\infty dx \, x^n \, \mathcal{W}_m(x; a_r) = |K_n^m(a_r)|^2 \quad \text{with} \quad \mathcal{W}_m(x; a_r) = \pi \mathcal{N}_m^2(x; a_r) \, \omega_m(x; a_r).$$

(3.12)

Therefore, the weight function ω_m is related to the undetermined moment distribution $\mathcal{W}_m(x; a_r)$, which is the solution of the Stieltjes moment problem with the moments given by $|K_n^m(a_r)|^2$. In order to use Mellin transformation, we can rewrite (3.12) as

$$\int_0^\infty dx \, x^{n+m} \, g_m(x; a_r) = |K_n^m(a_r)|^2, \quad \text{where} \quad g_m(x; a_r)$$

$$= \pi \mathcal{N}_m^2(x; a_r) x^{-m} \, \omega_m(x; a_r) .$$

(3.13)

By performing the variable change $n + m \to s - 1$, Eq. (3.13) becomes:

$$\int_0^\infty dx \, x^{s-1} g_m(x; a_r) = |K_s^m(a_r)|^2 .$$

(3.14)

Comparing this relation with the Meijer's G-function and the Mellin inversion theorem [28]

$$\int_0^\infty dx \, x^{s-1} G_{p,q}^{m,n} \left(\alpha x \, \middle| \, \begin{matrix} a_1, \ldots, a_n; a_{n+1}, \ldots, a_p \\ b_1, \ldots, b_m; b_{m+1}, \ldots, b_q \end{matrix} \right)$$

$$= \frac{1}{\alpha^s} \frac{\displaystyle\prod_{j=1}^m \Gamma(b_j + s) \prod_{j=1}^n \Gamma(1 - a_j - s)}{\displaystyle\prod_{j=m+1}^q \Gamma(1 - b_j - s) \prod_{j=n+1}^p \Gamma(a_j + s)},$$

(3.15)

we see that if $|K_s^m(a_r)|^2$ in the above relation can be expressed in terms of Gamma functions, then $g_m(x; a_r)$ can be identified as the Meijer's G- function.

3.4 Thermal Statistics

In quantum mechanics, the density matrix, generally denoted by ρ, is an important tool for characterizing the probability distribution on the states of a physical system. For example, it is useful for examining the physical and chemical properties of a system (see [6, 31] and the references listed therein). Consider a quantum gas of the system in the thermodynamic equilibrium with a reservoir at temperature T, which satisfies a quantum canonical distribution. The corresponding normalized density operator is given, in the Hilbert space $H_m := span \, |\Psi_{n+m}\rangle_{n \geq 0}$, as

$$\rho^{(m)} = \frac{1}{Z} \sum_{n=0}^{\infty} e^{-\beta E_n} |\Psi_{n+m}\rangle \langle \Psi_{n+m}|, \qquad (3.16)$$

where, in the exponential, E_n is the eigen-energy, and the partition function Z is taken as the normalization constant.

The diagonal elements of $\rho^{(m)}$, essential for our purpose, also known as the Q-distribution or Husimi's distribution, are derived in the PA-SIPCS basis as

$$_m \Big\langle z; a_r |\rho^{(m)}|z; a_r \Big\rangle_m = \frac{\mathcal{N}_m^2(|z|^2; m)}{Z} \sum_{n=0}^{\infty} \frac{|z|^{2n}}{|K_n^m(a_r)|^2} e^{-\beta E_n}. \qquad (3.17)$$

The normalization of the density operator leads to

$$Tr\rho^{(m)} = \int_{\mathbb{C}} d^2z \, \omega_m(|z|^2; a_r) \, _m\langle z; a_r |\rho^{(m)}|z; a_r\rangle_m = 1. \qquad (3.18)$$

The diagonal expansion of the normalized canonical density operator over the PA-SIPCS projector is

$$\rho^{(m)} = \int_{\mathbb{C}} d^2z \, \omega_m(|z|^2; a_r)|z; a_r\rangle_m \, P(|z|^2) \, _m\langle z; a_r|, \qquad (3.19)$$

where the P-distribution function $P(|z|^2)|$ satisfying the normalization to unity condition

$$\int_{\mathbb{C}} d^2z \, \omega_m(|z|^2; a_r) P(|z|^2) = 1 \qquad (3.20)$$

must be determined.

Thus, given an observable $\mathcal{O}$, one obtains the expectation value, i.e., the thermal average given by

$$\langle \mathcal{O} \rangle_m = Tr(\rho^{(m)}\mathcal{O}) = \int_{\mathbb{C}} d^2z \, \omega_m(|z|^2; a_r) P(|z|^2) \, _m\langle z; a_r |\mathcal{O}|z; a_r\rangle_m. \qquad (3.21)$$

One can check that for a PA-SIPCS (3.4) the expectation values of the operator $N := B_+ B_-$ [38] are:

$$\langle N \rangle = \mathcal{N}_m^2(|z|^2; a_r) \sum_{n=0}^{\infty} E_{n+m} \frac{|z|^{2n}}{|K_n^m(a_r)|^2}, \quad \Big\langle N^2 \Big\rangle$$

$$= \mathcal{N}_m^2(|z|^2; a_r) \sum_{n=0}^{\infty} E_{n+m}^2 \frac{|z|^{2n}}{|K_n^m(a_r)|^2}. \qquad (3.22)$$

Using (3.22), the pseudo-thermal expectation values of the operator N and of its square N^2, given by $\langle N \rangle^{(m)} = Tr(\rho^{(m)} N)$ and $\langle N^2 \rangle^{(m)} = Tr(\rho^{(m)} N^2)$, respectively, allow to obtain the thermal intensity correlation function as follows:

$$(g^2)^{(m)} = \frac{\langle N^2 \rangle^{(m)} - \langle N \rangle^{(m)}}{\left(\langle N \rangle^{(m)} \right)^2}. \tag{3.23}$$

Then, we deduce the thermal analogue of the Mandel parameter

$$Q^{(m)} = \langle N \rangle^{(m)} \left[(g^2)^{(m)} - 1 \right]. \tag{3.24}$$

4 Pöschl–Teller Potential

Consider the family of potentials

$$V_{l,l'}(x) = \begin{cases} \dfrac{1}{4a^2} \left[\dfrac{l(l-1)}{\sin^2 u(x)} - \dfrac{l'(l'-1)}{\sin^2 u(x)} \right] - \dfrac{(l+l')^2}{4a^2}, & u(x) = \dfrac{x}{2a}, \quad 0 < x < \pi a \\ \infty, \quad x \le 0, x \ge \pi a \end{cases} \tag{4.1}$$

of continuously indexed parameters l, l'. This class of potentials called Pöschl–Teller potentials of first type (PT-I), intensively studied in [4, 9, 15, 23], is closely related to other classes of potentials, widely used in molecular physics, namely: (1) the symmetric Pöschl–Teller potentials well ($l = l' \ge 1$), (2) the Scarf potentials $\frac{1}{2} < l' \le 1$ [35], (3) the modified Pöschl–Teller potentials which can be obtained by replacing the trigonometric functions by their hyperbolic counterparts [13, 33], (4) the Rosen-Morse potential which is the symmetric modified Pöschl–Teller potentials [34].

Let us define the corresponding Hamiltonian operator $H_{l,l'}$ with the action

$$H_{l,l'} \phi := \left(-\frac{d^2}{dx^2} + \frac{1}{4a^2} \left[\frac{l(l-1)}{\sin^2(x/2a)} - \frac{l'(l'-1)}{\sin^2(x/2a)} \right] - \frac{(l+l')^2}{4a^2} \right) \phi$$
$$\text{for} \quad \phi \in \mathcal{D}_{H_{l,l'}} \tag{4.2}$$

in the suitable Hilbert space $\mathcal{H} = L^2((0, \pi a), dx)$. $\mathcal{D}_{H_{l,l'}}$ is the domain of definition of $H_{l,l'}$. We consider here the case where $l, l' \ge 3/2$, then the operator $H_{l,l'}$ is in the limit point case at both ends $x = 0, \pi a$, therefore it is essentially self-adjoint. In this case (see [4, 9] for more details) the Pöschl–Teller Hamiltonian can be defined as the self-adjoint operator $H_{l,l'}$ in $L^2([0, \pi a], dx)$ acting as in (4.2), on the dense domain

$$\mathcal{D}_{H_{l,l'}} = \left\{ \phi \in AC^2(0, \pi a) | \left(\frac{1}{4a^2} \left[\frac{l(l-1)}{\sin^2(x/2a)} - \frac{l'(l'-1)}{\sin^2(x/2a)} \right] - \frac{(l+l')^2}{4a^2} \right) \right.$$

$$\phi \in L^2([0, \pi a], dx),$$

$$\left. \phi(0) = \phi(\pi a) = 0 \right\}. \tag{4.3}$$

with $AC^2(0, \pi a) = \{ \phi \in ac^2(0, \pi a) : \phi' \in \mathcal{H} \}$, where $ac^2(0, \pi a)$ denotes the set of absolutely continuous functions with absolutely continuous derivatives.

PT-I potentials are SUSY and fulfill the property of shape invariance [10]. Their superpotentials are:

$$W(x; l, l') = -\frac{1}{2a} \left[l \cot(u(x)) - l' \tan(u(x)) \right]. \tag{4.4}$$

One can define the first differential operators A, $A^{\dagger}$ that factorize the Hamiltonian operator in (4.2) as:

$$A := \frac{d}{dx} + W(x; l, l'), \quad A^{\dagger} := -\frac{d}{dx} + W(x; l, l') \tag{4.5}$$

with the domains:

$$\mathcal{D}(A) = \{ \psi \in ac[0, \pi a], \quad (\psi' + W(x; l, l')\psi) \in \mathcal{H} \} \tag{4.6}$$

$$\mathcal{D}(A^{\dagger}) = \left\{ \phi \in ac[0, \pi a] \mid \exists \tilde{\phi} \in \mathcal{H} : [\psi(x)\phi(x)]_0^a = 0, \right.$$

$$\left. \langle A\psi, \phi \rangle = \left\langle \psi, \tilde{\phi} \right\rangle, \forall \psi \in \mathcal{D}(A) \right\} \tag{4.7}$$

with $A^{\dagger}\phi = \tilde{\phi}$. The partner potentials $V_{1,2}$ satisfy the following shape invariance relation:

$$V_2(x, l, l') = V_1(x, l+1, l'+1) + \frac{1}{a^2}(l + l' + 1). \tag{4.8}$$

The potential parameters $a_1 \equiv (l, l')$ and $a_2 \equiv (l + 1, l' + 1)$ are related as

$$a_2 = a_1 + 2, \tag{4.9}$$

while the remainder in the shape invariant condition (2.14) is $\mathcal{R}(a_1) = \frac{1}{a^2}(l+l'+1)$. Then the products in terms of the quantity $\mathcal{R}(a_s)$ in the numerator and denominator of the coefficient $K_n^m(a_r)$, see Eq. (3.5), can be read, respectively, as:

$$\prod_{k=m+1}^{n+m} \left(\sum_{s=k}^{n+m} \mathcal{R}(a_s) \right) = \lambda^{2n} \frac{\Gamma(n+1)\Gamma(2n+2m+2\rho)}{\Gamma(n+2m+2\rho)} \tag{4.10}$$

$$\prod_{k=1}^{m} \left(\sum_{s=k}^{n+m} \mathcal{R}(a_s) \right) = \lambda^{2m} \frac{\Gamma(n+m+1)\Gamma(n+2m+2\rho)}{\Gamma(n+1)\Gamma(n+m+2\rho)} \tag{4.11}$$

where we set $\lambda = \frac{1}{a}$ and $2\rho = l + l'$, $\rho \geq 3/2$. The explicit form of the expansion coefficient $K_n^m(a_r)$ depends on the choice of the functional $\mathcal{Z}_j$.

4.1 First Choice of the Functional $\mathcal{Z}_j$

First we define the functional $\mathcal{Z}_j$ as $\mathcal{Z}_j = e^{-i\alpha \mathcal{R}(a_1)}$, then we obtain

$$\prod_{k=m}^{n+m-1} \mathcal{Z}_{j+k} = e^{i\alpha E_n}, \quad E_n = \lambda^2 n(n+2\rho). \tag{4.12}$$

Inserting this relation and the results (4.10) and (4.11) in (3.5), we obtain the expansion coefficient as:

$$K_n^m(a_r) = \lambda^{n-m} \sqrt{\frac{\Gamma(n+1)^2 \, \Gamma(2n+2m+2\rho) \, \Gamma(n+m+2\rho)}{\Gamma(n+m+1)\Gamma(n+2m+2\rho)^2}} \, e^{i\alpha E_n}. \tag{4.13}$$

(i) Normalization

The normalization factor, in terms of the generalized hypergeometric functions $_3F_4$, can readily be deduced from (3.6) as:

$$\mathcal{N}_m(|z|^2; a_r) = \left[\frac{\lambda^{2m} \Gamma(m+1)\Gamma(2m+2\rho)}{\Gamma(m+2\rho)} \, _3F_4 \left(\begin{matrix} m+1, 2m+2\rho, 2m+2\rho & ; \\ 1, m+\rho, m+2\rho, m+\rho+1/2 & ; \end{matrix} \frac{|az|^2}{4} \right) \right]^{-1/2}. \tag{4.14}$$

In terms of Meijer's G-function, we have:

$$\mathcal{N}_m(|z|^2; a_r) = \left[\frac{\lambda^{2m} \Gamma(m+\rho)\Gamma(m+\rho+\frac{1}{2})}{\Gamma(2m+2\rho)} \right.$$

$$G_{3,5}^{1,3}\left(-\frac{|az|^2}{4}\left|\begin{array}{l}-m, 1-2m-2\rho, 1-2m-2\rho\\0,0,1-m-\rho,1-m-2\rho,1/2-m-\rho\end{array}\right.\right)\right]^{-1/2}. \quad (4.15)$$

The explicit form of these PA-SIPCS are:

$$|z;a_r\rangle_m = \mathcal{N}_m(|z|^2;a_r)\lambda^m \sum_{n=0}^{\infty}\sqrt{\frac{\Gamma(n+m+1)\Gamma(n+2m+2\rho)^2}{\Gamma(n+m+2\rho)\Gamma(2n+2m+2\rho)}}\frac{(az)^n}{n!}|n+m\rangle \quad (4.16)$$

defined on the whole complex plane. For $m=0$, we recover the expansion coefficient and the normalization factor obtained in [2] for the generalized SIPCS:

$$K_n^0 = \lambda^n\sqrt{\frac{\Gamma(n+1)\Gamma(2\rho+2n)}{\Gamma(2\rho+n)}}e^{i\alpha E_n} = h_n(a_r),$$

$$\mathcal{N}_0(|z|^2;a_r) = \left[{}_1F_2\left(2\rho;\rho,\rho+1/2;\frac{|az|^2}{4}\right)\right]^{-1/2} = \mathcal{N}(|z|^2;a_r). \quad (4.17)$$

(ii) Non-orthogonality

The inner product of two different PA-SIPCS $|z;a_r\rangle_m$ and $\left|z';a_r\right\rangle_{m'}$ follows from Eq. (3.8):

$${}_{m'}\left\langle z';a_r\right|z;a_r\rangle_m = \chi(z',z,m,m',\rho)\,{}_3F_4\left(\begin{array}{l}m+1,2m+2\rho,m+m'+2\rho\\m-m'+1,m+2\rho,m+\rho,m+\rho+\frac{1}{2}\end{array};\frac{a^2z'^*z}{4}\right)$$

where $\chi(z',z,m,m',\rho) = \mathcal{N}_{m'}(|z'|^2;a_r)\,\mathcal{N}_m(|z|^2;a_r)\,z'^{*(m-m')}\lambda^{(m+m')}$
$\frac{\Gamma(m+1)\Gamma(m+m'+2\rho)}{\Gamma(m-m'+1)\Gamma(m+2\rho)}$.

(iii) Overcompleteness

The non-negative weight function $\omega_m(|z|^2;a_r)$ is related to the function g_m satisfying (3.13):

$$\int_0^{\infty}dx\,x^{n+m}\,g_m(x;a_r) = \xi(x,n,m,\rho)$$

$$\frac{\Gamma(n+1)^2\Gamma(n+m+2\rho)\Gamma(n+m+\rho)\Gamma(n+m+\rho+\frac{1}{2})}{\Gamma(n+m+1)\Gamma(n+2m+2\rho)^2} \quad (4.18)$$

where x stands for $|z|^2$, $\xi(x,n,m,\rho) = \lambda^{2(n-m)}\dfrac{2^{(2n+2m+2\rho)}}{2\sqrt{\pi}}$ and $\omega_m = \dfrac{x^m g_m(x;a_r)}{\pi \mathcal{N}_m^2(x;a_r)}$. After variable change $n+m \to s-1$ and using the Mellin inversion theorem in terms of Meijer's G-function (3.15), we deduce:

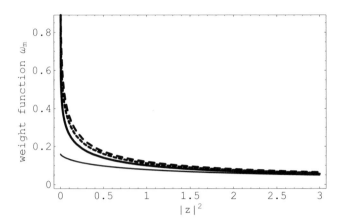

Fig. 1 Plots of the weight function (4.19) of the PA-SIPCS (4.16) versus $|z|^2$ with the potential parameters $\rho = 2, \lambda = 1$, for different values of the photon-added number m with $m = 0$ (thin solid line), $m = 1$ (solid line), $m = 2$ (dot line), and $m = 3$ (dashed line)

$$
\omega_m(|z|^2; a_r) = \frac{1}{2\pi \sqrt{\pi}} \frac{|z|^{2m}}{\mathcal{N}_m(|z|^2; a_r)^2} \lambda^{-2(1+2m)} 2^{2(1-\rho)}
$$

$$
G_{3,5}^{5,0} \left(\frac{|az|^2}{4} \middle| \begin{matrix} 0, -1+2\rho+m, -1+2\rho+m \\ -m, -m, 2\rho-1, -1+\rho, -1/2+\rho \end{matrix} \right). \quad (4.19)
$$

The weight function (4.19) is positive for the parameter $\rho > 0$ as shown in Figure 1, where the curves are represented for $\rho = 2$ and for $m = 0, 1, 2, 3$. All the functions are positive for $x = |z|^2 \in \mathbb{R}_+$ and tend asymptotically to the measure of the conventional CS ($m = 0$). The measure has a singularity at $x = 0$ and tends to zero for $x \to \infty$.

(iv) Thermal statistics

Consider the normalized density operator expression

$$
\rho^{(m)} = \frac{1}{Z} \sum_{n=0}^{\infty} e^{-\beta E_n} |n+m\rangle \langle n+m| \quad (4.20)
$$

in which the exponent βE_n is re-cast as follows: $\beta E_n = \beta\lambda^2 \left[n^2 + 2n\rho \right] = An^2 - B_\rho n$ where $A = \beta\lambda^2$, $B_\rho = -2\beta\rho\lambda^2$. Then, the energy exponential can be expanded in the power series (see, e.g., [32]) such that

$$
e^{-\beta E_n} = e^{-An} \left[\sum_{k=0}^{\infty} \frac{(B_\rho)^k}{k!} n^{2k} \right] = \left\{ \sum_{k=0}^{\infty} \frac{(B_\rho)^k}{k!} \left(\frac{d}{dA} \right)^{2k} \right\} \left(e^{-A} \right)^n
$$

$$
= \exp\left[B_\rho \left(\frac{d}{dA} \right)^2 \right] \left(e^{-A} \right)^n. \quad (4.21)
$$

Thereby,

$$\rho^{(m)} = \frac{\exp\left[B_\rho\left(\frac{d}{dA}\right)^2\right]}{Z} \sum_{n=0}^{\infty} \left(e^{-A}\right)^n |n+m\rangle\langle n+m|. \tag{4.22}$$

From (4.16) and (4.22), we get, in terms of Meijer's G functions, the Q-distribution or Husimi distribution:

$$\begin{aligned}
{}_m\langle z; m|\rho^{(m)}|z; m\rangle_m = {} & \frac{\Gamma(2m+2\rho)}{\Gamma(m+\rho)\Gamma(m+\rho+1/2)\,2^{2(m+\rho-1/2)}}\frac{\Gamma(\frac{1}{2})}{} \frac{\exp\left[B_\rho\left(\frac{d}{dA}\right)^2\right]}{Z} \times \\
& \times \frac{G^{1,3}_{3,5}\left(-\frac{(a|z|)^2}{4}e^{-A}\left|\begin{array}{l}-m,\,1-2m-2\rho,\,1-2m-2\rho;\\0;\,0,\,1-m-\rho,\,1-m-2\rho,\,1/2-m-\rho\end{array}\right.\right)}{G^{5,0}_{3,5}\left(-\frac{(a|z|)^2}{4}\left|\begin{array}{l}-m,\,1-2m-2\rho,\,1-2m-2\rho;\\0;\,0,\,1-m-\rho,\,1-m-2\rho,\,1/2-m-\rho\end{array}\right.\right)}.
\end{aligned} \tag{4.23}$$

The angular integration achieved, taking $x = |z|^2$, the condition (3.18) supplies

$$\begin{aligned}
\mathrm{Tr}\rho^{(m)} = {} & \frac{\exp\left[B_\rho\left(\frac{d}{dA}\right)^2\right]}{Z}\frac{2^{2[1-2\rho-m]}}{\lambda^{2(1+m)}}\int_0^{\infty} dx\,x^m \\
& G^{1,3}_{3,5}\left(-\frac{|a|^2}{4}xe^{-A}\left|\begin{array}{l}-m,\,1-2m-2\rho,\,1-2m-2\rho;\\0;\,0,\,1-m-\rho,\,1-m-2\rho,\,1/2-m-\rho\end{array}\right.\right) \times \\
& \times G^{5,0}_{3,5}\left(\frac{|a|^2}{4}x\left|\begin{array}{l};\,0,\,-1+2m+\rho,\,-1+2\rho+m\\-m,\,-m,\,2\rho-1,\,-1+\rho,\,-1/2+\rho;\end{array}\right.\right).
\end{aligned} \tag{4.24}$$

Then, the integral of Meijer's G-function product provides the partition function

$$Z = \frac{2^{4(1-\rho)}}{(\lambda|a|)^{2(1+m)}}\exp\left[B_\rho\left(\frac{d}{dA}\right)^2\right]\sum_{n=0}^{\infty}\left(e^{-A}\right)^n. \tag{4.25}$$

From (3.19), using the result $\langle n+m|\rho^{(m)}|n+m\rangle = \frac{1}{Z}\exp\left[B_\rho\left(\frac{d}{dA}\right)^2\right]\left(e^{-A}\right)^n$ and setting $\bar{n}_A = (e^A - 1)^{-1}$, we get the following integration equality:

$$\begin{aligned}
& \frac{1}{\bar{n}_A+1}\left(\frac{\bar{n}_A}{\bar{n}_A+1}\right)^n \frac{\Gamma(n+1)^2\Gamma(n+m+2\rho)\Gamma(2n+2m+2\rho)}{\Gamma(n+m+1)\Gamma(n+2m+2\rho)^2}\frac{\sqrt{\pi}\lambda^{4(1+m)}}{2^{5-6\rho}|a|^{2(n-m-1)}} \\
& = \int_0^{\infty} dx\,x^{n+m}\,P(x)\times \\
& G^{5,0}_{3,5}\left(\frac{|a|^2}{4}x\left|\begin{array}{l};\,0,\,-1+2m+\rho,\,-1+2\rho+m\\-m,\,-m,\,2\rho-1,\,-1+\rho,\,-1/2+\rho;\end{array}\right.\right).
\end{aligned}$$

After performing the exponent change $n + m = s - 1$ in order to get the Stieltjes moment problem, we arrive at the P-function as

$$P(|z|^2) = \frac{1}{\bar{n}_A} \left(\frac{\bar{n}_A + 1}{\bar{n}_A} \right)^m \frac{\lambda^2 |a|^{2(m+1)}}{2^{4(1-\rho)}}$$

$$\times \frac{G_{3,5}^{5,0} \left(\frac{\bar{n}_A+1}{\bar{n}_A} \frac{|az|^2}{4} \middle| \begin{array}{c} ; 0, -1+2\rho+m, -1+2\rho+m \\ -m, -m, 2\rho-1, -1+\rho, -1/2+\rho; \end{array} \right)}{G_{3,5}^{5,0} \left(\frac{|az|^2}{4} \middle| \begin{array}{c} ; 0, -1+2\rho+m, -1+2\rho+m \\ -m, -m, 2\rho-1, -1+\rho, -1/2+\rho; \end{array} \right)} \tag{4.26}$$

which obeys the normalization to unity condition (3.20).

Then, the diagonal representation of the normalized density operator in terms of the PA-SIPCS projector (3.19) takes the form

$$\rho^{(m)} = \frac{1}{\bar{n}_A} \left(\frac{\bar{n}_A + 1}{\bar{n}_A} \right)^m \frac{\lambda^2 |a|^{2(m+1)}}{2^{4(1-\rho)}} \int_{\mathbb{C}} d^2 z \, \omega_m(|z|^2; a_r) |z; a_r\rangle_m \mathfrak{S}_{3,5}^{5,0}(|z|^2; \bar{n}_A) \, _m\langle z; a_r| \tag{4.27}$$

with $\mathfrak{S}_{3,5}^{5,0}(|z|^2, \bar{n}_A)$—the Meijer's G-function quotient given in (4.26). Using the relations (4.26), (4.27), and the definition (3.21), the pseudo-thermal expectation values of the operator N and its square are given by

$$\langle N \rangle^{(m)} = \left(\frac{|a|^2}{4} \right)^{m+1} \frac{1}{\lambda^{2(m-1)}} \frac{m(m+2\rho)}{(m+1)(m+1+2\rho)}$$

$$\times \left[1 + \left(\frac{1}{m+1} + \frac{1}{m+1+2\rho} \right) \bar{n} + \frac{1}{(m+1)(m+1+2\rho)} \right.$$

$$\left. \times \left(\frac{\bar{n}}{1 - e^{-\beta}} + \bar{n}^2 \right) \right] \tag{4.28}$$

$$\langle N^2 \rangle^{(m)} = \left(\frac{|a|^2}{4} \right)^{m+1} \frac{1}{\lambda^{2(m-2)}} \left[\frac{m(m+2\rho)}{(m+1)(m+1+2\rho)} \right]^2 \times \left\{ 1 + 2 \left(\frac{1}{m+1} + \frac{1}{m+1+2\rho} \right) \bar{n} + \right.$$

$$+ \left(\frac{1}{(m+1)^2} + \frac{1}{(m+1+2\rho)^2} + \frac{4}{(m+1)(m+1+2\rho)} \right) \left(\frac{\bar{n}}{1 - e^{-\beta}} + \bar{n}^2 \right) +$$

$$+ 2 \left(\frac{1}{(m+1)^2(m+1+2\rho)} + \frac{1}{(m+1)(m+1+2\rho)^2} \right) \left[\frac{\bar{n}}{(1 - e^{-\beta})^2} + \frac{4\bar{n}^2}{1 - e^{-\beta}} + \bar{n}^3 \right] +$$

$$+ \frac{1}{(m+1)^2(m+1+2\rho)^2} \left[\frac{\bar{n}}{(1 - e^{-\beta})^3} + \frac{11\bar{n}^2}{(1 - e^{-\beta})^2} + \frac{11\bar{n}^3}{1 - e^{-\beta}} + \bar{n}^4 \right] \right\} \tag{4.29}$$

where $\bar{n} = (e^{-\beta} - 1)^{-1}$. Thereby,

$$(g^2)^{(m)} = 1 + \left\{ \left(\frac{1}{(m+1)} + \frac{1}{(m+1+2\rho)} \right)^2 \frac{\bar{n}}{(1-e^{-\beta})} + \right.$$
$$\left(\frac{1}{(m+1)^2(m+1+2\rho)} + \frac{1}{(m+1)(m+1+2\rho)^2} \right) \left[\frac{2\bar{n}}{(1-e^{-\beta})^2} + \frac{6\bar{n}^2}{1-e^{-\beta}} \right] +$$
$$\left. \frac{1}{(m+1)^2(m+1+2\rho)^2} \left[\frac{\bar{n}}{(1-e^{-\beta})^3} + \frac{10\bar{n}^2}{(1-e^{-\beta})^2} + \frac{9\bar{n}^3}{1-e^{-\beta}} \right] \right\}$$
$$\times \frac{1}{\left(\frac{|a|^2}{4} \right)^{-(m+1)} \lambda^{2m} \left(\langle N \rangle^{(m)} \right)^2} - \frac{1}{\left(\frac{|a|^2}{4} \right)^{-\frac{m+1}{2}} \lambda^m \langle N \rangle^{(m)}}. \tag{4.30}$$

Then, the thermal analogue of the Mandel parameter is given by

$$Q^{(m)} = \left(\frac{|a|^2}{4} \right)^{-\frac{m+1}{2}} \lambda^m \langle N \rangle^{(m)} \left[(g^2)^{(m)} - 1 \right]$$
$$= \left\{ \left(\frac{1}{(m+1)} + \frac{1}{(m+1+2\rho)} \right)^2 \frac{\bar{n}}{(1-e^{-\beta})} + \right.$$
$$\left(\frac{1}{(m+1)^2(m+1+2\rho)} + \frac{1}{(m+1)(m+1+2\rho)^2} \right) \left[\frac{2\bar{n}}{(1-e^{-\beta})^2} + \frac{6\bar{n}^2}{1-e^{-\beta}} \right] +$$
$$\left. \frac{1}{(m+1)^2(m+1+2\rho)^2} \left[\frac{\bar{n}}{(1-e^{-\beta})^3} + \frac{10\bar{n}^2}{(1-e^{-\beta})^2} + \frac{9\bar{n}^3}{1-e^{-\beta}} \right] \right\}$$
$$\times \frac{1}{\left(\frac{|a|^2}{4} \right)^{-\frac{m+1}{2}} \lambda^m \langle N \rangle^{(m)}} - 1. \tag{4.31}$$

4.2 Second Choice of the Functional $\mathcal{Z}_j$

We now take $\mathcal{Z}_j = \sqrt{g(a_1; \kappa, \kappa) g(a_1; \kappa, 0)} \, e^{-i\alpha \mathcal{R}(a_1)}$ with κ a real constant and where we use the auxiliary function [2] $g(a_j; c, d) = c a_j + d$, c and d being real constants. From the potential parameter relations (4.9) we obtain:

$$\prod_{k=m}^{n+m-1} g(a_{j+k}; c, d) = 2c^n \frac{\Gamma(n+m+\frac{a_1}{2}+j-1+d/2c)}{\Gamma(m+\frac{a_1}{2}+j-1+d/2c)}. \tag{4.32}$$

Setting $a_1 = 2\rho$, we have:

$$\prod_{k=m}^{n+m-1} \mathcal{Z}_{j+k} = \left[\kappa^{2n} \frac{\Gamma(2n+2m+2\rho)}{\Gamma(2m+2\rho)} \right]^{\frac{1}{2}} e^{-i\alpha E_n} \tag{4.33}$$

with the eigen-energy E_n given by (4.12). Inserting Eqs. (4.33), (4.10), and (4.11) in the expansion coefficient (3.5), we obtain

$$K_n^m(a_r) = \left[\frac{1}{\kappa^{2m}} \frac{\Gamma(n+1)^2 \, \Gamma(n+m+\nu+1)\Gamma(2m+\nu+1)}{\Gamma(n+m+1)\Gamma(n+2m+\nu+1)^2} \right]^{\frac{1}{2}} e^{i\alpha E_n}, \quad (4.34)$$

where we assume $\lambda = \frac{1}{a} = \kappa$ and $\rho = \frac{\nu}{2} + \frac{1}{2}, \nu \geq 1$ in (4.10) and (4.11). For $m = 0$, we recover the coefficient h_n in [2]:

$$K_n^0(a_r) = \left[\frac{\Gamma(n+1)\Gamma(\nu+1)}{\Gamma(n+\nu+1)} \right]^{\frac{1}{2}} e^{i\alpha E_n} = h_n(a_r) . \quad (4.35)$$

(i) Normalization

The normalization factor in terms of hypergeometric and Meijer's G-functions is

$$\mathcal{N}_m(|z|^2; a_r) = \kappa^{2m} \Gamma(m+1) \frac{\Gamma(2m+\nu+1)}{\Gamma(m+\nu+1)} \left[{}_3F_2 \left(\begin{matrix} m+1, 2m+\nu+1, 2m+\nu+1 \; ; \\ 1, m+\nu+1 \end{matrix} \; ; \; |z|^2 \right) \right]^{-\frac{1}{2}}$$

$$(4.36)$$

$$\mathcal{N}_m(|z|^2; a_r) = \left[\frac{\kappa^{2m}}{\Gamma(2m+\nu+1)} G_{3,3}^{1,3} \left(-|z|^2 \left| \begin{matrix} -m, -2m-\nu, -2m-\nu \\ 0, 0, -m-\nu \end{matrix} \right. \right) \right]^{-\frac{1}{2}} . \quad (4.37)$$

The explicit form of the PA-SIPCS, defined for $|z| < 1$, is provided by:

$$|z; a_r\rangle_m = \mathcal{N}_m(|z|^2; a_r) \sum_{n=0}^{\infty} \sqrt{\kappa^{2m} \frac{\Gamma(n+m+1)\Gamma(n+2m+\nu+1)^2}{\Gamma(n+1)^2\Gamma(n+m+\nu+1)\Gamma(2m+\nu+1)}} z^n e^{-i\alpha E_n} |n+m\rangle .$$

$$(4.38)$$

For $m = 0$, we recover the normalization factor

$$\mathcal{N}_0(|z|^2; a_r) = {}_1F_0(\nu+1; -; |z|^2)^{-\frac{1}{2}} = (1 - |z|^2)^{-1/2-\nu/2} = \mathcal{N}(|z|^2; a_r) \quad (4.39)$$

obtained in [2]. For $m = 0$, the PA-SIPCS is reduced to the SIPCS

$$|z; a_r\rangle = (1 - |z|^2)^{(\nu+1)/2} \sum_{n=0}^{\infty} \sqrt{\frac{\Gamma(n+\nu+1)}{\Gamma(\nu+1)\Gamma(n+1)}} e^{-i\alpha E_n} |\Psi_n\rangle \quad (4.40)$$

obtained in [2] and in [15] as CS of Klauder-Perelomov's type for the PT-I.

(ii) Non-orthogonality

The inner product of two different PA-SIPCS $|z; a_r\rangle_m$ and $|z'; a_r\rangle_{m'}$ is given by:

$$_{m'}\langle z'; a_r | z; a_r\rangle_m = \chi(z', z, m, m', \nu) \, _3F_2 \left(\begin{array}{c} m+1, m+m'+\nu+1, 2m+\nu+1 \, ; \\ m-m'+1, m+\nu+1 \end{array} \; ; z'^*z \right)$$

where

$$\chi(z', z, m, m', \nu) = \mathcal{N}_{m'}(|z'|^2; a_r) \, \mathcal{N}_m(|z|^2; a_r) \, \frac{z'^{*(m-m')} \kappa^{(m+m')}}{\sqrt{\Gamma(2m+\nu+1)\Gamma(2m'+\nu+1)}} \times$$

$$\times \frac{\Gamma(m+1)\Gamma(m+m'+\nu+1)\Gamma(2m+\nu+1)}{\Gamma(m-m'+1)\Gamma(m+\nu+1)} \, e^{i\alpha(E_n - E_{n+m-m'})}.$$

(iii) Overcompleteness

Following the steps of Sect. 3.3, we obtain the weight-function of the PA-SIPCS (4.38) as

$$\omega_m(|z|^2; a_r) = \frac{1}{\pi} G_{3,3}^{1,3} \left(-|z|^2 \left| \begin{array}{c} -m, -2m-\nu, -2m-\nu \\ 0, 0, -m-\nu \end{array} \right. \right) G_{3,3}^{3,0} \left(|z|^2 \left| \begin{array}{c} m, 2m+\nu, 2m+\nu \\ 0, 0, m+\nu \end{array} \right. \right).$$

(4.41)

We recover, for $m = 0$, the result:

$$\omega_0(|z|^2; a_r) = \frac{\Gamma(\nu+1)}{\pi} \, _1F_0(\nu+1; -; |z|^2) \, G_{1,1}^{1,0} \left(|z|^2 \left| \begin{array}{c} ; \nu \\ 0 ; \end{array} \right. \right) = \frac{\nu}{\pi}(1-|z|^2)^{-2}$$

(4.42)

obtained in [2] for the corresponding ordinary SIPCS.

(iv) Thermal statistics

Since the eigen-energy E_n (4.12) is the same as before, we start by maintaining the relations (4.20)–(4.22). From (4.38) and (4.22), we get, in terms of Meijer's G functions, the Q-distribution or Husimi distribution

$$_m\langle z; m | \rho^{(m)} | z; m\rangle_m = \frac{\exp\left[B_\rho \left(\frac{d}{dA} \right)^2 \right] G_{3,3}^{1,3} \left(-|z|^2 e^{-A} \left| \begin{array}{c} -m, -2m-\nu, -2m-\nu; \\ 0; 0, -m-\nu \end{array} \right. \right)}{Z \, G_{3,3}^{1,3} \left(-|z|^2 \left| \begin{array}{c} -m, -2m-\nu, -2m-\nu; \\ 0; 0, -m-\nu \end{array} \right. \right)}.$$

(4.43)

The angular integration achieved, taking $x = |z|^2$, the condition (3.18) supplies

$$
\mathrm{Tr}\rho^{(m)} = \frac{\exp\left[B_\rho\left(\frac{d}{dA}\right)^2\right]}{Z} \int_0^\infty dx\, x^m G_{3,3}^{3,0}\left(|z|^2 \left|\begin{array}{c} ;\, m, 2m+\nu, 2m+\nu \\ 0, 0, m+\nu; \end{array}\right.\right) \times
$$
$$
G_{3,3}^{1,3}\left(-|z|^2 e^{-A} \left|\begin{array}{c} -m, -2m-\nu, -2m-\nu; \\ 0;\, 0, -m-\nu \end{array}\right.\right).
\tag{4.44}
$$

Then, the properties of the integral of Meijer's G-function product provide the expression of the partition function $Z = \exp\left[B_\rho\left(\frac{d}{dA}\right)^2\right]\sum_{n=0}^{\infty}\left(e^{-A}\right)^n$. From (3.19), taking $\bar{n}_A = (e^A - 1)^{-1}$, we get the following integration equality

$$
\frac{1}{\bar{n}_A + 1}\left(\frac{\bar{n}_A}{\bar{n}_A + 1}\right)^n \frac{\Gamma(n+1)^2 \Gamma(n+m+\nu+1)}{\Gamma(n+m+1)\Gamma(n+2m+\nu+1)^2}
$$
$$
= \int_0^\infty dx\, x^n P(x) G_{3,3}^{3,0}\left(|z|^2 \left|\begin{array}{c} ;\, m, 2m+\nu, 2m+\nu \\ 0, 0, m+\nu; \end{array}\right.\right).
$$

Finally, we arrive at the P-function:

$$
P(|z|^2) = \frac{1}{\bar{n}_A}\frac{G_{3,3}^{3,0}\left(\frac{\bar{n}_A+1}{\bar{n}_A}|z|^2 \left|\begin{array}{c} ;\, m, 2m+\nu, 2m+\nu \\ 0, 0, m+\nu; \end{array}\right.\right)}{G_{3,3}^{3,0}\left(|z|^2 \left|\begin{array}{c} ;\, m, 2m+\nu, 2m+\nu \\ 0, 0, m+\nu; \end{array}\right.\right)}
\tag{4.45}
$$

which obeys the normalization to unity condition (3.20).

Then, the diagonal representation of the normalized density operator in terms of the PA-SIPCS projector (3.19) takes the form

$$
\rho^{(m)} = \frac{1}{\bar{n}_A}\int_{\mathbb{C}} d^2z\, \omega_m(|z|^2; a_r)|z; a_r\rangle_m \mathfrak{S}_{3,3}^{3,0}(|z|^2; \bar{n}_A)\, {}_m\langle z; a_r|
\tag{4.46}
$$

with $\mathfrak{S}_{3,3}^{3,0}(|z|^2, \bar{n}_A)$—the Meijer's G-function quotient given in (4.45). Using the relations (4.45), (4.46), and the definition (3.21), the pseudo-thermal expectation values of the operator N and its square are given by

$$
\langle N\rangle^{(m)} = \kappa^2 m(m+\nu+1)\left[1 + \left(\frac{1}{m+1} + \frac{1}{m+\nu+2}\right)\bar{n}\right.
$$
$$
\left. + \frac{1}{(m+1)(m+\nu+2)}\left(\frac{\bar{n}}{1-e^{-\beta}} + \bar{n}^2\right)\right]
\tag{4.47}
$$

$$\langle N^2 \rangle^{(m)} = \kappa^4 m^2 (m + \nu + 1)^2 \left\{ 1 + 2 \left(\frac{1}{m+1} + \frac{1}{m+\nu+2} \right) \bar{n} + \right.$$
$$\left(\frac{1}{(m+1)^2} + \frac{1}{(m+\nu+2)^2} + \frac{4}{(m+1)(m+\nu+2)} \right) \left(\frac{\bar{n}}{1-e^{-\beta}} + \bar{n}^2 \right) +$$
$$2 \left(\frac{1}{(m+1)^2(m+\nu+2)} + \frac{1}{(m+1)(m+\nu+2)^2} \right) \left(\frac{\bar{n}}{(1-e^{-\beta})^2} + \frac{4\bar{n}^2}{1-e^{-\beta}} + \bar{n}^3 \right) +$$
$$\left. \frac{1}{(m+1)^2(m+\nu+2)^2} \left[\frac{\bar{n}}{(1-e^{-\beta})^3} + \frac{11\bar{n}^2}{(1-e^{-\beta})^2} + \frac{11\bar{n}^3}{1-e^{-\beta}} + \bar{n}^4 \right] \right\}.$$

$$(4.48)$$

Thereby,

$$(g^2)^{(m)} = 1 + \left\{ \left(\frac{1}{(m+1)} + \frac{1}{(m+\nu+2)} \right)^2 \frac{\bar{n}}{(1-e^{-\beta})} + \left(\frac{1}{(m+1)^2(m+\nu+2)} \right. \right.$$
$$+ \frac{1}{(m+1)(m+\nu+2)^2} \right) \left[\frac{2\bar{n}}{(1-e^{-\beta})^2} + \frac{6\bar{n}^2}{1-e^{-\beta}} \right] + \frac{1}{(m+1)^2(m+\nu+2)^2}$$
$$\left. \left[\frac{\bar{n}}{(1-e^{-\beta})^3} + \frac{10\bar{n}^2}{(1-e^{-\beta})^2} + \frac{9\bar{n}^3}{1-e^{-\beta}} \right] \right\} \frac{1}{\left(\langle N \rangle^{(m)} \right)^2} - \frac{1}{\langle N \rangle^{(m)}}.$$

$$(4.49)$$

Then, the thermal analogue of the Mandel parameter is given by

$$Q^{(m)} = \langle N \rangle^{(m)} \left[(g^2)^{(m)} - 1 \right] = \left\{ \left(\frac{1}{(m+1)} + \frac{1}{(m+\nu+2)} \right)^2 \frac{\bar{n}}{(1-e^{-\beta})} + \right.$$
$$\left(\frac{1}{(m+1)^2(m+\nu+2)} + \frac{1}{(m+1)(m+\nu+2)^2} \right) \left(\frac{2\bar{n}}{(1-e^{-\beta})^2} + \frac{6\bar{n}^2}{1-e^{-\beta}} \right) +$$
$$\left. \frac{1}{(m+1)^2(m+\nu+2)^2} \left[\frac{\bar{n}}{(1-e^{-\beta})^3} + \frac{10\bar{n}^2}{(1-e^{-\beta})^2} + \frac{9\bar{n}^3}{1-e^{-\beta}} \right] \right\} \frac{1}{\langle N \rangle^{(m)}} - 1.$$

$$(4.50)$$

5   Concluding Remarks

In this paper, we have shown the use of the shape invariant potential method to construct generalized CS for photon-added particle systems under Pöschl–Teller potentials. These states have fully been characterized and discussed from both mathematics and physics points of view. This algebro-operator method can be exploited to investigate a large class of solvable potentials.

Acknowledgements This work is supported by TWAS Research Grant RGA No. 17-542 RG/MATHS/AF/AC_G -FR3240300147. The ICMPA-UNESCO Chair is in partnership with the Association pour la Promotion Scientifique de l'Afrique (APSA), France, and Daniel Iagolnitzer Foundation (DIF), France, supporting the development of mathematical physics in Africa.

References

1. G.S. Agarwal, K. Tara, Nonclassical properties of states generated by the excitations on a coherent state. Phys. Rev A. **43**, 492 (1991); G.S. Agarwal, K. Tara, Nonclassical character of states exhibiting no squeezing or sub-Poissonian statistics. Phys. Rev. A **46**, 485 (1992)
2. A.N.F. Aleixo, A.B. Balantekin, An algebraic construction of generalized coherent states for shape-invariant potentials. J. Phys. A Math. Gen. **37**, 8513 (2004)
3. S.T. Ali, J.-P. Antoine, J.-P. Gazeau, *Coherent States, Wavelets, and Their Generalizations.* Theoretical and Mathematical Physics, 2nd edn. (Springer, New York, 2014)
4. J.-P. Antoine, J.-P. Gazeau, P.M. Monceau, J.R. Klauder, K.A. Penson, Temporally stable coherent states for infinite well and Pöschl-Teller potentials. J. Math. Phys. **42**, 2349 (2001)
5. I. Aremua, J.-P. Gazeau, M.N. Hounkonnou, Action-angle coherent states for quantum systems with cylindric phase space, J. Phys. A Math. Gen. **45**, 335302 (2012)
6. I. Aremua, M.N. Hounkonnou, E. Baloïtcha, Coherent states for Landau Levels: algebraic and thermodynamical properties. Rep. Math. Phys. **45**(2), 247 (2015)
7. A.B. Balantekin, Algebraic approach to shape invariance. Phys. Rev. A **57**(6), 4188 (1998); A.B. Balantekin, M.A. Cândido Ribeiro, A.N.F. Aleixo, Algebraic nature of shape-invariant and self-similar potentials. J. Phys. A Math. Gen. **32**, 2785 (1999); E.D. Filho, M.A. Cândido Ribeiro, Generalized ladder operators for shape-invariant potentials. Phys. Scr. **64**(6), 548 (2001)
8. M. Ban, Photon statistics of conditional output states of lossless beam splitter. J. Mod. Opt. **43**, 1281 (1996)
9. H. Bergeron, P. Siegl, A. Youssef, New SUSYQM coherent states for Pöschl-Teller potentials: a detailed mathematical analysis. J. Phys. A Math. Theor. **45**, 244028 (2012)
10. F. Cooper, A. Khare, U. Sukhatme, Supersymmetry and quantum mechanics. Phys. Rep. **251**, 267 (1995)
11. J.W. Dabrowska, A. Khare, U.P. Sukhatme, Explicit wavefunctions for shape-invariant potentials by operator techniques. J. Phys. A Math. Gen. **21**, L195 (1988)
12. M. Dakna, T. Anhut, T. Opatrny, L. Knöll, D.-G.Welsch, Generating Schrödinger-cat-like states by means of conditional measurements on a beam splitter. Phys. Rev. A. **55**, 3184 (1997); M. Dakna, L. Knöll, D.-G. Welsch, Photon-added state preparation via conditional measurement on abeam splitter. Opt. Commun. **145**, 309 (1998)
13. C. Daskaloyannis, Generalized deformed oscillator corresponding to the modified Pöschl-Teller energy-spectrum. J. Phys. A Math. Gen. **25**, 2261 (1992)
14. R. Dutt, A. Khare, U.P. Sukhatme, Exactness of supersymmetric WKB spectrum for shape invariant potentials. Phys. Lett. B **181**, 295 (1986)
15. A.H. El Kinani, M. Daoud, Coherent states à la Klauder-Perelomov for the Pösch-Teller potentials. Phys. Lett. A **283**, 291 (2001); M. Daoud, Photon-added coherent states for exactly solvable Hamiltonians. Phys. Lett. A **305**, 135 (2002)
16. T. Fukui, N. Aizawa, Shape-invariant potentials and associated coherent states. Phys. Lett. A **180**, 308 (1993)
17. L.E. Gendenshtein, Derivation of exact spectra of the Schrödinger equation by means of supersymmetry. JETP Lett. **38**, 356 (1983)
18. R.J. Glauber, The quantum theory of optical coherence. Phys. Rev. **130**, 2529 (1963); R.J. Glauber, Coherent and incoherent states of the radiation field. Phys. Rev. **131**, 2766 (1963)
19. M.N. Hounkonnou, K. Sodoga, Generalized coherent states for associated hypergeometric-type functions, J. Phys. A Math. Gen. **38**, 7851 (2005)
20. M.N. Hounkonnou, K. Sodoga, E. Azatassou, Factorization of Sturm-Liouville operators: solvable potentials and underlying algebraic structure, J. Phys. A Math. Gen. **38**, 371 (2005)
21. M.N. Hounkonnou, E.B. Ngompe Nkouankam, On $(p, q, \mu, \nu, \phi_1, \phi_2)$-generalized oscillator algebra and related bibasic hypergeometric functions. J. Phys. A Math. Theor. **40**, 8835 (2007); M.N. Hounkonnou, E.B. Ngompe Nkouankam, New $(p, q; \mu, \nu, f)$-deformed states. J. Phys. A Math. Theor. **40**, 12113 (2007); M.N. Hounkonnou, E.B. Ngompe Nkouankam, (q, ν)-deformation of generalized basic hypergeometric states. J. Phys. A Math. Theor. **42**, 065202 (2008)

22. M.N. Hounkonnou, E.B. Ngompe Nkouankam, Generalized hypergeometric photon-added and photon-depleted coherent states. J. Phys. A Math. Theor. **42**, 025206 (2009)
23. M.N. Hounkonnou, S. Arjika, E. Baloïtcha, Pöschl-Teller Hamiltonian: Gazeau-Klauder type coherent states, related statistics, and geometry. J. Math. Phys. **55**, 123502 (2014)
24. L. Infeld, T.E. Hull, The factorization method. Rev. Mod. Phys. **23**, 28 (1951)
25. A. Khare, U.P. Sukhatme, New shape invariant potentials in supersymmetry quantum mechanics. J. Phys. A Math. Gen. **26**, L901 (1991)
26. J.R. Klauder, K.A. Penson, J.-M. Sixdeniers, Constructing coherent states through solutions of Stieltjes and Hausdorff moment problems. Phys. Rev. A **64**, 013817 (2001)
27. L. Mandel, E. Wolf, *Optical Coherence and Quantum Optics* (Cambridge University Press, Cambridge, 1995)
28. A.M. Mathai, R.K. Saxena, *Generalized Hypergeometric Functions with Applications in Statistics and Physical Sciences*. Lecture Notes in Mathematics, vol. 348 (Springer, Berlin, 1973)
29. K.A. Penson, A.I. Solomon, New generalized coherent states. J. Math. Phys. **40**, 2354 (1999)
30. A.M. Perelomov, *Generalized Coherent States and Their Applications* (Springer, Berlin, 1986)
31. D. Popov, Photon-added Barur-Girardello coherent states of the pseudoharmonic oscillator. J. Phys. A Math. Gen. **35**, 7205 (2002)
32. D. Popov, I. Zaharie, S.H. Dong, Photon-added coherent states for the Morse oscillator. Czech. J. Phys. **56**, 157 (2006); D. Popov, Some properties of generalized hypergeometric thermal coherent states. Electron. J. Theor. Phys. **3**(11), 123 (2006)
33. G. Pöschl, E. Teller, Bemerkungen zur Quantenmechanik des Anharmonischen Oszillators. Z. Phys. **83**, 143 (1933)
34. N. Rosen, P.M. Morse, On the vibrations of polyatomic molecules. Phys. Rev. **42**, 210 (1932)
35. F.L. Scarf, New soluble energy band problem. Phys. Rev. **112**, 1137 (1958)
36. E. Schrödinger, The continuous transition from micro- to macro-mechanics. Naturwiss **14**, 664 (1926)
37. J.-M. Sixdeniers, K.A. Penson, On the completeness of photon-added coherent states. J. Phys. A. Math. Gen **34**, 2859 (2001)
38. K. Sodoga, M.N. Hounkonnou, I. Aremua, Photon-added coherent states for shape invariant systems. Eur. Phys. J. D **72**, 105 (2018)
39. G. Teschl, *Mathematical Methods in Quantum Mechanics: With Application to Schrödinger Operators*. Graduate Studies in Mathematics, vol. 157, 2nd edn. (American Mathematical Society, Providence, RI, 1999)

On the Fourier Analysis for L^2 Operator-Valued Functions

Mawoussi Todjro and Yaogan Mensah

Abstract We endow the set of square integrable operator-valued functions on a locally compact group with a pre-Hilbert module structure and define the ρ-Fourier transform for such functions. We also describe the Fourier transform of Hilbert-Schmidt operator-valued function on compact groups.

Keywords Fourier transform · Hilbert module · Hilbert-Schmidt operator · Topological groups

1 Introduction

Square integrable functions (L^2-functions) play an important rôle in mathematics, Physics, and Engineering. In Quantum Mechanics the L^2-setting allows one to apply the Hilbert space methods. In the context of the analysis on locally compact groups, square integrable functions allows one to define, for instance, the regular representations of a group. Also with L^2-functions one can obtain the Fourier transform as an isometry [2, 5].

In order to handle efficiently problems involving vector valued functions/ measures it may be interesting to deepen the L^2-analysis for such functions/ measures. This paper brings its contribution in that direction.

Our work here put together the following concepts: topological group, bounded operator, square integrability, and pre-Hilbert module. More precisely we are interested in $L^2(G, \mathfrak{B}(\mathcal{H}))$ the space of square integrable bounded operator-valued functions on locally compact groups.

In Sect. 2, we recall some basic definitions and facts which we may need and also we fix some notations. In the Sect. 3 we construct a pre-Hilbert module structure on $L^2(G, \mathfrak{B}(\mathcal{H}))$. Section 4 is devoted to the Fourier transform of Hilbert-Schmidt

M. Todjro · Y. Mensah (✉)
Department of Mathematics, University of Lomé, Lomé, Togo

© Springer Nature Switzerland AG 2018   423
T. Diagana, B. Toni (eds.), *Mathematical Structures and Applications*,
STEAM-H: Science, Technology, Engineering, Agriculture,
Mathematics & Health, https://doi.org/10.1007/978-3-319-97175-9_18

operator-valued functions on a compact group. In the last section we construct a transform of Fourier type related to a representation of the group in a Hilbert module over a $\mathfrak{B}(\mathcal{H})$.

2 Preliminaries

We recall in this section some basic definitions and facts that we may need.

A *topological group* is a group together with a topology such that the map $G \times G \rightarrow G, (x, y) \mapsto xy^{-1}$ is continuous. A locally compact group is a topological group whose neutral element e (hence all elements) has a compact neighborhood. It turns out that each locally compact group G has a unique (up to a positive constant) left invariant measure called the Haar measure of G. Then one can perform integration against this measure of functions defined on G.

A *representation* π of the group G in a (complex) Hilbert space $\mathcal{H}_\pi$ is an homomorphism of G into $Aut(\mathcal{H}_\pi)$, the space of invertible bounded operators on $\mathcal{H}_\pi$. The dimension of π is by definition the dimension of $\mathcal{H}_\pi$. If for each $x \in G$, $\pi(x)$ is unitary, then the representation π is said to be *unitary*. Two unitary representations $(\pi_1, \mathcal{H}_{\pi_1})$ and $(\pi_2, \mathcal{H}_{\pi_2})$ of G are said to be *equivalent* if there exists a unitary isomorphism . $T : \mathcal{H}_1 \rightarrow \mathcal{H}_2$ such that

$$T \circ \pi_1(x) = \pi_2(x) \circ T, \ \forall x \in G. \tag{2.1}$$

A closed subspace $\mathcal{K}$ of $\mathcal{H}_\pi$ is said to be π-*invariant* if $\pi(x)v \subset \mathcal{K}$ for all $x \in G$.

If the only invariant subspaces of $\mathcal{H}_\pi$ are $\{0\}$ and $\mathcal{H}_\pi$, then π is said to be *irreducible*.

Through this paper we denote by $\widehat{G}$ the set of all the equivalence classes of continuous irreducible unitary representations of G.

For a Hilbert space $\mathcal{H}$ we denote by $\mathfrak{B}(\mathcal{H})$ the space of bounded operators on $\mathcal{H}$. The set $\mathfrak{B}(\mathcal{H})$ is endowed with the operator norm

$$\|T\| := \sup\{\|T\psi\| : \psi \in \mathcal{H}, \|\psi\| = 1\}. \tag{2.2}$$

We will also consider the set $\mathfrak{B}_2(\mathcal{H})$ of Hilbert-Schmidt operators on $\mathcal{H}$. These are bounded operators T such that $\sum_i \|Te_i\|^2 < \infty$ where $(e_i)_i$ is an arbitrary basis of $\mathcal{H}$. One may show that

$$\|T\|_2 = (\sum_i \|Te_i\|^2)^{\frac{1}{2}} = Tr(A^*A) \tag{2.3}$$

is a complete norm on $\mathfrak{B}_2(\mathcal{H})$ which is derived from the inner product

$$\langle S, T \rangle = Tr(S^*T). \tag{2.4}$$

Let $f : G \rightarrow \mathfrak{B}(\mathcal{H})$ be a strong measurable function. One says that f is square integrable (Bochner integral) with respect to the left Haar measure dx of G if

$\int_G \|f(x)\|^2 dx < \infty$. We denote by $L^2(G, \mathfrak{B}(\mathcal{H}))$ the set of all the (equivalence classes of) square integrable $\mathfrak{B}(\mathcal{H})$-valued functions on G. One can easily verify that $L^2(G, \mathfrak{B}(\mathcal{H}))$ is a vector space over $\mathbb{C}$ and it is endowed with the norm

$$\|f\| = \left(\int_G \|f(x)\|^2 dx \right)^{\frac{1}{2}}. \tag{2.5}$$

For more details on the integration of vector valued functions with respect to a scalar measure, we refer to [3] or [4].

In the next section we construct a pre-Hilbert module structure on $L^2(G, \mathfrak{B}(\mathcal{H}))$.

A pre-Hilbert module over a C^*-algebra $\mathcal{A}$ is a complex vector space E which is also a right $\mathcal{A}$-module such that there is a map

$$E \times E \to \mathcal{A}, \ (X, Y) \mapsto \langle X, Y \rangle$$

with the following properties. For $X, Y, Z \in E, \lambda \in \mathbb{C}, a \in \mathcal{A}$,

1. $\langle X, \lambda Y + Z \rangle = \lambda \langle X, Y \rangle + \langle X, Z \rangle$
2. $\langle X, Ya \rangle = \langle X, Y \rangle a$
3. $\langle Y, X \rangle = \langle X, Y \rangle^*$
4. $\langle X, X \rangle$ is a positive element in $\mathcal{A}$
5. $\langle X, X \rangle = 0 \Rightarrow X = 0$.

The equality

$$\|X\| = \|\langle X, X \rangle\|^{\frac{1}{2}} \tag{2.6}$$

defines a norm on E. One has the Cauchy-Schwarz inequality

$$\|\langle X, Y \rangle\| \leqslant \|X\| \|Y\|. \tag{2.7}$$

If E is complete with respect to the norm (2.6), then E is called a Hilbert $\mathcal{A}$-module or a Hilbert C^*-module over $\mathcal{A}$. See [6, 7] for more details.

3 A Pre-Hilbert Module Structure on $L^2(G, \mathfrak{B}(\mathcal{H}))$

For $f \in L^2(G, \mathfrak{B}(\mathcal{H}))$ and $T \in \mathfrak{B}(\mathcal{H})$, define

$$(fT)(x) = f(x)T, \ x \in G, \tag{3.1}$$

where $f(x)T$ is the composition of the operator T by the operator $f(x)$.

Proposition 3.1 *The map* $\theta : L^2(G, \mathfrak{B}(\mathcal{H})) \times \mathfrak{B}(\mathcal{H}) \to L^2(G, \mathfrak{B}(\mathcal{H})), \ (f, T) \mapsto \theta(f, T) = fT$ *is a right action of* $\mathfrak{B}(\mathcal{H})$ *on* $L^2(G, \mathfrak{B}(\mathcal{H}))$.

Proof Let us firstly show that θ is well-defined. Assume $f \in L^2(G, \mathfrak{B}(\mathcal{H}))$ and $T \in \mathfrak{B}(\mathcal{H})$. Then

$$
\begin{aligned}
\int_G \|(fT)(x)\|^2 dx &= \int_G \|f(x)T\|^2 dx \\
&\leqslant \int_G \|f(x)\|^2 \|T\|^2 dx \\
&= \|T\|^2 \int_G \|f(x)\|^2 dx < +\infty
\end{aligned}
$$

So $\theta(f, T) \in L^2(G, \mathfrak{B}(\mathcal{H}))$ and therefore θ is well defined.

On the other hand, for $S, T \in \mathfrak{B}(\mathcal{H})$ and $x \in G$ we have

$$
[(fS)T](x) = (f(x)S)T = f(x)(ST) = [f(ST)](x).
$$

Hence $(fS)T = f(ST)$.   □

Let us consider the map $\langle \cdot, \cdot \rangle : L^2(G, \mathfrak{B}(\mathcal{H})) \times L^2(G, \mathfrak{B}(\mathcal{H})) \rightarrow \mathfrak{B}(\mathcal{H})$ defined by

$$
(f, g) \mapsto \langle f, g \rangle = \int_G f(x)^* g(x) dx \tag{3.2}
$$

where $f(x)^*$ denotes the adjoint of the operator $f(x)$.

Proposition 3.2 *The space* $\left(L^2(G, \mathfrak{B}(\mathcal{H})), \langle \cdot, \cdot \rangle\right)$ *is a pre-Hilbert module over* $\mathfrak{B}(\mathcal{H})$.

Proof Let $f, g, h \in L^2(G, \mathfrak{B}(\mathcal{H}))$, $\lambda \in \mathbb{C}$ and $T \in \mathfrak{B}(\mathcal{H})$.

1.

$$
\begin{aligned}
\langle f, \lambda g + h \rangle &= \int_G f(x)^* (\lambda g(x) + h(x)) dx \\
&= \int_G f(x)^* \lambda g(x) dx + \int_G f(x)^* h(x) dx \\
&= \langle f, \lambda g \rangle + \langle f, h \rangle \\
&= \lambda \langle f, g \rangle + \langle f, h \rangle.
\end{aligned}
$$

2.

$$
\begin{aligned}
\langle f, gT \rangle &= \int_G f(x)^* (gT)(x) dx \\
&= \int_G f(x)^* (g(x)T) dx
\end{aligned}
$$

$$= \left(\int_G f(x)^* g(x) dx \right) T$$

$$= \langle f, g \rangle T.$$

3.

$$\langle f, g \rangle = \int_G f(x)^* g(x) dx$$

$$= \int_G \left(g(x)^* f(x) \right)^* dx$$

$$= \left(\int_G g(x)^* f(x) dx \right)^*$$

$$= \langle g, f \rangle^*$$

4. $\langle f, f \rangle = \int_G f(x)^* f(x) dx \geqslant 0$ since for all $x \in G$, $f(x)^* f(x)$ is a positive operator on $\mathfrak{B}(\mathcal{H})$.

5. $\int_G f(x)^* f(x) dx = 0$ implies $f(x)^* f(x) = 0$ a.e. Therefore $f = 0$ a.e.

It results that $L^2(G, \mathfrak{B}(\mathcal{H}))$ is a pre-Hilbert $\mathfrak{B}(\mathcal{H})$-module. $\qquad\square$

4 The Fourier Transform of Functions in $L^2(G, \mathfrak{B}_2(\mathcal{H}))$

In this section we assume that the group G is compact. We denote by $\widehat{G}$ its unitary dual, that is the set of equivalence classes of all the unitary representations of G. We consider L^2-functions on G with values in the class of Hilbert-Schmidt operators. The space $L^2(G, \mathfrak{B}_2(\mathcal{H}))$ is a Hilbert space under the inner product

$$\langle f, g \rangle = \int_G Tr(f(x)^* g(x)) dx. \tag{4.1}$$

The Fourier transform for $f \in L^2(G, \mathfrak{B}_2(\mathcal{H}))$ is the collection $(\widehat{f}(\pi))_{\pi \in \widehat{G}}$ of sesquilinear maps $\widehat{f}(\pi) : \mathcal{H}_\pi \times \mathcal{H}_\pi \to \mathfrak{B}_2(\mathcal{H})$ defined by

$$\widehat{f}(\pi)(\xi, \eta) = \int_G \langle \pi(x)^* \xi, \eta \rangle f(x) dx, \quad \xi, \eta \in \mathcal{H}_\pi. \tag{4.2}$$

Since G is compact, each representation of G is of finite dimension. For $\pi \in \widehat{G}$ we denote by d_π the dimension of $\mathcal{H}_\pi$ and we fix a basis $(\xi_1^\pi, \cdots, \xi_{d_\pi}^\pi)$ of $\mathcal{H}_\pi$. We set

$$u_{ij}^\pi(x) = \langle \pi(x) \xi_j^\pi, \xi_i^\pi \rangle \tag{4.3}$$

Then applying [1, Lemma 4.7] we have

$$f \in L^2(G, \mathfrak{B}_2(\mathcal{H})) \Rightarrow f(x)$$

$$= \sum_{\pi \in \widehat{G}} d_\pi \sum_{i=1}^{d_\pi} \sum_{j=1}^{d_\pi} u_{ij}^\pi(x) \widehat{f}(\pi)(\xi_j^\pi, \xi_i^\pi), \ \forall x \in G. \qquad (4.4)$$

We define $\mathcal{S}_2(\widehat{G}, \mathfrak{B}_2(\mathcal{H}))$ as the set of all the $(\phi(\pi))_{\pi \in \widehat{G}} \in \prod_{\pi \in \widehat{G}} \mathcal{S}(\mathcal{H}_\pi \times \mathcal{H}_\pi, \mathfrak{B}_2(\mathcal{H}))$ such that

$$\sum_{\pi \in \widehat{G}} d_\pi \sum_{i=1}^{d_\pi} \sum_{i=1}^{d_\pi} \|\phi(\pi)(\xi_i^\pi, \xi_j^\pi)\|_2^2 < \infty \qquad (4.5)$$

Here $\mathcal{S}(\mathcal{H}_\pi \times \mathcal{H}_\pi, \mathfrak{B}_2(\mathcal{H}))$ is the set of sesquilinear maps from $\mathcal{H}_\pi \times \mathcal{H}_\pi$ into $\mathfrak{B}_2(\mathcal{H})$. From [1, Theorem 4.8], one obtains that the Fourier transform $f \to \widehat{f}$ is an isometry from $L^2(G, \mathfrak{B}_2(\mathcal{H}))$ onto $\mathcal{S}_2(\widehat{G}, \mathfrak{B}_2(\mathcal{H}))$.

5   The ρ-Fourier Transform for $L^2(G, \mathfrak{B}(\mathcal{H}))$ Functions

Let E be a Hilbert module over $\mathfrak{B}(\mathcal{H})$. We denote by $\mathcal{L}(E)$ the set of adjointable operators on E.

Let $\rho : G \to \mathcal{L}(E)$ be a strongly continuous representation of G on E with $\rho \in L^2(G, \mathcal{L}(E))$.

Proposition 5.1 *For all $X, Y \in E$, we have $\langle \rho(\cdot)X, Y \rangle \in L^2(G, \mathfrak{B}(\mathcal{H}))$ and $\langle \rho(\cdot)^* X, Y \rangle \in L^2(G, \mathfrak{B}(\mathcal{H}))$.*

Proof Let $X, Y \in E$; we have:

$$\int_G \|\langle \rho(x)X, Y \rangle\|^2 dx \leqslant \int_G \|\rho(x)X\|^2 \|Y\|^2 dx$$

(the Cauchy-Schwarz inequality (2.7))

$$= \|Y\|^2 \int_G \|\rho(x)X\|^2 dx$$

$$\leqslant \|Y\|^2 \int_G \|\rho(x)\| \|X\|^2 dx$$

$$= \|X\|^2 \|Y\|^2 \int_G \|\rho(x)\|^2 dx < +\infty.$$

The second affirmation is proved similarly.   □

We obviously denote by $L^1(G, \mathfrak{B}(\mathcal{H}))$ the set of integrable functions $f : G \to \mathfrak{B}(\mathcal{H})$.

If $f \in L^2(G, \mathfrak{B}(\mathcal{H}))$, then the map $\langle \rho(\cdot)^* X, Y \rangle f(\cdot)$ is integrable. One can easily prove the following proposition.

Proposition 5.2 Let $f \in L^2(G, \mathfrak{B}(\mathcal{H}))$. Then the map

$$(X, Y) \mapsto \int_G \langle \rho(x)^* X, Y \rangle f(x) dx$$

is well-defined and is sesquilinear from $E \times E$ into $\mathfrak{B}(\mathcal{H})$.

From Proposition 5.2 and following [8] we set the following definition of the map $\mathcal{F}^\rho$ the well-definedness of which is ensured by the Proposition 5.2.

Definition 5.3 The ρ-Fourier transform of $f \in L^2(G, \mathfrak{B}(\mathcal{H}))$ is the $\mathfrak{B}(\mathcal{H})$-valued sesquilinear map $\mathcal{F}^\rho f : E \times E \to \mathfrak{B}(\mathcal{H})$ defined by

$$\mathcal{F}^\rho f(X, Y) = \int_G \langle \rho(x)^* X, Y \rangle f(x) dx, \quad (X, Y) \in E \times E. \tag{5.1}$$

In the sequel we assume that E is of finite dimension n. Let $(X_1, \cdots, X_n)$ be a basis of E. Assume also the following orthogonal relations:

$$\int_G \langle \rho(x)^* X_k, X_l \rangle \langle \rho(x) X_j, X_i \rangle dx = \delta_{kj} \delta_{li} \mathbb{I}_{\mathcal{H}}, \tag{5.2}$$

where $\mathbb{I}_{\mathcal{H}}$ designates the identity operator on $\mathcal{H}$.

Definition 5.4 The $\mathfrak{B}(\mathcal{H})$-valued functions $U_{ij}^\rho : x \mapsto \langle \rho(x) X_j, X_i \rangle$ defined on G will be called ρ- coefficients.

The ρ-Fourier transform of a function $f \in L^2(G, \mathfrak{B}(\mathcal{H}))$ is expressed in terms of ρ-coefficients in the following theorem.

Proposition 5.5 Let $f \in L^2(G, \mathfrak{B}(\mathcal{H}))$. Then there exists a family $(T_{ij})_{1 \leqslant i, j \leqslant n}$ of bounded operators of such that

$$\mathcal{F}^\rho f = \sum_{i=1}^n \sum_{j=1}^n T_{ij} \mathcal{F}^\rho U_{ij}^\rho.$$

Proof Let $X, Y \in E$. We write X and Y on the basis $(X_1, \cdots, X_n)$.

$$X = \sum_{k=1}^n \beta_k X_k, \quad Y = \sum_{l=1}^n \gamma_l X_l.$$

Let $f \in L^2(G, \mathfrak{B}(\mathcal{H}))$, then $\mathcal{F}^\rho f(X, Y) = \sum_{k=1}^n \sum_{l=1}^n \overline{\beta_k} \gamma_l \mathcal{F}^\rho f(X_k, X_l)$.

On the other hand, we have:

$$\mathcal{F}^{\rho} U_{ij}^{\rho}(X, Y) = \int_G \langle \rho(x)^* X, Y \rangle \langle \rho(x) X_j, X_i \rangle dx$$

$$= \sum_{k=1}^{n} \sum_{l=1}^{n} \overline{\beta_k} \gamma_l \int_G \langle \rho(x)^* X_k, X_l \rangle \langle \rho(x) X_j, X_i \rangle dx$$

$$= \sum_{k=1}^{n} \sum_{l=1}^{n} \overline{\beta_k} \gamma_l \delta_{kj} \delta_{li} \mathbb{I}_{\mathcal{H}}$$

$$= \overline{\beta_j} \gamma_i \mathbb{I}_{\mathcal{H}}$$

Hence

$$\mathcal{F}^{\rho} f(X, Y) = \sum_{j=1}^{n} \sum_{i=1}^{n} \overline{\beta_j} \gamma_i \mathcal{F}^{\rho} f(X_j, X_i)$$

$$= \sum_{j=1}^{n} \sum_{i=1}^{n} \mathcal{F}^{\rho} f(X_j, X_i)(\overline{\beta_j} \gamma_i \mathbb{I}_H)$$

$$= \sum_{j=1}^{n} \sum_{i=1}^{n} \mathcal{F}^{\rho} f(X_j X_i) \mathcal{F}^{\rho} U_{ij}^{\rho}(X, Y)$$

We set $T_{ij} = \mathcal{F}^{\rho} f(X_j, X_i)$. Hence

$$\mathcal{F}^{\rho} f(X, Y) = \sum_{i=1}^{n} \sum_{j=1}^{n} T_{ij} \mathcal{F}^{\rho} U_{ij}^{\rho}(X, Y). \qquad \Box$$

Acknowledgements The co-author Yaogan Mensah would like to thank the Third World Academy of Science (TWAS) for its financial support (Fellowship for Research and Advanced Training, FR Number: 3240257263). He also thanks Prof. K. B. Sinha from the Jawaharlal Nehru Centre for Advanced Scientific Research, Bangalore (India) for introducing him to some aspects of operator theory.

References

1. V.S.K. Assiamoua, A. Olubummo, Fourier-Stieltjes transforms of vector valued measures on compact groups. Acta Sci. Math. **53**, 301–307 (1989)
2. A. Deitmar, *A First Course in Harmoncic Analysis* (Springer, Berlin, 2012)
3. J. Diestel, J. Uhl, *Vector Measures*. Mathematical Surveys, vol. 15 (American Mathematical Society, Providence, 1977)

4. N. Dinculeanu, *Integration on Locally Compact Spaces* (Noorhoff International Publishing, Leyden, 1974)
5. E. Hewitt, K.A. Ross, *Abstract Harmonic Analysis*, vol. II (Springer, New York-Berlin-Heidelberg, 1970)
6. E.C. Lance, *Hilbert C*-Modules, A Toolkit for Operator Algebraists* (Cambridge University Press, Cambridge, 1995)
7. V.M. Manuilov, E.V. Troitsky, *Hilbert C*-Modules* (American Mathematical Society, Providence, 2005)
8. Y. Mensah, Y.M. Awussi, Ad-Fourier-Stieltjes transform of vector measures on compact Lie groups. Glob. J. Pure Appl. Math. **7**(4), 375–381 (2011)

Electrostatic Double Layers in a Magnetized Isothermal Plasma with two Maxwellian Electrons

Odutayo Raji Rufai

Abstract Finite amplitude nonlinear ion-acoustic double layers are discussed in a magnetized plasma consisting of warm isothermal ion fluid and two Boltzmann distributed electron species by assuming the charge neutrality condition at equilibrium. The model is compatible with the evolution of negative potential double layer structures in the auroral acceleration region. The model predicts maximum electric field amplitude of about ~ 30 mV/m, which is within the satellite measurements in the auroral acceleration region of the Earth's magnetosphere.

Keywords Electrostatic waves · Double layers · Warm ions · Hot and cool electrons · Quasineutrality condition · Auroral acceleration region

1 Introduction

S3-3 and Viking satellite observations have frequently indicated the propagation of nonlinear electrostatic wave structures along the auroral magnetic field lines of the Earth's Magnetosphere [1–3]. Ergun et al. [4] reported the observation of large-amplitude electromagnetic structures from the Fast Auroral Snapshot (FAST) satellite called "fast solitary waves." Further evidence of nonlinear electrostatic structures in the presence of two electron components in the auroral plasma has been received from [3].

Various theoretical models have been developed to describe the observed solitary wave and double layer structures at different regions of Earth's magnetosphere (for example [5, 6]). Recently, several theoretical analyses [7–11] have been done to explain the oblique propagation of finite amplitude nonlinear electrostatic waves in the auroral acceleration region of the Earth's magnetosphere with two electron

O. R. Rufai (✉)
Department of Physics and Astronomy, University of the Western Cape, Bellville, Cape Town, South Africa
e-mail: orufai@uwc.ac.za

© Springer Nature Switzerland AG 2018
T. Diagana, B. Toni (eds.), *Mathematical Structures and Applications*,
STEAM-H: Science, Technology, Engineering, Agriculture,
Mathematics & Health, https://doi.org/10.1007/978-3-319-97175-9_19

components. In hot adiabatic ion temperature, Rufai et al. [8] presented the evolution of nonlinear electrostatic solitary waves and double layers in a magnetized space plasma with cool and hot Boltzmann electron population.

The present investigation extends the model of [7] by including a warm isothermal ion to investigate the characteristics of finite amplitude electrostatic double layer structures in a magnetized auroral plasma with two Maxwellian electrons.

2 Theoretical Analysis

Consider a collisionless, magnetized plasma consisting of warm isothermal ions $(N_i, T_i \neq 0)$ and two distinct groups of Boltzmann distributed electron species, cool electron (N_c, T_c) and hot electron (N_h, T_h), respectively. Assuming the plasma is embedded in an external magnetic field $\mathbf{B}_0 = B_0 \hat{\mathbf{z}}$, where $\hat{\mathbf{z}}$ is the unit vector along the z-axis. Then, the electron densities are

$$N_c = N_{c0} \exp\left(\frac{e\phi}{T_c}\right), \tag{1}$$

$$N_h = N_{h0} \exp\left(\frac{e\phi}{T_h}\right) \tag{2}$$

where ϕ is the potential and N_c, N_h are the cool and hot electron densities.

The dynamics of the warm isothermal ion fluid are governed by the hydrodynamic (continuity, momentum, and pressure) equations

$$\frac{\partial N_i}{\partial t} + \nabla.(N_i V_i) = 0, \tag{3}$$

$$\frac{\partial V_i}{\partial t} + V_i \nabla.V_i = -\frac{e\nabla.\phi}{m_i} + e\frac{\mathbf{V}_i \times \mathbf{B}_0}{m_i c} \tag{4}$$

and

$$\frac{\partial P_i}{\partial t} + \mathbf{V}_i \nabla.P_i + \gamma P_i \nabla.\mathbf{V}_i = 0, \tag{5}$$

where N_i, $\mathbf{V}_i$, and m_i are the number density, fluid velocity, and mass of the ions, e is the magnitude of the electron charge, c is the speed of the light in vacuum, and the ion pressure P_i is given by the balance pressure equation (5). Further, the ion pressure can be written as

$$P_i = P_{i0} \left(\frac{N_i}{N_{i0}}\right)^{\gamma}, \tag{6}$$

where $\gamma = \frac{(N+2)}{N}$ is adiabatic index [8, 11]. N is the number of degrees of freedom. For magnetized isothermal ions $N = 1$, hence $\gamma = 3$ and the ion pressure at equilibrium is $P_{i0} = N_{i0}T_i$.

At equilibrium, the charge neutrality assumption (i.e., $N_i = N_c + N_h$) is valid for low-frequency electrostatic wave studies [8, 10, 11].

The three-species auroral plasma is normalized as follows: the quasi-neutrality condition at equilibrium is given by $N_{i0} = N_{c0} + N_{h0} = N_0$. Density is normalized by total density N_0, velocity is normalized by the effective ion-acoustic speed $c_s = (T_{eff}/m_i)^{1/2}$, $\psi = e\phi/T_{eff}$ is the normalized electrostatic potential distance by effective ion Larmor radius, $\rho_i = c_s/\Omega$, time t normalized by inverse of ion gyro-frequency Ω^{-1}, ($\Omega = eB_0/m_i c$). The temperature ratio is defined as $\tau = T_c/T_h$, cool electron density ratio $f = N_{c0}/N_0$, where $N_{j0}(j = c, h, i)$, $T_{eff} = T_c/(f + (1 - f)\tau)$ is the effective electron temperature, $\alpha_c = T_{eff}/T_c$, $\alpha_h = T_{eff}/T_h$, $\sigma = T_i/T_{eff}$, where T_i is the ion temperature. Also, $\alpha = k_x/k = \sin\theta$, $\gamma = k_z/k = \cos\theta$ (θ is the propagating angle between $\mathbf{k} = (k_x, 0, k_z)$ and $\mathbf{B}_0$, in which $k = \sqrt{k_x^2 + k_z^2}$).

In order to obtain the nonlinear localized traveling wave solution, the normalized set of Eqs. (1)–(5) can be transformed in a stationary moving frame with position $\xi = (\alpha x + \beta z - Mt)/M$, where M is the Mach number. Then, solving for perturbed densities with the quasi-neutrality condition using appropriate boundary conditions for solitary wave solutions (namely, $n_i \to 1$, $\psi \to 0$, and $d\psi/d\xi \to 0$ at $\xi \to \pm\infty$) to obtain

$$\frac{1}{2}\left(\frac{d\psi}{d\xi}\right)^2 + V(\psi, M) = 0 \tag{7}$$

where $V(\psi, M)$ is the Sagdeev potential, given by

$$V(\psi, M) = -\left[\frac{A(\psi, M) + B(\psi, M) + C(\psi, M)}{D(\psi, M)}\right] \tag{8}$$

where

$$A(\psi, M) = -\frac{M^4(1 - n_i)^2}{2n_i^2} - M^2(1 - \beta^2)\psi + M^2 H(\psi), \tag{9}$$

$$B(\psi, M) = \frac{\sigma M^2}{2}(2n_i + 1)(n_i - 1)^2$$

$$+ \frac{M^2 \sigma \beta^2}{2n_i}(n_i + 2)(n_i - 1)^2 - \frac{\beta^2 \sigma^2}{2}(1 - n_i^3)^2, \tag{10}$$

$$C(\psi, M) = \beta^2 \sigma H(\psi) - \frac{\beta^2}{2}H^2(\psi) - \frac{(M^2 + \sigma n_i^4)\beta^2}{n_i}H(\psi), \tag{11}$$

$$D(\psi, M) = \left(1 - \left(\frac{M^2}{n_i^3} - 3\sigma n_i\right)n_i'\right)^2, \tag{12}$$

$$n_i = fe^{\alpha_c \psi} + (1 - f)e^{\alpha_h \psi}, \tag{13}$$

$$n_i' = \alpha_c fe^{\alpha_c \psi} + \alpha_h(1 - f)e^{\alpha_h \psi} \tag{14}$$

and

$$H(\psi) = \frac{f}{\alpha_c}(e^{\alpha_c \psi} - 1) + \frac{1 - f}{\alpha_h}(e^{\alpha_h \psi} - 1). \tag{15}$$

Equation (7) represents an energy integral for a particle moving with velocity $d\psi/d\xi$ in a potential $V(\psi, M)$.

In order to obtain the solitary wave solution from the energy integral Eq. (7), the Sagdeev potential (ψ, M) has to satisfy the following usual soliton conditions : $V(\psi, M) = 0, dV(\psi, M)/d\psi = 0, d^2V(\psi, M)/d\psi^2 < 0$ at $\psi = 0$; $V(\psi, M) = 0$ at some $\psi = \psi_m$, and $V(\psi, M) < 0$ for $0 < |\psi| < |\psi_m|$. For double layer solutions, an additional condition $dV(\psi, M)/d\psi = 0$ at $\psi = \psi_m$ (ψ_m is maximum amplitude) must be satisfied.

Since $f\alpha_c + (1 - f)\alpha_h = 1$, the analysis of the second derivative of the Sagdeev potential $V(\psi, M)$ [7, 9, 12], which has to be negative at origin, namely, $d^2V(\psi, M)/d\psi^2 < 0$ at $\psi = 0$, shows that the nonlinear solutions can be found to exist within the Mach number range, $M_1 < M < M_2$, where $M_1 = \beta\sqrt{1 + 3\sigma}$ is the critical Mach number and $M_2 = \sqrt{1 + 3\sigma}$ is the upper Mach number of which no solution can be found beyond this range.

The energy integral Eq. (7) and Sagdeev potential Eq. (8) will be numerically computed for the nonlinear electrostatic double layer structures with the plasma parameters such as Mach number M, density ratio f, wave obliqueness β, ion temperature ratio σ, and electron temperature ratio τ. Figure 1 shows the variation of the Sagdeev potential with the real electrostatic potential for different values of ion temperature ratio $\sigma = 0.01$ and 0.05 and wave obliqueness $\beta = 0.80523$ and 0.80955. The fixed parameters are $f = 0.1, \tau = 0.04$ and $M = 1.00$.

The curves in Fig. 2 show the variation of the normalized electrostatic double layer potential ψ against ξ for the fixed plasma parameters, $f = 0.1, \sigma = 0.01$, and $\beta = 0.966$.

3   Conclusion

In this paper, the effect of finite ion temperature on the existence of nonlinear finite amplitude ion-acoustic waves in a magnetized three-component plasma making up of a warm isothermal ion fluid and Boltzmann distribution of cool and hot electron species has been investigated using the Sagdeev pseudo-potential technique. The model is consistent with the existence of negative potential (compressive)

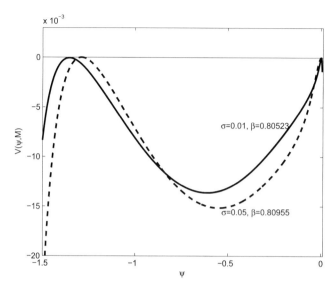

Fig. 1 Sagdeev potential, $V(\psi, M)$ vs normalized electrostatic potential ψ, for $M = 1.00$, $f = 0.1$, $\tau = 0.04$

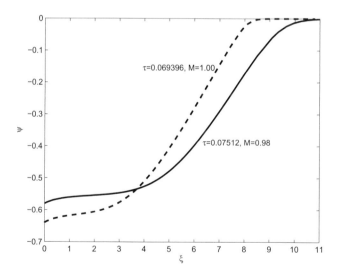

Fig. 2 Electrostatic potential ψ vs ξ for $f = 0.1$, $\sigma = 0.01$, $\beta = 0.966$

double layer structures obliquely propagating along the magnetic field lines of the Earth's auroral acceleration region.

Related to the Viking Spacecraft observations in the auroral zone of the Earth's magnetosphere [3] recorded the electrostatic wave electric field amplitude of less than $100\,\mathrm{mV/m}$, width $\approx 100\,\mathrm{m}$, pulse duration $\approx 20\,\mathrm{ms}$, and velocities of about

10–50 km/s. For the ion temperature in the auroral region, $\sigma = 0.01$, electron density $\tau = 0.069396$, Mach number $M = 1.00$ and wave obliqueness $\beta = 0.966$, the maximum double layer electric field generated comes out to be 30.09 mV/m, with the corresponding width, pulse duration, and velocity of about 148.7 m, 5.74 ms, and 25.90 km/s, respectively. These results agree with the spacecraft measurements.

References

1. M.Temerin, K. Cerny, W.Lotko and F. S. Mozer, Phys. Rev. Lett., **48** 1175, (1982).
2. R. Boström et al., Phys.Rev.Lett., **61** 82–85. (1988).
3. M. Berthomier et al., J. Geophys. Res.,**103** A3 4261–4270, (1998).
4. R. E. Ergun et al., Geophys. Res. Lett., **25** 12 2041–2044, (1998).
5. R. Bharuthram and P. K. Shukla, Phys. Fluids, **29** 10, (1986).
6. T. K. Baluku, M. A. Hellberg and F. Verheest, EPL, **91** 15001, (2010).
7. O. R. Rufai, R. Bharuthram, S. V. Singh and G. S. Lakhina, Phys. Plasmas, **19** 122308, (2012).
8. O. R. Rufai, R. Bharuthram, S. V. Singh and G. S. Lakhina, Commun Nonlin. Sci. Numer. Simulat., **19** 1338–1346, (2014).
9. O. R. Rufai, Phys. Plasmas **22** 052309, (2015).
10. O. R. Rufai, R. Bharuthram, S. V. Singh and G. S. Lakhina, Adv. Space Res., **57** 813–820, (2016).
11. O. R. Rufai, R. Bharuthram, S. V. Singh and G. S. Lakhina, Phys. Plasmas **23** 032309, (2016).
12. O. R. Rufai and R. Bharuthram, Phys. Plasmas **23** 092306, (2016).

Star Products, Star Exponentials, and Star Functions

Akira Yoshioka

Dedicated to Professor Norbert Hounkonnou on the occasion of his 60th birthday.

Abstract We give a brief review on non-formal star products and star exponentials and star functions (Omori et al., Deformation of expressions for elements of an algebra, in *Symplectic, Poisson, and Noncommutative Geometry*. Mathematical Sciences Research Institute Publications, vol. 62 (Cambridge University Press, Cambridge, 2014), pp. 171–209; Deformation of expressions for elements of algebra, arXiv:1104.1708v1[math.ph]; Deformation of expressions for elements of algebras (II), arXiv:1105.1218v2[math.ph]). We introduce a star product on polynomials with a deformation parameter $\hbar > 0$. Extending to functions on complex space enables us to consider exponential element in the star product algebra, called a star exponential. By means of the star exponentials we can define several functions called star functions in the algebra, with some noncommutative identities. We show certain examples.

Keywords Star product · Star exponential · Star function

1 Star Product on Polynomials

First we start by considering well-known star product, the Moyal product. Also we introduce typical star products, that is, normal product and anti-normal product. These products are mutually isomorphic with explicit isomorphisms, which are given by changing orderings in physics.

A. Yoshioka (✉)
Tokyo University of Science, Tokyo, Japan
e-mail: yoshioka@rs.kagu.tus.ac.jp

© Springer Nature Switzerland AG 2018
T. Diagana, B. Toni (eds.), *Mathematical Structures and Applications*,
STEAM-H: Science, Technology, Engineering, Agriculture,
Mathematics & Health, https://doi.org/10.1007/978-3-319-97175-9_20

439

Next we define a class of star products containing the Moyal, normal, and anti-normal products as examples. The product is defined on the space of complex polynomials and is given as power series with respect to a positive parameter $\hbar$.

Also we discuss isomorphisms between these star products. We describe that these objects naturally give rise to a bundle of star product algebras over space of complex symmetric matrices.

1.1 Moyal Product

The Moyal product is a well-known example of star product [2, 4].

For polynomials f, g of variables $(u_1, \ldots, u_m, v_1, \ldots, v_m)$, we define a biderivation $\left(\overleftarrow{\partial_v} \cdot \overrightarrow{\partial_u} - \overleftarrow{\partial_u} \cdot \overrightarrow{\partial_v} \right)$ by

$$f \left(\overleftarrow{\partial_v} \cdot \overrightarrow{\partial_u} - \overleftarrow{\partial_u} \cdot \overrightarrow{\partial_v} \right) g = \sum_{j=1}^{m} \left(\partial_{v_j} f \, \partial_{u_j} g - \partial_{u_j} f \, \partial_{v_j} g \right)$$

Here the overleft arrow $\overleftarrow{\partial}$ indicates that the partial derivative is acting on the polynomial on the left and the overright arrow indicates the right.

Then the Moyal product $f *_O g$ is given by the power series of the biderivation $\left(\overleftarrow{\partial_v} \cdot \overrightarrow{\partial_u} - \overleftarrow{\partial_u} \cdot \overrightarrow{\partial_v} \right)$ such that

$$f *_O g = f \exp \frac{i\hbar}{2} \left(\overleftarrow{\partial_v} \cdot \overrightarrow{\partial_u} - \overleftarrow{\partial_u} \cdot \overrightarrow{\partial_v} \right) g = f \sum_{k=0}^{\infty} \frac{1}{k!} \left(\frac{i\hbar}{2} \right)^k \left(\overleftarrow{\partial_v} \cdot \overrightarrow{\partial_u} - \overleftarrow{\partial_u} \cdot \overrightarrow{\partial_v} \right)^k g$$

$$= fg + \frac{i\hbar}{2} f \left(\overleftarrow{\partial_v} \cdot \overrightarrow{\partial_u} - \overleftarrow{\partial_u} \cdot \overrightarrow{\partial_v} \right) g + \frac{1}{2!} \left(\frac{i\hbar}{2} \right)^2 f \left(\overleftarrow{\partial_v} \cdot \overrightarrow{\partial_u} - \overleftarrow{\partial_u} \cdot \overrightarrow{\partial_v} \right)^2 g$$

$$+ \cdots + \frac{1}{k!} \left(\frac{i\hbar}{2} \right)^k f \left(\overleftarrow{\partial_v} \cdot \overrightarrow{\partial_u} - \overleftarrow{\partial_u} \cdot \overrightarrow{\partial_v} \right)^k g + \cdots \tag{1.1}$$

where $\hbar$ is a positive number. Although the Moyal product is defined as a formal power series of bidifferential operators, this becomes a finite sum on polynomials. One can check the associativity of the product directly, hence we have

Proposition 1.1 The Moyal product is well-defined on polynomials, and associative.

Other typical star products are normal product $*_N$, anti-normal product $*_A$ given similarly with replacing biderivations, respectively, by

$$f *_N g = f \exp i\hbar \left(\overleftarrow{\partial_v} \cdot \overrightarrow{\partial_u} \right) g, \quad f *_A g = f \exp -i\hbar \left(\overleftarrow{\partial_u} \cdot \overrightarrow{\partial_v} \right) g$$

These are also well-defined on polynomials and associative.

By direct calculation we see easily

Proposition 1.2

(i) *For these star products, the generators* $(u_1, \ldots, u_m, v_1, \ldots, v_m)$ *satisfy the canonical commutation relations*

$$[u_k, v_l]_{*_L} = -i\hbar \delta_{kl}, \quad [u_k, u_l]_{*_L} = [v_k, v_l]_{*_L} = 0, \quad (k, l = 1, 2, \ldots, m)$$

where $*_L$ *stands for* $*_O$, $*_N$, *and* $*_A$.

(ii) *Then the algebras* $(\mathbb{C}[u, v], *_L)$ $(L = O, N, A)$ *are mutually isomorphic and isomorphic to the Weyl algebra.*

The algebra isomorphisms have explicit expressions. For example, the algebra isomorphism

$$I_O^N : (\mathbb{C}[u, v], *_O) \rightarrow (\mathbb{C}[u, v], *_N)$$

is given by the power series of the differential operator such as

$$I_N^O (f) = \exp\left(-\frac{i\hbar}{2} \partial_u \partial_v\right)(f) = \sum_{l=0}^{\infty} \frac{1}{l!} \left(\frac{i\hbar}{2}\right)^l (\partial_u \partial_v)^l (f) \tag{1.2}$$

Other isomorphisms are given in similar forms (cf. [12]).

Remark 1.3 *We remark here that these isomorphisms are well-known as ordering problem in physics* [1].

1.2 Star Product

Using complex matrices we generalize biderivations and we define a star product on complex domain in the following way.

Let Λ be an arbitrary $n \times n$ complex matrix. We consider a biderivation

$$\overleftarrow{\partial_w} \Lambda \overrightarrow{\partial_w} = (\overleftarrow{\partial_{w_1}}, \cdots, \overleftarrow{\partial_{w_n}}) \Lambda (\overrightarrow{\partial_{w_1}}, \cdots, \overrightarrow{\partial_{w_n}}) = \sum_{k,l=1}^{n} \Lambda_{kl} \overleftarrow{\partial_{w_k}} \overrightarrow{\partial_{w_l}} \tag{1.3}$$

where $(w_1, \cdots, w_n)$ is a generator of polynomials.

Now we define a star product similar to (1.1) by

Definition 1.4

$$f *_\Lambda g = f \exp\left(\frac{i\hbar}{2} \overleftarrow{\partial_w} \Lambda \overrightarrow{\partial_w}\right) g \tag{1.4}$$

Remark 1.5 ([13])

(i) *The star product* $*_\Lambda$ *is a generalization of the products* $*_L$ $(L = O, N, A)$. *Actually*

- *if we put* $\Lambda = \begin{pmatrix} 0 & -1_m \\ 1_m & 0 \end{pmatrix}$, *then we have the Moyal product*

- *if* $\Lambda = 2\begin{pmatrix} 0 & 0 \\ 1_m & 0 \end{pmatrix}$, *then we have the normal product and*

- *if* $\Lambda = 2\begin{pmatrix} 0 & -1_m \\ 0 & 0 \end{pmatrix}$, *then the anti-normal product*

(ii) *If* Λ *is a symmetric matrix, the star product* $*_\Lambda$ *is commutative.*

Then similarly as before we see easily

Theorem 1.6 *For an arbitrary* Λ, *the star product* $*_\Lambda$ *is well-defined on polynomials, and associative.*

1.3 Equivalence and Geometric Picture of Weyl Algebra

In this section, we take Λ as a special class of matrices in order to represent Weyl algebra (cf. [5, 10]).

Let K be an arbitrary $2m \times 2m$ complex symmetric matrix. We put a complex matrix

$$\Lambda = J + K$$

where J is a fixed matrix such that

$$J = \begin{pmatrix} 0 & -1_m \\ 1_m & 0 \end{pmatrix}$$

Since Λ is determined by the complex symmetric matrix K, we denote the star product by $*_K$ instead of $*_\Lambda$.

We consider polynomials of variables $(w_1, \cdots, w_{2m}) = (u_1, \cdots, u_m, v_1, \cdots, v_m)$. By easy calculation one obtains

Proposition 1.7

(i) *For a star product* $*_K$, *the generators* $(u_1, \ldots, u_m, v_1, \ldots, v_m)$ *satisfy the canonical commutation relations*

$$[u_k, v_l]_{*_K} = -i\hbar\delta_{kl}, \quad [u_k, u_l]_{*_K} = [v_k, v_l]_{*_K} = 0, \quad (k, l = 1, 2, \ldots, m)$$

(ii) *Then the algebra* $(\mathbb{C}[u,v], *_K)$ *is isomorphic to the Weyl algebra, and the algebra is regarded as a polynomial representation of the Weyl algebra.*

Equivalence As in the case of typical star products, we have algebra isomorphisms as follows.

Proposition 1.8 *For arbitrary* $(\mathbb{C}[u,v], *_{K_1})$ *and* $(\mathbb{C}[u,v], *_{K_2})$ *we have an algebra isomorphism* $I_{K_1}^{K_2} : (\mathbb{C}[u,v], *_{K_1}) \to (\mathbb{C}[u,v], *_{K_2})$ *given by the power series of the differential operator* $\partial_w(K_2 - K_1)\partial_w$ *such that*

$$I_{K_1}^{K_2}(f) = \exp\left(\tfrac{i\hbar}{4}\partial_w(K_2 - K_1)\partial_w\right)(f)$$

where $\partial_w(K_2 - K_1)\partial_w = \sum_{kl}(K_2 - K_1)_{kl}\partial_{w_k}\partial_{w_l}$.

By a direct calculation we have

Theorem 1.9 *Isomorphisms satisfy the following chain rule:*

1. $I_{K_3}^{K_1}I_{K_2}^{K_3}I_{K_1}^{K_2} = Id, \quad \forall K_1, K_2, K_3$
2. $\left(I_{K_1}^{K_2}\right)^{-1} = I_{K_2}^{K_1}, \quad \forall K_1, K_2$

Remark 1.10

1. *By Proposition 1.8 we see the algebras* $(\mathbb{C}[u,v], *_K)$ *are mutually isomorphic and isomorphic to the Weyl algebra. Hence we have a family of star product algebras* $\{(\mathbb{C}[u,v], *_K)\}_K$ *where each element is regarded as a polynomial representation of the Weyl algebra.*
2. *The above equivalences are also valid for star products* $*_\Lambda$ *and* $*_{\Lambda'}$ *for arbitrary* Λ, Λ' *with a common skew symmetric part. More precisely, let* $\tilde{J}$ *by an arbitrary* $n \times n$ *skew-symmetric matrix, and for any* $n \times n$ *symmetric matrices* K, K' *we consider* $\Lambda = \tilde{J} + K$ *and* $\Lambda' = \tilde{J} + K'$. *Then* $I_K^{K'}$ *gives an algebra isomorphism* $I_K^{K'} : (\mathbb{C}[w], *_\Lambda) \to (\mathbb{C}[w], *_{\Lambda'})$, *where* $w = (w_1, \ldots, w_n)$ *is the generator of polynomials.*

According to the previous theorem, we introduce an infinite dimensional bundle and a connection over it and using parallel sections of this bundle we have a geometric picture (cf. [14]) for the family of the star product algebras $\{(\mathbb{C}[u,v], *_K)\}_K$.

Algebra Bundle We set $\mathcal{S} = \{K\}$ the space of all $2m \times 2m$ symmetric complex matrices. We consider a trivial bundle over $\mathcal{S}$ with fiber the star product algebras

$$\pi : E = \Pi_{K \in \mathcal{S}}(\mathbb{C}[u,v], *_K) \to \mathcal{S}, \quad \pi^{-1}(K) = (\mathbb{C}[u,v], *_K).$$

Then Proposition 1.8 shows that each fiber $(\mathbb{C}[u,v], *_K)$ is isomorphic to the Weyl algebra and any fibers of the bundle are mutually isomorphic by the intertwiners $I_{K_1}^{K_2}$.

Connection and Parallel Sections For a curve $C : K = K(t)$ in the base space S starting from $K(0) = K$, we define a parallel translation of a polynomial $f \in (\mathbb{C}[u, v], *_K)$ by

$$f(t) = \exp \tfrac{i\hbar}{4} \partial_w (K(t) - K) \partial_w (f).$$

It is easy to see $f(0) = f$. By differentiating the parallell translation we have a connection of this bundle such that

$$\nabla_X f(K) = \tfrac{d}{dt} f(t)|_{t=0}(K) = \tfrac{i\hbar}{4} \partial_w X \partial_w f(K)|_{t=0}, \quad X = \dot{K}(t)|_{t=0}$$

where $f(K)$ is a smooth section of the bundle E.

We set $\mathcal{P}$ the space of all parallel sections of this bundle, namely, f is an element of $\mathcal{P}$ iff $f(K_2) = I_{K_1}^{K_2} f(K_1)$ for any $K_1, K_2 \in S$. Since $I_{K_1}^{K_2}$ are algebra isomorphisms, namely it holds for sections f, g

$$I_{K_1}^{K_2}(f(K_1) *_{K_1} g(K_1)) = \left(I_{K_1}^{K_2}(f(K_1)) \right) *_{K_2} \left(I_{K_1}^{K_2}(g(K_1)) \right),$$

we have a star product $f * g$ for the parallel sections $f, g \in \mathcal{P}$ by setting

$$f * g (K) = f(K) *_K g(K)$$

Then we have

Theorem 1.11

(i) *The space of the parallel sections $\mathcal{P}$ consists of the sections such that $\nabla_X f = \tfrac{i\hbar}{4} \partial_w X \partial_w f = 0, \ \forall X$.*
(ii) *The space $\mathcal{P}$ is canonically equipped with the star product $*$, and the associative algebra $(\mathcal{P}, *)$ is isomorphic to the Weyl algebra.*

Remark 1.12 *The algebra $(\mathcal{P}, *)$ is regarded as a geometric realization of the Weyl algebra.*

2 Extension to Functions

We consider to extend the star products $*_\Lambda$ for an arbitrary complex matrix Λ from polynomials to functions (cf. [9]).

2.1 Star Product on Certain Holomorphic Function Space

For ordinary smooth functions, the star products $*_\Lambda$ are not necessarily well-defined, e.g., convergent in general. However, we can discuss star products by restricting the product to certain class of smooth functions. Although there may be many classes for such functions, we consider the following space of certain entire functions in this note (cf. [6]).

Semi-norm Let f(w) be a holomorphic function on $\mathbb{C}^n$. For a positive number p, we consider a family of semi-norms $\{|\cdot|_{p,s}\}_{s>0}$ given by

$$|f|_{p,s} = \sup_{w \in \mathbb{C}^n} |f(w)| \exp(-s|w|^p), \quad |w| = \sqrt{|w_1|^2 + \cdots + |w_n|^2}.$$

Space We put

$$\mathcal{E}_p = \{f : \text{entire} \mid |f|_{p,s} < \infty, \forall s > 0\}$$

With the semi-norms the space $\mathcal{E}_p$ becomes a Fréchet space.

As to the star products, we have for any matrix Λ

Theorem 2.1

(i) *For $0 < p \leq 2$, $(\mathcal{E}_p, *_\Lambda)$ is a Fréchet algebra. That is, the product converges for any elements, and the product is continuous with respect to this topology.*

(ii) *Moreover, for any Λ' with the common skew symmetric part with Λ, $I_\Lambda^{\Lambda'} = \exp(\frac{i\hbar}{4}\partial_w(\Lambda' - \Lambda)\partial_w)$ is well-defined algebra isomorphism from $(\mathcal{E}_p, *_\Lambda)$ to $(\mathcal{E}_p, *_{\Lambda'})$. That is, the expansion converges for every element, and the operator is continuous with respect to this topology.*

(iii) *For $p > 2$, the multiplication $*_\Lambda : \mathcal{E}_p \times \mathcal{E}_{p'} \to \mathcal{E}_p$ is well-defined for p' such that $\frac{1}{p} + \frac{1}{p'} = 2$, and $(\mathcal{E}_p, *_\Lambda)$ is a $\mathcal{E}_{p'}$-bimodule.*

In this topology, the parameter $\hbar$ can be taken as $\hbar \in \mathbb{C}$.

3   Star exponentials

Since we have a complete topological algebra, we can consider exponential elements in the star product algebra $(\mathcal{E}_p, *_\Lambda)$. (cf. [13]).

3.1   Definition

For a polynomial H_*, we want to define a star exponential $e_*^{t\frac{H_*}{i\hbar}}$. However, except special cases, the expansion $\sum_n \frac{t^n}{n!}\left(\frac{H_*}{i\hbar}\right)^n$ is not convergent, so we define a star exponential by means of a differential equation.

Definition 3.1 The star exponential $e_*^{t\frac{H_*}{i\hbar}}$ is given as a solution of the following differential equation

$$\frac{d}{dt}F_t = H_* *_\Lambda F_t, \quad F_0 = 1. \tag{3.1}$$

3.2   Examples

We are interested in the star exponentials of linear and quadratic polynomials. For these, we can solve the differential equation and obtain explicit form. For simplicity, we consider $2m \times 2m$ complex matrices Λ with the skew symmetric part $J = \begin{pmatrix} 0 & -1_m \\ 1_m & 0 \end{pmatrix}$. We write $\Lambda = J + K$ where K is a complex symmetric matrix.

First we remark the following. For a linear polynomial $l = \sum_{j=1}^{2m} a_j w_j$, we see directly an ordinary exponential function e^l satisfies

$$e^l \notin \mathcal{E}_1, \quad \in \mathcal{E}_{1+\epsilon}, \quad \forall \epsilon > 0.$$

Then put a Fréchet space

$$\mathcal{E}_{p+} = \cap_{q > p} \mathcal{E}_q$$

Linear Case

Proposition 3.2 *For* $l = \sum_j a_j w_j = <\boldsymbol{a}, \boldsymbol{w}>$, $a_j \in \mathbb{C}$, *we have*

$$e_*^{t(l/i\hbar)} = e^{t^2 \boldsymbol{a} K \boldsymbol{a}/4i\hbar} e^{t(l/i\hbar)} \in \mathcal{E}_{1+}$$

Quadratic Case   (Cf. [8]).

Proposition 3.3 *For* $Q_* = \langle \boldsymbol{w} A, \boldsymbol{w} \rangle_*$ *where* A *is a* $2m \times 2m$ *complex symmetric matrix,*

$$e_*^{t(Q_*/i\hbar)} = \frac{2^m}{\sqrt{\det(I - \kappa + e^{-2t\alpha}(I + \kappa))}} e^{\frac{1}{i\hbar} \langle \boldsymbol{w} \frac{1}{I - \kappa + e^{-2t\alpha}(I+\kappa)} (I - e^{-2t\alpha}) J, \boldsymbol{w} \rangle}$$

where $\kappa = KJ$ *and* $\alpha = AJ$.

Remark 3.4 *The star exponentials of linear functions are belonging to* $\mathcal{E}_{1+}$ *then the star products are convergent and continuous. But for a quadratic polynomial* Q_*, *it is easy to see*

$$e_*^{t(Q_*/i\hbar)} \in \mathcal{E}_{2+}, \quad \notin \mathcal{E}_2$$

and hence star exponentials $\{e_*^{t(Q_*/i\hbar)}\}$ *are difficult to treat. Some anomalous phenomena happen.* (cf. [7]).

Remark 3.5 *Beside solving differential equation, we also construct star exponential of quadratic polynomial by so-called path-integral method. Namely, we divide the time interval by a positive integer* N. *Consider commutative exponential functions* $e^{(t/N)(Q/i\hbar)}$, $N = 1, 2, \ldots$, *and take* N-multiple products $e^{(t/N)(Q/i\hbar)}$ $*$

$\cdots * e^{(t/N)(Q/i\hbar)}$. *Then taking a limit we have* (cf. [3])

$$e_*^{t(Q*/i\hbar)} = \lim_{N \to \infty} e^{(t/N)(Q/i\hbar)} * \cdots * e^{(t/N)(Q/i\hbar)}$$

4   Star Functions

There are many applications of star exponential functions (cf. [11–14]). In this note we show examples using linear star exponentials.

In what follows, we consider the star product for the simple case where Λ has only one nonzero entry

$$\Lambda = \begin{pmatrix} \rho & 0 \\ 0 & 0 \end{pmatrix}, \quad \rho \in \mathbb{C}$$

Then we see easily that the star product is commutative and explicitly given by $p_1 *_\Lambda p_2 = p_1 \exp\left(\frac{i\hbar\rho}{2} \overleftarrow{\partial}_{w_1} \overrightarrow{\partial}_{w_1}\right) p_2$. This means that the algebra is essentially reduced to the space of functions of one variable w_1. Thus, we consider functions $f(w)$, $g(w)$ of one variable $w \in \mathbb{C}$ and we consider a commutative star product $*_\tau$ with complex parameter τ such that

$$f(w) *_\tau g(w) = f(w) e^{\frac{\tau}{2} \overleftarrow{\partial}_w \overrightarrow{\partial}_w} g(w)$$

4.1   Star Hermite Function

Recall the identity

$$\exp\left(\sqrt{2}tw - \tfrac{1}{2}t^2\right) = \sum_{n=0}^{\infty} H_n(w) \frac{t^n}{n!}$$

where $H_n(w)$ is an Hermite polynomial. We remark here that

$$\exp\left(\sqrt{2}tw - \tfrac{1}{2}t^2\right) = \exp_*(\sqrt{2}tw_*)_{\tau=-1}$$

Since $\exp_*(\sqrt{2}tw_*) = \sum_{n=0}^{\infty}(\sqrt{2}tw_*)^n \frac{t^n}{n!}$, we have $H_n(w) = (\sqrt{2}tw_*)^n_{\tau=-1}$. We define $*$-Hermite function by

$$H_n(w, \tau) = (\sqrt{2}tw_*)^n, \quad (n = 0, 1, 2, \cdots)$$

with respect to $*_\tau$ product. Then we have

$$\exp_*(\sqrt{2}t w_*) = \sum_{n=0}^{\infty} H_n(w, \tau) \frac{t^n}{n!}$$

Trivial identity $\frac{d}{dt} \exp_*(\sqrt{2}t w_*) = \sqrt{2}w * \exp_*(\sqrt{2}t w_*)$ yields at every $\tau \in \mathbb{C}$ the identity

$$\frac{\tau}{\sqrt{2}} H_n'(w, \tau) + \sqrt{2}w H_n(w, \tau) = H_{n+1}(w, \tau), \quad (n = 0, 1, 2, \cdots).$$

The exponential law $\exp_*(\sqrt{2}s w_*) * \exp_*(\sqrt{2}t w_*) = \exp_*(\sqrt{2}(s+t) w_*)$ yields at every $\tau \in \mathbb{C}$ the identity

$$\sum_{k+l=n} \frac{n!}{k!l!} H_k(w, \tau) *_\tau H_l(w, \tau) = H_n(w, \tau).$$

4.2   Star Theta Function

In this note we consider the Jacobi's theta functions by using star exponentials as an application.

A direct calculation gives

$$\exp_{*_\tau} i \, tw = \exp(i \, tw - (\tau/4)t^2)$$

Hence for $\Re\tau > 0$, the star exponential $\exp_{*_\tau} ni \, w = \exp(ni \, w - (\tau/4)n^2)$ is rapidly decreasing with respect to integer n and then we can consider summations for τ satisfying $\Re\tau > 0$

$$\sum_{n=-\infty}^{\infty} \exp_{*_\tau} 2ni \, w = \sum_{n=-\infty}^{\infty} \exp\left(2ni \, w - \tau n^2\right) = \sum_{n=-\infty}^{\infty} q^{n^2} e^{2ni \, w}, \quad (q = e^{-\tau})$$

This is Jacobi's theta function $\theta_3(w, \tau)$. Then we have expression of theta functions as

$$\theta_{1*_\tau}(w) = \frac{1}{i} \sum_{n=-\infty}^{\infty} (-1)^n \exp_{*_\tau}(2n+1)i \, w, \qquad \theta_{2*_\tau}(w) = \sum_{n=-\infty}^{\infty} \exp_{*_\tau}(2n+1)i \, w$$

$$\theta_{3*_\tau}(w) = \sum_{n=-\infty}^{\infty} \exp_{*_\tau} 2ni \, w, \qquad \theta_{4*_\tau}(w) = \sum_{n=-\infty}^{\infty} (-1)^n \exp_{*_\tau} 2ni \, w$$

Remark that $\theta_{k*_\tau}(w)$ is the Jacobi's theta function $\theta_k(w, \tau)$, $k = 1, 2, 3, 4$, respectively. It is obvious by the exponential law

$$\exp_{*_\tau} 2i\, w *_\tau \theta_{k*_\tau}(w) = \theta_{k*_\tau}(w) \quad (k = 2, 3)$$

$$\exp_{*_\tau} 2i\, w *_\tau \theta_{k*_\tau}(w) = -\theta_{k*_\tau}(w) \quad (k = 1, 4)$$

Then using $\exp_{*_\tau} 2i\, w = e^{-\tau} e^{2i\, w}$ and the product formula directly we have

$$e^{2i\, w - \tau} \theta_{k*_\tau}(w + i\,\tau) = \theta_{k*_\tau}(w) \quad (k = 2, 3)$$

$$e^{2i\, w - \tau} \theta_{k*_\tau}(w + i\,\tau) = -\theta_{k*_\tau}(w) \quad (k = 1, 4)$$

4.3   *-Delta Functions

Since the $*_\tau$-exponential $\exp_*(it w_*) = \exp(it w - \frac{\tau}{4}t^2)$ is rapidly decreasing with respect to t when $\Re\tau > 0$, then the integral of $*_\tau$-exponential

$$\int_{-\infty}^{\infty} \exp_*(it(w-a)_*)\, dt = \int_{-\infty}^{\infty} \exp_*(it(w-a)_*)dt = \int_{-\infty}^{\infty} \exp(it(w-a) - \frac{\tau}{4}t^2)dt$$

converges for any $a \in \mathbb{C}$. We put a star δ-function

$$\delta_*(w - a) = \int_{-\infty}^{\infty} \exp_*(it(w - a)_*)dt$$

which has a meaning at τ with $\Re\tau > 0$. It is easy to see for any element $p_*(w) \in \mathcal{P}_*(\mathbb{C})$,

$$p_*(w) * \delta_*(w - a) = p(a)\delta_*(w - a), \quad w_* * \delta_*(w) = 0.$$

Using the Fourier transform we have

Proposition 4.1

$$\theta_{1*}(w) = \frac{1}{2} \sum_{n=-\infty}^{\infty} (-1)^n \delta_*(w + \frac{\pi}{2} + n\pi)$$

$$\theta_{2*}(w) = \frac{1}{2} \sum_{n=-\infty}^{\infty} (-1)^n \delta_*(w + n\pi)$$

$$\theta_{3*}(w) = \tfrac{1}{2} \sum_{n=-\infty}^{\infty} \delta_*(w + n\pi)$$

$$\theta_{4*}(w) = \tfrac{1}{2} \sum_{n=-\infty}^{\infty} \delta_*(w + \tfrac{\pi}{2} + n\pi).$$

Now, we consider the τ with the condition $\Re\tau > 0$. Then we calculate the integral and obtain $\delta_*(w - a) = \frac{2\sqrt{\pi}}{\sqrt{\tau}} \exp\left(-\tfrac{1}{\tau}(w - a)^2\right)$. Then we have

$$\theta_3(w, \tau) = \tfrac{1}{2} \sum_{n=-\infty}^{\infty} \delta_*(w + n\pi) = \sum_{n=-\infty}^{\infty} \frac{\sqrt{\pi}}{\sqrt{\tau}} \exp\left(-\tfrac{1}{\tau}(w + n\pi)^2\right)$$

$$= \frac{\sqrt{\pi}}{\sqrt{\tau}} \exp\left(-\tfrac{1}{\tau}\right) \sum_{n=-\infty}^{\infty} \exp\left(-2n\tfrac{1}{\tau}w - \tfrac{1}{\tau}n^2\tau^2\right)$$

$$= \frac{\sqrt{\pi}}{\sqrt{\tau}} \exp\left(-\tfrac{1}{\tau}\right) \theta_{3*}(\tfrac{2\pi w}{i\tau}, \tfrac{\pi^2}{\tau}).$$

We also have similar identities for other $*$-theta functions by the similar way.

Acknowledgements The author would like to thank JSPS KAKENHI Grant Number JP15K04856 for its financial support.

References

1. G.S. Agarwal, E. Wolf, Calculus for functions of noncommuting operators and general phase-space methods in quantum mechanics I. Mapping Theorems and ordering of functions of noncommuting operators. Phys. Rev. D **2**, 2161–2186 (1970)
2. F. Bayen, M. Flato, C. Fronsdal, A. Lichnerowicz, D. Sternheimer, Deformation theory and quantization. I. Deformations of symplectic structures. Ann. Phys. **111**(1), 61–110 (1978)
3. T. Matsumoto, A. Yoshioka, Path integral for star exponential functions of quadratic forms, in *AIP Conference Proceedings on Geometry, Integrability and Quantization*, Varna, 2002, vol. 956 (Coral Press Scientific Publishing, Sofia, 2003), pp. 115–122
4. J.E. Moyal, Quantum mechanics as a statistical theory. Proc. Camb. Philol. Soc. **45**, 99–124 (1949)
5. H. Omori, Toward geometric quantum theory, in *From Geometry to Quantum Mechanics.* Progress in Mathematics, vol. 252 (Birkhäuser, New York, 2007), pp. 213–251
6. H. Omori, Y. Maeda, N. Miyazaki, A. Yoshioka, Deformation quantization of Frécht-Poisson algebras: Convergence of the Moyal product, in *Proceedings of Conference Mosh'e Flato 1999, Mathematical Physics Studies* vol. 22 (Kluwer Academic Publishers, Dordrecht/Boston/London, 2000), pp. 233–245
7. H. Omori, Y. Maeda, N. Miyazaki, A. Yoshioka, Strange phenomena related to ordering problems in quantizations. J. Lie Theory **13**(2), 481–510 (2003)
8. H. Omori, Y. Maeda, N. Miyazaki, A. Yoshioka, Star exponential functions as two-valued elements, in *The Breadth of Symplectic and Poisson Geometry.* Progress in Mathematics, vol. 232 (Birkhäuser Boston, 2005), pp. 483–492

9. H. Omori, Y. Maeda, N. Miyazaki, A. Yoshioka, Orderings and non-formal deformation quantization. Lett. Math. Phys. **82**, 153–175 (2007)
10. H. Omori, Y. Maeda, N. Miyazaki, A. Yoshioka, Geometric objects in an approach to quantum geometry, in *From Geometry to Quantum Mechanics*. Progress in Mathematics, vol. 252 (Birkhäuser, New York, 2007), pp. 303–324
11. H. Omori, Y. Maeda, N. Miyazaki, A. Yoshioka, Deformation of expressions for elements of an algebra, in *Symplectic, Poisson, and Noncommutative Geometry*. Mathematical Sciences Research Institute Publications, vol. 62 (Cambridge University Press, Cambridge, 2014), pp. 171–209
12. H. Omori, Y. Maeda, N. Miyazaki, A. Yoshioka, Deformation of expressions for elements of algebra, arXiv:1104.1708v1[math.ph]
13. H. Omori, Y. Maeda, N. Miyazaki, A. Yoshioka, Deformation of expressions for elements of algebras (II), arXiv:1105.1218v2[math.ph]
14. A. Yoshioka, A family of star products and its application, in *AIP Conference Proceedings of the XXVI Workshop on Geometrical Mathods in Physics*, Bialowieza, 2007, vol. 956 (American Institute of Physics, Merville, 2007), pp. 115–122

Index

© Springer Nature Switzerland AG 2018
T. Diagana, B. Toni (eds.), *Mathematical Structures and Applications,*
STEAM-H: Science, Technology, Engineering, Agriculture,
Mathematics & Health, https://doi.org/10.1007/978-3-319-97175-9

Printed in the United States
By Bookmasters